# Lecture Notes in Computer Science

*Commenced Publication in 1973*
Founding and Former Series Editors:
Gerhard Goos, Juris Hartmanis, and Jan van Leeuwen

Eric Andres  Guillaume Damiand
Pascal Lienhardt (Eds.)

# Discrete Geometry for Computer Imagery

12th International Conference, DGCI 2005
Poitiers, France, April 13-15, 2005
Proceedings

Volume Editors

Eric Andres
Guillaume Damiand
Pascal Lienhardt
Université de Poitiers, Laboratoire S.I.C.
Bât. SP2MI, Teleport 2, Bvd Marie et Pierre Curie, B.P. 30179
86962 Futuroscope Chasseneuil Cedex, France
E-mail: {andres, damiand, lienhardt}@sic.univ-poitiers.fr

Library of Congress Control Number: Applied for

CR Subject Classification (1998): I.4, I.3.5, G.2, I.6.8, F.2.1

ISSN      0302-9743
ISBN-10   3-540-25513-3 Springer Berlin Heidelberg New York
ISBN-13   978-3-540-25513-0 Springer Berlin Heidelberg New York

Springer is a part of Springer Science+Business Media

springeronline.com

© Springer-Verlag Berlin Heidelberg 2005
Printed in Germany

Typesetting: Camera-ready by author, data conversion by Scientific Publishing Services, Chennai, India
Printed on acid-free paper      SPIN: 11414292      06/3142      5 4 3 2 1 0

# Preface

In 2005, the twelfth edition of the conference Discrete Geometry for Computer Imagery was held in Poitiers, France, April 13–15, 2005. The conference was organized by the laboratory SIC ("Signal, Image, Communications") of the University of Poitiers, Centre National de la Recherche Scientifique and the Technical Committee 18 of the International Association for Pattern Recognition. DGCI 2005 was sponsored by the Faculty of Science, the University of Poitiers, the Conseil Général de la Vienne and the Region of Poitou-Charentes. The aim of the conference was to present recent advances in both theoretical aspects and applications in discrete geometry.

This year's conference was organized in combination with the 5th Workshop on Graph-Based Representations in Pattern Recognition, April 11–13, 2005 also organized in Poitiers. The workshop GbR aims at using graph-based structures in image analysis. There is a strong connection between the community interested in the GbR workshop and the discrete geometry community. For this reason, for the first time, both the workshop and the DGCI conference were organized in the same place, with a common session of four papers, two submitted to GbR and two to DGCI.

The DGCI conference attracted again for this edition many excellent papers, with 53 submitted papers from 21 countries. After careful reviewing by two and sometimes three reviewers, 36 papers were accepted, from which 22 were selected for oral presentation and 14 as posters. These contributions were regrouped into topics: applications, discrete topology, discrete hierarchical geometry, discrete tomography, object properties, recognition, and reconstruction, uncertain geometry, and visualization.

The program was completed by invited lectures from internationally known speakers: As first invited speaker we had Achille Braquelaire, who spoke about 2D images and planar maps. His contribution can be read in the GbR conference proceedings, also published by LNCS 3434. As second invited speaker, Peter Veerlaert spoke about uncertain geometry. Finally Jean-Pierre Guédon presented the Mojette Transform.

Many people contributed to the renewed success of this conference and we would like to thank them all, and in particular the authors who submitted papers and the invited speakers for their contributions. A special thanks also to the Steering Committee, and to the Program Committee whose members reviewed so many papers in a very efficient way.

Finally, we thank all the participants and we hope that they enjoyed their stay in Poitiers.

April 2005

Éric Andrès
Guillaume Damiand
Pascal Lienhardt

# Organization

DGCI 2005 was organized by the Signal, Image, Communications Laboratory of the University of Poitiers in France. The conference venue was the IFMI building on the Futuroscope campus ground in Chasseneuil du Poitou. The conference was sponsored by the International Association for Pattern Recognition (IAPR).

## Conference Chairs

Éric Andres      SIC, Poitiers, France
Guillaume Damiand      SIC, Poitiers, France
Pascal Lienhardt      SIC, Poitiers, France

## Steering Committee

Gabriella Sanniti di Baja      Italy
Achille Braquelaire      France
Gunilla Borgefors      Sweden
Jean-Marc Chassery      France
Annick Montanvert      France
Gabor Szekely      Switzerland

## Programme Committee

Reneta Barneva      Gilles Bertrand      Valentin E. Brimkov
David Coeurjolly      Michel Couprie      Leila De Floriani
Isabelle Debled-Rennesson      Ulrich Eckhardt      Oscar Figueiredo
Christophe Fiorio      Atsushi Imiya      Pieter Jonker
Ron Kimmel      Nahum Kiryati      Christer O. Kiselman
Walter G. Kropatsch      Jacques-Olivier Lachaud      Gregoire Malandain
Remy Malgouyres      Serge Miguet      Ingela Nyström
Klette Reinh      Pierre Soille      Edouard Thiel

# Table of Contents

## Applications

Increasing Interconnection Network Connectivity for Reducing
Operator Complexity in Asynchronous Vision Systems
*Valentin Gies, Thierry M. Bernard* ............................. 1

Geometric Robot Mapping
*Rolf Lakaemper, Longin Jan Latecki, Xinyu Sun, Diedrich Wolter* .... 11

Discrete Geometry Applied in Hard Real-Time Systems Validation
*Gaëlle Largeteau, Dominique Geniet, Éric Andrès* ................. 23

## Discrete Hierarchical Geometry

Hierarchical Watersheds Within the Combinatorial Pyramid Framework
*Luc Brun, Myriam Mokhtari, Fernand Meyer* ..................... 34

Optimal Design of 2D/3D Hierarchical Content-Based Meshes for
Multimedia
*Işil Celasun, Rupen Melkisetoğlu, A. Murat Tekalp* ................ 45

Receptive Fields for Generalized Map Pyramids: The Notion of
Generalized Orbit
*Carine Grasset-Simon, Guillaume Damiand, Pascal Lienhardt* ....... 56

Resolution Pyramids on the FCC and BCC Grids
*Robin Strand, Gunilla Borgefors* ................................. 68

## Discrete Tomography

The Mojette Transform: The First Ten Years
*JeanPierre Guédon, Nicolas Normand* ........................... 79

On the Stability of Reconstructing Lattice Sets from X-Rays Along
Two Directions
*Andreas Alpers, Sara Brunetti* ................................. 92

Reconstruction of Decomposable Discrete Sets from Four Projections
*Péter Balázs* ................................................ 104

A Tomographical Characterization of L-Convex Polyominoes
    *Giusi Castiglione, Andrea Frosini, Antonio Restivo, Simone Rinaldi* . .  115

Computerized Tomography with Digital Lines and Linear Programming
    *Fabien Feschet, Yan Gérard* . . . . . . . . . . . . . . . . . . . . . . . . . . . . . . . . . . . .  126

A Discrete Modulo $N$ Projective Radon Transform for $N \times N$ Images
    *Andrew Kingston, Imants Svalbe* . . . . . . . . . . . . . . . . . . . . . . . . . . . . . . . .  136

Two Remarks on Reconstructing Binary Vectors from Their Absorbed
Projections
    *Attila Kuba, Gerhard J. Woeginger* . . . . . . . . . . . . . . . . . . . . . . . . . . . . . .  148

How to Obtain a Lattice Basis from a Discrete Projected Space
    *Nicolas Normand, Myriam Servières, JeanPierre Guédon* . . . . . . . . . . .  153

## Discrete Topology

Local Characterization of a Maximum Set of Digital $(26, 6)$-Surfaces
    *Jose C. Ciria, Angel de Miguel, Eladio Domínguez,*
    *Angel R. Francés, Antonio Quintero* . . . . . . . . . . . . . . . . . . . . . . . . . . . . .  161

Algorithms for the Topological Watershed
    *Michel Couprie, Laurent Najman, Gilles Bertrand* . . . . . . . . . . . . . . . .  172

The Class of Simple Cube-Curves Whose MLPs Cannot Have Vertices
at Grid Points
    *Fajie Li, Reinhard Klette* . . . . . . . . . . . . . . . . . . . . . . . . . . . . . . . . . . . . . .  183

Computation of Homology Groups and Generators
    *Samuel Peltier, Sylvie Alayrangues, Laurent Fuchs,*
    *Jacques-Olivier Lachaud* . . . . . . . . . . . . . . . . . . . . . . . . . . . . . . . . . . . . . . .  195

Inclusion Relationships and Homotopy Issues in Shape Interpolation
for Binary Images
    *Javier Vidal, Jose Crespo, Victor Maojo* . . . . . . . . . . . . . . . . . . . . . . . . .  206

## Object Properties

Discrete Bisector Function and Euclidean Skeleton
    *Michel Couprie, Rita Zrour* . . . . . . . . . . . . . . . . . . . . . . . . . . . . . . . . . . . .  216

Pixel Queue Algorithm for Geodesic Distance Transforms
    *Leena Ikonen* . . . . . . . . . . . . . . . . . . . . . . . . . . . . . . . . . . . . . . . . . . . . . . . .  228

Analysis and Comparative Evaluation of Discrete Tangent Estimators
  *Jacques-Olivier Lachaud, Anne Vialard, François de Vieilleville* ...... 240

Surface Volume Estimation of Digitized Hyperplanes Using Weighted
Local Configurations
  *Joakim Lindblad* ............................................... 252

Rectification of the Chordal Axis Transform and a New Criterion for
Shape Decomposition
  *Lakshman Prasad* ............................................... 263

## Reconstruction and Recognition

Generalized Functionality for Arithmetic Discrete Planes
  *Valerie Berthé, Christophe Fiorio, Damien Jamet* .................. 276

Complexity Analysis for Digital Hyperplane Recognition in Arbitrary
Fixed Dimension
  *Valentin E. Brimkov, Stefan S. Dantchev* ......................... 287

An Elementary Algorithm for Digital Line Recognition in the General
Case
  *Lilian Buzer* ................................................... 299

Supercover Model and Digital Straight Line Recognition on Irregular
Isothetic Grids
  *David Coeurjolly* ............................................... 311

Discrete Epipolar Geometry
  *Masatoshi Hamanaka, Yukiko Kenmochi, Akihiro Sugimoto* .......... 323

Local Point Configurations of Discrete Combinatorial Surfaces
  *Yukiko Kenmochi, Yusuke Nomura* ............................... 335

Reversible Polygonalization of a 3D Planar Discrete Curve: Application
on Discrete Surfaces
  *Isabelle Sivignon, Florent Dupont, Jean-Marc Chassery* ............ 347

## Uncertain Geometry

Uncertain Geometry in Computer Vision
  *Peter Veelaert* ................................................. 359

Optimal Blurred Segments Decomposition in Linear Time
*Isabelle Debled-Rennesson, Fabien Feschet, Jocelyne Rouyer-Degli* .... 371

Shape Preserving Digitization of Binary Images After Blurring
*Peer Stelldinger, Ullrich Köthe* .................................. 383

# Visualization

A Low Complexity Discrete Radiosity Method
*Pierre Y. Chatelier, Rémy Malgouyres* ........................... 392

A Statistical Approach for Geometric Smoothing of Discrete Surfaces
*Bertrand Kerautret, Achille Braquelaire* ......................... 404

Arbitrary 3D Resolution Discrete Ray Tracing of Implicit Surfaces
*Nilo Stolte* ..................................................... 414

**Author Index** .................................................. 427

# Increasing Interconnection Network Connectivity for Reducing Operator Complexity in Asynchronous Vision Systems

Valentin Gies and Thierry M. Bernard

ENSTA, 32 Bd Victor 75015, Paris, France
contact@vgies.com,
http://www.ensta.fr/uer/uei/eng/index.html

**Abstract.** Due to the restriction of SIMD mode to local operations in VLSI massively parallel vision chips, using programmable connections and asynchronous communications are key ingredients to support regional computations. Asynchronism implies using combinatorial multi-input operators having an important hardware cost. To reduce it, we propose to use a connection network having a connectivity level greater than the mesh being mapped. This solution allows to use only 2-inputs asynchronous operators having a reduced hardware cost in each pixel. Examples and results will be presented on the examples of the regional sum algorithm computed over a 4-connectivity squared mesh connected with a 6-connectivity interconnection network, and the regional sum computed over a 6-connectivity squared mesh connected with a 8-connectivity interconnection network.

## 1 Introduction

An artificial retina is an image sensor with a processing element (PE) in each pixel. Such VLSI circuits are also called "vision chips" [1]. Motivated by the low power implementation of vision applications, we focus our research [2] on digital programmable artificial retinas (PAR), for which the PE is a tiny digital processor called the pixellic processor. The latter allows the on-site processing of data from the pixel or its neighbors, according to instructions provided by an external program.

The basic operating mode of a PAR is the SIMD mode (Single Instruction Multiple Data) : at a given time, the same instruction is simultaneously executed by each pixellic processor. SIMD mesh arrays for image processing were popular in the eighties as they allow the efficient implementation of local and shift-invariant operators (linear filtering, mathematical morphology, ...). But they were later abandoned due to several drawbacks. Nowadays, SIMD processing has come back into favor within commercial microprocessors in order to cope with frequency and power consumption limitations. While PARs fully benefit from the SIMD low power advantages, they are much less subject to SIMD

E. Andres et al. (Eds.): DGCI 2005, LNCS 3429, pp. 1–10, 2005.
© Springer-Verlag Berlin Heidelberg 2005

drawbacks than the mesh arrays of the eighties. Still, the SIMD mode is only well adapted to low-level vision.

Rather than processed images produced by low-level vision operators, a PAR should ideally output image descriptors, which can only result from higher levels of vision. These descriptors are based on regions resulting from a segmentation of the image. These regions need efficient regional operators for manipulating it. In contrast with neighbor-to-neighbor communications used in PARs for low-level vision, regional operators need to communicate between sparse and distant pixels.

Programmable neighbor-to-neighbor connections [3] allow to implement data-dependant communication networks within the SIMD framework, but with very poor synchronous performances. In the synchronous case, communication speed is limited to "one pixel farther per clock cycle".

Suppressing the above drawback leads to use asynchronous instead of synchronous communication. Thus PARs have to feature programmable connections and asynchronous communications and computations to efficiently handle regions.

This paper first presents some existing solutions for computing an exemplary asynchronous regional task, the "regional sum" in a 4-connectivity network using a dedicated asynchronous adder in each pixel. Since this adder cost is prohibitive for very large scale implementations, we propose a new communication network based on 6-connectivity reducing the hardware cost of asynchronous operators by reducing their necessary inputs to the minimum possible. Algorithm for installing the communication network is presented, a hardware cost comparison is proposed.

## 2     Linear Bit-Serial Multi-input Adder

Computing a regional sum implies to collect data from all the pixels of the region. Collecting and adding these data in one chosen place implies moving data on long distances. In order to overcome this problem, data must be added locally. To do this, a possibility is to chain pixels with an adder operator inside the pixel. The operator will have to add the binary value provided by the preceding pixel, and the local value. For digital sum computation, bits have to be processed one after the other, from less significant bit to most significant bit. In this case, one also has to sum the carry stored in each pixel during the computation of the preceding bit sum. Finally, the local operator has to be an adder able to compute the sum of 3 binary inputs. The operator used is a full adder.

The principle of the global addition is explained in fig.1.This solution has been proposed and implemented by in [4] using dynamically reconfigurable chains of pixels set by external programming.

### 2.1     Sum Algorithm

In each processor, the full adder inputs are connected to local binary data (internal bit and carry) and to the preceding full adder less significant output bit (usually called the sum bit). The least significant output bit is connected to the

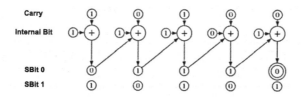

**Fig. 1.** Linear bit-serial multi-input adder process

next full adder input in the chain. This bit can also be seen as the parity of the number of 1's input to the full adder in that pixel. By associativity of the parity operator, the result at the end of the chain can be interpreted as the parity of the number of 1's among all local binary inputs in the chain (fig.1). This value is also the least significant bit of the sum of local binary data of the whole region. To start the regional sum algorithm, the least significant bit of the operand in each pixel is placed in the internal bit while all carries are reset. The first step is to run the global combinatorial sum computation and to get the least significant bit of the sum in the adder at the end of the chain (displayed as a double circled *Sbit0* in fig.1). The second step is to move the most significant bit (called *SBit1* in fig.1) of each full adder in the carry bit. Besides, local values corresponding to the next bit of the operands are loaded in the internal bit of the pixel.

Then, the process is iterated to produce each bit of the sum, as the output of the processor in the pixel at the end of the chain.

This algorithm computes the regional sum in $N$ combinatorial operations where $N$ is the number of bits required to represent the sum. These combinatorial operations are executed in a synchronous sequence. Since the regional sum operator is based on the ripple propagation, from pixel to pixel of parity information from the beginning to the end of the chain, we consider it as an *asynchronous* operator.

## 2.2    Limitations

Although this implementation allows to compute the regional sum quickly, the main problem is that a chain is a linear structure (cf. fig.2), and it is impossible to cover arbitrary connected sets of pixels with chains. Figure 3 shows a simple example of this impossibility. In such a situation, a tree-based bit-serial multi-input adder is needed instead of a linear one.

**Fig. 2.** Linear adder

**Fig. 3.** Tree adder

# 3    Tree Bit-Serial Multi-input Adder

The asynchronous algorithm used to compute a sum over an arbitrary shaped region is an extension of the linear bit-serial multi-input one presented in previous section. The main difference is the tree structure of the global adder. Such a structure has been implemented in the Associative Mesh of Orsay [5] [6].

What is a tree? It is a direct acyclic graph [7]. That means there is no loop (it is impossible to find 2 points connected by more than one direct subgraph) and there is only one root in it. The acyclic property is needed for using non idempotent associative operators (such as sum) on a graph [8]. The root is be used to collect the sum information computed on the graph.

Every pixel in the region is connected to this root through a spanning tree (fig. 3). As a consequence, inputs of the adder are connected to local binary data (internal bit and carry) and to *all* the directly preceding full adder least significant output bits in the tree. Consecutively, in 4-connectivity, each pixel different from the root can be connected to up to 3 neighbors as input of the adder, the fourth one being necessarily connected to the output of the adder. Taking into account 1 local binary data (both binary local data are added synchronously before the asynchronous sum), a total of 4 binary inputs are needed for the local adder. A Wallace tree analysis shows that 1 full adders and 2 half adder are needed to perform this task in a combinatorial way.

The algorithm used to compute the sum is very similar to the one described in section 2.1. The only difference lies in the number of inputs of the adders.

At this point, one main issue is still remaining : How to install a spanning tree over a region using a fast enough procedure regardless of the region shape?

## 3.1    Asynchronous Spanning Tree Installation

A spanning tree cannot be settled efficiently in a synchronous way, because the number of steps of the algorithm would grow linearly with the geodesic diameter, and it would take a long time. So, we have to perform this task in an asynchronous way.

The asynchronous algorithm used is the following one. At initialization, all connections between pixels of a same region are established. All pixels are inactive and a root is chosen, deterministically or at random. Then the root is activated, and communicate its state to the neighbor pixels. Each activated pixel keeps in memory the connection through which it was activated, and forward its active state to its neighbors. This process propagates asynchronously throughout the region until all pixels are active. The oriented spanning tree is obtained looking at the unique connections used for the activation of each pixel.

As explained before, a spanning tree is a direct acyclic graph, that means each pixel must have only one antecedent. During the algorithm, a pixel may have to choose between 2 or more antecedents if they want to activate the considered pixel at the same time. For this reason a 4 inputs arbiter is needed in each pixel.

## 3.2    Tree Bit-Serial Adder Asynchronous Hardware Cost

The different asynchronous components needed to perform regional sum computation and spanning tree installation have been defined before. According to the specification above, the elementary processor asynchronous part is composed of a 4-input arbiter (32 transistors), a 4-input adder(44 transistors), and 6 programmable connections (necessary for choosing 3 out of 4 inputs). The hardware cost of the asynchronous part is finally 82 transistors. Such an important cost is worth being reduced for a VLSI implementation.

# 4    Network Topology for Regional Sum Computation over a 4-Connectivity Region

Asynchronous dedicated operators used for sum computation are expensive mostly because they have 4 inputs. To cut this transistor expense, a lighter structure is desirable. When looking at an example of computing network, we notice that most of the cells exploits their 4-input convergent operators as simple 1-input operators only. This is a waste of resources. Is there any way to better distribute the network, that allows the use of k-input convergent operators with $k$ smaller than 4?

First, let's recall that $k = 1$ corresponds to 1-input operators and is therefore insufficient. What about using a 2-input asynchronous operator? Let's call it a *2-input convergent operator*. Let's consider a computation network over a region with $m$ pixels. Connecting these $m$ pixels together in a tree structure requires exactly $m - 1$ operators, convergent or not. How to settle them among $m$ pixels? Only one 2-input convergent operator in each pixel could be enough. There are 2 main issues for implementing such a structure. The first one is the network topology needed to implement it, the second one is how to install the spanning tree.

## 4.1    Network Topology

We recall that a 4-connectivity network combined with the use of 2-input convergent operators is insufficient to set-up a network over an arbitrary shaped region. Let's consider a cross-shaped region of 5 pixels, such a region is an example of this impossibility (Fig. 5). A solution is to increase the connectivity level used. An 8-connectivity interconnection network could be an obvious solution to the problem, allowing vertical, horizontal and both diagonal connections. However, the hardware cost would be rather expensive. A better solution is to use 6-connectivity and 2-input convergent operators. First, let's show that this solution fits our needs. For this, let's consider the pixel matrix as a hexagonal mesh (Fig. 4). Thanks to hexagonal mesh properties, an arbitrary pixel configuration can be connected with only 2-input convergent operators. For example, installation of a spanning tree using only 2-input convergent operators over a 5 pixels cross-shaped region is proposed (Fig. 5),whereas it was impossible to map in 4-connectivity. 6-connectivity is the lowest connectivity level allowing the connection of an arbitrary shaped region into a tree structure with only binary

**Fig. 4.** Transformation from square to hexagonal mesh

**Fig. 5.** Spanning tree over a cross-shaped region

operators, and it leads to the lowest possible hardware cost using asynchronous computation operators.

### 4.2 Asynchronous Spanning Tree Installation

Using 4-inputs convergent operators, installation of a spanning tree is a rather simple task. Starting from a fully connected network, a signal propagates from the tree root through the network until all pixels have been reached (cf. 3.1). Using 2-inputs convergent operators, this task is much more difficult because at the initialization of the algorithm, each pixel can be connected to 2 pixels only, and not to all its neighbors. Connecting all the neighbors is something simplifying the construction of the spanning tree but fortunately it is not really necessary. One as only to ensure that propagation starting from one pixel will reach all the other pixels of the region. That means every pixel of the region has to be connected to all other pixels. Such a region is called a strongly connected component (SCC). The issue is how to build a SCC in 6-connectivity using only 2-inputs convergent operators.

**Algorithm Principles.** A way to solve this problem is to connect all the boundary pixels of the region into a clockwise oriented chain and then to connect all the pixels not connected yet and the boundary rings together. Fig.6 shows the original region on the top, and its corresponding hexagonal representation after the initialization of the connections on the bottom. As explained before, pixel inputs are connected to a maximum of 2 other pixels. According to this connection method, boundary pixels are connected in a SCC (a ring is a simple SCC), and other pixels are added to the SCC thanks to bi-directional connections, this ensuring them to be part of the SCC.

The final step of the proposed method is to extract a spanning tree from the SCC by propagating a token from the root as explained before in section 3.1.

**Algorithm for Connecting a SCC.** The algorithm used is very simple, and can performed in a very cheap and fast synchronous way. Initialization of the connections can be done by only considering 6 local pixel configurations as described

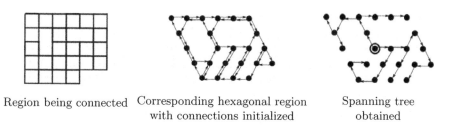

Region being connected   Corresponding hexagonal region   Spanning tree
with connections initialized   obtained

**Fig. 6.** Example of spanning tree installation using a 6-connectivity interconnection network and 2-inputs convergent operators over a 4-connectivity region

in Fig. 7. Configurations 1 to 6 are used for connecting boundary pixels. Configurations 5 and 6 also allow to connect all other pixels diagonally. In the different configurations of Fig. 7., the pixel to connect is double-circled. Black pixels are pixels belonging to the region while white ones are pixels outside the region.

An important fact is the non-isotropy of the local transformation. Configuration 5 and 6 are not rotated versions of configurations 1 and 2 or 3 and 4. This is a consequence of mapping a 4-connectivity square mesh onto an hexagonal network. Instead of configurations 5 and 6, using a $2\pi/3$ rotated versions of configurations 1 and 2 would lead to connect diagonal configurations of pixels not connected in a 4-connectivity squared mesh. Actually, the diagonal connection, not present in 4-connectivity, is used here for establishing all the non-boundary connections.

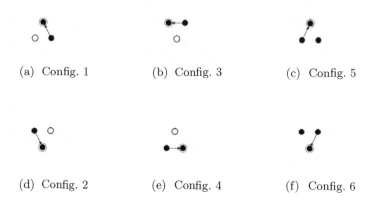

(a)  Config. 1          (b)  Config. 3          (c)  Config. 5

(d)  Config. 2          (e)  Config. 4          (f)  Config. 6

**Fig. 7.** Local configurations for SCC initialization in 4-connectivity using a 6-connectivity interconnection network

**Validity of the Algorithm.** Having presented the principles and operation of the algorithm, let's demonstrate its validity. By construction, all pixels are connected in a same SCC. By construction, all pixels of a same region are connected into one SCC. The only point to check is that pixel inputs do not have to be connected to more than 2 neighbor pixels, to allow the use of 2-inputs operators.

As shown in Fig.7, each configuration sets-up one input connection. We have to verify that if a pixel neighborhood corresponds to 2 configurations, all the other configurations are false. For this let's consider the mutual exclusions of the configurations. Configuration 1 excludes configuration 3 and 5. Configuration 3 excludes 1 and 5, and configuration 5 excludes 1 and 3. Finally only one odd-numbered configuration can be true at one time. It is the same for even configuration. Only one even configuration can be true for a given neighborhood. There are no exclusions between odd and even configurations. Finally, a pixel neighborhood can only match one even and one odd configuration. That means a maximum of 2 configurations can be valid at one time, and maximum 2 input connections will be set-up in the pixel.

**Performance of the Algorithm.** The proposed algorithm for initializing the SCC can be performed very efficiently in a synchronous non-iterative way. This allows using this algorithm on a massively parallel synchronous machine having only limited resources for synchronous computation. There is no hardware dedicated to the spanning tree initialization task, which means that the reduction of hardware cost due to the use of 2-input convergent operators does not imply additional costs for installing the spanning tree.

### 4.3    Hardware Reduction

Using 6-connectivity connections over a squared mesh allows to reduce the hardware cost dedicated to asynchronous regional sum to one 2-input arbiter (8 transistors) and one 3-inputs adder (20 transistors) in each pixel. 3 inputs are necessary for the adder : one for the local bit, and two for the neighbor connections. However, the necessary number of programmable connections increases. 2 out of 6 neighbors have to be connected at one time. Consequently, the minimal number of programmable connections necessary is 5 connections for each input. Finally, the number of transistors needed is $8 + 20 + 10 = 38$ transistors.

44 transistors are saved in each pixel by using a 6-connectivity topology. This leads to a reduction of 54% of the asynchronous hardware expense. However, the algorithmic capabilities are remaining the same.

## 5    Network Topology for Regional Sum Computation over a 6-Connectivity Region

### 5.1    Network Topology

Increasing the connectivity level of the interconnection network for reducing hardware cost of regional computation using only 2-input convergent operators

**Fig. 8.** Transformation from squared to hexagonal 8-connectivity mesh

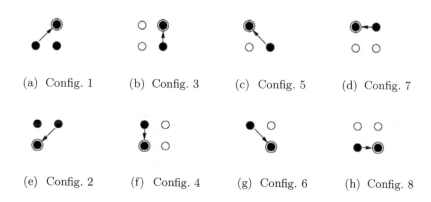

| (a) Config. 1 | (b) Config. 3 | (c) Config. 5 | (d) Config. 7 |

| (e) Config. 2 | (f) Config. 4 | (g) Config. 6 | (h) Config. 8 |

**Fig. 9.** Local configurations for SCC initialization in 6-connectivity using a 8-connectivity interconnection network

can be extended to 6-connectivity meshes. A 8-connectivity interconnection network is used. The interconnection network will be represented in a squared mesh because as shown in Fig. 8, in a squared mesh, a 8-connectivity neighborhood is more regular ( central symmetry) than in a hexagonal mesh (only 2 axial symmetries).

The principles of the algorithm remain the same as in 4-connectivity. Boundary pixels are connected into a clockwise oriented chain and non-boundary pixels are connected to the SCC formed by the boundary ring using a synchronous algorithm (cf. Fig. 9).There are 8 configurations to consider for setting the interconnection network. All theses configurations are used for connecting boundary pixels. Configurations 1 and 2 are also used for connecting non-boundary pixels to the boundary ring. With the proposed algorithm, an 6-connectivity network can be connected using 2-input convergent operators.

## 5.2    Hardware Reduction

Using an 8-connectivity interconnection network over a 6-connectivity hexagonal mesh allows to reduce the hardware cost dedicated to asynchronous regional sum to one 2-input arbiter (8 transistors) and one 3-inputs adder (20 transistors) in each pixel. 3 inputs are necessary for the adder : one for the local bit, and two for the neighbor connections. Considering interconnections, 2 out of 8 neighbors have to be connected at one time. Consequently, the minimal number of programmable connections necessary is 7 connections for each input. Finally, the number of transistors needed is $8 + 20 + 14 = 42$ transistors.

Using 7-input asynchronous adders and 6-input arbiters leads to an hardware cost of 156 transistors (72 for the adder and 84 for the arbiter). Finally, 114 are saved in each pixel by using a 8-connectivity interconnection network. This leads to a dramatic reduction of 73% of the asynchronous hardware expense, without lowering algorithmic capabilities.

## 6    Conclusion

We first presented implementations for regional sum computation through subsets of pixels in the image. After evaluating the hardware cost of dedicated operators needed for computing the sum asynchronously, we proposed using a 6-connectivity interconnection network and 2-input operators for reducing the hardware cost of asynchronism in 4-connectivity squared meshes. This leads to a important reduction of more than half the original transistor cost. The reduction is even more important (73%) when using an 8-connectivity interconnection network and 2-input convergent operators for mapping 6-connectivity region. Such a reduction offers an opportunity for achieving a dense large scale implementation of this circuit. Such an implementation is now on its way, and will lead to a vision chip allowing to perform medium level image processing.

## References

1. Moini, A.: Vision Chips. Kluwer Academic Publishers, ISBN: 0-7923-8664-7 (2000)
2. Paillet, F., Mercier, D., Bernard, T.: Second generation programmable artificial retina. In: IEEE ASIC/SOC Conf. (1999) 304–309
3. Li, H., Stout, Q.: Reconfigurable Massively Parallel Computers. Prentice-Hall, Englewood Cliffs, NJ (1991)
4. Komuro, T., Kagami, S., Ishikawa, M.: A dynamically reconfigurable simd processor for a vision chip. IEEE Journal of Solid-State Circuits **39** (2004) 265–268
5. Merigot, A.: Associative nets: A graph-based parallel computing net. IEEE Transactions on Computers **46** (1997) 558–571
6. Dulac, D., Mohammadi, S., Merigot, A.: Implementation and evaluation of a parallel architecture using asynchronous communications. In: CAMP. (1995) 106–111
7. Ducourthial, B., Merigot, A.: Graph embedding in the associative mesh model. Technical Report TR-96-02 (1996)
8. Ducourthial, B., Mérigot, A.: Parallel asynchronous computations for image analysis. Proceedings of the IEEE **90** (2002) 1218–1228

# Geometric Robot Mapping

Rolf Lakaemper[1], Longin Jan Latecki[1], Xinyu Sun[2],
and Diedrich Wolter[3]

[1] Temple University, Philadelphia, PA 19122, USA
{latecki, lakamper}@temple.edu
[2] Texas A&M University, College Station, TX 77840, USA
xsun@math.tamu.edu
[3] Universität Bremen, Bremen, Germany
dwolter@informatik.uni-bremen.de

**Abstract.** The purpose of this paper is to present a technique to create a global map of a robot's surrounding by converting the raw data acquired from a scanning sensor to a compact map composed of just a few generalized polylines (polygonal curves). To merge a new scan with a previously computed map of the surrounding we use an approach that is composed of a local geometric process of merging similar line segments (termed Discrete Segment Evolution) of map and scan with a global statistical control process. The merging process is applied to a dataset gained from a real robot to show its ability to incrementally build a map showing the environment the robot has traveled through.

**Keywords:** Robot Mapping, Polygon Merging, Polygon Simplification, Perceptual Grouping.

## 1 Introduction

Imagine a scenario where a robot explores an unknown terrain. The goal is to acquire in real time a global overview map integrating all measurements collected by the robot. Here we deal with measurements obtained by 2D range sensors, called scans, that represent partial top views of the robots environment. Building a global overview map from scans is a typical scenario for rescue robots, where the overview knowledge, in the form of a global map, is particularly important to localize victims in catastrophe scenarios (e.g., in collapsed buildings) and to ensure that the whole target region has been searched [4]. Since odometry information under such conditions is very unreliable, we assume that it is not available. Also landmarks are ambiguous.

The whole process from reception of raw scanning data to the final map refers to the problem of simultaneous localization and mapping (SLAM) in robotics. The proposed approach addresses two main problems in SLAM stated in [10].

E. Andres et al. (Eds.): DGCI 2005, LNCS 3429, pp. 11–22, 2005.
© Springer-Verlag Berlin Heidelberg 2005

1. The measurement errors are statistically dependent, since errors in control accumulate over time, and they affect the way future sensor measurements are interpreted.
2. The second complicating aspect of the robot mapping problem arises from the high dimensionality of the entities that are being mapped, which leads to serious runtime and storage problems.

We address the problem of measurement errors being statistically dependent with a new process of map merging that is based on geometric local process of line segment merging with a global statistical control.

The second problem arises from the fact that in most mapping approaches the objects of which maps are built are simply points. These are either directly scan reflection points or point landmarks, e.g., [2] and [10]. In some approaches simple geometric features, especially line segments [7, 9, 1] are used. However, the maps are still composed of a huge numbers of them, since these approaches do not provide any mechanisms to incrementally reduce the number of building blocks, which can be line segments of simply points. Consequently, the obtained maps are composed of thousands or even millions of points or line segments. An example of such a map is shown in Fig. 1(a). It is composed of 144400 points and obtained by alignment of 400 scans. It is then clear that such maps lead to serious runtime and storage problems, e.g., it is impossible to map larger environments and to perform loop closing in real time. In our map representation, we simply do not run into the second problem. Our representation is built of higher level objects, which are line segments and generalized polylines, and we have an explicit process, called Discrete Segment Evolution, that reduces the number of line segments to a minimal number required to represent the mapped environment. An example map obtained by our approach is shown in Fig. 1(b). This map was obtained from the same scan data as the map in (a), and it is composed of only about 50 line segments (which amounts to about 100 endpoints). Videos illustrating our incremental mapping results can be viewed on http://knight.cis.temple.edu/~shape/robot/.

A nice probabilistic framework to construct a global map from scan data is presented in [8]. However, this framework is based on the assumption that the uncertainty of scan points' positions is known. Due to the dependence of laser scan measurements on surface characteristics of scanned objects, e.g., glass-like surface, brick wall, and metal surface, this assumption is not satisfied in our example of rescue robots. We approach the problem of constructing a global map using the principles of perceptual grouping, which look for geometric structures in the data without any assumptions about the error characteristics [6].

## 2   Robot Mapping

In this section we introduce some notation regarding the system used by a robot to create its global map of its surroundings, and we summarize the main steps performed at each iteration of the algorithm, i.e., on the arrival of a new scan.

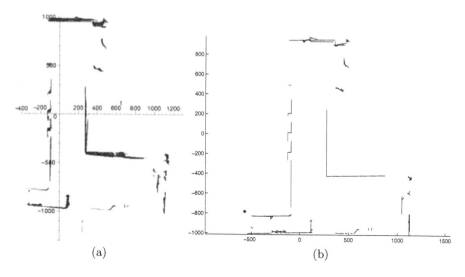

(a)                                              (b)

**Fig. 1.** (a) A global map obtained by alignment of 400 scans is composed of 144400 points. (b) A global map obtained by the proposed map merging algorithm is composed of about 50 line segments (100 endpoints). Both maps are obtained from the same laser range data set showing a hallway at the Univ. of Bremen

The output is a global map that represents a top view of the environment using a small number of polylines. Fig. 1(b) illustrates such a map. For comparison, a global map obtained by alignment only is shown in Fig. 1(a).

The proposed algorithm merges the newest laser range scan $S_t$ at time $t$ with a global map $G_{t-1}$ built from previous scans 0 to $t-1$. The global map is produced incrementally, which means that at every time $t$ we have a ready to use and a very simple global map of the environment. This is very important for all navigation and mapping tasks. Both global map $G_t$ and scan $S_t$ are composed of *generalized polylines*.

A *generalized polyline* is a set of line segments, having a specific ordering, whose vertices may or may not be connected. Observe that a classical definition of a polyline (polygonal curve) requires that the endpoints of consecutive segments coincide. Generalized polylines result naturally when scan points are approximated with line segments, which is our first processing step of the input range data. By dropping the constraint that a polyline be composed of line segments whose vertices are connected, we do not introduce additional noise that would result from connecting these vertices. The usage of generalized polylines is particularly important in the polyline merging and shape similarity algorithms described below.

Our first processing step (approximation of scan points with line segments) is followed by the segment grouping step. We form an ordered list of segments by minimizing the sum of the distances of their endpoints. Finally, if the endpoints of consecutive segments are too far apart, we split the list into sublists. Thus, generalized polylines are sublists of this list.

To create a global map $G$, we start with the first global map $G_1$ being equal to the first scan $S_1$. Henceforth, assuming we have created the global map $G_{t-1}$ at time $t - 1$ and a new scan $S_t$ has arrived, $G_t$ is created in the following three steps:

**Correspondence:** We use a very simple and common approach to establish correspondence between global map $G_{t-1}$ and a new scan $S_t$. We position the new scan $S_t$ at the pose of the previous scan $S_{t-1}$ that has been aligned to the previous global map $G_{t-2}$ (in the process of construction of the actual global map $G_{t-1}$). The pose is the position and rotation angle in the coordinates of the global map. Then the correspondence is establish by mapping each segment endpoint of $S_t$ to a closest point in $G_{t-1}$ and the same with roles of $S_t$ and $G_{t-1}$ interchanged.

This process of correspondence computation works fine if there is continuity in the robot pose, i.e., robot pose changes only slightly from scan to scan. If the continuity assumption is not satisfied, we use shape similarity to establish the correspondence [5] and [11]. However, this approach is outside the scope of this paper.

**Alignment:** The current scan $S_t$ is rotated and translated until a minimum distance is found between the corresponding points. Then the closest points are found again, and the whole process is repeated until it stabilized. This algorithm to align scan $S_t$ and map $G_{t-1}$ is called Iterative Closest Point (ICP), and is described in [3].

**Merging:** This is the main contribution of this paper and its detailed discussion follows in Section 2.1. The output of alignment overlays the actual scan on the global map, but the surfaces of the same objects are still represented by separate polylines. The goal of merging is to represent surfaces of the same objects by single polylines.

## 2.1   Merging

Merging is the task of combining similar segments taken from two aligned maps to form new segments in a joint map. The similarity between pairs of segments is modeled following principles of perceptual grouping. In the case of incremental building of a global map, the task of merging is to combine similar line segments of the new scan, $S_t$, and the previous global map, $G_{t-1}$, to form new segments that define a new and current global map $G_t$. We assume that $S_t$ has been aligned to $G_{t-1}$. Merging consists of two steps, which integrate the new information contained in $S_t$ with the previous global map to produce $G_t$. The two steps are restrictive pairing and simplification.

**Restrictive Pairing:** The result of pairing can intuitively be understood as *visual average* of $S_t$ and $G_{t-1}$. Since our goal is to combine the information from both maps, we allow only pairing of segments from different maps. Therefore, we define two classes of line segments, class $C_1$ is a list consisting of segments from $G_{t-1}$ and class $C_2$ is a list consisting of segments of $S_t$.

Although the final task of merging is to decrease the number of line segments by combination, the pairing step goes into the opposite direction: it might create many new segments, which will be simplified in the second step called simplification. Pairing can be compared to a pencil drawing technique known as sketching, e.g., used for cartoon drawings: to find the final outline of an object, it is first approximated by a larger number of light strokes, giving the eye the opportunity to imagine and select the correct position.

The process creates all possible pairs of line segments that are sufficiently similar, taking one segment from $C_1$ and one from $C_2$. Pairing of two segments from the same class is not allowed.

The similarity of line segments $L_1$ and $L_2$ is measured with the cost function $C(L_1, L_2, ad)$ (defined below), where $ad$ is an angular direction given by global statistics (defined below). If $C(L_1, L_2, ad)$ is below a given threshold, we create a new line segment $ms(L_1, L_2, ad)$ (defined below) that is visually close to $L_1$ and $L_2$. Figure 2 shows an example of restricted pairing. The newly created segments must follow the main directions, i.e., they are only allowed to have angles of $0°, +60°, -60°$ with the $x$ axis.

A single line segment $L_1$ can create many *children* line segments $ms(L_1, L_2^i, ad)$ by pairing with segments $L_2^i$ for $i = 1, ..., n$. We need to allow a single segment to pair with more than one segment from the other class, since we do not know the exact segment correspondence. It might be that the correct shape feature is created by the pairing with a second partner. However, to limit the computational complexity, each line segment is allowed to create only a small number of segments (at most 3 children segments are allowed in our implementation).

We remove all line segments from $C_1$ and $C_2$ that were parents of at least one new segment. We denote the resulting lists by $C_1'$ and $C_2'$. We denote with $A_t$ a list of all resulting children together with $C_1'$ and $C_2'$. Formally, the output of restricted pairing is defined as ($T_p$ is a pairing cost threshold):

$$A_t = C_1' \cup C_2' \cup \{ms(L_1, L_2, ad) : C(L_1, L_2, ad) < T_p, L_1 \in C_1, L_2 \in C_2\}.$$

Restrictive pairing may create some small artifacts in addition to features present in the reality, such as parallel segments in Fig. 2(b). The artifacts may be introduced, since we do not know the exact correspondence of line segments, and therefore, must allow a single line segment to pair with many line segments in the other class. We therefore need a cleaning process to remove these artifacts, or, in analogy to the sketching example mentioned above, a process that selects or creates an appropriate precise set of strokes based on the approximation. This process is called *simplification* and it is the second step in the merging process. Its result is illustrated in Fig. 2(c).

**Simplification:** The input is the joint map $A_t$ created by restrictive pairing. Simplification can be viewed as cleaning process after pair creation to create a smaller set of possibly new segments, being the final merging result. To use the

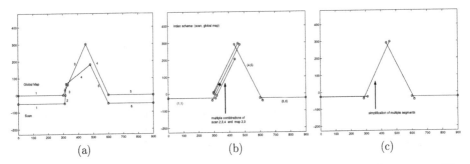

**Fig. 2.** The figure (b) is obtained from (a) by the proposed pair creation process. The index pairs in (b) refer to parent segments in (a). (c) shows the final result obtained by simplification of (b). The newly created segments in (b) and (c) must follow the main directions of $0°, +60°, -60°$ with the $x$ axis

sketch analogy again, the simplification process creates a single line by visually averaging the approximating bundle of strokes. Pairs of line segments are merged together to form new segments using the same merging process $ms$ and cost function $C$ but with different constraints. Simplification is done without any class restriction, i.e., a line segment in $A_t$ can pair with any other segment in $A_t$.

The main difference is that the simplification process has a global control mechanism: We iteratively merge a pair whose merging cost $C$ is the lowest at each pass. More precisely, the segment pair $L_1, L_2$ with lowest cost $C(L_1, L_2, ad)$, for one of the main angular directions $ad$, is merged in each pass. This means that $ms(L_1, L_2, ad)$ is inserted to list $A_t$, and $L_1, L_2$ are removed from $A_t$. Thus, each line segment can have at most one child. The process stops when the lowest cost is above a threshold $T_s$. The resulting simplified version of $A_t$ is the new global map $G_t$. Fig. 3(c) shows the simplified version of line segments in (b). Observe that both merged shape features from (a) are preserved (the straight line and the tent). This is acceptable, since given the input as in Fig. 3(a), we

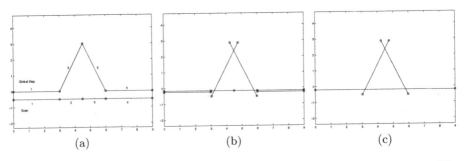

**Fig. 3.** The figure (b) is obtained from (a) by the proposed pair creation process. The index pairs in (b) refer to parent segments in (a). (c) shows the result simplification of (b). The newly created segments in (b) and (c) have to follow the main directions of $0°, +60°, -60°$ with the $x$ axis

cannot decide which feature is the right one. This decision can be, however, made after merging several consecutive scans.

The main angular directions for simplification are computed with the same global statistics as in pairing process. This allows us to cope with accumulative errors. Section 2.2 gives more details on global statistics.

## 2.2     Main Directions

The main directions $ad$ are obtained as significant peaks in the direction histogram of line segments (angles with the $x$ axis) in $G_{t-1}(tm)$, where $G_{t-1}(tm)$ is a global map $G_{t-1}$ restricted to line segments created in at most $tm$ time steps ago. $tm$ is a *time memory* factor. This restriction implies that only part of global map recently created determines the main directions. Therefore, the choice of directions is dynamic: the set of main directions $ad$ is created with respect to the most recent part of the global map. This dynamic process assures the robustness of the algorithm with respect to main directions, e.g., for indoor environments, while simultaneously being flexible enough to react to changing or even non present main directions, e.g., outdoor or natural environments.

Each line segment in the restricted global map $G_{t-1}(tm)$ contributes to the bin representing its direction with the weight given by its length. This histogram is cyclic, e.g., a window of size 5 around bin 1 contains bins 179, 180, 1, 2, 3.

An important feature of our approach is the interaction between main directions and merging: Let us assume we have a significant number of segments in $S_t$, all having a similar direction that is significantly different from main directions present in $G_{t-1}(tm)$, i.e., these segments do not define a main direction yet. For example, this is the case if the robot starts to perceive a new surface.

Since the new segments follow a significantly different direction, the cost function $C$ assures that they create no children, because the pairing cost is above the threshold $T_p$, i.e., the transformation of the parent segments to child segments is too expensive. Thus, the original segments will remain in $A_t$. Since $A_t$, after simplification, becomes the new global map $G_t$, the new segments will lead to a peak, and will open a new main direction in the direction histogram. Consequently, pairing and simplification in the new direction will be allowed. A real example will be given in Section 3.

The direction histogram represents the statistical distribution of line segment directions. This statistical control provides a solution to the problem of cumulative errors (Problem 1 in the introduction). Cumulative errors introduce systematic distortions in the directions of line segments that accumulate slowly. The main issue is that accumulative errors do not lead to peaks in the direction histogram, and consequently, appearing line segments are correctly mapped to the existing main directions. On the other hand, as we have just described, if a surface of a new object is oriented into a new direction, it will lead to a peak in the direction histogram after a few scans of the surface have been acquired. This solution is based on quantization of the angular directions. Thus, a new main direction is created in the direction of the new surface if the difference between the new direction and the existing main directions is larger then the quantization factor, which is determined by the thresholds $T_p$ and $T_s$. The fact that the direction histogram provides a solution to the problem of cumulative errors is also true for other histogram-based approaches, e.g., [9].

## 2.3   New Line Segment Creation and Cost Function

This section will describe the merging function $ms$ and the associated cost function $C$ used for the pairing and simplification process. The merging function is the most important module in the merging system, since it is responsible for the creation of the segments finally seen in the new map. Given a pair of line segments, $L_1$ and $L_2$, and the angular direction $ad$, it computes a merged segment $ms(L_1, L_2, ad)$ with the angular direction $ad$. The cost function $C(L_1, L_2, ad)$ is responsible for the filtering step: it produces the basic values for the decision if a created segment will be accepted or rejected as a member of the approximated map $A_t$ or the new global map $G_t$. It measures the similarity of $L_1$ and $L_2$ in the context of the main direction $ad$.

The geometric intuition of the presented merging process and, in particular, the definition of merging cost $C(L_1, L_2, ad)$ is based on cognitively motivated principles of perceptual grouping. We followed the approach presented in [6] on grouping line segments to form longer line segments. It states that proximity of endpoints, parallelism, and collinearity are the main geometric relations that influence the perceptual grouping of line segments. Our setting is slightly different, since we merge two line segments only with respect to a given main direction $ad$. Therefore, we developed a new cost function. As mentioned above, the usage of main directions is necessary to cope with cumulative errors.

Before we explain the meaning of the perceptual grouping principles in our setting, we need to introduce one more concept of a straight line $ld$ that follows one of the main directions $ad$. Let two line segments $L_1$ and $L_2$ and a main angular direction $ad$ be given. A first step in our cost computation is to position a line following direction $ad$ between two line segments $L_1$ and $L_2$. The straight line $ld$ with direction $ad$ is positioned between $L_1$ and $L_2$ so that the equation

$$d_1 \cdot l_1 = d_2 \cdot l_2$$

is satisfied (see Figure 4), where $l_i$ is the length of segment $L_i$ and $d_i$ is the distance of the midpoint of $L_i$ to line $ld$ for $i = 1, 2$. The **merged segment** $ms(L_1, L_2, ad)$ is defined by the convex hull of the projections of $L_1$ and $L_2$ on line $ld$. It is the segment from $P_1'$ to $P_4'$ in Figure 4.

Now we can explain the meaning of the perceptual grouping principles in our setting.

- **Parallelism:** The greater the angles between $L_1$ and $L_2$, and between $L_1$ and $L_2$ and $ms(L_1, L_2, ad)$, the greater the cost of merging them together. Likewise, the angle difference of longer segments have more weight than shorter ones.

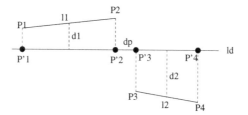

**Fig. 4.** Geometric illustration of the cost function $C(L_1, L_2, ad)$

- **Collinearity:** The greater the distance of the endpoints of $L_1$ and $L_2$ from the target line $ld$, the higher the value of the cost function.
- **Proximity:** The greater the distance between the projections of $L_1$ and $L_2$ on line $ld$, the higher the value of the cost function.

Finally we can define the new cost function that integrates the three perceptual grouping principles. Given the line $ld$, the cost of merging $L_1$ and $L_2$ to $ms(L_1, L_2, ad)$ is defined by the following measure that incorporates our realization of the perceptual grouping principles (see Figure 4):

$$C(L_1, L_2, ad) =$$
$$\frac{lr\left((l_1 + 1)\frac{d(P_1,ld)+d(P_2,ld)}{1+K\cos(a_1)} - l_1\frac{d(P_1,ld)+d(P_2,ld)}{1+K}\right)}{1 + K\cos(a_{12})} +$$
$$\frac{lr\left((l_2 + 1)\frac{d(P_3,ld)+d(P_4,ld)}{1+K\cos(a_2)} - l_2\frac{d(P_3,ld)+d(P_4,ld)}{1+K}\right)}{1 + K\cos(a_{12})}$$

where $a_1$ is the angle between $L_1$ and $ld$, $a_2$ is the angle between $L_2$ and $ld$, $a_{12}$ is the angle between $L_1$ and $L_2$, $P_1, P_2$ are endpoints of $L_1$, $P_3, P_4$ are endpoints of $L_2$, $d(P_i, ld)$ is the distance between point $P_i$ and line $ld$. The constant $K$ depends on the metric units used, and need s to be adjusted to obtain a balance between angular and metric units. The length ratio

$$lr = \frac{l(ms(L_1, L_2, ad))}{l(p(L_1)) + l(p(L_2))}$$

is the quotient of the length of the merged line segment $ms(L_1, L_2, ad)$ to the sum of the length of the projections $p(L_1)$ and $p(L_2)$ of line segments of $L_1$ and $L_2$ on line $ld$.

## 3   Implementation Details

### 3.1   Merging

**Pairing:** As described in Section 2.1, the merging consists of pairing followed by simplification. The pairing step creates a rough approximation of the new map,

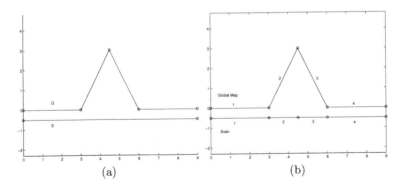

(a)                                    (b)

**Fig. 5.** New vertices are inserted in $S$ in (a) as projections of vertices in $G$. The vertex insertion is necessary to make the correspondence of line segments possible

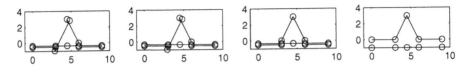

**Fig. 6.** Input map is from Figure 5(b). Pairing with different threshold $T_p$ and global direction control with main directions of 0, $+62°$, $-62°$. From left to right we have $T_p$ = 0.05, 0.04, 0.03, 0.01

and it must take into account features in both the scan and the global map, even if a certain feature is only present in one of them. An example is given in Fig. 5, where the triangle feature is only present in the global map $G$. To permit appropriate combinations with segments of this feature, the straight segment of the scan $S$ is split up to create segments corresponding to the feature segments of the global map. In order to do so a correspondence between endpoints of line segments is established. For every endpoint $E$ in $G$, we find a closest point $p(E)$ in $S$. If distance from $E$ to $p(E)$ is below a predefined threshold and $p(E)$ is not an existing end point in $S$, then $p(E)$ is inserted to $S$, splitting an existing line segment in $S$ into two collinear line segments (that meet at $p(E)$). We perform the same for every endpoint $E$ in $S$. The maps modified this way are the input to pair building.

The creation of descendants in the pairing process is dependent on the pairing threshold $T_p$. Fig. 6 shows the output of the pairing with global direction control of the input shown in Fig. 5 with different thresholds $T_p$, leading to different constellations: The leftmost map consists of newly created segments only. In the second figure the triangle segments could only pair once. The rightmost figure finally did not change the input; the threshold was too low to create any new segments.

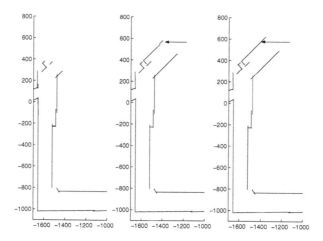

**Fig. 7.** Bremen hallway experiment: a new main direction is detected in the right figure and used for merging

## 3.2    Main Directions

The following section shows an example for the interaction between the merging process and the dynamic detection of main directions. The basic idea is not to merge segments with a cost above $T_p$ or $T_s$ respectively, but to keep them in their original position/direction Such segments can be described as outliers with respect to the current main directions $ad$. If these outliers become dominant, they will open a new main direction and will be merged. This process can be observed in Fig. 7 showing a section of an experiment with real laser finder data collected in a hallway at the University of Bremen. The robots' position is in the upper left corner, the robot moves up right and scans backward, i.e., down left.

In the left figure the robot approaches a $45^o$ corner, the current main directions consist of $0^o$ and $90^o$ angles with the $x$-axis. New $45^o$ segments are not simplified but transferred unchanged into the new global map. New $45^o$ segments still do not have enough weight to create a new direction in the center figure. Additionally the $0^o$ walls are out of sight, and therefore, the histogram loses this entry (which does not effect any segments in this situation). The only main direction remaining is $90^o$. In the right figure, new $45^o$ segments finally create a peak in the direction histogram. Hence the new main directions are now $90^o$ and $45^o$. Consequently, the new segments are merged together and simplified following the new $45^o$ direction.

## 4    Conclusions and Future Work

We presented a novel approach to robot mapping, based on an iterative process that merges similar line segments of subsequent scans. The process is locally controlled by a similarity of line segments, which is motivated by principles of perceptual grouping. Additionally the process is controlled by global statistics, giving emphasis to the main directions present in the map to eliminate effects of accumulated errors of local scans. The experiments performed on real robot data show that the interaction between local and global control is able to successfully adapt to changes in directions without any pre-knowledge of the environment to build the environmental map given by subsequent scans. The merging process can be useful for any application where (visual) simplification of sets of line segments is needed; further experiments will include simplification of images gained by edge detection and hand drawn sketches for stroke recognition.

## Acknowledgment

This work was supported in part by the National Science Foundation under grant INT-0331786 and grant R3 [Q-Shape] in the framework of the SFB/TR 8 Spatial Cognition from German Research Foundation (DFG). We would like to thank Sebastian Thrun for a helpful discussion regarding the topic of this paper. Thomas Röfer is acknowledged for providing scan data.

## References

1. Ingemar J. Cox. Blanche: Position estimation for an autonomous robot vehicle. In Ingemar J. Cox and G.T. Wilfong, editors, *Autonomous Robot Vehicles*, pages 221–228. Springer-Verlag, 1990.
2. G. Dissanayake, P. Newman, S. Clark, H.F. Durrant-Whyte, and M. Csorba. A solution to the simultaneous localization and map building (SLAM) problem. *IEEE Transactions of Robotics and Automation*, 2001.
3. A. Fitzgibbon. Robust registration of 2d and 3d point sets. In *Proc. British Machine Vision Conference, volume II, Manchester, UK*, pages 411–420, 2001.
4. A. Jacoff, E. Messina, and J. Evans. Performance evaluation of autonomous mobile robots. *Industrial Robot: An Int. Journal*, 29(3), 2002.
5. Longin Jan Latecki, Rolf Lakämper, and Diedrich Wolter. Shape similarity and visual parts. In *Proceedings of the 11th International Conference on Disrecte Geometry for Computer Imagery (DGCI), Naples, Italy*, November 2003.
6. D. G. Lowe. Three-dimensional object recognition from single two-dimensional images. *Artificial Intelligence*, 31:355–395, 1987.
7. F. Lu and E. Milios. Robot pose estimation in unknown environments by matching 2D range scans. *Journal of Intelligent and Robotic Systems*, 1997.
8. S. T. Pfister, S. I. Roumeliotis, and J. W. Burdick. Weighted line fitting algorithms for mobile robot map building and efficient data representation. In *ICRA*, 2003.
9. T. Röfer. Using histogram correlation to create consistent laser scan maps. In *Proceedings of the IEEE International Conference on Robotics Systems (IROS-2002)*, 2002.
10. S. Thrun. Robotic mapping: A survey. In G. Lakemeyer and B. Nebel, editors, *Exploring Artificial Intelligence in the New Millenium*. Morgan Kaufmann, 2002.
11. D. Wolter and L. J. Latecki. Shape matching for robot mapping. In Chengqi Zhang, Hans W. Guesgen, and Wai K. Yeap, editors, *Proc. of 8th Pacific Rim Int. Conf. on Artificial Intelligence*, Auckland, New Zealand, August 2004.

# Discrete Geometry Applied in Hard Real-Time Systems Validation

Gaëlle Largeteau[1], Dominique Geniet[1], and Éric Andrès[2]

[1] LISI, Université de Poitiers & ENSMA, Téléport 2 - 1 avenue Clément Ader BP
40109 86961 Futuroscope Chasseneuil cédex, France.
[2] IRCOM-SIC, SP2MI, BP 30179, 86962 Futuroscope Cedex, France

**Abstract.** Off-line validation of hard real-time systems usually stands on state based models. Such approaches always deal with both space and time combinatorial explosions. This paper proposes a discrete geometrical approach to model applications and to compute operational feasability from topological properties. Thanks to this model, we can decide the feasability of real-time synchronous systems composed of periodic tasks, sharing resources, running on multiprocessor architectures. This method avoids state enumeration and therefore limits both space and time explosion: computing an automaton model takes at least 2 hours for a real application instead of at most 1 second using discrete geometry.

**Keywords:** Real-time, operational validation, multiprocessors, resource sharing, geometrical modeling.

## 1   Introduction

In a real-time system, the correctness of a computation depends on both the logical results of the computation and the time when results are produced. Time constraints are called strict if not respecting them involves irreparable consequences on the system safety. In this case, the system is called *hard* [But97]. On the opposite, if not respecting deadlines keeps the system safe, the system is called *soft*. In this study, we only consider hard real-time systems where time constraints are strict.

A real-time system is a task set: each task is a process designed to react to an external incoming event. The systems we study use resources and run on centralized multiprocessor architectures. All processors are identical; tasks are preemptive and can move from a processor to another one at any time. Each task $\tau_i$ is specified by time characteritics: its first activation date $r_i$, its deadline $D_i$, its period $T_i$, and its execution time $C_i$ [LL73]. We assume that tasks are periodic and not reentrant: $\forall i \in [1, n], D_i \leq T_i$.

The operational validation of a real-time system is reached by proving that no task misses deadline, i.e by proving that there exists at least one time valid scheduling sequence for the system. This proof is obtained by feasability conditions or simulations. Validation is performed off-line for systems sharing re-

E. Andres et al. (Eds.): DGCI 2005, LNCS 3429, pp. 23–33, 2005.

sources which run on multiprocessor architectures, since there exists no necessary and sufficient feasability condition in this case [Mok83]. Off-line methods are usually based on state models (Petri nets, automata)[ALU94][CHO96]. Each transition is associated with the same duration and constraints are expressed as numbers of transitions. Therefore time is discrete in these models. In [LG02], we have defined an implicit timed model, based on finite automata, that enumerates states of systems and therefore involves both space and time combinatorial explosion (about 2 h. 30 min. for 1000 states and 5000 transitions).

Observing the graphs of the automata we obtained in this approach, we have conceived a new model, based on discrete geometry, that is presented in this work. Our goal is to reduce notably both time and space combinatorial explosion in the validation process. In this model, we associate each task with a geometrical figure which only depends on time characteristics r, C, D, T. Geometrical operations (extrusion, cartesian product, intersection) allow to model concurrency and synchronization in a geometrical way. This model makes state enumeration implicit and therefore decreases both space and time combinatorial explosion while keeping a strong expression capacity (about 0.2 s. for 1000 discrete points). We define the geometrical model for a single task in section 2. In section 3, we present compositional operations to integrate both parallelism and synchronization in the model. Section 4 is dedicated to the presentation of a software implementing this modeling process.

## 2    Model Definitions

### 2.1    A Two Dimensional Discrete Model to Represent Single Tasks

A task is usually modeled by an automaton (see figure 1). Each transition of this automaton is associated with a duration of one time unit (time is discrete). A task state is defined by the execution progress $x=C(t)$ of the task and the total time $t$ since the system activation. Therefore, we consider, for each task, a two-dimensional space $(t, c)$: $t$ adresses absolute time and $c$ adresses $x$.

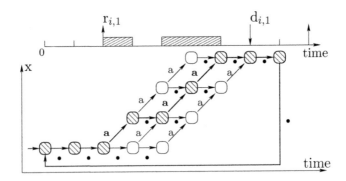

**Fig. 1.** Automaton model for task $\tau$ (r=2,C=3,D=5,T=7)

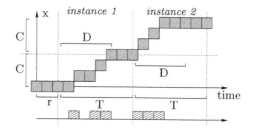

**Fig. 2.** A trace of task $\tau$ (r=2,C=3,D=5,T=7)

A task execution corresponds to the successive executions of its instances. At time $t$, $k$ instances of task $\tau$ (r,C,D,T) are completed (corresponding to $k \times C$ time units of CPU owning) and the $(k+1)^{\text{th}}$ instance is running ($C_{k+1}(t)$ time units of past CPU owning). So, at time $t$, $C(t)=k \times C + C_{k+1}(t)$ time units have been executed. During its execution, an instance of task $\tau$ goes through several states whether it owns the CPU or not: the initial state $e_0$ of $\tau$'s first instance is $(r, 0)$, its last state is $(r + T, C)$. The final state of any instance is the first state of the next instance. The final state $e_k$ of the $k^{\text{th}}$ instance of $\tau$ is $(e_{k-1} + (T, C))$. We denote $r^k = r + k \times T$ the activation date of the $k^{\text{th}}$ task's instance of $\tau$.

A task execution is then totally defined by the set of all instance states. This set is the graph of a function in space $(t, c)$. This function is called "trace". Figure 2 is a discrete representation of the automaton sequence of figure 1. Note that for $\tau$, this function is not unique.

Let us now characterize traces. Since tasks cannot be parallelized, a task cannot be in more than one state at once and its state can not be undefined. Therefore, a task trace is a mapping between time and task execution progress. This mapping is an increasing function: either the task is progressing during execution; or the task is suspended and its execution progress state keeps the same value. Moreover, during any time interval [t,t+1], no task can progress more than one step in one time unit since it is the maximal CPU time that can be allocated to the task for its execution during this interval.

**Definition 1.** *We call **trace** of task $\tau$ an increasing mapping* $Tr(\tau)$:

$$Tr_\tau : \mathbb{Z}^+ \to \mathbb{Z}^+ \atop t \quad \to Tr_\tau(t) \quad such\ that\ \forall t \geq 0,\ Tr_\tau(t + 1) \in \{Tr_\tau(t), Tr_\tau(t) + 1\}.$$

Task $\tau$ must deal with its temporal characterisation (r,C,D,T): this property imposes geometrical constraints on execution traces of $\tau$ (see Figure 3). Some task traces are compatible with task operational charateristics: they are "valid task traces". Others are not compatible: they are unvalid.

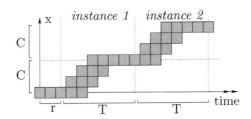

**Fig. 3.** Discrete model for two instances of task $\tau$ (r=2,C=3,D=5,T=7)

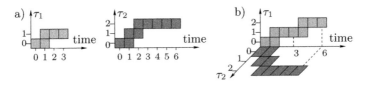

**Fig. 4.** Examples: $\tau_1$ (r=0, C=1,D=2,T=3), $\tau_2$ (r=0,C=2,D=3,T=6)

**Definition 2.** A **valid trace** of $\tau$ (r,C,D,T) is an execution trace $Tr_\tau V$ such that:

$$
\begin{aligned}
Tr_\tau V([0,r]) &= \{0\} \\
Tr_\tau V([r^k,r^k+D]) &= [(k-1)\times C,k\times C] \\
Tr_\tau V([r^k+D,r^{k+1}]) &= \{k\times C\}
\end{aligned}
$$

We call $TV(\tau)$ the set of $\tau$'s valid traces. If $\tau$ misses no deadline, no instance of $\tau$ runs on time intervals [0,r] and $[r^k+D,r^{k+1}]$. Therefore, traces $Tr_\tau V$ are constant functions over these intervals. The equation $Tr_\tau V(r^k+D)=k\times C$ means that the execution requirement $C_i$ of the $k^{\text{th}}$ instance of $\tau_i$ is completed before its deadline. Such a trace is a valid trace. We can then characterize the set $\Omega(\tau)$ which collects all points of valid traces.

**Definition 3.** The validity space $\Omega(\tau)$ of a task $\tau$ is:

$$
\Omega(\tau) = \bigcup_{\psi \in TV(\tau)} \{(t,\psi(t))\}.
$$

We note $T(\Omega(\tau))$ for $TV(\tau)$.

Figure 3 presents the validity space for the two first instances of a task. Each valid state of $\tau$ is associated with a point of $\mathbb{Z}^2$.

## 2.2   Concurrency Modeling : $(n+1)$ Dimensional Discrete Model

Let $\Gamma=(\tau_i)_{i\in[1,n]}$ be a set of tasks, designed to run concurrently. The state of $\Gamma$ at time t is defined by the states of all $\tau_i$. At time t, $\Gamma$ is valid if and only if all tasks of $\Gamma$ are valid.

**Definition 4.** *A valid state $P$ of a system $\Gamma$ at time $t$ is defined by:*

$$P = (t, x_1, ..., x_n) / \ \forall i \in [1, n], \ (t, x_i) \in \Omega(\tau_i)$$

Therefore, the space of $\Gamma$'s valid states is (n+1)-dimensional (see Figure 4).

$$\Omega(\Gamma) = \{(t, x_1, ..., x_n) / \ \forall i \in [1, n] \ (t, x_i) \in \Omega(\tau_i)\}$$

## 2.3    Computing $\Omega(\Gamma)$

We consider a system $\Gamma$ of n tasks.

**Geometrical Basic Notions [And03]:**
Two discrete points $p$ and $q$ in $n$ dimensions are $k$-**neighbour**, for $0 \le k \le n$, if $\forall 1 \le i \le n$, $|p_i - q_i| \le 1$ and if $k \le n - \displaystyle\sum_{i=1}^{n} |p_i - q_i|$. A $k$-**path** in a discrete object $A$ is a discrete $A$ point list such that two consecutive points of this list are $k$-neighbour (a task trace is a 0-path in 2-dimension). If there exists a $k$-path in a discrete object $A$, $A$ is said $k$-connected. A $k$-component is a maximal $k$-connected discrete object.

**Definition 5.** *Let $\mathcal{I} = \{i_1, ..., i_{|\mathcal{I}|}\}$ $(i_1 < i_2 < ... < i_{|\mathcal{I}|})$. We define the injection operation $\mathcal{J}_{\mathcal{I},n}$ in the following way: $\mathcal{J}_{\mathcal{I},n} : \ \mathbb{Z}^{|\mathcal{I}|} \to \mathbb{Z}^{n+1}$*

$$(x_1, ..., x_{|\mathcal{I}|}) \to \overbrace{(0, ..., 0, x_1, 0, ..., 0, x_j, 0, ..., 0, x_{|\mathcal{I}|}, 0, ..., 0)}^{n+1}$$
$$\phantom{(x_1, ..., x_{|\mathcal{I}|}) \to (0, ..., 0, }i_1 \phantom{x_1, 0, ..., 0, }i_j \phantom{0, ..., 0, }i_{|\mathcal{I}|}$$

Notations:
- $\mathcal{J}_{i,n} = \mathcal{J}_{\{1,i+1\},n}$: $(a,b) \to (a,0,..,0,b,0...,0)$
  $\phantom{\mathcal{J}_{i,n} = \mathcal{J}_{\{1,i+1\},n}: (a,b) \to (}1 \phantom{,0,..,0,}i+1$
- $\Lambda_{\mathcal{I},n}$ is the following cartesian product:

$$\Lambda_{\mathcal{I},n} = \mathbb{Z}^{i_1-1} \times \{0\} \times \mathbb{Z}^{i_1-i_2-1} \times \{0\} \times ... \times \mathbb{Z}^{i_{|\mathcal{I}|}-i_{|\mathcal{I}|-1}-1} \times \{0\} \times \mathbb{Z}^{n+1-i_{|\mathcal{I}|}}.$$

- $\Lambda_{i,n}$ is the cartesian product $\Lambda_{i,n} = \{0\} \times \mathbb{Z}^{i-1} \times \{0\} \times \mathbb{Z}^{n-i} = \Lambda_{\{1,i+1\},n}$.

**Definition 6.** *We define the interleaved cartesian product $\mathcal{J}_{\mathcal{I},n}^{\Lambda} : \mathbb{Z}^{|\mathcal{I}|} \to \mathcal{P}(\mathbb{Z}^{n+1})$*
*$(x_1, ..., x_{|\mathcal{I}|}) \to \{\mathcal{J}_{\mathcal{I},n}((x_1, ..., x_{|\mathcal{I}|})) + \lambda, \ \lambda \in \Lambda_{\mathcal{I},n}\}$*

Notations:
- $\mathcal{J}_{i,n}^{\Lambda} = \mathcal{J}_{\{1,i+1\},n}^{\Lambda} = \{\mathcal{J}_{i,n}((a,b)) + \lambda, \ \lambda \in \Lambda_{i,n}\}$.

**Definition 7.** *We call "concurrent product" (denoted $\otimes$) of $\Omega(\tau_1)$ and $\Omega(\tau_2)$ the set $\Omega(\tau_1) \otimes \Omega(\tau_2) = \mathcal{J}_{1,n}^{\Lambda}(\Omega(\tau_1)) \cap \mathcal{J}_{2,n}^{\Lambda}(\Omega(\tau_2))$.*

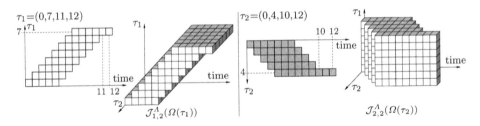

**Fig. 5.** Example: $\mathcal{J}_{1,2}^{\Lambda}(\Omega(\tau_1))$ and $\mathcal{J}_{2,2}^{\Lambda}(\Omega(\tau_2))$

**Remarks**:

-The operation $\otimes$ is associative, therefore we can generalize the notation:

$$\Omega(\tau_1, ..., \tau_n) \;=\; \bigotimes_{i=1}^{i=n} \Omega(\tau_i)$$

- $\forall\, \mathcal{J} \in \mathcal{J}_{i,n}^{\Lambda}((a,b))$, $\forall k \in [1, i-1] \cup [i+1, n]$, $\exists\, x_k \in \mathbb{Z}$ such that:

$$\mathcal{J} \;=\; (\mathbf{a}, x_1, ..., x_{i-1}, \mathbf{b}, x_{i+1}, ..., x_n)$$

- Projection $\Pi_i$: $(y_1,...,y_{i+1},...,y_{n+1}) \rightarrow (y_1, y_{i+1})$ is a reverse operation of $\mathcal{J}_{i,n}^{\Lambda}$:
$\Pi_i(\mathcal{J}_{i,n}^{\Lambda}(a,b)) = \{(a,b)\}$.

The following theorem says that the $(n+1)$ dimensional discrete model is obtained as an intersection of our interleaved cartesian product. It provides a direct algorithm.

**Theorem 1.** $\Omega(\Gamma) \;=\; \bigotimes_{i=1}^{i=n} \Omega(\tau_i)$.

**Proof**: Let us show that $\bigotimes_{i=1}^{i=n} \Omega(\tau_i) \subset \Omega(\Gamma)$:

Let P=$(t, x_1, x_2, ....x_n) \in \bigotimes_{i=1}^{i=n} \Omega(\tau_i)$. The definition of $\otimes$ gives:

P=$(t, x_1, x_2, ....x_n) \in \bigcap_{i \in [1,n]} \left( \mathcal{J}_{i,n}^{\Lambda}(\Omega(\tau_i)) \right)$. Since P belongs to an intersection
set, we get: $\forall i \in [1, n]$, $P \in \mathcal{J}_{i,n}^{\Lambda}(\Omega(\tau_i))$. Using the two-dimensional projection
on $\tau_i$-space $(t, \tau_i)$: $\forall i \in [1, n]$, $\Pi_i(P) \in \Pi_i(\mathcal{J}_{i,n}^{\Lambda}(\Omega(\tau_i))) = \Omega(\tau_i)$. And then
$\forall i \in [1, n]$, $(t, x_i) \in \Omega(\tau_i)$. Therefore, $P \in \Omega(\Gamma)$.

Let us show that $\Omega(\Gamma) \subset \bigotimes_{i=1}^{i=n} \Omega(\tau_i)$:

Let P=$(t, x_1, x_2, ....x_n) \in \Omega(\Gamma)$. We consider the projection of P on each
plane $(t, \tau_i)$: $\forall i \in [1, n]$ $\Pi_i(P) \in \Omega(\tau_i)$. We then consider $\mathcal{J}_{i,n}^{\Lambda}$ for each $\Pi_i$ (P):
$\forall i \in [1, n]$ $\mathcal{J}_{i,n}^{\Lambda}(\Pi_i(P)) \subset \mathcal{J}_{i,n}^{\Lambda}(\Omega(\tau_i))$. Finally, we consider the intersection of
all sets we have obtained:

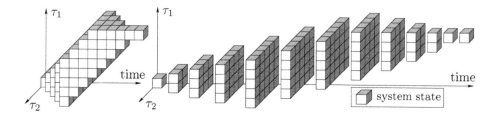

**Fig. 6.** Example: geometrical modeling for a two tasks system, $\Omega(\tau_1) \otimes \Omega(\tau_2)$

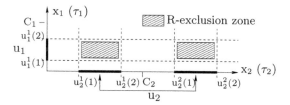

**Fig. 7.** Example: both $\tau_1$ and $\tau_2$ use resource R

$$\bigcap_{i\in[1,n]}\left(\mathcal{J}_{i,n}^{\Lambda}(\Pi_i(P))\right) \subset \bigcap_{i\in[1,n]}\left(\mathcal{J}_{i,n}^{\Lambda}(\Omega(\tau_i))\right) \text{ Since } \bigcap_{i\in[1,n]}\left(\mathcal{J}_{i,n}^{\Lambda}(\Pi_i(P))\right) =$$

$$\{(t,x_1,x_2,....x_n)\} = \{P\}, \text{ we get: } P \in \bigotimes_{i=1}^{i=n} \Omega(\tau_i). \qquad \square$$

## 2.4   Synchronization Integration

In real-time systems, tasks use critical resources and the processor number is limited. In next parts, we integrate these constraints in our modelisation.

**Resource Sharing**

Resources must be used in mutual exclusion : only one task can use resource R at time t. Therefore, some states of $\Omega(\Gamma)$ become invalid from the resource sharing point of view. On a geometric point of view, the $k^{\text{th}}$ instance of a task corresponds to the task execution progress interval $[C_i \times k, C_i \times k + C_i]$. We define $u^k(1)$ and $u^k(2)$ such that this instance uses R during the execution time interval: $]u_i^k(1),u_i^k(2)[ \subset [C_i \times k, C_i \times k + C_i]$.

We denote by $u_i$ the time interval union $\bigcup_{k \in \mathbb{N}} ]u_i^k(1),u_i^k(2)[$, corresponding to all resource requirement intervals for all instances of $\tau_i$ (see Figure 7).

Since a task cannot own the resource if another task already uses it, two tasks cannot be in R critical section at the same time. Then if $P=(t,x_1,...,x_n)$ and $x_i \in u_i$, we must get $\forall j \neq i, x_j \notin u_j$. Therefore, for a valid state $P=(t,x_1,...,x_n)$ of $\Gamma$, the following property stands: $|\{x_i/x_i \in u_i\}| \leq 1$. The validity space of $\Gamma$ respecting resource R sharing is (see figure 8):

$$\Omega_R(\Gamma) = \{(t, x_1, ..., x_n) \in \Omega(\Gamma)/|\{x_i/x_i \in u_i\}| \leq 1\}$$

Here, we deal only with one resource R. For many resources, since resources are independant, the technique can be applied by induction. Let $\Gamma_R = (\tau_j)_{j \in \mathcal{I}_R}$ be the set of tasks sharing resource R (we note $\mathcal{I}_R$ the set of indices of tasks sharing R). The states of $\Gamma_R$ which are associated with a simultaneous use of R are unvalid. The R-exclusion zone $\Omega(R)$ of R collects all states of $\Gamma_R$ corresponding R misuses. $\Omega(R)$ is part of the subspace associated with all tasks sharing R, since it implies the simultaneous run of at least two of these tasks.

**Definition 8.** *The R-exclusion zone is (see Figure 7):*

$$\Omega(R) = \{(x_i)_{i \in \mathcal{I}_R}/ |\{x_i/x_i \in u_i\}| > 1\}.$$

A state of $\Omega(R)$ only concerns tasks of $\Gamma_R$. A state s=$(t, x_1, ..., x_n)$ is unvalid if $\Pi_{\Gamma_R}(s) \in \Omega(R)$. Therefore, the set $\eta_R = \{(t, x_1, ..., x_n)/t \in \mathbb{Z}, |\{x_i/x_i \in u_i\}| > 1\}$ of unvalid states can be obtained thanks to a concurrent cartesian product and an extrusion operation.

**Theorem 2.** $\eta_R = Extr\left(\mathbb{Z}, \mathcal{J}_{\mathcal{I}_R,n}^{\Lambda}(\Omega(R))\right).$

**Proof**: This theorem comes directly from the definitions of $\Omega$(R), the extrusion operation and interleaved cartesian product operation. $\Omega$(R) collects all unvalid states of $\Gamma_R$: it is a $|\mathcal{I}_R|$-dimensional object. The interleaved cartesian product associates these states with all possible states of $\Gamma \setminus \Gamma_R$. Then all states of $\Gamma$ that are unvalid in the R-sharing point of view are reached. $\mathcal{J}_{\mathcal{I}_R,n}^{\Lambda}(\Omega(R))$ is an n-dimensional object. Now one must integrate that these states are always unvalid. This is done by extruding this objet following the time direction ($\mathbb{Z}$). This operation collects all R-sharing unvalid states of $\Gamma$. Then we get $\eta_R = Extr\left(\mathbb{Z}, \mathcal{J}_{\mathcal{I}_R,n}^{\Lambda}(\Omega(R))\right).$ □

All states of $\eta_R$ are not valid from the resource sharing point of view. Valid states of the application are then in $\Omega(\Gamma)$ but not in $\eta_R$.

**Theorem 3.** $\Omega_R(\Gamma) = \Omega(\Gamma) \setminus \eta_R.$

Therefore, the set of valid traces including resource sharing is:

$$T(\Omega_R(\Gamma)) = \{\psi \in T(\Omega(\Gamma))/ \forall t \in \mathbb{Z}^+, \psi(t) = (x_1, ..., x_n), (t, x_1, ..., x_n) \in \Omega_R(\Gamma)\}.$$

**Processor Sharing**
While building $\Omega_R(\Gamma)$, we have not considered the number of processors. However, this parameter makes each trace $\omega$ in $T_R(\Gamma)$ valid or unvalid according to the minimal number of processors useful to execute $\omega$.

During an execution, the number of active processors is constant between two consecutive context switches. To decide the validity of a trace, we only have to look at it at context switch times. Let us note by q the scheduling quantum. If there are $k$ running tasks between two given context switches $a$ and $b$, the trace is called $k$-concurrent between $a$ and $b$.

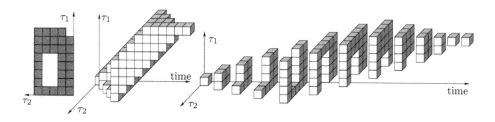

**Fig. 8.** Geometrical model of a system including resource sharing

**Definition 9.** *A trace $\psi \in T(\Omega_R(\Gamma))$ is k-concurrent between two context switch times $i \times q$ and $(i+1) \times q$ if and only if: $\psi(i \times q) = (i \times q, (x_i)_{i \in [1,n]})$ and $\psi((i+1) \times q) = ((i+1) \times q, (y_i)_{i \in [1,n]}) \Rightarrow \sum_{i \in [1,n]} |x_i - y_i| \leq k \times q$.*

We said that discrete points $\psi(i \times q)$ and $\psi((i + 1) \times q)$ are $k$-concurrent. This definition of $k$-concurrency in a $n + 1$ dimensional space corresponds to the definition of the $n - k$-neighbourhood. A $k$-concurrent trace is then a $n - k$-path in $\Omega_R(\Gamma)$. We denote by $T_{R,k}(\Gamma)$ the set of $k$-concurrent traces of $T(\Omega_R(\Gamma))$.

**Definition 10.** *A set $\Omega_R(\Gamma)$ is k-concurrent if there exists at least one k-concurrent trace $\psi$ in $T_R(\Gamma)$.*

**Remark:** If a set $\Omega_R(\Gamma)$ is $k$-concurrent, then it is a $n - k$-component and there exists at least one valid sheduling sequence for $\Gamma$ on a $k$ processor architecture.

## 2.5 Feasability Decision

For a system running on a $k$ processor architecture and sharing a resource $R$, a valid scheduling is a $k$-concurrent trace in $T\Omega_R(\Gamma)$. The feasability decision is reached by evaluating the predicate: $T_{R,k}(\Gamma) \neq \emptyset$.

## 3 Implementation

We have developed the software GemSMARTS (Geometric Scheduling Modeling and Analysis of Real-Time Systems) which computes the set $\Omega_R(\Gamma)$. We have tested a discrete data structure implemented through classical matrices. Since we avoid enumeration, we get more efficient computing times : using automata, models are obtained after at least 2 hours instead of 1 second at most using discrete geometry.

Figure 9 shows $\Omega_R(\Gamma)$ for a two-tasks system sharing a resource.

While avoiding enumeration in the discrete model, we reach very efficient computation time. As a comparison, for a seven task system sharing four resources, computing the automaton model takes more than 2 hours while the computation of the discrete modele last less than 1 second.

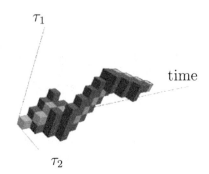

**Fig. 9.** System geometrical model:$\tau_1$=(0,7,10,12), $\tau_2$=(0,3,6,6)

## 4    Conclusion

Validity spaces are useful to model hard real-time systems running on multi-processor architectures and sharing resources. Feasability of task systems and optimal numbers of processors can be computed thanks to the k-concurrency concept.

Feasability is usually decided using state based model and model checkers [LG02]. Using validity spaces involves a noteworthy improvement: the time saved on a time free automata model is about 85%. The data structure we have tried (matrices) is useful to implement our model and developed algorithms are already more efficient than the previous ones (using automata) although they are far from being optimized. We have reached a limited time complexity depending mainly on the number of tasks whereas classical models follow complexities which depend on time. This method allows to drive back both space and time explosion.We are studying both space and time complexities for optimized version of the geometrical algorithms.

Ongoing works concern definitions of both topological and geometrical properties to precisely characterize scheduling sequences. We plan, for example, to define topological properties for validate on-line classical scheduling (RM, ED, and so on), in order to propose multiprocessor versions of on-line validation techniques integrating resource sharing.

## References

[ALU94]  R.Alur, D.Dill: A theory of timed automata. Theoretical Computer Science, vol. 126, pp 183-235.(1994).

[And03]  E.Andres: Discrete linear objects in dimension n: the standard model. Graphical Models 65(1-3), pp 92-111. (2003).0

[But97]  G. Buttazzo: Hard real-time computing systems: Predictable scheduling algorithms and applications. Kluwer Academic Publishers. (1997).

[CHO96]  A.Choquet-geniet, D.Geniet, F.Cottet: Exhaustive computation of scheduled task execution sequences of real-time application. Proc FTRTFT'96.(1996)

[LL73]    C.L. Liu et J.W. Layland : Scheduling algorithms for multiprogramming in real-time environnement. Journal of the ACM, vol 20, pp 46-61.(1973)

[LG02]    G.Largeteau, D.Geniet: Term Validation of Distributed Hard Real-time Applications. Conference on Implementation and Application of Automata. (2002).

[Mok83]   A.K.Mok: Fundamental design problems for the hard real-time environments. Ph.D. MIT.(1983).

# Hierarchical Watersheds Within the Combinatorial Pyramid Framework

Luc Brun[†], Myriam Mokhtari[†], and Fernand Meyer[‡]

[†] GreyC CNRS UMR 6072, Équipe Image - Ensicaen,
6, Boulevard du Maréchal Juin,
14050 CAEN Cedex - France
{luc.brun, myriam.brun}@greyc.ensicaen.fr
[‡] Centre de Morphologie Mathématique (CMM),
35, rue Saint Honoré,
77305 Fontainebleau Cedex - France
Fernand.Meyer@cmm.ensmp.fr

**Abstract.** Watershed is one of the most popular tool defined by mathematical morphology. The algorithms which implement the watershed transform generally produce an over segmentation which includes the right image's boundaries. Based on this last assumption, the segmentation problem turns out to be equivalent to a proper valuation of the saliency of each contour. Using such a measure, hierarchical watershed algorithms use the edge's saliency conjointly with statistical tests to decimate the initial partition. On the other hand, Irregular Pyramids encode a stack of successively reduced partitions. Combinatorial Pyramids consitute the latest model of this family. Within this framework, each partition is encoded by a combinatorial map which encodes all topological relationships between regions such as multiple boundaries and inclusion relationships. Moreover, the combinatorial pyramid framework provides a direct access to the embedding of the image's boundaries. We present in this paper a hierarchical watershed algorithm based on combinatorial pyramids. Our method overcomes the problems connected to the presence of noise both within the basins and along the watershed contours.

## 1    Introduction

Segmentation and contour extraction are important tasks in image analysis. Among the multitude of methods, the watershed transformation [17, 12, 14, 8, 4] arises as a popular image segmentation algorithm. This method usually based on the gradient of the image provides a partition of the image into a set of basins corresponding to local minima of the gradient and a set of watershed pixels. These pixels may be roughly understood as the borders of the basins. Using a flooding process [17] watershed pixels are defined as the places where water coming from several basins merges. Watershed algorithms presents the main advantage of providing closed curves leading to a proper definition of regions. A well known drawback of the watershed algorithms is the over segmentation

E. Andres et al. (Eds.): DGCI 2005, LNCS 3429, pp. 34–44, 2005.

often produced by these methods (e.g. [17]). Since the contours appear to be correct, the over segmentation problem turns out to be equivalent to a proper valuation of the saliency of each contour. This contour's saliency is generally used conjointly with an homogeneity criteria in order to derive a hierarchy of partitions.

This hierarchy of partitions may be encoded using Irregular Pyramids [13, 10, 3]. These data structures encode each partition as a graph whose nodes and edges respectively correspond to regions and region's adjacencies. Usual Irregular Pyramids [13] are made of a stack of simple graphs (i.e. graphs without multiple edges nor self-loops). Within this framework several contours between two regions are encoded by a single edge which thus simply encodes the existence of at least one contour between the two regions. However, within the hierarchical watershed framework the contours of the partition play a major role in the decimation process. The explicit encoding of each contour of the partition by one edge requires thus to encode an irregular pyramid made of non simple graphs. Such enriched graphs may be created using the Dual graph reduction scheme [10]. Within this framework, the reduction operation is performed in two steps: First, the contraction of a set of edges identifies a set of vertices. This operation may create redundant edges such as empty self-loops or double edges [10]. These redundant edges are characterized in the dual of the graph and removed by a set of edge removals. Applied to the watershed transform such a reduction scheme provides a graph where each vertex encodes a basin and each edge corresponds to one contour between two basins.

Combinatorial Pyramids inherit all the useful properties from the dual graph pyramids with several additional advantages: Firstly, within the combinatorial pyramid framework the dual graph may be implicitly encoded and thus updated. This property allows to decrease both the memory and computational time requirements. Secondly, combinatorial pyramids preserve the local orientation of edges around vertices and faces. This last property is used to retrieve efficiently the set of points encoding a contour.

The aim of this paper is to present one implementation of a hierarchical watershed algorithm within the combinatorial pyramid framework. The paper is thus organized as follows: We first present the main features of combinatorial pyramids (Section 2). Then, the specific advantages of this model within this framework are illustrated by a new hierarchical watershed construction scheme using specific features of combinatorial pyramids (Section 3).

## 2   Combinatorial Pyramids

A combinatorial pyramid corresponds to a stack of successively reduced combinatorial maps where the initial combinatorial map $G_0$ usually encodes a 4 connected planar sampling grid. A combinatorial map $G = (\mathcal{D}, \sigma, \alpha)$ may be understood as an encoding of a planar graph. The construction of a combinatorial map from a plane graph is as follows: first edges are split into a set of half-edges called darts, the set of darts being denoted by $\mathcal{D}$. Two darts sharing the same edge are connected by the involution $\alpha$ which maps each of the two darts to the other one. The vertices of the graph are encoded by the permutation $\sigma$ whose cycles correspond to the sequence of darts encountered when turning counter-clockwise around each vertex. Each vertex of the graph is thus encoded by one cycle of the permutation $\sigma$. In the same way each edge of the graph is encoded by one cycle of $\alpha$. In what follows, the cycles of $\sigma$ and $\alpha$ containing a dart $d$ will be respectively denoted by $\alpha^*(d)$ and $\sigma^*(d)$. An introduction to combinatorial maps and Combinatorial Pyramids may be found in [2, 3].

As in the dual graph pyramid scheme [10] (Section 1) the two operations used to reduce combinatorial maps within the pyramid are the contraction and the removal. In order to preserve the number of connected components of the initial combinatorial map, we forbid the removal of bridges and the contraction of self-loops. Such contractions may be avoided by using a contraction kernel defined as a forest of the initial combinatorial map. As mentioned in the introduction of this paper the contraction operation may create redundant edges such as empty self loops and double edges. A contraction kernel is thus followed by a removal kernel removing the eventual empty-self loops and double edges. A reduction step in the pyramid involves thus the application of 2 kernels : One contraction kernel and one removal kernel. Note that, while a contraction kernel is application dependent, the removal kernel is automatically defined from one combinatorial map. Indeed, within our reduction scheme a contraction kernel specifies a set of regions to be merged while the removal kernel is restrained to the removal of redundant edges.

Given an initial combinatorial map $G_0$ encoding the 4 connected planar sampling grid and a sequence of contraction or removal kernels $K_1, \ldots, K_n$ each reduced combinatorial map $G_i = (\mathcal{D}_i, \sigma_i, \alpha_i)$ may be build from $G_{i-1} = (\mathcal{D}_{i-1}, \sigma_{i-1}, \alpha_{i-1})$ and the kernel $K_i$ [3]. Note that we have $\mathcal{D}_i = \mathcal{D}_{i-1} - K_i$. The set of darts of any reduced combinatorial map is thus included in the initial set of darts $\mathcal{D}_0$. The resulting pyramid is usually stored explicitly as a sequence of successively reduced combinatorial maps $(G_1, \ldots, G_n)$. However, we have shown [2, 3] that within the combinatorial pyramid framework all the kernels and all the reduced combinatorial maps may be encoded efficiently by storing for each initial dart in $G_0$, the maximal level where this dart survives in the pyramid and the operation applied at each level. This implicit encoding may be performed by :

1. one function *state* from $\{1, \ldots, n\}$ to the 2 states $\{Contracted, Removed\}$ which specifies the type of each kernel.
2. one function *level* defined for all darts in $\mathcal{D}_0$ such that $level(d)$ is equal to the maximal level where $d$ survives:

$$\forall d \in \mathcal{D}_0 \; level(d) = Max\{i \in \{1, \ldots, n+1\} \,|\, d \in \mathcal{D}_{i-1}\}$$

a dart $d$ surviving up to the top level has thus a level equal to $n + 1$.

Given the function level, each kernel $K_i$ may be efficiently retrieved as the set of darts whose level equals to $i$. Moreover, any reduced combinatorial map may be retrieved from this implicit encoding in a time proportional to the total length of the boundaries encoded by this combinatorial map [2, 3].

The explicit encoding of the pyramid $(G_0, \ldots, G_n)$ may thus be replaced by $(G_0, level, sate)$. Moreover, if the initial combinatorial map $G_0$ encodes a planar sampling grid, the permutations $\sigma_0$ and $\alpha_0$ may be implicitly encoded using any convention on the numbering of darts. The pyramid may thus be simply encoded by $(\mathcal{D}_0, level, state)$. On the other hand, the current top level combinatorial map is frequently accessed during the construction of the pyramid. We thus decided to store additionally a combinatorial map encoding the top level of the pyramid. This combinatorial is updated at each level during the construction of the pyramid. Our encoding of the pyramid is thus defined by $(G_n, \mathcal{D}_0, level, state)$ where $G_n$ denotes the current top level combinatorial map. This choice allows an efficient construction scheme of the pyramid while avoiding the explicit encoding of all the intermediate combinatorial maps.

Moreover, we have shown [2] that using the two functions *level* and *state* we can associate to each edge $\alpha_i^*(d)$ an ordered sequence of 1-cells [18] (also denoted cracks or linels) which encodes the embedding of the edge, i.e. the boundary between the two regions associated to the vertices $\sigma_i^*(d)$ and $\sigma_i^*(\alpha_i(d))$. The sequence of linels of one contour is retrieved in a time proportional to its length [2].

The construction of a combinatorial pyramid is performed by successive simplifications of the top level combinatorial map. From this point of view, combinatorial pyramids may be compared to other topological data structures [6, 9]. However, combinatorial pyramids differ from these alternative encodings on two points : Firstly, the implicit encoding provided by the function *level* allows us to encode the whole sequence of reduced combinatorial maps rather than the top level one. Secondly, alternative data structures [6, 9] encode the geometry of the partition thanks to an additional geometrical model cooperating with the topological one in order to provide a full description of the partition. Using combinatorial pyramids, the geometrical embedding of the partition is provided without additional memory requirements by the function *level*.

# 3    Hierarchical Watersheds with Combinatorial Pyramids

Within the combinatorial pyramid framework, the initial combinatorial map is usually associated to the 4 connected sampling grid. Given a $m \times n$ grey level image, we build an initial combinatorial map $G_0$ encoding the $m \times n$ sampling grid and we store within each vertex the grey value (or altitude) of the associated pixel. This vertex's altitude is the basic feature used to compute the watershed transform on $G_0$.

## 3.1    Building the Initial Watershed Partition

Several methods [17, 5, 8] have been proposed to build the basins of a graph. The topological watershed method designed by Bertrand and Couprie [5] produces a grey level image $W$ whose minima encode the basins. The construction of a contraction kernel from such an image may be performed by computing a spanning tree [10] which covers each basin. The union of all trees forms the contraction kernel. Using Meyer's [12] or Vincent [17] algorithms the basins are built iteratively using a flooding process. The main property satisfied by these algorithms are :

1. the assignment of a vertex to a label (watershed or basin) is performed only once,
2. each vertex marked as belonging to a basin is adjacent to at least one vertex already aggregated to this basin.

   Starting from an empty kernel, condition 2 insures that for each vertex adjacent to a basin we can find one edge connecting it to this basin. We can thus add this edge to the contraction kernel. Moreover, the contraction kernel may contain a loop only if one vertex is aggregated twice to a same basin which is refused by condition 1. The contraction kernel can thus, in this case, be built in parallel with the watershed transform.

   Using any of the above methods we can thus build a contraction kernel $K_1$ whose trees span each basin of $G_0$. The contraction of $K_1$ contracts each of these trees into a single vertex. Since each vertex of $G_0$ contains the altitude of the associated pixel, we can compute during the contraction process the minimal altitude of each tree and store the resulting value within the contracted vertex. Each vertex of $G_1$ associated to a basin stores thus the minimal altitude of this basin.

   The kernel $K_1$ is followed by a removal kernel $K_2$ in order to remove redundant edges(Section 2). Let us denote by $G_2$ the combinatorial map obtained from the successive applications of $K_1$ and $K_2$. Since the kernel $K_2$ does not imply any merge of vertices, the vertice's values computed during the contraction step remain unchanged. Moreover, since the trees of $K_1$ span only the basin of $G_0$ the vertices of $G_2$ correspond either to basins or to watershed pixels.

## 3.2   Building a Partition into Basins

Hierarchical watershed algorithms are generally based on a partition of the image into a set of basins. However, watershed algorithms produce a partition of the image into a set of basins *and* a set of watershed pixels each of these pixels being encoded by one vertex in $G_2$. The explicit encoding of watershed vertices induces two types of problems within this framework: First of all if two basins are separated by a thin watershed line the adjacency between the two basins is not encoded by a single edge but by a sequence of two edges encoding for each watershed vertex its adjacency to the two basins. Secondly, watershed vertices may form thick connected components [17, 14] where many watershed vertices are incident to 0 or 1 basin. In such a case, the adjacency between the basins surrounding such a component and thus the existence and location of the contours between the basins is relative to a labeling of the watershed vertices to the different basins.

Two recent algorithms [11, 9] have been proposed to encode an image partition defined by pixel's boundaries. These two approaches encode a sequence of pixels defining a boundary between two basins by a single edge. However, each approach suffers of different drawbacks. The method presented by Marchadier [11] must pre-process the boundary pixels in order to avoid some configurations. This last step modifies the partition without taking into account the image's content. The method present by Köethe [9] may violate some basic topological properties by contracting basins into single points. Finally, using either of these methods boundary pixels do not belong to any basin. Some well known properties of an image partition into 4 or 8 connected regions may thus be violated. For example, the method presented by Köethe encodes 4 connected basins but may produce partitions with more the 4 basins incident to a same point.

To overcome these drawbacks we designed [4] an algorithm which aggregates the watershed vertices to the basins using a flooding process. This algorithm ensures that each watershed vertex aggregated to a basin may be connected to the minimum of this basin by an always descending path. Moreover, this algorithm satisfies the same conditions than Meyer's and Vincent's algorithms (conditions 1 and 2 Section 3.1). We can thus build a contraction kernel $K_3$ during the aggregation process. As previously the contraction kernel is followed by a removal kernel $K_4$. The final combinatorial map is denoted by $G_4$.

The above method is similar to the minima extension presented by Bertrand [1]. However, both methods differ on the following point: Roughly speaking, the greedy algorithm presented by Bertrand preserves the minimal altitude one as to climb to connect two adjacent basins. This method allows to attach a global pass value value to each couple of adjacent basins. Our method [4] preserves the minimal altitude one has to climb to connect two adjacent basins *while* passing by one watershed pixel. The aim of this method is to attach one pass value to each elementary element of the border between two adjacent basins (see below).

## 3.3   From Watershed Values to Linel's Pass Values

The combinatorial map $G_4$ encodes a partition of the image into a set of basins. Each edge between two vertices of $G_4$ encodes a contour between two basins and may be associated to a sequence of linels encoding the embedding of the associated boundary (Section 2). Each linel along the contour separates two pixels belonging to each basin. Moreover, since each basin is initially surrounded by watershed pixels, at least one of these two pixels was initially marked as a watershed. Let us consider a linel $l$ between two basins $B_1$ and $B_2$ of $G_4$ separating two pixels $P$ and $Q$ belonging respectively to $B_1$ and $B_2$. If $P$ and $Q$ were both initially marked as watershed pixels, there is by construction [4](Section 3.2) two descending paths from $P$ to the minimum of $B_1$ and from $Q$ to the minimum of $B_2$. If one of the two pixels, say $P$, was not initially marked as a watershed we can induce from the construction scheme of the basins [14] that $P$ is connected to the minimum of $B_1$ by an always descending path. The maximum of the altitudes $h(P)$ and $h(Q)$ represents thus the minimal altitude one has to reach to connect the minima of $B_1$ and $B_2$ while passing by $P$ and $Q$. This value is associated to each linel and called a linel's pass value. These valuated linels correspond intuitively to the values of the watershed pixels along the contours. However the aggregation of the watershed vertices to the basins and the transfer of the watershed pixels altitudes to the linel's pass values allows us to overcome the two drawbacks mentioned in Section 3.2.

## 3.4   From Linel's Pass Values to Edge's Pass Values

Given an edge $\alpha_4^*(d)$ of $G_4$ let us consider the function $Pv(t)$ which encodes the sequence of linel's pass values encountered along the contour associated to $\alpha_4^*(d)$. The symbol $t$ may be understood as the rank of the linel along the contour while $Pv(t)$ represents the pass value of the associated linel. The value usually determined from the function $Pv$ within the hierarchical watershed framework [14] is its minimum. Such a value may be associated to each edge of the combinatorial map $G_4$. However, the minimal linel's pass value along a contour is sensitive to the noise which may be present along it. Moreover, this choice does not take into account the distribution of $Pv$ and thus the saliency of the minimum.

In order to overcome this last drawbacks we propose to measure the saliency of the different minima of the function $Pv$ using the following decomposition: If the function $Pv$ contains less then a given number (fixed to 5 in our experiments) of samples we consider that no reliable values on the saliency of the minima may be defined and we fix the edge's pass value to the minimum of the function $Pv$. Otherwise, we use the volumic filters defined by Vachier [16] to compute the saliency of the different minima as follows:

Given an edge $\alpha_i^*(d)$ of the current top level combinatorial map $G_i$, we consider the function $Pv$ associated to $\alpha_i^*(d)$ as a 1D relief which is progressively flooded. When two 1D basins $b_1$ and $b_2$ merge along a maxima $m$ the volume of $b_1$ and $b_2$ are computed by:

$$\forall j \in \{1, 2\} \quad vol(b_j) = \sum_{t \in b_j} m - Pv(t) \tag{1}$$

The two basins $b_1$ and $b_2$ are then filled up to the altitude $m$ and the process continues on the updated signal. This process stops when the signal has only one minimum left. Note that our method is based on a family of leveling functions. Indeed, the signal used at step $i$ of our algorithm is defined as $Pv_i(t) = \psi^i(Pv(t))$ where $\psi^i$ is the $i^{th}$ iteration of the leveling operator [16] $\psi$ which merges the basins separated by the lowest maxima and fills them up to the altitude of this maxima.

Given the set $\{b_1, \ldots, b_n\}$ of 1D basins merged by our method we define the global pass value of the contour as the minimal altitude of the basins with the greatest volume:

$$pass\_value(\alpha_i^*(d)) = \min_{j \in \{1, \ldots, n\}} \{Depth(b_j) \mid Vol(b_j) = \max_{k \in \{1, \ldots, n\}} Vol(b_k)\} \tag{2}$$

where $Depth(b_i)$ and $Vol(b_i)$ denote respectively the minimal altitude of $b_i$ and its volume (equation 1).

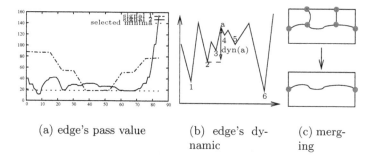

(a) edge's pass value          (b) edge's dy-          (c) merg-
                                   namic                    ing

**Fig. 1.** (a) Two different signals with a same edge's pass value. (b) Computation of the dynamic of the edge $a$ on a 1D example. (c) Enlargement of a contour

Intuitively, this choice corresponds to a measure of the saliency of each minimum by the volume of the associated basin and a selection of the minimum of greatest volume. Note that, in practical applications, the basin of maximal volume is generally uniquely defined and the $Min$ operator in equation 2 becomes useless.

Fig. 1(a) shows two signals with a same pass value. The signal 1 which has only one minimum is valuated by the value of this minimum. On the other hand, the small gaps at the beginning of signal 2 are not selected since the last minimum has a higher altitude but a maximal volume.

### 3.5   From Edge's Pass Values to Edge's Dynamics

The computation of the edge's pass values allows us to reduce the influence of noise along the contour by affecting to each contour its more significant minimum. However, the contrast between two basins is relative both to the pass values of their common contours and to the minima of the two basins. In order to reduce the influence of noise inside the basins which may induce the presence of many non significant basins, we use the contour's dynamic introduced by Najman [15]. Intuitively, the dynamic of a contour is defined by a flooding process which progressively merges all basins. The dynamic of each edge is then defined as the maximal difference between the edge's pass value and the depth of the two basins which merge along the contour. An illustration of the computation of the edge's dynamics on a 1D signal is provided in Fig. 1(b). Our algorithm floods thus progressively the current combinatorial map by merging at each step the two basins separated by the edge with the lowest pass value. The dynamic of the edge is then computed and we store within the basin with the higher altitude a pointer to the remaining basin. For example, in Fig. 1, before the flooding of edge $a$, the basins 3 and 4 points respectively towards the basins 2 and 5. After the merge of edge $a$, basin 5 points to the basin 2. These pointers allows us to retrieve for each basin the deeper basin to which it has been merged in order to compute the edge dynamic. This set of pointers defines a forest within the set of basins, the root of each tree being retrieved in almost constant time using union-find operations [7].

Note that the computation of the dynamics is based on the edge's pass value (Section 3.4) rather than the minimum of each contour. This difference influence both the computation of the dynamic at each step and the flooding process which is based on the edge values. The edge's dynamics computed by our algorithm are thus different from the ones computed using the contour's minima.

## 3.6    From Edge's Dynamic to Hierarchical Segmentation

Within the hierarchical watershed framework, the edge's dynamics are usually computed once and combined with an other homogeneity criteria to merge progressively the different basins. This approach suffers from two main drawbacks: First of all, as mentioned in Section 1 the edge's dynamics are often used to reduce the over segmentation of the image produced by watershed algorithms. Due to the over segmentation, many contours of the partition are initially composed of a small number of linels (e.g. 4 or 5). The reliability of a global value from a such reduced sample of data is difficult to state (Section 3.4). Secondly, the edge's dynamics are not updated according to the updates of the partition and may thus contain unreliable values all along the reduction process. However, after each sequence of merge operations, the removal of redundant edges (Section 2) either removes a contour or enlarges it by a concatenation with other contours (Fig. 1(c)). Therefore, the length of a contour in the pyramid is an increasing function of the level and the problems connected with the presence of very short contours tends to disappear as we go up in the hierarchy. In order to overcome the drawbacks connected with the poor reliability of the edge's dynamics at the first levels of the pyramid our method update the edge's pass values and edge's dynamics after each sequence of contraction and removal operations. More precisely, our method iterates the following steps:

1. Initialization step: Compute the edge's pass value and goto step 3,
2. Update the pass value of edges adjacent to a merged region,
3. Compute edge's dynamics,
4. Build a contraction kernel containing the edges with the lowest dynamic ; apply the contraction kernel and remove redundant edges. If more than one region left goto step 2.

Step 2 corresponds to a lazy programming. Indeed, since the computation of an edge's pass value requires only features of the associated contour, we can ensure that an edge not adjacent to a region merged at the previous step keeps its pass value. Step 3 performs the operations described in Section 3.5. Note that, after the first iteration some vertices do not encode a single basin but a set of merged basins. In this case, the minimal altitude of the vertex is defined as the minimal altitude of the merged basins. Step 4 builds a contraction kernel from the set of edges with a low pass value. Note that this set of edges may defines loops in the current combinatorial map. In this case one of the edge of each loop is not added to the kernel in order to respect the forest requirement of a contraction kernel(Section 2). However, this case is rare in practical cases and the contraction kernel generally include all the edges with the lowest dynamic.

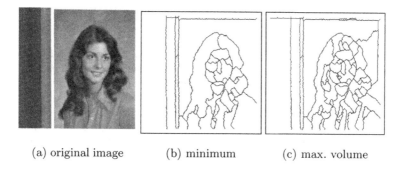

(a) original image          (b) minimum          (c) max. volume

**Fig. 2.** Two segmentations using different edge's pass values

Fig. 2(b) and (c) shows two levels of two pyramids built by valuating edges respectively with the minimal value of the contour and the edge pass value as defined in Section 3.4. The levels in each pyramid have been selected such as the white bar on the left of Fig. 2(a) forms only one region at the level above. Much more meaningful details are preserved in Fig. 2(c) which thus better fit to the intuitive notion of contour's saliency. This phenomena is due to the edge's pass value which do not take into account minima with a small volume within the profile of the contours. Note that the operations used to obtain Fig. 2(b) may be performed without our hierarchical data structure (using e.g. [15]) while Fig. 2(c) is obtained using both the geometrical and topological features of Combinatorial Pyramids.

## 4    Conclusion

We have presented in this paper a new hierarchical watershed method based on the edge's dynamic. The different partitions of the hierarchy are encoded within the combinatorial pyramid framework. The main advantages of combinatorial pyramids within this framework are the encoding of each contour by one edge and the efficient retrieval of each contour's embedding as a sequence of linels. We used these properties to define a new edge's pass value which allows us to overcome the noise which may be present within the contours. The presence of noise within the basins is corrected using edge's dynamics based on the edge's pass values previously computed. In future studies we are planing to combine the edge's dynamic with statistical tests on the content of the regions. More studies should also be undertaken on the valuation of the minimal value of a contour.

## References

[1] G. Bertrand. Some properties of topological greyscale watersheds. In *procs. SPIE Vision Geometry XII*, volume 5300, pages 182–191, 2004.
[2] L. Brun. *Traitement d'images couleur et pyramides combinatoires*. Habilitation à diriger des recherches, Université de Reims, 2002.

[3] L. Brun and W. Kropatsch. Combinatorial pyramids. In Suvisoft, editor, *IEEE International conference on Image Processing (ICIP)*, volume II, pages 33–37, Barcelona, September 2003. IEEE.

[4] L. Brun, P. Vautrot, and F. Meyer. Hierarchical watersheds with inter-pixel boundaries. In *Image Analysis and Recognition: International Conference ICIAR 2004, Part I*, pages 840–847, Proto (Portugal), 2004. Springer Verlag Heidelberg (LNCS).

[5] M. Couprie and G. Bertrand. Topological grayscale watershed transformation. In *SPIE Vision Geometry VI Proceedings*, volume 3168, pages 136–146, 1997.

[6] G. Damiand. *Définition et étude d'un modèle topologique minimal de représentation d'images 2d et 3d*. PhD thesis, Université des Sciences et Techniques du Languedoc, Décembre 2001.

[7] C. Fiorio and J. Gustedt. Two linear time Union-Find strategies for image processing. *Theoretical Computer Science*, 154(2):165–181, 5 Feb. 1996.

[8] R. Glantz and W. Kropatsch. Plane embedding of dually contracted graphs. In *Discrete Geometry for Computer Imager DGCI'2000*, Lecture Notes in Computer Science. Springer, Berlin Heidelberg, New York, 2000. In Press.

[9] U. Köthe. Deriving topological representations from edge images. In *Geometry, Morphology, and Computational Imaging, 11th Intl. Workshop on Theoretical Foundations of Computer Vision, LNCS, Springer Verlag*, volume 2616, pages 320–334, 2003.

[10] W. G. Kropatsch and H. Macho. Finding the structure of connected components using dual irregular pyramids. In *Cinquième Colloque DGCI*, pages 147–158. LLAIC1, Université d'Auvergne, ISBN 2-87663-040-0, September 1995.

[11] J. Marchadier, D. Arquès, and S. Michelin. Thinning grayscale well-composed images. *Pattern Recognition Letters*, 25:581–590, 2004.

[12] F. Meyer. Topographic distance and watershed lines. *Signal Processing*, (38):113–125, 1994.

[13] A. Montanvert, P. Meer, and A. Rosenfeld. Hierarchical image analysis using irregular tessellations. *IEEE Transactions on Pattern Analysis and Machine Intelligence*, 13(4):307–316, APRIL 1991.

[14] L. Najman and M. Couprie. Watershed algorithms and contrast preservation. In *Discrete geometry for computer imagery*, volume 2886, pages 62–71. LNCS, Springer Verlag, 2003.

[15] L. Najman and M. Schmitt. Geodesic saliency of watershed contours and hierarchical segmentation. *IEEETPAMI*, 18(2):1163–1173, December 1996.

[16] C. Vachier and F. Meyer. A morphological scale-space approach to image segmentation based on connected operators. In *Workshop on Mathematics and Image applications*, Paris, September 2000.

[17] L. Vincent and P. Soille. Watersheds in digital spaces : an efficient algorithm based on immersion simulations. *IEEE Transactions on Pattern Analysis and Machine Intelligence*, 13(6):583–598, 1991.

[18] J. Webster. Cell complexes, oriented matroids and digital geometry. *Theoretical Computer Science*, 305(1–3):491–502, Aug. 2003.

# Optimal Design of 2D/3D Hierarchical Content-Based Meshes for Multimedia

Işıl Celasun[1], Rupen Melkisetoğlu[1], and A. Murat Tekalp[2]

[1] Dept. of Electronics and Communication Eng.,
Istanbul Technical University, Maslak,
Istanbul, 34469, Turkey
[2] Dept. of Electrical Eng. and Center for Electronic Imaging Systems,
Univ. of Rochester, P.O.Box 270126,
Rochester, NY 14627, USA

**Abstract.** This paper proposes and compares methods for designing hierarchical 2D meshes for representation of object-based video and hierarchical 3D meshes for 3D objects used in telemedicine and multimedia applications. The same approach has been applied both in 2D and 3D but with different constraints. This representation consists of a hierarchy of Delaunay meshes, obtained by recursive simplification of the initial fine level-of-detail mesh geometry. There is no guarantee of an optimal mesh in 3D that uses a specific given set of node points whereas in 2D it is guaranteed that there is a unique 2D Delaunay mesh which uses all the node points for a specific set. To solve this problem an optimized alpha value is used in 3D Delaunay triangulation in the proposed algorithm. Mesh simplification entails removal of mesh nodes to reduce the level of detail. The selection of nodes to be removed is achieved by associating a cost with each mesh node. The Delaunay topology constraint on each mesh level not only helps to design meshes with desired geometric properties, but also enables efficient compression of the mesh data for multimedia applications.

## 1 Introduction

Dynamic mesh representation of video objects has recently been proposed for object-based video coding (by warped motion compensation), storage (e.g. video database query by motion) and manipulation (e.g. video animation). A 2D triangular mesh is initially designed on the first video object plane of a video object sequence, and subsequently tracked by motion estimation techniques. Hence, a 2D dynamic mesh compactly represents the shape and motion of a video object. We have recently proposed a hierarchical mesh-based representation of object-based video [1]. Hierarchical mesh representation has attracted attention recently, because it provides rendering at various level of detail (quality scalability), and allows progressive/scalable transmission of the mesh geometry and motion. Also, the hierarchical representation leads to improved motion track-

E. Andres et al. (Eds.): DGCI 2005, LNCS 3429, pp. 45–55, 2005.

ing performance [1]. This paper discusses optimal hierarchical construction of content-based 2D and 3D dynamic meshes.

A hierarchy of fine-to-coarse meshes can be obtained by repeated removal of some detail information from an initial mesh. A 2D mesh partitions the image domain into triangular patches, the vertices of which are referred to as *node points*. The straight-line segments between node points are referred to as edges. The *degree* of a node is the number of incident edges to the node. In the initial fine level-of-detail mesh design, node points are placed at salient edge and corner points of the boundary and interior of a video object plane. The number of nodes in the initial mesh can be found using the number of points detected by a corner detection algorithm. The initial mesh topology is constructed using constrained Delaunay triangulation of the node points, where the boundary edge segments serve as constraints. The Delaunay property of the initial mesh is preserved during mesh simplification by the constraint that only non-adjacent nodes can be removed, i.e., removed nodes must form an *independent* set. Each removed node leaves a hole in the mesh which is retriangulated using the Delaunay criterion.

Three-dimensional (3-D) polygonal meshes have been popular in computer graphics to describe the geometry (structure) of world objects. They have been employed to view objects from different angles and/or to render photorealistic synthetic images by texture mapping [2]. Three-dimensional meshes are an elementary building block of the Virtual Reality Modeling Language (VRML), a standard for storing and interacting with graphics objects and virtual worlds over the World Wide Web.

Hierarchical representation of 3D meshes have attracted attention because it: 1) provides rendering at various levels of detail (quality scalability); 2) allows progressive/scalable transmission or storage of the object geometry and motion information. Scalability means that terminals of different complexity can extract data of different quality levels from this single bit stream. Hierarchical representation of 3-D meshes has been addressed in computer graphics for adaptive level-of-detail (LOD) rendering of 3-D objects. A wavelet-based multiresolution mesh approximation was proposed in [3, 4, 5]. Meshes of tetrahedra have many applications, including interpolation, rendering, compression, and numerical methods such as the finite element method. Most such applications demand more than just a triangulation of the object or domain being rendered or simulated. To ensure accurate results, the tetrahedra must be "well-shaped", having small aspect ratios or bounds on their smallest and largest angles [6]. In this paper, we propose a new hierarchy of 3D Delaunay meshes and we only remove vertices in the fine to coarse design strategy. We do not reposition vertices nor edges which increase efficient use of bandwidth when compared to other methods mentioned so far.

In Section 2 and 3, algorithms for 2D/3D mesh simplification with their proper design parameters respectively are explained. In Section 4, experimental results for proposed decimation algorithms and conclusions are given.

## 2  Two Dimensional Hierarchical Mesh Representation and Design

A hierarchy of fine to coarse level meshes $M_\ell$ is defined, with $\ell = 0, 1, \ldots, L-1$, where each coarser level mesh is obtained by removal of some detail information from the finer level mesh. Thus, a hierarchical decomposition of the mesh data into $L$ levels is obtained, consisting of a base level and $L-1$ enhancement levels, suitable for progressive transmission of a 2D dynamic mesh. The base level mesh contains a coarse version $M_{L-1}$ of the original mesh, while each enhancement level contains detail information such that finer versions $M_{L-2}, \ldots, M_0$ of the mesh can be reconstructed, where $M_0$ is the original mesh. Hierarchical mesh representation and mesh simplification have been addressed in computer graphics for adaptive level-of-detail rendering of 3D objects. Hierarchical mesh modeling of video has been addressed previously in the context of uniform topology only.

### 2.1  Delaunay Hierarchy

Here, the decomposition of the initial mesh is restricted such that each mesh $M_\ell$ reconstructed at level $\ell$ has Delaunay topology. Thus, no topology data has to be encoded on any level of the decomposition. Only mesh node locations need to be encoded.

A simplification step starts by removing an *independent set* of nodes from the mesh at a certain level. An independent set of nodes is a set of nodes among which no two nodes are connected to each other by an edge. Edges that are incident to a removed node are removed as well. Each removed node of the mesh leaves a hole in the mesh which is to be retriangulated using the Delaunay criterion. It has been noted by de Berg et al. [7] that, after locally retriangulating the areas around removed independent nodes, the global triangulation remains Delaunay. The nodes that were removed in this step form the detail information to be encoded in an enhancement layer. This removal step is iterated a number of times, thus defining $L$ versions of the mesh from fine to coarse. This strategy leads to a well-defined hierarchy of triangles and Delaunay meshes for all levels $\ell = 0, \ldots, L-1$, in which part of the topology is preserved going from one hierarchical level to the next. The approach represents the middle ground between enforcing a strict triangular hierarchy as one extreme, and ignoring topology preservation completely by picking nodes to be removed freely as the other extreme. It is naturally desirable to retain important nodes while going from one level to the next, such that essential mesh features are not removed. In this paper, we determine the importance of a node adaptively using image-based and shape-based criteria. The nodes on a 2D mesh boundary, convey essential object *shape* information.

### 2.2  Removal of Mesh Boundary Nodes

A sequential simplification algorithm is used to remove boundary nodes going from one hierarchy level to the next. The algorithm uses a distance parameter

$D_{max}$ to control the error in the polygonal boundary shape, which is increased from level to level. A candidate approximating edge segment, or chord, is drawn between an initial boundary node point and another node point, the chord node. Initially, the chord node is the second node on the boundary with respect to the initial node. For each node between the initial boundary node and the chord node, the distance $d$ from the candidate node to the chord is computed and compared to $D_{max}$. If $d$ is smaller than $D_{max}$ for all candidate node points, the next node on the boundary becomes the new chord node and a new chord is drawn. If $d$ is greater than or equal to $D_{max}$ for one or more candidate nodes, the candidate node with maximum distance $d$ is retained and all nodes between this node and the initial node become removable nodes. The node that is retained becomes the initial node in the next step. Note that only an independent subset of the removable nodes is actually removed, so as to conform to the hierarchy discussed in the previous subsection. These steps are repeated until all nodes on the boundary have been processed.

### 2.3    Removal of Mesh Interior Nodes

The set of interior nodes to be removed while going from one hierarchy level to the next can be determined using both spatial and temporal image-based criteria. That is, one desires to retain nodes with locally important motion activity, as well as nodes that are salient in terms of "edgeness" and "cornerness" measures. The following measure of the saliency of node $n$ is used:

$$C_n = |I_x(\vec{p}_n)|^2 + |I_y(\vec{p}_n)|^2 + \Gamma(\vec{p}_n) \tag{1}$$

where $\vec{p}_n(x_n, y_n)$ is the location of node $n$, $I_x(\vec{p}_n)$ and $I_y(\vec{p}_n)$ are partial derivatives of the intensity image at $\vec{p}_n$, and $\Gamma(\vec{p}_n)$ is the response of a corner detector at $\vec{p}_n$, based on a non-linear measure of similarity between the image intensities in a neighborhood around $\vec{p}_n$. Here, the intensity image refers to the initial VOP of the video object from which the mesh was derived. Using this importance measure, interior nodes are removed using an iterative greedy algorithm. Initially, all nodes that were connected directly to a removed boundary node are marked. Then, during each iteration, the unmarked node with the smallest value of the importance measure is removed and all nodes connected directly to this node are marked. Repeat until there are no more unmarked nodes. Note that the marking of nodes ensures that the nodes that are removed form an independent set. We allow a node to be removed only if its degree (number of incident edges) is at most 6. One can still ensure that a certain percentage of nodes will be removed.

## 3    Three Dimensional Hierarchical Mesh Representation and Design

### 3.1    Initial Fine Detail Mesh

The 3D Delaunay triangulation is defined as the triangulation that satisfies the Delaunay criterion for $n$-dimensional simplexes (in this case $n=3$ and the simplexes are tetrahedra). This criterion states that a circumsphere of each simplex in a triangulation contains only the $n+1$ defining points of the simplex. It has been proven that two dimensional Delaunay triangulation satisfies an "optimal" triangulation, but in 3D Delaunay triangulation the situation is not so, since a measurement for optimality in 3D is not agreed on in the literature.

**Computation of the "$\alpha$" Parameter.** A graph can be defined as $\mathcal{G} = (\mathcal{V}, \mathcal{E})$. Here $\mathcal{V}$ is the set of vertices, $\mathcal{V}=\{v_0, v_1, \ldots, v_1, \ldots, v_{v-1}\}$, and $\mathcal{E}$ is the set of edges, $\mathcal{E}=\{e_0, e_1, \ldots, e_i, \ldots, e_{E-1}\}$. The $\alpha$ parameter used in the 3-D Delaunay tetrahedralization tool specifies the radius of the circumsphere of each tetrahedron. Only tetrahedrons lying within this circumsphere are allowed. If the $\alpha$ value is zero, the output of the tetrahedralizator is a convex hull. Usually the output is a tetrahedral mesh, but if $\alpha$ distance value is not to be set as zero, then output data consists of tetrahedra, triangles, edges, and vertices lying within the $\alpha$ radius. Optimal $\alpha$ value has to be computed in order to get a good approximation of the original image. The average tetrahedron edge length ($l_e$) of the convex hull is used in the proposed decimation method to determine this optimal value. Although the convex hull may contain edges of large length when connecting end nodes of the volume data, experiments show that the average edge length gives us a proper $\alpha$ value. In fact, to preserve all tetrahedra present in the 3-D mesh, maximum circumradius of them has to be used to determine $\alpha$ value, which in practice is expressed as average value when convex hull is employed. Again these "ill-conditioned" tetrahedra force us to choose average edge length instead of circumradius in $\alpha$ value determination.

When used a *global* $\alpha$, some regions in the mesh, as pointed in Figure 3 (a), (c), (e) cannot be filled. The reason is that, an optimal Delaunay triangulation in 3D using all given node points cannot be guaranteed 3D which is different from 2D case [6, 7]. These vertices cannot be located in the spheres by using global $\alpha$ value in the Delaunay triangulation topology. There may be non connected regions since $\alpha$ value does not guarantee a connected mesh. The $\alpha$ variable in the *local* Delaunay topology, is obtained by vertex density value calculated by scanning the number of neighbours of a vertex in a region. Optimization is realized consecutively by bounding the alpha variable with a minimum and maximum value. Region with a density of vertices with constrained Delaunay triangulation, and the boundaries of these regions are kept and after the application of the same process for other density value regions, these regions boundaries are connected. So, generally a Delaunay mesh is formed by connected submeshes which are constrained Delaunay triangulated.

**Boundary Extraction Algorithm.** The boundary extraction algorithm for 3-D images represented by the Delaunay tetrahedralization, uses the sum of the solid angles at every vertex to determine whether the vertex is on the boundary or not. The solid angle at the vertex $v_i$ of the tetrahedron $\mathcal{T}(v_0, v_1, v_2, v_3)$ is defined to be the surface area formed by projecting each point of the face not containing $v_i$ to the unit sphere centered at $v_i$. For each tetrahedron, the solid angle at each vertex is calculated. A vertex is said to be on the boundary, if the sum of the solid angles of every tetrahedron the vertex belongs to at this node, is less than the surface area of a unit sphere, $4\pi$. If the sum of the solid angles equals $4\pi$, the vertex has to be in the interior of the image. It is obvious from the definition of the solid angle of a tetrahedron, that there is no possibility for the sum of the solid angles to be greater than the surface area of the unit sphere, provided that the tetrahedrons do not intersect. The solid angle $(\Psi)$ of a tetrahedron is as:

$$(\Theta_{e_1 e_2} : \text{dihedral angle at edge } e_1 e_2) \quad \Psi_A = (\Theta_{AB}) + (\Theta_{AC}) + (\Theta_{AD}) - \pi \quad (2)$$

where the dihedral angle $(\Theta)$ at the edge $e_{AB}$ is the angle between the intersection of the two faces containing the edge $e_{AB}$ and a plane perpendicular to this edge:

$$\Theta_{AB} = \pi - |\arccos(\text{face}_{ABC} \bullet \text{face}_{ABD})| \quad (3)$$

A sequential simplification algorithm is again used as in 2D case for 3D too to remove boundary vertices going from one hierarchy level to the next.

**Mesh Interior Simplification.** Importance function assigns an importance value for each vertex in the mesh. If the vertex is an important vertex that should not be removed, his importance value is also large. The importance $(IP_1)$ value for an interior node is defined as the ratio of the sum of its neighbors' tetrahedral volume to its tetrahedron volume. This $IP_1$ stress on the connectivity of the mesh and try to retain detailed regions' vertices of the volume data. $Degree(n)$ is criterion of the connectivity of relevant vertex $v_n$. It represents how many edges are connected to the vertex. It is obvious that more smaller the $volume(n)$ and more larger the sum of $volume(i)$ is more higher the importance value of relevant vertex because small volume represents a more detailed region and vertex in this region is more important. The interior vertex remove algorithm is a simple Greedy-type algorithm that removes the vertex having the smallest importance value among the unprocessed ones, and keeps its neighbor vertices.

The importance value for an interior vertex, is defined as the multiplication of his degree with the ratio of the total volume of the tetrahedrons formed by its neighbors to its own volume. If the volume of a tetrahedron related to a vertex, is smaller then the volumes of the tetrahedrons related to its neighbors, this means that this vertex contains a detail

$$IP(n) = Degree(n) * \frac{\sum_{i=1}^{neighbors-1} volume(i)}{volume(n)} \quad (4)$$

information and removing will be deletion of this detail information.

# 4    Experimental Results and Conclusions

## 4.1    Results for Design of 2D Meshes

Initial fine level-of-detail content-based meshes were designed on initial Bream and Akiyo video object planes, as illustrated in Figure 1 (a) and Figure 2 (a) respectively. These fine level-of-detail meshes contain 165 nodes and 210 nodes, respectively. Successively coarser level-of-detail meshes, obtained using the simplification methods are illustrated in Figure 1 (b), (c), (d) and (e) for Bream,

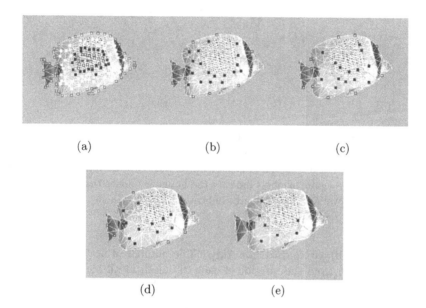

(a)                    (b)                    (c)

(d)                    (e)

**Fig. 1.** (a) Illustrates six node layers of the original Bream mesh used in mesh simplification, by successive empty and filled squares. In (b), (c), (d) and (e), the hierarchical mesh for four consecutive levels of Bream is shown overlaid onto the original video object plane, where empty square symbols indicate the boundary nodes that will be removed at that hierarchy level and filled squares indicate the interior nodes to be removed at that level

**Table 1.** Number of removed boundary nodes and number of removed interior nodes at each level of the mesh hierarchy for the initial Bream video object plane

| Hier. level | $D_{max}$ | # rem. boundary | # rem. interior |
|:---:|:---|:---:|:---:|
| 1 | 0.8997 | 8 | 18 |
| 2 | 1.6246 | 10 | 14 |
| 3 | 1.8342 | 6 | 11 |
| 4 | 2.0613 | 0 | 7 |

**Table 2.** Number of removed boundary nodes and number of removed interior nodes at each level of the mesh hierarchy for the initial Akiyo video object plane

| Hier. Level | $D_{max}$ | # rem. boundary | # rem. interior |
|---|---|---|---|
| 1 | 0.4662 | 4 | 32 |
| 2 | 1.3083 | 6 | 23 |
| 3 | 3.3609 | 6 | 19 |

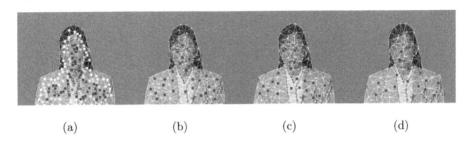

(a)              (b)              (c)              (d)

**Fig. 2.** (a) Illustrates six node layers of the original Akiyo mesh used in mesh simplification, by successive empty and filled squares. In (b), (c) and (d), the hierarchical mesh for three consecutive levels of Akiyo is shown overlaid onto the original video object plane, where empty square symbols indicate the boundary nodes that will be removed at that hierarchy level and filled squares indicate the interior nodes to be removed at that level

and in Figure 2 (b), (c) and (d) for Akiyo. The number of nodes removed from the boundary and from the interior of each 2D mesh at all hierarchy levels are summarized in Tables 1 and 2, for the Bream and Akiyo sequences, respectively.

### 4.2    Results for Design of 3D Meshes

Experimental results related to boundary and interior vertex simplification are given for "Cat" and "Engine" data. The proposed algorithm works in general for volume data but it can also provide successful results for surface data in terms of minimized number of vertices for a good mesh quality. Boundary node simplification is accomplished for $D_{max} = 2.0$ for "Cat". "Cat" data has a boundary consisted of concave and convex regions so its boundary has to be processed carefully for its global shape information. That is why a low $D_{max}$ value is chosen for "Cat". The stopping criterion is such that no more removed vertices on its boundary is applied. Global $\alpha$ value and $IP$ are used for "Engine" and "Cat" in Figures 3-5. As seen from Figure 3, there are some gaps (nonconnected regions) at the output the rendered object. To get rid of this, local $\alpha$ values are used as in Figure 3-5 with $IP$ calculation. Although number of vertices are very small in Figure 4 and 5, good results in terms of visual information and node simplification are obtained as given in Table 3 and 4 where two test results are also added.

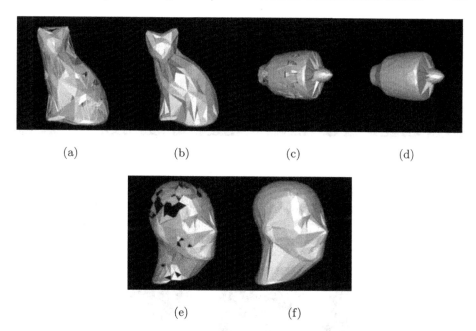

**Fig. 3.** "Results for 3D Delaunay meshes using global $\alpha$ are given in (a), (c), (e) and for 3D Delaunay meshes using local $\alpha$ s are given in (b), (d), (f) respectively for "Cat", "Engine", "Mannequin" data

Mesh quality criteria can be provided by a "Volume Test" and an "Angle Test": First, ratio of inner radius $(r)$ to outer $(R)$ one for a tetrahedron cannot pass 1/3. In a sliver $r \to 0$ and $R \to \infty$, thus $\frac{3r}{R} \cong 0$. Using this ratio as a mesh quality measure, we can say that this mesh performance criterion is between 0 and 1. By calculating this ratio for each tetrahedron and taking the average value, we can determine the overall quality of the mesh at hand. This is called a "Volume Test". Second, absolute value of the deviation of each angle of each triangle in the 3D mesh from 60 is calculated. Average of the deviation angles of all triangles in the 3D mesh is taken and "Angle Test" is thus calculated.

For both 2D and 3D hierarchical meshed only vertices have to be known, triangles are determined by Delaunay triangulation. The main difficulty with Delaunay triangulation in 3D is that its optimality is not proven. An optimal choice of alpha value controls this problem. In the proposed method, one only needs the vertices so a high transmission rate is possible for both 2D and 3D applications. Removal of maximum number of independent vertices will then provide high compression ratios by preserving mesh and thus rendered object quality.

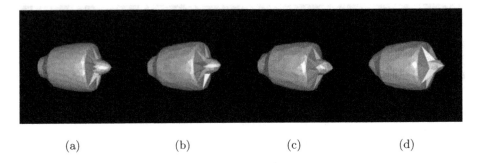

(a)                    (b)                    (c)                    (d)

**Fig. 4.** (a) Original "Engine"graphics model with 426 vertices -3D Delaunay Mesh $\ell_0$ (b) 3D Delaunay Mesh $\ell_1$ hierarchy level 1 with 366 vertices (c) 3D Delaunay Mesh $\ell_2$ hierarchy level 2, with 291 vertices (d) 3D Delaunay Mesh $\ell_3$ hierarchy level 3 with 235 vertices

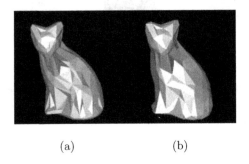

(a)                    (b)

**Fig. 5.** (a) Original graphics model "Cat" with 328 vertices,-3D Delaunay Mesh $\ell_0$ (b) 3D Delaunay Mesh $\ell_3$ hierarchy level 3 with 238 vertices

**Table 3.** Performance of the hierarchical design algorithm for "Engine" data as given in Figure 4 with IP calculation and local $\alpha$ values (V: vertices, B.V.: Boundary vertices)

|                   | V   | B.V. | Volume Test | Angle Test |
|-------------------|-----|------|-------------|------------|
| Original          | 426 | 189  | 0.419       | 24.04      |
| Hierarchy Level 1 | 366 | 161  | 0.419       | 23.85      |
| Hierarchy Level 3 | 291 | 153  | 0.437       | 22.92      |
| Hierarchy Level 5 | 235 | 144  | 0.462       | 22.09      |

**Table 4.** Performance of the hierarchical design algorithm for "Cat" data as given in Figure 5 with IP calculation and local $\alpha$ values (V: vertices, B.V.: Boundary vertices)

|                   | V   | B.V. | Volume Test | Angle Test |
|-------------------|-----|------|-------------|------------|
| Original          | 328 | 149  | 0.284       | 28.02      |
| Hierarchy Level 1 | 268 | 120  | 0.284       | 27.91      |
| Hierarchy Level 2 | 238 | 120  | 0.294       | 27.45      |

**Table 5.** Local $\alpha$ values for "Engine" and "Cat" data for different hierarchy levels (E:Engine, C:Cat, $\ell_i$:level $i$)

| E $\ell_0$ | E $\ell_1$ | E $\ell_3$ | E $\ell_5$ | C $\ell_0$ | C $\ell_1$ |
|---|---|---|---|---|---|
| 0.78 | 0.81 | 0.87 | 1.08 | 0.11 | 0.12 |
| 1.34 | 1.37 | 1.48 | 0.94 | 0.18 | 0.13 |
| 1.33 | 1.36 | 0.98 | 1.60 | 0.16 | 0.19 |
| 1.23 | 1.25 | 1.11 | 1.44 | 0.12 | 0.19 |
| 0.95 | 0.96 | 1.24 | 1.25 | 0.15 | 0.17 |
| 1.03 | 0.88 | 1.39 | | 0.14 | 0.12 |
| 1.13 | 1.15 | | | 0.18 | 0.16 |
| 0.86 | 1.05 | | | 0.15 | 0.17 |
| | | | | 0.17 | 0.14 |
| | | | | 0.12 | 0.15 |
| | | | | 0.13 | |

# References

1. van Beek, P., Tekalp, A. M., Zhuang, N., Celasun, I., Xia, M.: Hierarchical 2-D mesh representation, tracking, and compression for object-based video. IEEE Trans. on CSVT 9, (1999)
2. Hearn D., and Baker M. P.: Computer Graphics, 2nd ed. Englewood Cliffs, NJ: Prentice Hall, (1997)
3. P. S., Heckbert, M., Garland: Multiresolution Modeling for Fast Rendering, in Proc. Graphics Interface '94, Banff, Alta. (May 1994)
4. Hoppe H.: Progressive Meshes, in Computer Graphics-Proc. SIGGRAPH '96, (1996), 99–108
5. Eck M., DeRose T., Duchamp T., Stuetzle W.: Multiresolution Analysis of Arbitrary Meshes, in Computer Graphics-Proc. SIGGRAPH '95, (1995), 173–182
6. Shewchuck, J. R.: Tetrahedral mesh generation by Delaunay refinement, in Proceedings of the fourteenth annual symposium on computational geometry, Association for Computing Machinery, (June 1998), 86–95
7. de Berg, M., van Kreveld, M. , Overmars, M., Schwarzkopf, O.: Computational Geometry-Algorithms and Applications, Springer, 1997

# Receptive Fields for Generalized Map Pyramids: The Notion of Generalized Orbit

Carine Grasset-Simon, Guillaume Damiand, and Pascal Lienhardt

SIC, FRE-CNRS 2731 - Université de Poitiers,
bât. SP2MI, Bvd M. et P. Curie,
BP 30179, 86962 Futuroscope Chasseneuil Cedex - France
{simon, damiand, lienhardt}@sic.univ-poitiers.fr

**Abstract.** A pyramid of $n$-dimensional generalized maps is a hierarchical data structure. It can be used, for instance, in order to represent an irregular pyramid of $n$-dimensional images. A pyramid of generalized maps can be built by successively removing and/or contracting cells of any dimension. In this paper, we define generalized orbits, which extend the classical notion of receptive fields. Generalized orbits allow to establish the correspondence between a cell of a pyramid level and the set of cells of previous levels, the removal or contraction of which have led to the creation of this cell. In order to define generalized orbits, we extend, for generalized map pyramids, the notion of connecting walk defined by Brun and Kropatsch.

**Keywords.** Irregular pyramids, generalized maps, generalized map pyramids, connecting walks, generalized orbits.

## 1 Introduction

For image analysis, it can be useful for some applications to segment an image at different levels. According to the application, some informations appear more clearly at some levels. An image pyramid corresponds to several segmentation levels of an image; levels 0 corresponds to the original image, the following levels correspond to the successive segmentations of this image. Many works deal with 2D image regular pyramids (cf. e.g. [1]) or 2D image irregular pyramids (cf. e.g. [2, 3, 4]).

Most image processing algorithms need to extract informations from images, for instance the adjacency between regions of images (e.g. an algorithm of segmentation by region aggregation). Order is another interesting notion. For example, it can be useful to retrieve the order of the edges which compose the boundary of a 2D region, or in 3D, to know the order of volumes and faces around an edge or a vertex.

Many definitions of irregular pyramids are based upon graphs [3, 5]. More recently, Brun and Kropatsch [6, 7, 8, 9] have studied 2D combinatorial map pyramids, since combinatorial maps allow to represent the whole topological in-

E. Andres et al. (Eds.): DGCI 2005, LNCS 3429, pp. 56–67, 2005.

formation of subdivisions of orientable surfaces without boundary (for instance the order information is generally not represented by graphs).

3D and 4D images (time being the $4^{th}$ dimension) are now usual images. So, we want to extend the previous works for any dimension, by defining pyramids of generalized maps [10]. The $n$-dimensional generalized maps (or $n$-G-maps) represent the topology of $n$-dimensional quasi-manifolds [11], orientable or not, with or without boundary. We have chosen generalized maps since their definition is homogeneous for all dimensions. So, we can easily define generic operations and algorithms.

A pyramid of $n$-G-maps can be constructed in the following way. Given an $n$-G-map which represents for instance an image, each level of the pyramid is deduced from the previous level by applying simultaneously removals and/or contractions of cells of any dimension[1].

It is essential for many applications, to establish the correspondence between a cell at a given level, and the set of cells of previous levels the removal or the contraction of which has led to the creation of this cell. In 2D for instance, it can be useful to associate a face with the corresponding region of a lower level. In particular, it allows to retrieve any information contained in a lower pyramid level. The notion of receptive field has been introduced in order to establish this correspondence between regions of different levels. First, this notion has been defined in the context of graph hierarchy [3], in the following way: the receptive field of a cell of level $n$ is the set of all pixels of level 0 which "compose" this cell. More recently, Brun and Kropatsch define the notion of receptive field of a dart for 2D combinatorial map pyramid [12]. This notion is based on the notion of connecting walk between a surviving dart and its successor at a given level, which is the set of darts which separate these two darts at the previous level. The notion of reduction window generalizes that notion of connecting walks for any levels.

The main result presented in this paper is the definition of generalized orbits of $n$-G-map pyramids which makes it possible to associate any cell of any dimension of a given level with the set of corresponding cells of any lower level. This definition is based upon a generalization of the connecting walk notion, which is itself based upon the operation of "simultaneous removals and contractions of cells" presented in [13].

The notion of orbit is a classical one for combinatorial and generalized maps. It allows to define cells as set of darts, darts being the unique type of elements defining maps (c.f. section 2). We generalize the notion of receptive field for any orbit (i.e. any cell) and any levels by defining generalized orbit. With this notion, we retrieve for $n$-cells the union of $n$-cells at a lower level which have been "merged" in a unique $n$-cell in a upper level.

This paper is organized in the following way. We give in section 2 a brief recall about pyramids of $n$-dimensional generalized maps. The connecting walk

---

[1] Note that 2D combinatorial map pyramids as defined by Brun and Kropatsch are built in a particular way: each odd (resp. even) level is deduced from the previous one by contracting (resp. removing) edges.

notion is defined in section 3, and generalized orbits are defined in section 4. At last, we conclude and give some perspectives in section 5.

## 2 Recall: Pyramids of $n$-Dimensional Generalized Maps

An $n$-dimensional generalized map is a set of abstract elements, called darts, together with applications defined on these darts (c.f. figure 1):

**Definition 1 ($n$-G-map).** *Let $n \geq 0$. An $n$-dimensional generalized map $G$ (or $n$-G-map) is defined by $G = (D, \alpha_0, \ldots, \alpha_n)$ where:*

1. *$D$ is a finite set of darts;*
2. *$\forall i, 0 \leq i \leq n, \alpha_i$ is an* involution[2] *on $D$;*
3. *$\forall i, j, 0 \leq i < i + 2 \leq j \leq n, \alpha_i \alpha_j$ is an involution.*

The $n$-G-maps represent the topology of subdivided objects, more precisely the topology of quasi-manifolds (see [11]). Cells are implicitly represented as subset of darts:

**Definition 2 ($i$-cell).** *Let $G$ be an $n$-G-map, $d$ be a dart and $i \in N = \{0, \ldots, n\}$. The $i$-cell incident to $d$ is the orbit[3]*
$$<>_{N-\{i\}} (d) = < \alpha_0, \ldots, \alpha_{i-1}, \alpha_{i+1}, \ldots, \alpha_n > (d).$$

Figure 1 illustrates the notions of generalized map and $i$-cell. Intuitively, an $i$-cell is the set of all darts which can be reached starting from $d$, by using any combination of all involutions except $\alpha_i$. The set of $i$-cells is a partition of the set of darts $D$, for each $i$ between 0 and $n$. Two cells are disjoint when their intersection is empty, i.e. when no dart is shared by the cells. More precisions about $n$-G-maps are provided in [11] and [14].

In order to define $n$-G-map pyramids, Damiand and Lienhardt have defined the operation of "simultaneous removals and contractions of cells of any dimension" [13] which allows to contract and remove a set of cells of any dimension in a simultaneous way. The formal definition of this operation is:

**Definition 3 (Simultaneous Removal and Contraction of Cells of Any Dimension).**
*Let $G = (D, \alpha_0, \ldots, \alpha_n)$ be an $n$-G-map, $R_0, \ldots, R_{n-1}$ be sets of 0-cells, $\ldots$, $(n-1)$-cells to be removed and $C_1, \ldots, C_n$ be sets of 1-cells, $\ldots$, $n$-cells to be contracted. Let $R = \cup_{i=0}^{n-1} R_i$ and $C = \cup_{i=1}^{n} C_i$. Two preconditions have to be satisfied: cells are disjoint (i.e. $\forall c, c' \in C \cup R, c \cap c' = \emptyset$), and "the degree of each cell is equal to 2", i.e.:*

---

[2] An involution $f$ on a finite set $S$ is a one to one mapping from $S$ onto $S$ such that $f = f^{-1}$.

[3] Let $\{\Pi_0, \ldots, \Pi_n\}$ be a set of permutations on D. The orbit of an element $d$ relatively to this set of permutations is $< \Pi_0, \ldots, \Pi_n > (d) = \{\Phi(d), \Phi \in < \Pi_0, \ldots, \Pi_n >\}$, where $< \Pi_0, \ldots, \Pi_n >$ denotes the group of permutations generated by $\Pi_0, \ldots, \Pi_n$.

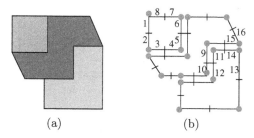

(a)                                        (b)

**Fig. 1.** (a) A subdivision of a surface. (b) The corresponding 2-G-map. Darts are represented by (numbered) black segments. Two darts in relation by $\alpha_0$ share a little vertical segment (ex. darts 1 and 2). Two darts in relation by $\alpha_1$ share a same point (ex. darts 2 and 3). Two distinct darts in relation by $\alpha_2$ are parallel and close to each other (ex. darts 9 and 11); otherwise, the dart is its own image by $\alpha_2$ (ex. dart 2). The vertex incident to dart 14 is $< \alpha_1, \alpha_2 > (14) = \{13, 14, 15, 16\}$, the edge incident to dart 9 is $< \alpha_0, \alpha_2 > (9) = \{9, 10, 11, 12\}$, and the face incident to dart 4 is $< \alpha_0, \alpha_1 > (4) = \{1, 2, 3, 4, 5, 6, 7, 8\}$

- $\forall i,\ 0 \leq i \leq n - 2,\ \forall d \in R_i,\ d\alpha_{i+1}\alpha_{i+2} = d\alpha_{i+2}\alpha_{i+1}$
- $\forall i,\ 2 \leq i \leq n,\ \forall d \in C_i,\ d\alpha_{i-1}\alpha_{i-2} = d\alpha_{i-2}\alpha_{i-1}$

Let $SD_i = (R_i \cup C_i)\alpha_i - (R_i \cup C_i)\ \forall i, 0 \leq i \leq n$ *(it is the set of surviving darts "neighbour" of removed and contracted cells). The resulting n-G-map is* $G' = (D', \alpha'_0, \ldots, \alpha'_n)$ *defined by:*

- $D' = D - (C \cup R);$
- $\forall i, 0 \leq i \leq n,\ \forall d \in D' - SD_i,\ d\alpha'_i = d\alpha_i;$
- $\forall i, 0 \leq i \leq n,\ \forall d \in SD_i,\ d\alpha'_i = d' = d(\alpha_i\alpha_{k_1})\ldots(\alpha_i\alpha_{k_p})\alpha_i,$ *where p is the smallest integer such that* $d' \in SD_i,$ *and* $\forall j, 1 \leq j < p,$ *if* $d_j = d(\alpha_i\alpha_{k_1})\ldots(\alpha_i\alpha_{k_{j-1}})\alpha_i \in R_i$ *then* $k_j = i+1$ *else* $(d_j \in C_i)$ $k_j = i - 1.$

Figure 2 illustrate this operation.

A pyramid of $n$-dimensional generalized maps (or $n$-G-map pyramid) is a hierarchical data structure [10]. Each level is an $n$-G-map, deduced from the previous level by applying the general operation of removal and/or contraction of cells. The choice of the removed or contracted cells depends on the application (we assume here that this choice is the result of an external process). An $n$-G-map pyramid can be defined in the following way (see figure 3-a):

**Definition 4 ($n$-G-map Pyramid).** *Let* $n,\ m \geq 0.$ *An* $m + 1$ *level pyramid* $\mathcal{P}$ *of n-dimensional generalized maps is defined by* $\mathcal{P} = \{G^k\}_{0 \leq k \leq m}$ *where:*

1. *for each* $k,\ 0 \leq k \leq m,\ G^k = (D^k, \alpha_0^k, \ldots, \alpha_n^k)$ *is an n-G-map;*
2. *for each level* $k,\ 0 < k \leq m,$

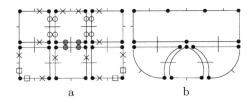

a                              b

**Fig. 2.** An example of simultaneous removal and contraction of cells of different dimensions. (a) A 2-G-map where the darts 0-removed, 1-removed and 1-contracted are respectively marked by empty squares, circles and gray disks. The surviving "neighbour" darts are marked by crosses. (b) The 2-G-map obtained by application of the operation

- $R^{k-1} = \bigcup_{i=0}^{n-1} R_i^{k-1}$ *(resp.* $C^{k-1} = \bigcup_{i=1}^{n} C_i^{k-1}$*) is a set of cells of* $G^{k-1}$. *These cells are disjoint, and their degrees equals to 2.*
- $G^k$ *is deduced from* $G^{k-1}$ *by simultaneously removing* $R^{k-1}$ *and contracting* $C^{k-1}$.

In the following, $\mathcal{P}$ denotes an $n$-G-map pyramid composed of $m + 1$ levels numbered from 0 to $m$. Level $k$ refers to $n$-G-map $G^k$. A dart keeps its name when it is not suppressed (for instance dart 1 of level 0 and 1 in figure 3-b). So, $D^{k+1} \subseteq D^k$ for any $k$, $0 \leq k < m$: each dart appears for the first time in the first pyramid level; if a dart belongs to the $k^{th}$ pyramid level, it does not belong to a cell which is removed or contracted in a precedent level. $Lev_d$ denotes the last level in which dart $d$ exists. At last, note that for a given dimension $i$ and a given level $k$, each dart $d$ in $D^k$ is either a dart which belongs to a removed $i$-cell ($d \in R_i^k$) or to a contracted $i$-cell ($d \in C_i^k$), or a dart which is a "neighbour" of a contracted or removed $i$-cell ($d \in SD_i^k$), or an other dart ($d \notin (R_i^k \cup C_i^k \cup SD_i^k)$).

## 3     Connecting Walks

The notion of connecting walk has been defined by Brun and Kropatsch for combinatorial map pyramids. More precisely, a connecting walk is the set of darts at a given level which separates a surviving dart and its successor in the next level. We extend this notion for $n$-G-map pyramids, for any two levels, and any dart of the pyramid. In the standard case (for surviving darts), the definition of connecting walk corresponds to that of Brun and Kropatsch. For all other darts of the pyramid (non surviving darts), this definition is extended in order to be able to define generalized orbits. A connecting walk is a sequence of darts in a lower level that separates two darts of a upper level. Intuitively, a connecting walk at a given level is obtained by concatenating connecting walks of the previous level concerned by removals or contractions. $Ch_{(i,a,b)}(d)$ denotes the connecting walk between levels $a$ and $b$ ($a \leq b$), for any dart $d \in D^a$ and dimension $i$.

First, assume $d \in D^b$ (a surviving dart). $Ch_{(i,a,b)}(d)$ is the sequence of darts at level $a$ separating dart $d$ and its neighbour $d\alpha_i^b$.

When $b = a + 1$, we get a definition near to that of Brun and Kropatsch. In fact, when $b = a + 1$, $Ch_{(i,a,a+1)}(d)$ is then the sequence of darts removed and contracted between levels $a$ and $a + 1$ linking $d$ and $d\alpha_i^{a+1}$. These darts are traversed when we define $d\alpha_i^{a+1}$ (cf. definition 3 of simultaneous removal and contraction).

When $b > a + 1$, $Ch_{(i,a,b)}(d)$ is the sequence of darts removed and contracted between levels $a$ and $b$ linking $d$ and $d\alpha_i^b$. Using the definition of removals and contractions of cells, we can iterate the previous process for each level. So, we can express $Ch_{(i,a,b)}(d)$ as a concatenation of walks examined between levels $a$ and $b - 1$. We obtain a recursive definition of $Ch_{(i,a,b)}(d)$.

In the particular case where there is no dart between $d$ and its neighbour for $\alpha_i^b$ at level $b - 1$, that is to say where $d$ is not the neighbour for $\alpha_i^{b-1}$ of a contracted or removed dart, then $Ch_{(i,a,b)}(d)$ is equal to $Ch_{(i,a,b-1)}(d)$.

We can observe that $Ch_{(i,a,a)}(d)$ is composed of only one dart: $d\alpha_i^a$, since no darts at level $a$ separates $d$ and $d\alpha_i^a$.

Second, assume $d \notin D^b$ (a non surviving dart).

We have to extend the notion of connecting walk in order to define the notion of generalized orbit. When $d \in D^{b-1}$, the removal or contraction of the cell containing $d$ is directly concerned in the construction of level $b$, and has direct consequences in the definition of new orbits of this level. This is not the case when $d \notin D^{b-1}$. For these reasons, if $d \notin D^{b-1}$, $Ch_{(i,a,b)}(d)$ is the empty sequence, and if $d \in D^{b-1}$, $Ch_{(i,a,b)}(d)$ corresponds to the sequence of darts traversed from $d$ and applying the same rules than for the definition of removal and contraction. There are here two conditions to stop: when the last dart $d'$ belong to $D^b$, or when the last dart is $d$ (In the first case, $Ch_{(i,a,b)}(d)$ corresponds to a subsequence or to the inverse of a subsequence of $Ch_{(i,a,b)}(d')$. In the second case, $Ch_{(i,a,b)}(d)$ is a cycle).

See figure 3 for examples of these different cases of connecting walks. The notion of connecting walk is formally defined in the following way:

**Definition 5 (Connecting Walk).** *Let $i \in N$, $a$ and $b$ be such that $0 \leq a \leq b \leq m$.*
*For each dart $d \in D^a$, $Ch_{(i,a,b)}(d)$ is defined by:*
*if $b = a$: $Ch_{(i,a,b)}(d) = (d\alpha_i^a)$,*
*else if $b > lev_d + 1$: $Ch_{(i,a,b)}(d) = ()$,*
*else if $d \notin (SD_i^{b-1} \cup R_i^{b-1} \cup C_i^{b-1})$: $Ch_{(i,a,b)}(d) = Ch_{(i,a,b-1)}(d)$,*
*else: $Ch_{(i,a,b)}(d) = C = \left( Ch_{(k_1,a,b-1)}(d_1), \ldots, Ch_{(k_p,a,b-1)}(d_p) \right),$*

$\quad$ *where: $d_1 = d$,*

$\quad\quad$ *$\forall u, 1 \leq u < p$, $d_{u+1}$ is the last dart of $Ch_{(k_u,a,b-1)}(d_u)$,*

$\quad\quad$ *$\forall u, 1 \leq u \leq p$, $k_u = \begin{cases} i & \text{if } u \text{ is odd} \\ i+1 & \text{if } u \text{ is even and } d_u \in R_i^{b-1}, \\ i-1 & \text{if } u \text{ is even and } d_u \in C_i^{b-1}, \end{cases}$*

$\quad\quad$ *and $p$ is the smallest integer such that the last dart of $C$ is equal to $d$, or is a surviving dart.*

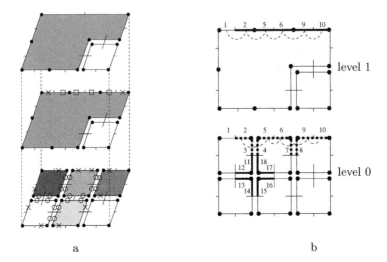

a                                      b

**Fig. 3.** (a) Example of 2-G-map pyramid composed of three levels. The darts 0-removed (resp. 1-removed) are marked by empty squares (resp. circles). (b) Two levels of the pyramid. The darts of connecting walk $Ch_{(0,1,2)}(1)$ are drawn thick in level 1. These darts are between darts 1 and 10, where $10 = 1\alpha_0^2$: $Ch_{(0,1,2)}(1) = (2,5,6,9,10) = (1\alpha_0^1, 1\alpha_0^1\alpha_1^1, 1\alpha_0^1\alpha_1^1\alpha_0^1, 1\alpha_0^1\alpha_1^1\alpha_0^1\alpha_1^1, 1\alpha_0^1\alpha_1^1\alpha_0^1\alpha_1^1\alpha_0^1)$. $Ech_{(0,1,2)}^\circ(1) = \{2,5,6,9\}$, and $Ech_{(0,1,2)}(1) = \{1,2,5,6,9,10\}$. The darts of connecting walk $Ch_{(0,0,2)}(1)$ are drawn dotted in level 0, and $Ch_{(0,0,2)}(1) = (2,3,4,5,6,7,8,9,10)$. $Ch_{(0,0,2)}(1)$ is the concatenation of $Ch_{(0,0,1)}(1)$, $Ch_{(1,0,1)}(2)$, $Ch_{(0,0,1)}(5)$, $Ch_{(1,0,1)}(6)$ and $Ch_{(0,0,1)}(9)$. Note that $Ch_{(0,0,1)}(5) = Ch_{(0,0,0)}(5) = (6)$. Darts 5 and 11 are non surviving darts at level 2: $Ch_{(0,0,2)}(5) = (6,7,8,9,10)$ is a subsequence of $Ch_{(0,0,2)}(1)$, and $Ch_{(1,0,1)}(11) = (12,13,14,15,16,17,18,11)$ is a cycle

From this definition, we deduce in a straightforward way an algorithm which computes a connecting walk with a cost $\Theta(n)$, $n$ being the number of darts of the connecting walk.

Given a connecting walk $Ch_{(i,a,b)}(d)$, we deduce two sets of darts: open connecting set $Ech_{(i,a,b)}^\circ(d)$ and closed connecting set $Ech_{(i,a,b)}(d)$ (c.f. figure 3-b). Intuitively, the first set corresponds to the interior of the connecting walk and the second set corresponds to the whole connecting walk, extremities included. The notion of generalized orbits is based upon these sets.

**Definition 6 (Open and Closed Connecting Sets).** *Let $i \in N$, $a$ and $b$ be such that $0 \leq a \leq b \leq m$.*
*For each dart $d \in D^a$:*

- *$Ech_{(i,a,b)}^\circ(d)$ is the set of darts of the connecting walk $Ch_{(i,a,b)}(d)$, the last dart excepted;*

- $Ech_{(i,a,b)}(d) = \begin{cases} \emptyset \text{ if } Ch_{(i,a,b)}(d) = () \\ Ech_{(i,a,b)}^\circ(d) \cup \{d, d'\} \text{ otherwise,} \\ \qquad \text{where } d' \text{ is the last dart of } Ch_{(i,a,b)}(d) \end{cases}$

## 4    Generalized Orbits

A possible use of $n$-G-map pyramid is the representation of an image segmented at several levels. For instance, an $n$-G-map pyramid can be used for representing an $n$D image in gray level which is segmented using a simple gray level distance as homogeneous criterion. Level 0 of the pyramid represents the initial image. Each xel is represented by an $n$-cell associated with a gray level. At the following level, neighbour regions which satisfy the homogeneity criterion are merged into a unique region. Merging is achieved by removing $(n - 1)$-cells which separate them (note that other operations are possible in order to simplify the boundary between two adjacent regions).

We can compute the gray level of an $n$-cell from the gray levels of all the $n$-cells at level 0 which correspond to this cell. For that we need to compute the set of $n$-cells at level 0 which correspond to a given $n$-cell.

More generally, let $a$ and $b$ be any two levels of an $n$-G-map pyramid $\mathcal{P}$ such that $0 \leq a \leq b \leq m$. Let $K \subseteq \{0, \ldots, n\}$ and let $O =<>_{(K)^b} (d)$ be an orbit of $G^b$, the $n$-G-map of the level $b$ of $\mathcal{P}$. The set of darts corresponding to $O$ at level $a$ is called *generalized orbit*. Informally, a generalized orbit is the set of orbits at level $a$ which are "merged" into the orbit at level $b$.

A generalized orbit can be computed in a sequential way (see figure 4). Let $GO$ be the generalized orbit associated to $O$. $GO$ is initialized by $GO_0 = \{d\}$. Then we repeat the two following phases: first we add the darts of the connecting walks of the darts which belong to $GO$ (i.e. we define $GO_{2p+1}$ as the union of $GO_{2p}$ and the set of all darts of closed connecting set $Ech_{(i,a,b)}(d')$ for all dimension $i \in K$ and all darts $d'$ of $GO_{2p}$); second, we add all darts of the orbits $<>_K$ of darts belonging to $GO$ (i.e. we define $GO_{2p+2}$ as the union of all orbits $<>_{(K)^a} (d')$ for all darts $d'$ of $GO_{2p+1}$). Since an $n$-G-map contains a finite number of darts, we can show that $q \geq 0$ exists, such that no dart is added to $GO_{q(K,a,b)}(d)$ by repeating the process. Generalized orbit $GO$ is defined as $GO_{q(K,a,b)}(d)$.

**Definition 7 (The Series $\left(GO_{p(K,a,b)}(d)\right)_{0 \leq p \leq q}$).** *Let $K \subseteq N$, let $a$ and $b$ be such that $0 \leq a \leq b \leq m$, and let $d \in D^b$.*
*We define the series $\left(GO_{p(K,a,b)}(d)\right)_{0 \leq p \leq q}$ in the following way:*

- $GO_{0(K,a,b)}(d) = \{d\}$;
- $\forall n > 0$:
  $GO_{2p-1(K,a,b)}(d) = \bigcup_{i \in K} \bigcup_{d' \in GO_{2p-2(K,a,b)}(d)} Ech_{(i,a,\min\{b,lev_{d'}+1\})}(d')$;
  $GO_{2p(K,a,b)}(d) =<>_{(K)^a} (GO_{2p-1(K,a,b)}(d))$.

Note that the even elements of this series are unions of orbits $<>_K$ (for instance, we can see in figure 5 that a generalized face is a set of faces). As we have said before, a property of this series is that it is convergent in a finite number of iterations. Indeed, it is increasing, bounded by $D^a$ and so stationary. We can thus define a generalized orbit as the limit of such a series.

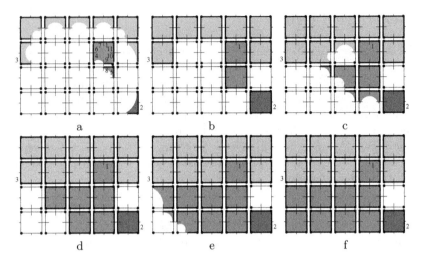

**Fig. 4.** Computing the generalized orbit $<>_K=<>_{\{0,1\}}$ (the face orbit) for darts 1, 2 and 3 in the pyramid of the figure 5. (a) $GO_{1(K,0,5)}$ for darts 1, 2 and 3, are represented in different gray levels. (b) to (f) $GO_{j(K,0,5)}$, for darts 1, 2 and 3, from $j=2$ to $j=6$, are represented in different gray levels. Note that $GO_{j(K,0,5)} = GO_{6(K,0,5)}$ for $j > 6$. For example, when we compute $GO_{1(K,0,5)}(1)$, we add the darts of $Ch_{(0,0,5)}(1) = (7,6,4)$ and $Ch_{(1,0,5)}(1) = (11,10,9,8,5)$

**Definition 8 (Generalized Orbit).** *Let $K \subseteq N$, let $a$ and $b$ be such that $0 \le a \le b \le m$, and let $d \in D^b$. The generalized orbit $GO_{(K,a,b)}(d)$ is defined as the limit of the series $\left(GO_{p(K,a,b)}(d)\right)_{0 \le p \le q}$.*

From definitions 7 and 8, we can deduce directly a first algorithm which computes a generalized orbit with a cost $\Theta(kn^2)$, where $n$ is the number of darts of this generalized orbit, and $k = card(K)$.

The problem of this algorithm is to consider all darts at each step, whereas it is not useful. In order to optimize this algorithm, we can remark that:

1. it is necessary to consider each dart of the orbit for all involutions of this orbit. So the lower bound is $\Theta(kn)$.
2. when we have added the darts of $Ch_{(i,a,b)}(d)$, it is not useful to consider the connecting walk for $\alpha_i$, of a dart of the interior of $Ch_{(i,a,b)}(d)$ since it is included in $Ch_{(i,a,b)}(d)$. Moreover a dart will not be considered another time for $\alpha_i$ since the intersection of two different connecting walks is empty.

So, we can propose an optimized algorithm that computes a generalized orbit with a cost $\Theta(kn)$.

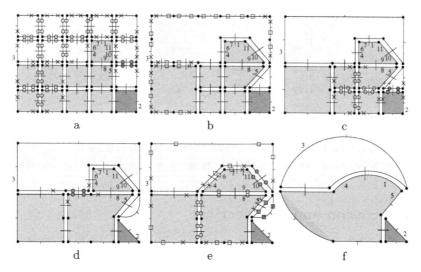

**Fig. 5.** A pyramid. a (resp. b, c, d, e, et f) represents level 0 (resp. 1, 2, 3, 4 et 5). Darts 0-removed (resp. 1-removed, 1-contracted and 2-contracted) are marked with empty squares (resp. circles, gray disks and gray squares). The faces of level 5 and the corresponding generalized faces in lower levels are represented in different gray level

Among the properties of generalized orbits we have:

- $\forall d \in D^a$, $GO_{(K,a,a)}(d) =<>_{(K)a}(d)$. Generalized orbits are an extension of orbits and we retrieve them when we consider generalized orbits at only one level;
- $GO_{(K,a,b)}(d)$ is a union of orbits $<>_K$ at level $a$ (by definition).

$n$-cells satisfy some additional properties. We can easily show that the $n$-cells of a level are necessarily the result of the "merging" of $n$-cells of the previous level by applying $(n-1)$-removals. All the other operations only modify the boundary of existing $n$-cells.

So the generalized orbit associated to an $n$-cell is the union of all $n$-cells of level $a$ which have been "merged" to construct it. We can note that even when $(n-1)$-removals have led to a disconnection (see figure 6), we retrieve for the generalized orbit associated to an $n$-cell, the set of $n$-cells at level $a$ which compose it. With this property, we can deduce that two generalized orbits associated with $n$-cells and to the same levels are either equal or disjoint.

These last properties are not necessarily true for other $i$-cells ($i \neq n$). But, in general, the interesting cells that contain information are $n$-cells and not the others. In order to get such properties for the other cells, it could be necessary to define other sets of darts and perhaps other generalized orbits.

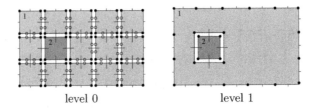

level 0                                level 1

**Fig. 6.** A 2-G-map pyramid composed of two levels. 1-removed darts are marked by circles. The general orbits associated with faces for darts 1 and 2 are represented in two gray levels. Note in the second level that the removal of edges has led to a disconnection

## 5    Conclusion and Perspectives

We have defined the notion of generalized orbit which extends that of receptive field. This notion, defined for graph pyramids and combinatorial maps pyramids, establishes a correspondence between a region at a given level and the corresponding set of regions at a lower level. The notion of generalized orbit is defined for $n$-G-map pyramids, which can be used in order to represent pyramids of $n$-dimensional images. Moreover, this notion is defined for any cells of any dimension, and between any two levels. The definition of this notion is based upon a generalization of the connecting walk notion, initially defined by Brun and Kropatsch.

Some properties of generalized orbits have been established. Among them, two concern only $n$-cells. In order to define similar properties for other cells, we are going to define other connecting walks and generalized orbits, and compare them for different cells of different dimensions.

Moreover we are conceiving operations for handling $n$-G-map pyramids. More precisely, we want to be able to modify a level of a pyramid, and to automatically compute the modifications of the upper and lower levels. Last, given a generalized orbit and an operation which modifies the pyramid, we want to optimize the computation of the modified generalized orbit, i.e. to directly deduce it without re-computing it from scratch.

## Acknowledgements

The authors wish to thank Luc Brun and Walter Kropatsch for their encouragements and help.

## References

1. Burt, P., Hong, T., Rosenfeld, A.: Segmentation and estimation of image region properties through cooperative hierarchical computation. IEEE Transactions on Systems, Man and Cybernetics **11** (1981) 802–809.

2. Meer, P.: Stochastic image pyramids. Computer Vision, Graphics and Image Processing **45** (1989) 269–294.
3. Montanvert, A., Meer, P., Rosenfeld, A.: Hierarchical image analysis using irregular tessellations. IEEE Transactions on Pattern Analysis and Machine Intelligence **13** (1991) 307–316.
4. Jolion, J., Montanvert, A.: The adaptive pyramid: A framework for 2d image analysis. Computer Vision, Graphics and Image Processing **55** (1992) 339–348.
5. Kropatsch, W.: Building irregular pyramids by dual-graph contraction. Vision, Image and Signal Processing **142** (1995) 366–374.
6. Brun, L., Kropatsch, W.: Irregular pyramids with combinatorial maps. In Amin, A., Ferri, F.J., Pudil, P., I nesta, F.J., eds.: Advances in Pattern Recognition, Joint IAPR International Workshops SSPR'2000 and SPR'2000. Volume 1451 of Lecture Notes in Computer Science., Alicante, Spain, Springer, Berlin Heidelberg, New York (2000) 256–265.
7. Brun, L., Kropatsch, W.: Introduction to combinatorial pyramids. In G. Bertrand, A. Imiya, R.K., ed.: Digital and Image Geometry. Volume 2243 of LNCS. Springer Verlag (2001) 108–127.
8. Brun, L., Kropatsch, W.: Combinatorial pyramids. In Suvisoft, ed.: IEEE International conference on Image Processing (ICIP). Volume II., Barcelona (2003) 33–37.
9. Brun, L., Kropatsch, W.: Contraction kernels and combinatorial maps. Pattern Recognition Letters **24** (2003) 1051–1057.
10. Grasset-Simon, C., Damiand, G., Lienhardt, P.: Pyramids of n-dimensional generalized map. Technical Report 2, SIC, Université de Poitiers (2004)
11. Lienhardt, P.: N-dimensional generalized combinatorial maps and cellular quasi-manifolds. In: International Journal of Computational Geometry and Applications. (1994) 275–324.
12. Brun, L., Kropatsch, W.: Receptive fields within the combinatorial pyramid framework. In: Discrete Geometry for Computer Imagery. Number 2301 in LNCS, Bordeaux, France (2002) 92–101.
13. Damiand, G., Lienhardt, P.: Removal and contraction for n-dimensional generalized maps. In: Discrete Geometry for Computer Imagery. Number 2886 in Lecture Notes in Computer Science, Naples, Italy (2003) 408–419.
14. Lienhardt, P.: Subdivisions of n-dimensional spaces and n-dimensional generalized maps. In: Proceedings of the fifth annual Symposium on Computational Geometry, Saarbruchen, West Germany (1989) 228–236.

# Resolution Pyramids on the FCC and BCC Grids

Robin Strand[1] and Gunilla Borgefors[2]

[1] Centre for Image Analysis, Uppsala University,
Lägerhyddsvägen 3, SE-75237 Uppsala, Sweden
[2] Centre for Image Analysis, Swedish University of Agricultural Sciences,
Lägerhyddsvägen 3, SE-75237 Uppsala, Sweden
{robin, gunilla}@cb.uu.se

**Abstract.** Partitionings on the face-centered cubic grid and the body-centered cubic grid that are suitable for resolution pyramids are found. The partitionings have properties similar to a partitioning that has been used for the resolution pyramids on the cubic grid. Therefore, they are well-suited for adapting methods to construct multiscale representations developed for the cubic grid. Multiscale representations of images are constructed using different methods.

## 1 Introduction

Three-dimensional images are usually captured into the cubic grid. Often, the images are computed from projections rather than captured directly. It is, however, possible to adjust image capturing techniques such as CT or MRI to produce images on other grids, such as the face-centered cubic (fcc) and the body-centered cubic (bcc) grids [1]. The sampling theorem allows a decrease of the number of grid points by a factor 1.41 on the bcc grid compared to the cubic grid without influencing the image representation/reconstruction quality, [1]. It has been demonstrated that the hexagonal grid in two dimensions is theoretically better than the square grid, [2]. For example, the hexagonal grid has a higher packing density than the square grid. In three dimensions, the fcc and bcc grids both have higher packing densities than the cubic grid. The fcc and bcc grids are the three-dimensional "equivalents" of the hexagonal grid, [3]. Since it is very easy to construct efficient data structures for the fcc and bcc grids, the only missing piece for allowing full use of the fcc and bcc grids in image processing is to construct efficient algorithms for analysis and processing.

Data structures based on resolution pyramids can be used to improve both the computation time and the quality in image analysis applications, e.g. matching [4] and segmentation [5]. When using the two dimensional square grid or the three dimensional cubic grid, partitioning is straightforward, [6]. The literature on resolution pyramids on the hexagonal grid is rich, e.g., [7, 8, 9, 10]. In this article, resolution pyramids on the fcc and bcc grids are examined.

E. Andres et al. (Eds.): DGCI 2005, LNCS 3429, pp. 68–78, 2005.

In the first part of the article, we calculate partitionings that are suitable for resolution pyramids on the fcc and bcc grids. We use conditions similar to those derived for the hexagonal grid in [7]. The most important property is that all levels in the pyramid should be represented by the same grid. The partitionings have many properties in common with the partitioning used on the cubic grid and are therefore well-suited for adapting methods to construct multiscale representations developed for the cubic grid. Two different methods are presented and illustrated using a running example.

## 2    The Grids

In [3], a *grid* $\mathbb{G}$ is defined as any set of points in $\mathbb{R}^n$. In this article, we will use three-dimensional grids on the form:

$$\mathbb{G}_T = \{T(x,y,z) : x,y,z = a\vec{v_1} + b\vec{v_2} + c\vec{v_3}, a,b,c \in \mathbb{Z}\}, \qquad (1)$$

where $T$ is an affine transformation. We refer to $\vec{v_1}$, $\vec{v_2}$, and $\vec{v_3}$ as base vectors. By using this definition, we do not require the origin to be in the grid. We will consider three types of grids:

**Definition 1.**
- *With vectors $\vec{v_1} = (1,0,0)$, $\vec{v_2} = (0,1,0)$, and $\vec{v_3} = (0,0,1)$, and a rigid transformation $T$ such that (1) is fulfilled, $\mathbb{G}_T$ is a cubic grid, denoted $\mathbb{Z}^3_T$.*
- *With vectors $\vec{v_1} = (1,0,1)$, $\vec{v_2} = (0,1,1)$, and $\vec{v_3} = (1,1,0)$, and a rigid transformation $T$ such that (1) is fulfilled, $\mathbb{G}_T$ is an fcc grid, denoted $\mathbb{F}_T$.*
- *With vectors $\vec{v_1} = (1,-1,1)$, $\vec{v_2} = (-1,1,1)$, and $\vec{v_3} = (1,1,-1)$, and a rigid transformation $T$ such that (1) is fulfilled, $\mathbb{G}_T$ is a bcc grid, denoted $\mathbb{B}_T$.*

If $T = I$ is the identity mapping, then we use the notation $\mathbb{G} = \mathbb{G}_T$.

For each grid point in an fcc grid, there are 12 face-neighbours and 6 vertex-neighbours, resulting in the 12- and 18-neighbourhoods, see Fig. 1(b),(c). In a bcc grid, the neighbours connected to a grid point are all face-neighbours, see Fig. 1(e),(f). However, there are two kinds of face-neighbours, resulting in the 8-

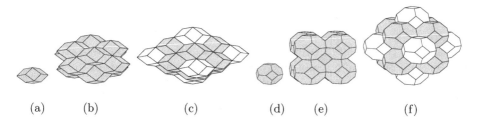

(a)         (b)              (c)              (d)       (e)              (f)

**Fig. 1.** A voxel (a) in fcc and its 12- and 18-neighbours, shown in (b) and (c) respectively. A voxel in bcc (d) and its 8- and 14-neighbours, shown in (e) and (f) respectively

and 14-neighbourhoods. The light-grey voxels in Fig. 1 are referred to as first-order neighbours and the white voxels as second-order neighbours. Throughout this article, the set of object grid points are visualized by their voxels (the Voronoi regions).

## 3      Tessellating the FCC and BCC Grids

### 3.1      Conditions for the Tessellations

In this section, a way to construct partitionings of the fcc and bcc grids is presented. Note that we are not concerned with tiles in continuous space, but with "patterns" of grid points. The goal is basically to cluster neighbouring grid points in such a way that the centroid of the cluster will form a grid point in the next level in a resolution pyramid, and these new grid points will form an appropiate grid.

There is no standard notation for these clusters. The term "aggregate" is used in [8] to denote a cluster in a resolution pyramid. The term "pattern" is used in [7]. In [9], the same term is used for a matching operation using resolution pyramids. A pattern is defined to be a vector of elements of the set $\{0, 1, D\}$, where $D$ indicates a "don't care" position. A centroid of a cluster is often referred to as *parent* and the elements in the cluster as *children*. We will adopt the notation used in [7] with a slight modification.

The tessellation of a grid is carried out by clustering grid points together in a *pattern*. Patterns in the hexagonal grid were examined in [7]. A *generator* $\mathbf{Q} \subset \mathbb{G}$ is a set of grid points, e.g., as in Fig. 2. By considering translations and rotations of the generator, a larger subset of the grid is covered. The set $\mathcal{P} = \{\mathbf{P}_i, i \in \mathcal{I}\}$, where $\mathcal{I}$ is a set of indices and $\mathbf{P}_i$ are the patterns, i.e., translations and rotations of $\mathbf{Q}$ can be constructed such that there is an index $i$ satisfying $\mathbf{P}_i = \mathbf{Q}$. The centroids of the patterns will form a new grid at the next level in the resolution pyramid.

**Definition 2.** *The set $\mathcal{P} = \{\mathbf{P}_i, i \in \mathcal{I}\}$ tessellates the grid $\mathbb{G}$ if the following is fulfilled:*

- $\mathbb{G} = \bigcup_{i \in \mathcal{I}} \mathbf{P}_i$.
- *For any two patterns $\mathbf{P}_i$ and $\mathbf{P}_j$, $\mathbf{P}_i \cap \mathbf{P}_j = \varnothing \iff i \neq j$.*

*Example 1.* Let $\mathbf{Q} = \{(0, 0, 0), (1, 1, 1)\}$ (see Fig. 2(b)) be a generator. A set of patterns $\mathcal{P}$ that tessellates $\mathbb{B}$ can be constructed by translating $\mathbf{Q}$ by integer multiples of $(2, 0, 0)$, $(0, 2, 0)$, and $(0, 0, 2)$. If $T$ is a translation from $(0, 0, 0)$ to the centroid of $\mathbf{Q}$ and a scaling by a factor 2, then the centroids of the patterns satisfy the conditions for being a cubic grid in Definition 1. The new grid is $\mathbb{Z}^3_T$.

### 3.2      Finding Admissible Patterns

If we can construct a set of patterns such that the grid formed by the centroids of the patterns $\mathbb{G}_T$ is of the same type as the original grid $\mathbb{G}$, then by repeating the

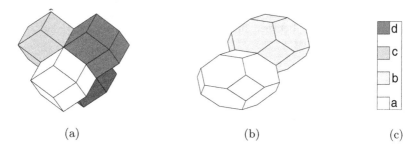

(a)                         (b)                         (c)

**Fig. 2.** Examples of patterns in (a) fcc and (b) bcc. The grid points are color-coded as in (c)

procedure recursively, a resolution pyramid with levels $\mathbb{G}, \mathbb{G}_T, \mathbb{G}_{T^2}, \ldots, \mathbb{G}_{T^m}$ is achieved. We will now calculate patterns that can be applied in such a recursive manner.

**Definition 3.** *Given a grid* $\mathbb{G}$*, a set of patterns* $\mathcal{P}$ *is admissible if it fulfills the following conditions:*

*C1* $\mathcal{P}$ *tessellates* $\mathbb{G}$*.*
*C2 The grid formed by the centroids of the patterns is of the same type as* $\mathbb{G}$*.*
*C3 For any pattern* $\mathbf{P}_i$*, there is a one-to-one translation mapping* $S$ *between the generator* $\mathbf{Q}$ *and the pattern* $\mathbf{P}_i$*,* $S : \mathbf{Q} \to \mathbf{P}_i$*.*
*C4 The pattern should be compact in the sense that it should be a connected set.*

Conditions **C1** and **C2** are necessary for a pattern to be used in a resolution pyramid where all levels form the same grid. The one-to-one mapping in **C3** will be used when labelling the grid points in the patterns. Moreover, **C3** will be used when searching for admissible patterns. Without condition **C4**, we might end up in a set of patterns that is not suitable for resolution pyramids; we want a grid point in one level to correspond to a set of connected grid points in the previous level. The patterns obtained by the generators in Fig. 2 satisfy all conditions but **C2**. In fact, both patterns form a cubic grid.

We denote the first level, the grid we start with, in the pyramid by $\mathbb{G}$. The second level, the grid formed by the centroids of patterns of the original grid, is denoted $\mathbb{G}_T$. By repeating this procedure, the resolution pyramid consisting of the levels $\mathbb{G}, \mathbb{G}_T, \mathbb{G}_{T^2}, \ldots, \mathbb{G}_{T^m}$ is achieved.

Suppose that each $\mathbf{P}_i$ consists of $n$ grid points. We denote by $\delta_{\mathbb{G}}$ the distance between two first-order neighbours in $\mathbb{G}$. For the fcc and bcc grids $\mathbb{F}$ and $\mathbb{B}$, $\delta_{\mathbb{F}} = \sqrt{2}$ and $\delta_{\mathbb{B}} = \sqrt{3}$. If the set of patterns satisfies **C1**, then the new grid, $\mathbb{G}_T$, will have $1/n$ as many grid points as the original, distributed over the same volume. If it also satisfies **C2**, the distance between first-order neighbours in $\mathbb{G}_T$ is $\delta_{\mathbb{G}_T} = \sqrt[3]{n}\delta_{\mathbb{G}}$. By **C1**, the patterns should also tessellate the grid, therefore $n$ must be an integer. We get the following condition:

**Table 1.** The grid points in $\mathbb{G}$ are sorted according to their distance to $\mathbf{0}$. This distance will be the distance between first-order neighbours in $\mathbb{G}_T$, $\delta_{\mathbb{G}_T}$. For each distance, condition (2) is tested

| fcc | | | bcc | | | cubic | | |
|---|---|---|---|---|---|---|---|---|
| $\delta_{\mathbb{G}_T}$ | $n$ | # points | $\delta_{\mathbb{G}_T}$ | $n$ | # points | $\delta_{\mathbb{G}_T}$ | $n$ | # points |
| $\sqrt{2}$ | 1 | 12 | $\sqrt{3}$ | 1 | 8 | 1 | 1 | 6 |
| 2 | $2^{\frac{3}{2}}$ | 6 | 2 | $\left(\frac{2}{3}\right)^{\frac{3}{2}}$ | 6 | $\sqrt{2}$ | $2^{\frac{3}{2}}$ | 12 |
| $\sqrt{6}$ | $3^{\frac{3}{2}}$ | 24 | $\sqrt{8}$ | $\left(\frac{8}{3}\right)^{\frac{3}{2}}$ | 12 | $\sqrt{3}$ | $3^{\frac{3}{2}}$ | 8 |
| $\sqrt{8}$ | 8 | 12 | $\sqrt{11}$ | $\left(\frac{11}{3}\right)^{\frac{3}{2}}$ | 24 | 2 | 8 | 6 |
| $\sqrt{10}$ | $5^{\frac{3}{2}}$ | 24 | $\sqrt{12}$ | 8 | 8 | $\sqrt{5}$ | $5^{\frac{3}{2}}$ | 24 |
| $\sqrt{12}$ | $6^{\frac{3}{2}}$ | 8 | 4 | $\left(\frac{16}{3}\right)^{\frac{3}{2}}$ | 6 | $\sqrt{6}$ | $6^{\frac{3}{2}}$ | 24 |
| $\sqrt{14}$ | $7^{\frac{3}{2}}$ | 48 | $\sqrt{19}$ | $\left(\frac{19}{3}\right)^{\frac{3}{2}}$ | 24 | $\sqrt{8}$ | $8^{\frac{3}{2}}$ | 12 |
| 4 | $8^{\frac{3}{2}}$ | 6 | $\sqrt{20}$ | $\left(\frac{20}{3}\right)^{\frac{3}{2}}$ | 24 | 3 | 27 | 30 |
| $\sqrt{18}$ | 27 | 36 | $\sqrt{24}$ | $8^{\frac{3}{2}}$ | 24 | $\sqrt{10}$ | $10^{\frac{3}{2}}$ | 24 |
| $\sqrt{20}$ | $10^{\frac{3}{2}}$ | 24 | $\sqrt{27}$ | 27 | 32 | $\sqrt{11}$ | $11^{\frac{3}{2}}$ | 24 |
| $\sqrt{22}$ | $11^{\frac{3}{2}}$ | 24 | $\sqrt{32}$ | $\left(\frac{32}{3}\right)^{\frac{3}{2}}$ | 12 | $\sqrt{12}$ | $12^{\frac{3}{2}}$ | 8 |

$$\left(\frac{\delta_{\mathbb{G}_T}}{\delta_{\mathbb{G}}}\right)^3 = n \in \mathbb{N} \tag{2}$$

In order to find the admissible patterns, we need to label the grid points in the patterns. We use the labels $\mathbf{a}, \mathbf{b}, \mathbf{c}, \ldots$ as in Fig. 2. Assume that the generator $\mathbf{Q}$ consists of $n$ grid points. Each element in $\mathbf{Q}$ is given a unique label. By **C3**, for any $i \in \mathcal{I}$, there is a one-to-one mapping involving only translation between $\mathbf{Q}$ and $\mathbf{P}_i$. Therefore, each grid point is identified by exactly one of the labels.

Using the one-to-one mapping, it is easy to see that the distance between the centroids of two patterns is equal to the distance between the $\mathbf{a}$-points in the patterns. Let $\mathbf{0}$ be the $\mathbf{a}$-point of the generator $\mathbf{Q}$. To find possible values of $n$, we sort the grid points by their distances to $\mathbf{0}$ and check if (2) is fulfilled. This is done by using $\delta_{\mathbb{G}_T} = d(\mathbf{0}, q)$, where $q$ is the $\mathbf{a}$-point in a neighbouring pattern. There might of course be many grid points at the same distance from $\mathbf{0}$. We want $q$ to correspond to a first-order neighbour in the new grid $\mathbb{G}_T$. Therefore, there must be at least as many grid points at distance $\delta_{\mathbb{G}_T}$ as the number of first-order neighbours in $\mathbb{G}$. For comparison, we include the cubic grid in Table 1.

In Table 1, we see that the smallest admissible patterns on $\mathbb{F}$ have distance $\sqrt{8}$ to the first-order neighbours. The generators are constructed by clustering 8 neighbouring grid points and checking that they satisfy **C1**–**C4**. A generator satisfying this is, e.g., $\mathbf{Q} = \{(0,0,0), (1,1,0), (1,-1,0), (1,0,-1), (1,0,1), (0,1,1), (0,1,-1), (2,0,0)\}$. On $\mathbb{B}$, the distance between first-order neighbours of the centroids is $\delta_{\mathbb{G}_T} = \sqrt{12}$. We get, e.g., the generator $\mathbf{Q} = \{(0,0,0), (1,1,1), (1,-1,1), (1,1,-1), (1,-1,-1), (2,0,0), (0,2,0), (0,0,2)\}$. Both these set of

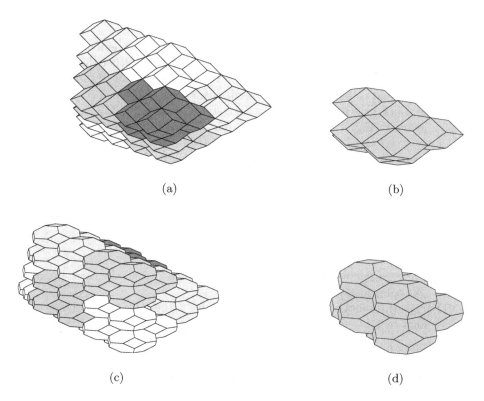

(a)                                                    (b)

(c)                                                    (d)

**Fig. 3.** A set of grid point in $\mathbb{G} = \mathbb{F}$, (a), and the corresponding set of grid points in $\mathbb{G}_T$, (b). Analogously on a bcc grid, (c) and (d)

patterns satisfy **C1–C4** and has the ratio (1 : 8), i.e., each pattern consists of 8 grid points. They are shown in Fig. 3. For both sets of patterns, if $T$ is a translation from $(0,0,0)$ to the centroid of **Q** and a scaling by a factor 2, then the centroids of the patterns satisfy the conditions for being an fcc/bcc grid in Definition 1.

Note that these patterns are not unique; there are 24 possibilities satisfying **C1–C4** in each case. These patterns are obtained by rotation of the generator around the $x$-, $y$-, and $z$-axes by multiples of $\pi/2$. This is in analogy with the patterns in the usual cubic grid. Often, resolution pyramids on the cubic grid is based on the generator $\{(0,0,0), (1,0,0), (0,1,0), (0,0,1), (1,1,0), (1,0,1), (0,1,1),$ $(1,1,1)\}$. However, any $2 \times 2 \times 2$ "cube" can be used as generator resulting in 8 possibilities.

From Table 1, the conclusion that none of the generators in Fig. 2 satisfy **C2** can be drawn. The reason is that there is no pattern with ratio (1 : 4) or (1 : 2) for which the centroids form an fcc or bcc grid.

## 4    Multiscale Representation of Binary Images

In this section, we will use the patterns obtained in the previous section to construct resolution pyramids. Each grid point is assigned either the value 1 or the value 0. In the figures, the value 1 will correspond to object and the value 0 will correspond to background. We need, given a set of 1:s in $\mathbb{G}_{T^i}$, some rules to decide which grid points should be assigned the value 1 in $\mathbb{G}_{T^{i+1}}$. Using the binary image in Fig. 4 as a running example, different rules will be examined. The images contains 18, 995 (fcc) and 18, 915 (bcc) object grid points. Observe that, in order to show the details in the different levels in the pyramid, the images are scaled. The *shifting grid method* and the *intermediate grey-level method* are generalizations of the methods in [11, 12].

### 4.1    Simple Rules

The logical rules OR and AND form the simplest pyramids.

**OR**  If any grid point in a pattern in $\mathbb{G}_{T^i}$ has value 1, then the corresponding grid point in $\mathbb{G}_{T^{i+1}}$ is set to 1.

**AND**  If all grid points in a pattern in $\mathbb{G}_{T^i}$ has value 1, then the corresponding point in $\mathbb{G}_{T^{i+1}}$ is set to 1.

The resolution pyramids obtained by using these rules on the objects in Fig. 4 are shown in Fig. 5. Note that for thin structures, the OR-rule is the only sensible choice, as otherwise the lower resolution levels will be empty.

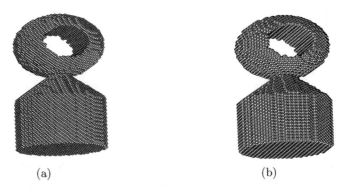

(a)                                            (b)

**Fig. 4.** A binary object on fcc (a) and bcc (b) used in the report

### 4.2    The Shifting Grid Method

In this method, the fact that there is an ambiguity when constructing the patterns is used. As mentioned in Section 3.2, there are 24 possible patterns all

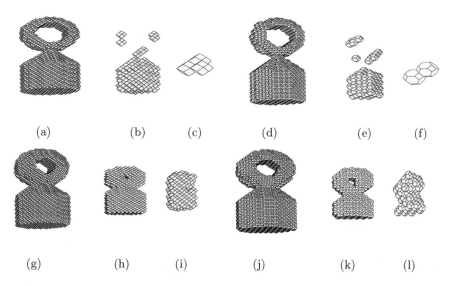

(a)        (b)        (c)        (d)        (e)        (f)

(g)        (h)        (i)        (j)        (k)        (l)

**Fig. 5.** Resolution pyramids consisting of $\mathbb{G}_T$–$\mathbb{G}_{T^3}$; (a)–(c) fcc, AND-rule; (d)–(f) bcc, AND-rule; (g)–(i) fcc, OR-rule; (j)–(l) bcc, OR-rule

giving the ratio $(1 : 8)$. The information obtained by considering the remaining 23 possible patterns can be used to construct a stable multiscale representation. The idea is to use all the possible patterns in 24 different grids, denoted $\mathbb{G}_{T_j^i}$, where the subscript $j \in \{1, \ldots, 24\}$ denotes which one of the 24 different patterns that is used. When constructing the $\mathbb{G}_{T_j^{i+1}}$ images from $\mathbb{G}_{T_j^i}$, we use the OR-rule to produce all the 24 possible images that can be constructed. To combine these to one single image $\mathbb{G}_{T^{i+1}}$, the AND-rule is used: if all grid points at a corresponding position in $\mathbb{G}_{T_j^{i+1}}, j = 1, \ldots, 24$ are 1:s, then the grid point in $\mathbb{G}_{T^{i+1}}$ is set to 1. This method is called "AND-ing the ORs" shifting grid method. Alternatively, by first using the AND-rule and then the OR-rule, a similar method called "OR-ing the ANDs" shifting grid method is obtained.

When implementing this method, it is not necessary to compute all 24 intermediate images. It is enough to scan the image once and for each grid point apply all the 24 patterns on the image in the previous level. By combining the OR and the AND-rule (or the AND- and the OR-rule), a more efficient shape preservation is obtained compared to the simple methods in Section 4.1. We get connected objects with a tunnel even in the lowest resolution.

Resolution pyramids of the objects in Fig. 4 using the AND-ing the ORs shifting grid method are shown in Fig. 6.

### 4.3   The Intermediate Grey-Level Method

As mentioned in the previous section, a scan through the image at level $i$ is enough to construct the image in level $i + 1$ using the shifting grid method. The idea resulting in the intermediate grey-level method is to simulate the construc-

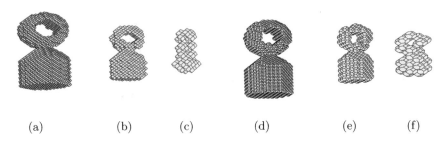

(a)              (b)              (c)              (d)              (e)              (f)

**Fig. 6.** Resolution pyramids consisting of $\mathbb{G}_T$–$\mathbb{G}_{T^3}$; (a)–(c) fcc, AND-ing the ORs shifting grid method; (d)–(f) bcc, AND-ing the ORs shifting grid method

tion of all possible patterns by using weights in a neighbourhood of each grid point. Suppose that **0** is included in all of the 24 generators. On the fcc grid, each first order neighbour is then included in 12 generators and each second order neighbour is included in 4 generators. On the bcc grid, each neighbour is included in 12 generators. These numbers will be the weights we use in the method. A mask is constructed consisting of the grid point and all neighbours of first and second order using these weights.

An intermediate grey-valued level $\mathbb{G}'_T$ is constructed by placing the weighted mask at each **a**-point in the patterns in $\mathbb{G}$ and adding the weights of the mask positioned on a grid point with value 1 in $\mathbb{G}$. Now, we need to binarize $\mathbb{G}'_T$ to achieve $\mathbb{G}_T$. The grey-level of a grid point in $\mathbb{G}'_T$ is denoted $\Sigma$. A low value of $\Sigma$ indicates a small number of neighbours with value 1. Thus, if $\Sigma$ has a low value, it would probably be set to 0 in the next level for most of the 24 patterns in an image using only one of the 24 patterns. If $\Sigma$ has a large value, it would probably be set to 1. An intermediate value of $\Sigma$ indicates that it would be set to 0 for some of the patterns and to 1 for some patterns. To decide whether $\Sigma$ should be set to 1 or 0, we state an extra criterion based on the values of the neighbours of $\Sigma$ in $\mathbb{G}'_T$. If $\Sigma$ is larger than the average value in the neighbourhood consisting of first- and second-order neighbours, then it is set to 1, otherwise to 0.

Criterion

If $0 \leq \Sigma < 64$ set $p$ to 0
If $64 \leq \Sigma \leq 128$ set $p$ to 1 if $\Sigma$ is larger than the average value in the first-order neighbourhood centered at $p$.
If $128 < \Sigma \leq 192$ set $p$ to 1

The thresholds that are used corresponds to the thresholds in [11]; the sum of the mask values on the cubic grid is 64 and the thresholds used are 21 and 43. The thresholds on the fcc and bcc grids are calculated by scaling these thresholds by using the sum of the mask values on the fcc and bcc grids, 192. The thresholds obtained by this procedure is 63 and 129. The method is, however, not very sensitive to changes of the threshold values. To get symmetric intervalls, we use the thresholds 64 and 128 instead. The values 60 and 124 or 68 and 132 would give similar results. Resolution pyramids using this method are shown in Fig. 7.

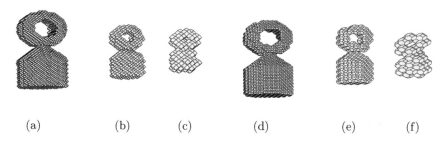

**Fig. 7.** Resolution pyramids consisting of $\mathbb{G}_T$–$\mathbb{G}_{T^3}$; (a)–(c) fcc, the intermediate grey-level image method; (d)–(f) bcc, the intermediate grey-level image method

Also in this case we get connected objects with a tunnel even in the lowest resolution level.

## 5    Conclusions

The most commonly used resolution pyramid on the hexagonal grid has ratio (1 : 7). One problem with this partitioning is that the grid is rotated by arctan $\sqrt{3/5}$ for each level in the resolution pyramid. If the image is represented by a hexagonal grid in each level, the image will be transformed by a rotation which makes the resolution pyramid not suited for, e.g., comparing images captured at different resolutions. In contrary, no rotation is needed for the partitionings of the fcc and bcc grids suggested in this article. The resolution pyramid on the cubic grid has the same ratio as the resolution pyramids in this article, which makes methods for multiscale representations on the cubic grid easy to adapt to the fcc and bcc grids.

As expected, the AND- and OR-rules for multiscale representation give poor results. When using the AND-rule, to many grid points are set to 0 and when using the OR-rule, too many grid points are set to 1. This implies that quality of the image (the topology preservation and the perceived shape) rapidly decreases when resolution is decreased. The perceived shape is better maintained using the shifting grid and the intermediate grey-level methods. A difference between the intermediate grey-level method and the shifting grid method is that when using the shifting grid method, each image obtained using the OR-rule is handled separately. This gives a slightly different result compared to when all possible patterns are handled simultanously using a mask. Also, when using the mask in the intermediate grey-level method, all possible positions of the patterns are considered, some of which might not be included in the shifting grid method. No one of these methods guarantees topology preservation of objects as changes of topology are unavoidable when the resolution is decreased. If we choose to maintain the connectedness of an object with a tunnel and decrease the resolution, then for each level of the resolution pyramid, the tunnel will be smaller than in the previous level. At some point, the tunnel will vanish. When the tunnels are big enough, the shifting grid and the intermediate grey-level methods have nice topology-preserving properties. They are also designed to remove the sensitivity to where the object is positioned in the image. It is also worth noticing that the intermediate grey-level method is faster, since a mask is used instead of applying 24 patterns at each grid point.

# References

1. Matej, S., Lewitt, R.M.: Efficient 3D grids for image reconstruction using spherically-symmetric volume elements. IEEE Transactions on Nuclear Science **42** (1995) 1361–1370

2. Bell, S.B.M., Holroyd, F.C., Mason, D.C.: A digital geometry for hexagonal pixels. Image and Vision Computing **7** (1989) 194–204

3. Herman, G.T.: Geometry of Digital Spaces. Birkhäuser, Boston (1998)

4. Borgefors, G.: Hierarchical chamfer matching: A parametric edge matching algorithm. IEEE Transactions on Pattern Analysis and Machine Intelligence **10** (1988) 849–865

5. Rezaee, M.R., van der Zwet, P.M.J., Lelieveldt, B.P.F., van der Geest, R.J., Reiber, J.H.C.: A multiresolution image segmentation technique based on pyramidal segmentation and fuzzy clustering. IEEE Transactions on Image Processing **9** (2000) 1238–1248

6. Rosenfeld, A.: Multiresolution image representation. In Levialdi, S., ed.: Digital Image Analysis. (1984) 18–28

7. Burt, P.J.: Tree and pyramid structures for coding hexagonally sampled binary images. Computer Graphics and Image Processing **14** (1980) 271–280

8. Lucas, D., Gibson, L.: Image pyramids and partitions. In: Proceedings $7^{th}$ international Conference on Pattern Recognition, Montreal. (1984) 230–233

9. Tanimoto, S.L., Crettez, J.P., Simon, J.C.: Alternative hierarchies for cellular logic. In: Proceedings $7^{th}$ international Conference on Pattern Recognition, Montreal. (1984) 236–239

10. Ahuja, N.: On approaches to polygonal decomposition for hierarchical image representation. Computer Vision, Graphics, and Image Processing **24** (1983) 200–214

11. Borgefors, G., Ramella, G., di Baja, G.S., Svensson, S.: On the multiscale representation of 2D and 3D shapes. Graphical Models and Image Processing **61** (1999) 44–62

12. Borgefors, G., di Baja, G.S., Svensson, S.: Multiresolution representation of shapes in binary images II: Volume images. In: Proceedings of Discrete Geometry for Computer Imagery (DGCI 1997), Montpellier, France. (1997) 75–86

# The Mojette Transform: The First Ten Years

JeanPierre Guédon and Nicolas Normand

Laboratoire IRCCyN, Team Image & Video Communications CNRS UMR 6795,
École polytechnique de l'Université de Nantes La Chantrerie Rue Christian Pauc,
BP 50609, F-44306 Nantes cedex 3
`firstname.lastname@polytech.univ-nantes.fr`

**Abstract.** In this paper the Mojette transforms class is described. After recalling the birth of the Mojette transform, the Dirac Mojette transform is recalled with its basic properties. Generalizations to spline transform and to nD Mojette transform are also recalled. Applications of the Mojette transform demonstrate the power of frame description instead of basis in order to match different goals ranging from image coding, watermarking, discrete tomography, transmission and distributed storage. Finally, new insights for the future trends of the Mojette transform are sketched.

## 1 Introduction

### 1.1 The Word "Mojette"

"Mojette" is a well known word in Poitiers; in old french, it describes the class of white beans. Until recently, these white beans were the standard tool for a child to start computing additions and subtractions. They were also used for computing the number of victories for card games like "Aluette" still played with middle age spanish cards. Mojette is the name of our transform to remember first that when only adds are invoked, the computations can be easily made, second that by sharing the information pot each player will get a part of it.

### 1.2 The First Papers

The first paper on the Mojette transform has been published in 1995 after one year of hard work. At the moment, we were looking for something that we still did not find: a discrete mathematical tool that can split the Fourier plane into radial and angular sectors. The initial application was the psychovisual image encoding that mimics the human channelized vision. Hopefully, what we found was so much practical that many applications did appeared later on. The first talk on the Mojette transform (denoted MT in the following) was published in SPIE Visual Communications and Image Processing [8]. The fact that a novel transform was welcomed by the community was of prime importance to continue this kind of research. Even if all the theory was not yet there, the structure of

E. Andres et al. (Eds.): DGCI 2005, LNCS 3429, pp. 79–91, 2005.

the transform was already explained by its first inverse version using a recursive implementation. The first laboratory talk was given on June $22^d$ 1995.

## 1.3   Related Works

The first two precursors of the Mojette transform were Myron Katz and Gabor Herman. They were both looking at discrete geometric tools to properly inverse the Radon transform at the beginning of the X-ray scan. The concept of discrete angles that we used later on in the MT was taken from Myron Katz' work [12]. Gabor Herman presented iterative algorithms to solve the equation $f = Rp$ where $f$ is the image, $R$ is a discrete Radon projection operator and $p$ is the set of projections obtained with a constant bin width [11]. In this regard, mixing both works gives the Mojette transform.

The mood for a discrete Radon transform use only came back in the nineties; except for pioneer works as Attila Kuba who was starting to see the generality of the problem using discrete geometry in 1984 [13]. Guédon [9, 7] starts with a spline version of the Radon transform in order to fill the gap between discrete and continuous points of view. L. Dorst added a major point by linking the Radon (or slope) transform to mathematical morphology [3] as independently we did obtained our first reconstruction theorem by this link to the two pixel structuring element. Notice that a big step has been made in 1999 when Jean-Marc Chassery did organize a workshop on discrete tomography that was using discrete geometry for mixing the kind of solutions for a given polyominos problem [2] with the complexity of others very close problems [5, 6]. However, it must be clear here that the Mojette transform does not exactly belongs to the "discrete tomography" community in the sense the word was defined by Larry Shepp in 1994, i.e. reconstructing an object from two to four projections.

A novel situation arises now, when many and very interesting papers are published to express different properties of transforms very close to the MT as [22] and unread papers come back to the community as [14].

## 1.4   Paper Organization

The paper is split into theoretical results and applications. The second section of the paper reviews the properties of the Mojette transform. The simple definition of the direct Mojette transform (that will be described by Dirac-Mojette afterwards) is first presented and its fast inversion (same order of complexity than the direct operator) follows. The strong relationships with mathematical morphology are then presented because they constitute both the core of the proofs of the reconstruction theorems we obtained and the geometric way to cope with the transform. The generalization to spline Mojette transform and other kernels are then presented to show the power of the relationships between continuous and discrete words. In other words, this link allows for translating various problems such as tomography into the discrete geometry world. Finally, extensions to higher dimensions demonstrate two notions : the order of complexity of the transform is still linear in the number of pixels, voxels, ixels (information elements) and linear into the number of hyperplanes. The third section present the variety of applications already found for the Mojette transform. The key point that is used here is the frame description notion (instead of the classical vectorial basis). Its first obvious success (because of its direct implementation) was to use the Mojette transform as a tool for multiple description which has evolved to a kind of new standard for communications[17]. A step further in this area was to add the hierarchical (or multiresolution) data description with the concept of multi-layer buffering. Another completely different application was to master the noise properties into the Mojette domain in order to allows the Mojette transform for tomographic reconstruction. Even if this sounds obvious from the Radon inheritance, this works is the last one we were able to perform. Back to the image domain, the Mojette transform has been successfully implemented to solve for new techniques as watermarking as well as texture analysis for image and video as presented in this section. The final part of the paper gives some insights for the next ten years.

## 2    Basic Mojette Transform Properties

### 2.1    Direct and Inverse Dirac-Mojette Transform

The mojette transform is derived from the Radon transform [20]. The 2-D transformed domain consists in projections (in the Radon sense) where each calculated element called a bin is the sum of ixel values. However, from an original block $f(k,l)$ of information elements (denoted ixel for information element), the Mojette transform gives a linear set of projections $proj_\theta(m)$ only for specific angles of the form $\theta = atan(q/p)$ where $p \in \mathbb{Z}, q \in \mathbb{Z}_+$ and are relatively prime $(GCD(p,q) = 1)$. The Mojette transform set is defined by $Mf(k,l)$ as a set of $I$ projections $M_{p,q}f(k,l)$, where $(k,l)$ belongs to the ixels block and $\Delta(m) = 1$ if $m = 0$, 0 otherwise, as follows:

$$Mf = \{M_{p_i,q_i}f, i = 1, \ldots, I\} = \{proj_{p_i,q_i}, i = 1, \ldots, I\} \qquad (1)$$
$$M_{p,q}f(k,l) = proj_{p,q}(m) = proj_\theta(m) = f(k,l)\Delta(m + qk - pl), \qquad (2)$$

A bin value is then simply the summation of every ixel intersecting the line $m = -qk + pl$. As described in Fig. 1, the major difference with the Radon transform is that the bin spacing on a projection depends on the projection angle.

The well-posed nature of this linear transform and its explicit discrete nature were the reason of the new name not to confuse with the numerous regularized versions of the Radon transform (ill-posed inverse operator). A direct consequence of this sampling is that the number of bins of the projection indexed by $(p,q)$ depends both of the projection angle and of the block dimensions. For instance, a $P \times Q$ rectangular block has its $(p,q)$ projection composed of $B$ bins with $B = (Q-1)|p| + (P-1)|q| + 1$. The order of complexity of the direct transform is obviously $O(IN)$ i.e. linear in the number of (p)ixels $N$ and linear in the number of projections $I$.

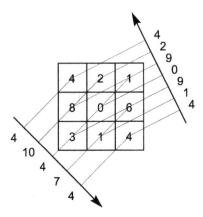

**Fig. 1.** Mojette Transform of a $3 \times 3$ f(k,l) lattice

The inverse transform can be implemented with the same order of complexity. This is an important characteristic of the Mojette transform which comes from the underlying mathematical morphology properties as described in the next paragraph. To get such a result, the reader has to pay attention to the projection of the "corners" of the shape: they exhibit some one to one ixel-bin relationships. Thus, these specific bins can be exactly backprojected in the image as soon as we keep for each bin the number of contributed ixels during the inverse transform. The tool for the proof of this algorithm (given in [15]) is now presented.

### 2.2    Mathematical Morphology

Reconstruction conditions were demonstrated for any convex image [15] by associating a projection angle (p,q) with a 2 pixels structuring element (2PSE) as a point couple $\{O, (p, q)\}$ giving the same direction axis. The set of ixels that are projected in one bin $m$ is the intersection of the image with the $(p, q) - line$ defined by: $-qk + pl = m$. If the image is convex, this set is connected relative to the $(p, q)$-neighborhood (all ixels are reachable from the others using $(p, q)$ displacements). Thus, an opening (an erosion followed by a dilation) with the 2PSE $\{O, (p, q)\}$ will empty the set if it is composed of exactly one ixel but will leave it unchanged in all other cases. The back projection in direction $(p, q)$, by removing all ixels in a one to one correspondence with a bin, results in the opening of the image with the 2PSE:

$$I' = I \circ \{O, (p, q)\} \ .$$

An ixel disappears from I only if it has no $(p, q)$-neighbor.

The final reconstruction result is not affected by the order bins are back projected. So the overall reconstruction can be viewed as an iterative series of openings with the 2PSEs:

$$I_{k+1} = I_k \circ \{O, (p_1, q_1)\} \circ \ldots \circ \{O, (p_I, q_I)\} \ .$$

structuring elements

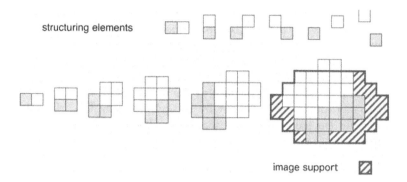

image support

**Fig. 2.** Reconstructibility of a convex image

The reconstruction stops if no ixels can be removed by any projection ($I_{k+1} = I_k$ or $I_{k+1} = \emptyset$). This occurs when every ixel in the image has at least one $(p,q)$-neighbor in every $(p,q)$ direction. The smallest non-null image with that property is built by the series of dilations with the 2PSEs.

The reconstructibility criteria of Katz finds here another expression only using dilations. A convex image is not reconstructible if and only if the dilation result by 2PSE is not included in the image support (even if one pixel is concerned). This general result is expressed in the following theorem by the equivalence of two statements.

**Theorem 1.** *Both propositions are equivalent*

- $f(k,l)$ defined on the convex $G$ is reconstructible by $\{proj_{p_i,q_i}, i = 1, \ldots, I\}$;
- $R$ constructed by $I$ dilations set $\{\{O, (p_i, q_i)\}, i = 1, \ldots, I\}$ is not included in $G$.

**Corollary 1.** *Both propositions are equivalent*

- $G$ is reconstructible by $\{proj_{p_i,q_i}, i = 1, \ldots, I\}$;
- the erosion of $G$ by $R$ is null.

This inversion criterion is an extension of Katz' criterion for rectangular images. In that particular case, $R$ can be included in the image if its width $(1 + \sum_i |p_i|)$ and height $(1 + \sum_i |q_i|)$ are respectively smaller than the width $P$ and height $Q$ of the image. Thus, for a $P \times Q$ image, a complete reconstruction is achievable if [12]:

$$P \leq \sum_i |p_i| \text{ or } Q \leq \sum_i |q_i| \tag{3}$$

## 2.3 Spline Mojette Transform

The previous Mojette transform is denoted Dirac-Mojette transform because the underlying 2D ixel representation is a Dirac field:

$$f(x,y) = \sum_k \sum_l f(k,l)\delta(x-k)\delta(y-l) \ . \tag{4}$$

However, very similar angles like $(p, q) = (1, 1)$ and $(p, q) = (30, 29)$ will do not have similar bin values: a direct inter-correlation computation between the two projections will be poor because the number of bins and the number of pixels contributing to each bin will be very different. However, Philippé and Guédon [19] have shown that the autocorrelation of the Mojette projection corresponds to the projection of the autocorrelation of the image.

Summing ixel intensities weighted by the relative length of the ray inside each ixel, defines the spline$_0$-Mojette transform. The corresponding continuous image is given by:

$$f(x, y) = \sum_k \sum_l f(k, l)\beta_0(x - k)\beta_0(y - l) \ , \tag{5}$$

where $\beta_0(x) = 1$ when $|x| < $ , $\beta_0() = $ and 0 elsewhere.

It can be simply shown that the spline$_0$-Mojette transform corresponds to the filtering of the Dirac-Mojette transform with a finite impulse response (FIR) filter. The FIR filter definition $w_{p,q}(m)$ corresponds to a trapezoidal shape, i.e. the projection of the shape of the ixel (a square in 2D) as described in [10].

$$M_0 f = [M_D f] * w_{p,q} \ , \tag{6}$$

Where the kernel $w$ can always be decomposed with small filters composed of unitary values. In other words only additions can be used for implementation of this convolution and the recursive filter that implements the inverse operator (going from spline$_0$ to Dirac Mojette data) will only use subtractions.

The generalization to higher spline order Mojette transform follows the same pattern. From a $\beta_n$ spline definition of the original function given as:

$$f(x, y) = \sum_k \sum_l f(k, l)\beta_n(x - k)\beta_n(y - l) \ , \tag{7}$$

with $\beta_n(x) = \beta_{n-1}(x) * \beta_0(x)$, the spline$_n$-Mojette transform is defined as:

$$M_n f(k, l) = [M_{n-1} f] * w_{p,q}(b) \ . \tag{8}$$

The main originating reason to use spline functions instead of Dirac fields is to complied with the generalization of the well-known Shannon-Wittaker sampling theorem obtained by Unser and AlDroubi [1, 23, 24, 25]. This theorem allows to combine the projection decomposition with other image needs like wavelet or geometric (e.g. rotation) operators.

The main use of spline Mojette transform is to introduce a controlled redundancy inside each projection via the kernel $w$. Direct applications to error correcting codes have already been obtained [21] in the presence of noise inside the projection data. This intra-redundancy can be managed at the same time that the inter projections redundancy to answer consistency of distributed information projections as shown in the applications.

## 2.4    nD Mojette Transform

Going from 2D results to 3D then $n$D is an indefectible temptation to generalize
the obtained results. This was done by still considering the Mojette transform as
giving $(n-1)$D hyperplanes from the $n$D discrete hypervolume. It can be easily
shown that from $N$ ixels distributed inside a convex set of dimension $n$ and a
choice of $I$ hyperplanes, the direct and inverse Mojette transform still exhibit a
complexity order of $O(IN)$. This is quite obvious for the direct transform. The
way of having this very low cost algorithm for the inverse transform is to use the
same mapping than previously used in 2D between ixels and bins. As a matter
of fact, this mapping is dimension independent.

Another result that can be seen as independent of the dimension is the discrete central slice theorem (CST). The continuous version of the CST has been
extensively used for Radon inversion as well as for image processing. It says that
the 1D Fourier transform of a projection of angle $\theta$ equals the slice of the 2D
Fourier transform of the image at angle $\theta + \pi/2$. The discrete version of this theorem was established in 2D by Dudgeon and Mersereau using the Z transform
[4]. It has been checked with the 2D Mojette transform and extended to higher
dimensions in a straightforward manner [26].

## 2.5    Mojette Transform and Redundancy

The Mojette transform matrix $M$ is only filled by 1 and 0 values. Only additions and subtractions are required for $M$ and $M^*$. It is also the case for $M^{-1}$
with the inverse Transform algorithm for exact values [15]. The matrix $M^*M$ is
Toeplitz-block-Toeplitz. Since only the additive group structure is needed for its
definition, replacing the natural addition by modulus addition will not change
any property of the transform. This can be done as soon as the initial values of
the ixels are quantized onto an interval $[0, 2b[$ where $b$ represents the number of
bits of an ixel. In this case, a bin will belong to the same interval (by modulus
addition), so do the reconstructed ixel. This represents an important matter for
not having an overbinary representation in the Mojette domain as well as to use
the same type of elements in both domains.

The computation of a redundancy factor as $Red = \#Bins/\#Ixels - 1$, gives
the true weight for managing a frame. However, this Red indicator does not
explain at all the stability of the transform under inconsistency (e.g. noise)
considerations. In such a case, the number of projections as well as the difference between the initial shape and the dilations of two ixel structuring el-

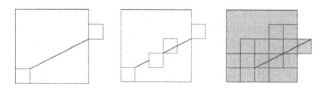

**Fig. 3.** Redundancy of three sets of projections under a $5 \times 5$ $f(k,l)$ lattice

ements (2ISE) will better explain the power of recovering inconsistency. For instance, for a $5 \times 5$ convex, the sets $S_0 = \{(5,3)\}$, $S_1 = \{(3,2),(2,1)\}$ and $S_2 = \{(1,0),(-1,1),(1,1),(2,1)\}$ all fulfill the reconstruction conditions and give Red$= 8/25$. The corresponding reconstructed shapes are illustrated in Fig. 3.

The importance of splitting the initial information into (projection) pieces can be pictured as in the previous figure or computed as the conditioning factor of the matrix $M * M$. In this example, when the first degenerated information set gets a false bin value it will not been detected at the reconstruction step, the second set will detect the error but not correct it as the third set does.

# 3    Applications

The power of the Mojette transform lies in it abilities to manage inconsistency. The first way to have inconsistent data is to have not enough data whereas it exists some intrinsic correlation inside the data as shown in the last paragraph of section two. In this case, only some partial exact descriptions of the information are available: this corresponds to the multiple description problem. This information can also already be sorted in an hierarchical manner and treated with the same formalism through the use of the Mojette transform. The second case is the shape description problem through some of its projections, where the Mojette transform becomes a geometric tool when allied with the mathematical morphology. The third case of inconsistent data arises when metadata is added inside the data as for the watermarking techniques. Then the two coexisting kinds of data must have different behaviors (one must be shown while the other should lie under the expressed data until a specific algorithm can reveal it. The fourth case is when enough projections are available but are corrupted by noise such that only an approximation can be computed as for tomography.

## 3.1    The Use of Redundancy: The Multiple Description Tool

The multiple description problem arises in the seventies and was studied by Ozarow and others as an information theory problem for telephony over two channels [16]. It has been developed for a decade for Internet transmission and now for distributed storage. The paradigm is as follows: given an initial information set, computes $I$ different descriptions such that for a lower number of descriptions there is still always an approximation that can be computed and used usually for real-time considerations.

Many kind of solutions have been employed ranging from Turbo-codes, Solomon codes, Tornado codes and of course Mojette projections. The major property of these codes is generally to be Minimum Description Separable (denoted MDS), that is without redundancy as for Solomon or Tornado codes or $(1 + \epsilon)$-MDS as for the Mojette codes demonstrated by Parrein [17]. The price of the redundancy can be easily paid to gain much more flexibility in the design of the code (which means to choose a pair of shape and set of projections) whereas Solomon codes must be only determined from Galois field $GF(2^n)$.

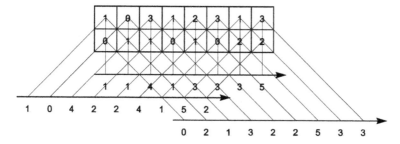

**Fig. 4.** An image support of height $Q = 2$ with directions $(0, 1)$, $(-1, 1)$ and $(1, 1)$. Any combination of two projections is sufficient to reconstruct the image

For the basic Mojette solution, a rectangular shape $P \times Q$ with $(P \gg Q)$ is employed in conjunction with a set of $I$ projections:

$$S_Q = \{(p_1, 1), (p_2, 1), \ldots, (p_Q, 1)\} \ .$$

With these particular directions $(\forall i, q_i = 1)$ the image is reconstructible with as least $Q$ projections among $I$ according to Katz' criterion Eq. 3 (the height of the smallest non-invertible image equals the number of received projections plus one). The value $P \times Q$ gives the amount of data that can be transferred at a time (available bandwidth), $I - Q$ is number of extra projections (descriptions) sent over the channel to protect against data losses.

With this simple image support, protection against losses is equal for all data. More precise control of the redundancy can be achieved with a concatenation of image supports of different heights, allowing to fit the protection level to the importance of the data: the less important the data, the higher the image support [18].

Finally, when the initial data exhibits some intra correlation, phantoms can be used to produce an approximated reconstruction from an iterated algorithm [19].

Distributed storage organization and management can follow the same path. Initial data are taken and dispatched after the Mojette operation giving $A$ projections. When a small fixed number $B(B \ll A)$ of different projections issued from the $A$ set are stored on each of the $S$ servers then the Mojette organization is made and gives the following management properties:

- if a server is attacked and destroyed by hackers, its content can be restored by any set of other servers as soon as it meets the Mojette reconstruction requirement,
- if a server content is stolen, its content can not be useful by itself (several servers have to be attacked at the same time),
- from the user side, the ask data will arrive from different sources and routes, which will decrease the network congestion and increase the trust in the data.

## 3.2    Discrete Tomography Tool

The continuous Radon transform described by :

$$proj(t, \theta) = R(t, \theta)[f(x, y)] = \int f(x, y)\delta(t - x\cos\theta + y\sin\theta)dxdy \ , \qquad (9)$$

represents the continuous function $f(x, y)$ by an infinite set of projections. The functional projection of $f(x, y)$ onto a spline space $\{\phi(x - k), k \in \mathbb{Z}\}$ leads to an interpolation equation:

$$f_\phi(x, y) = \sum_k \sum_l f(k, l).\phi(x - k).\phi(y - l) \ . \qquad (10)$$

When the discrete pixel grid is considered in the tomographic problem, the function $f(x, y)$ in Eq. 9 is replaced by $f_\phi(x, y)$ of Eq. 10. This leads (after inverting discrete and continuous sum signs) to a definition of the continuous projection from a discrete image and the spline interpolating function.

For tomographic purposes, $\phi$ is not taken as $\phi(x) = \delta(x)$ but rather as a spline function to express the "quality of the physical length" that can be recorded at the detector.

This Mojette sampling has been used both for direct spline FBP algorithm and for iterative conjugate gradient algorithm in the presence of noise with good results. The only remaining problem consists in the rebinning of the initial data recorded from the tomographic device.

## 3.3    Image Watermarking

Watermarking represents the way of transferring invisible data inside data. It can be text in text as in the private correspondence of George Sand with Alfred de Musset or more generally marks informations inside images or videos in order to carry the rights of the author and all the business steps allowing to sell information bytes. The invisibility notion is taken in the human sense, thus the locations to put this information into the carrier information must be given by psychovisual studies. The employed tool that implements the mixture of both information must be flexible enough both to use the given locations and be robust to image attacks as low-pass filtering, small rotation, and image compression for instance.

The Mojette transform in this context has produced different kinds of tools. The first one consists to watermark the projections instead of the image. This technique has also been lately develop by Macq and others in the Radon domain. The clue is to obtain an inconsistent set of informations from the marked projections in order to resist to rotation attacks.

The second tool is related to Normand's inversion algorithm that use pixel-bin correspondences at each step. By adding a controlled quantity of noise in the projections (in the proportion of the redundancy Red factor) the reconstruction becomes impossible because of the ill-posed properties of the Radon transform. However, if for each reconstructed pixel the choice can be made (at the coder step) to avoid corrupted bins then a unique reconstructing path can be determined and encoded as the clue to decode the image. In other words, a cryptographic scheme is obtained.

The first unsolved question lies in the possibility to decode the crypto-tattooed image thanks to the linearity of the transform even if the modulus addition is employed. As a matter of fact, this problem has to be solve thanks to both the Red factor and the possible decodable errors inside Mojette data : discrete group theory allied with geometry seems the right way to cope with this problem.

Another interesting question arises from the psychovisual locations able to hide a message. A soon as these locations are computed (with a respective strength for each pixel), the best set of projections can only be calculated by testing all the resulting phantoms (very large set) that can match the greater number of positions to assess the robustness of the watermarking. Going from this NP complete problem to a polynomial solution seems feasible using discrete shape descriptions and superimpositions.

We are confident that such a tool will be usefully employed in many other applications because of its intrinsic redundancy properties. However, it should be stressed again that the strong link between projections and discrete directions thus shape is today the only way to get mathematical tools to solve for problems. As soon as this projection structure disapear and is replaced by a simple set of bins, the tools complexity become NP complete instead of linear. This has been the case both for watermarking and for obtaining compression scheme from the right set of bins when each of them reconstructs an entire line of ixels.

## 4   Conclusion

This paper has matched the timing for the ten years of a new transform coming from the word of discrete geometry. It allows for reviewing both what has been done in the theory and to demonstrate that discrete geometry can be an important tool for network optimization or distributed storage. These trends must and will be continued and amplified for the next ten years.

## Acknowledgements

The authors would like to thank Jean-Marc Chassery for his indefectible help to promote the Mojette transform and Éric Andres and the DGCI board for their kind invitation to write this paper.

## References

1. Akram Aldroubi and Murray Unser. Families of wavelet transforms in connection with Shannon's sampling theory and the Gabor transform. In C. K. Chui, editor, *Wavelets: A Tutorial in Theory and Applications*, pages 509–528. Academic Press, 1992.

2. Alberto Del Lungo, Maurice Nivat, and Renzo Pinzani. The number of convex polyominoes reconstructible from their orthogonal projections. *Discrete Math.*, 157(1-3):65–78, 1996.
3. Leo Dorst and Rein van den Boomgaard. Morphological signal processing and the slope transform. *invited paper for Signal Processing*, 38:79–98, 1994.
4. Dan E. Dudgeon and Russell M. Mersereau. *Multidimensional Digital Signal Processing*. Prentice-Hall, 1984.
5. Richard Gardner. *Geometric Tomography*. Encyclopedia of Mathematics and its Applications. Cambridge University Press, 1995.
6. Peter Gritzmann and Maurice Nivat, editors. *Discrete Tomography: Algorithms and Complexity*, number 97042, Germany, jan 1997. Dagstuhl.
7. JeanPierre Guédon. *Les problèmes d'échantillonnage dans la reconstruction d'images à partir de projections.* PhD thesis, Université de Nantes, Novembre 1990.
8. Jeanpierre Guédon, Dominique Barba, and Nicole Burger. Psychovisual image coding via an exact discrete radon transform. In Lance T. Wu, editor, *VCIP'95*, pages 562–572, Taipei, Taiwan, may 1995. CORESA.
9. JeanPierre Guédon and Yves Bizais. Bandlimited and haar filtered back-projection reconstuction. *IEEE transaction on medical imaging*, 13(3):430–440, September 1994.
10. JeanPierre Guédon and Nicolas Normand. Spline mojette transform. application in tomography and communication. In *EUSIPCO*, sep 2002.
11. Gabor T. Herman and M. D. Altschuler. *Image Reconstruction from Projections - Implementation and Applications.* Topics in Applied Physics. Springer-Verlag New York, oct 1979.
12. Myron Katz. *Questions of uniqueness and resolution in reconstruction from projections.* Lecture Notes in Biomathematics. Springer-Verlag New York, dec 1978.
13. Attila Kuba. The reconstruction of two-directionally connected binary patterns from their two orthogonal projections. *Comput. Vision, Graphics. Image Process.*, (27):249–265, 1984.
14. F. Matus and J. Flusser. Image representation via a finite radon transform. *IEEE transaction on pattern analysis and machine intelligence*, 15(10):996–1006, oct 1993.
15. Nicolas Normand and Jeanpierre Guédon. La transformée mojette : une représentation redondante pour l'image. *Comptes-Rendus de l'Académie des Sciences*, pages 123–126, jan 1998.
16. L. Ozarow. On a source coding problem with two channels and three receivers. *Bell Sys. Tech. Journal*, 59:1909–1921, 1980.
17. Benoît Parrein, Nicolas Normand, and Jeanpierre Guédon. Multiple description coding using exact discrete radon transform. In *Data Compression Conference*, page 508, Snowbird, mar 2001. IEEE.
18. Benoît Parrein, Nicolas Normand, and Jeanpierre Guédon. Multimedia forward error correcting codes for wireless lan. *Annals of telecommunications*, (3-4):448–463, mar-apr 2003.
19. Olivier Philippé and Jeanpierre Guédon. Correlation properties of the mojette representation for non-exact image reconstruction. In ITG-Fachbericht Verlag, editor, *Proc. Picture Coding Symposium 97*, pages 237–241, Berlin, sep 1997.
20. Johan Radon. Über die bestimmung von funktionen durch ihre integralwerte längs gewisser mannigfaltigkeiten. *Ber. Ver. Sächs. Akad. Wiss. Leipzig, Math-Phys. Kl.*, 69:262–277, April 1917. In German. An english translation can be found in S. R. Deans: *The Radon Transform and Some of Its Applications*, app. A.

21. Benoît Souhard, Christian Chatellier, and Christian Olivier. Simulation d'une chaîne de communication adaptée à la transmission d'images fixes sur canal réel. In *CORESA*, Lyon, jan 2003.

22. Imants Svalbe and Andrew Kingston. Farey sequences and discrete radon transform projection angles. In *IWCIA'03*, Palermo, may 2003.

23. Michael Unser, Akram Aldroubi, and Murray Eden. Polynomial spline signal approximations: Filter design and asymptotic equivalence with shannon's sampling theorem. In *IEEE Transaction on Information theory*, volume 38, pages 95–103. IEEE, 1992.

24. Michael Unser, Akram Aldroubi, and Murray Eden. B-Spline signal processing: Part I - Theory. *IEEE Trans. Signal Process.*, 41(2):821–833, Feb 1993.

25. Michael Unser, Akram Aldroubi, and Murray Eden. B-Spline signal processing: Part II - Efficient design and applications. *IEEE Trans. Signal Process.*, 41(2):834–848, Feb 1993.

26. Pierre Verbert and Jeanpierre Guédon. An exact discrete backprojector operator. In *EUSIPCO*, Toulouse, 2002.

# On the Stability of Reconstructing Lattice Sets from X-Rays Along Two Directions

Andreas Alpers[1] and Sara Brunetti[2]

[1] Zentrum Mathematik, Technische Universität München,
Boltzmannstr. 3, D-85747 Garching bei München, Germany
`alpers@ma.tum.de`
[2] Dipartimento di Scienze Matematiche e Informatiche,
Universitá degli Studi di Siena, Pian dei Mantellini 44, 53100 Siena, Italy
`sara.brunetti@unisi.it`

**Abstract.** We consider the stability problem of reconstructing lattice sets from their noisy X-rays (i.e. line sums) taken along two directions. Stability is of major importance in discrete tomography because, in practice, these X-rays are affected by errors due to the nature of measurements. It has been shown that the reconstruction from noisy X-rays taken along more than two directions can lead to dramatically different reconstructions. In this paper we prove a stability result showing that the same instability result does not hold for the reconstruction from two directions. We also show that the derived stability result can be carried over by similar techniques to lattice sets with invariant points.

## 1 Introduction

The main problem in discrete tomography is to reconstruct lattice sets (or equivalently, binary pictures) from their X-rays, that is, from the number of its points lying along lines parallel to any set of prescribed directions. In practice, these X-rays are affected by noise. In [3] the authors investigate the stability of the problem of reconstruction from X-rays taken from $m \geq 3$ lattice directions. In this case they show that a small change in the data (of $2(m-1)$ measured in the $\ell_1$-norm) can lead to a dramatic change in the reconstruction. This instability persists even when the original is uniquely determined by its exact X-rays. This paper addresses the open question whether the reconstruction of lattice sets from X-rays along two directions is also an unstable task.

In this paper we analyze the case $m = 2$ with the same kind of data errors as in [3], i.e., where $2(m-1) = 2$. We show in Theorem 17 that lattice sets which are uniquely determined by their X-rays along two directions can only have stable reconstructions even if the X-rays are changed by 2 (in the $\ell_1$-norm). It still remains open what happens if there is a higher perturbation in the data. In any case, we show that a similar combinatorial reasoning leads to provable stability results (Theorem 25) when the requirement of uniqueness on the original set is weakened to the assumption that in every reconstruction there exists a suitable number of invariant points.

E. Andres et al. (Eds.): DGCI 2005, LNCS 3429, pp. 92–103, 2005.

Other kinds of mathematically proved stability results can be found in [1], [2], [5] and [6]. Reports on the stable behavior of reconstruction algorithms which incorporate a-priori knowledge can be found in [11] and [15]. For details on discrete tomography we refer to [8]. More about $(0, 1)$-matrices can be found in [4]. Invariant sets have been intensively studied by many authors (e.g. [7],[9],[13],[14]).

## 2    Notations and Statement of the Problem

The elements of $\mathcal{F}^2 = \{F \subset \mathbb{Z}^2 \ : \ F \text{ is finite}\}$ are called *lattice sets*, and the subspaces $\lin\{v\} \subseteq \mathbb{R}^2$, $v \in \mathbb{Z}^2 \setminus \{0\}$ are called *lattice directions*. For a lattice direction $S$, we define $\mathcal{A}(S) = \{w + S \ : \ w \in \mathbb{Z}^2\}$. The *(discrete) X-ray* of $F \in \mathcal{F}^2$ parallel to $S$ is the function $X_S F : \mathcal{A}(S) \rightarrow \mathbb{N}_0 = \mathbb{N} \cup \{0\}$ defined by $X_S F(T) = |F \cap T|$, for each $T \in \mathcal{A}(S)$.

In this paper we always consider X-rays from two directions $S_1, S_2$. By affine transformations of the grid we can assume in the following that $S_1 = \lin\{(1, 0)\}$ and $S_2 = \lin\{(0, 1)\}$, thus we speak of X-rays along horizontal and vertical lines, or even about row and column sums. It is clear that the X-rays can be assembled as a vector. Let two lattice sets $F_1, F_2$ be given. We will denote its vector containing the row sums by $R_i$ and the vector containing the column sums by $C_i$, $i = 1, 2$, respectively. Notice that we assume that the $i$-th entries of $R_1$ and $R_2$ refer to the same (horizontal) line, which can be achieved by inserting zero-entries. We make the same assumption on the entries of $C_1$ and $C_2$.

For $F_1 \in \mathcal{F}^2$ let $\mathcal{F}^2(F_1) := \{F_2 \in \mathcal{F}^2 : |F_1| = |F_2| \wedge ||R_1 - R_2||_1 + ||C_1 - C_2||_1 = 2\}$. So, we focus on the question: *Let $F_1 \in \mathcal{F}^2$, uniquely determined by $R_1$ and $C_1$. What is the sharpest upper bound on $\max_{F_2 \in \mathcal{F}^2(F_1)} |F_1 \triangle F_2|$?*

## 3    Preliminary Remarks

Since $F_1 \in \mathcal{F}^2$ is finite, it is contained in a rectangle whose rows and columns are non-empty. We refer to them also as the rows and the columns of $F_1$. If $p$ is a point of $F_1$ lying in the $i$th row and $j$th column, we write that $row(p) = i$ and $col(p) = j$. Sometimes we do not distinguish between the row itself and its index. We use the convention that the rows and columns are numbered starting from its left-upper corner (see Figure 1 (a)).

We can change the coordinates of each point in the rectangle by first permuting the columns so that $C_1$ is a non increasing vector, and then permuting the rows so that $R_1$ is also a non increasing vector. This is a one-to-one function on the points of the rectangle yielding a triangular shape when the set is uniquely determined by its X-rays (it is a *maximal matrix* with non increasing row sums, see [12]). So, we shall assume that $F_1$ has a triangular shape, like in Figure 1 (a).

The following remarks concern the X-ray errors. Proposing an X-ray error of 2 means that the error occurs on exactly two lines of a single direction. This follows from the assumption $|F_1| = |F_2|$. Indeed the sum of the X-ray values according to one direction equals the cardinality of the set to be reconstructed,

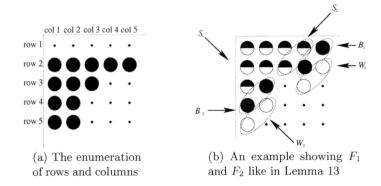

(a) The enumeration
of rows and columns

(b) An example showing $F_1$
and $F_2$ like in Lemma 13

**Fig. 1.** An illustration of sets $F_1$ (black points) and $F_2$ (white points). Points of $F_1 \cap F_2$ are colored half white and half black

and so if there is exactly one line where the error is $-1$ (which will in the following mean that there is one point more of $F_2 \setminus F_1$ than of $F_1 \setminus F_2$ on the line), then there exists exactly one line with a $+1$-error (meaning that there is one point more of $F_1 \setminus F_2$ than of $F_2 \setminus F_1$ on the line). Furthermore, by possibly rotating $F_1$ and $F_2$ we may assume that the error occurs along horizontal lines (rows). If $F_2$ contains a single point on a horizontal line, then we can assume without loss of generality that it is row 1.

*Remark 1.* In summary, we assume in the rest of the paper that $F_1$ is a maximal matrix with sorted rows (or equivalently, row and column sum vectors are monotone and uniquely determine $F_1$), $|F_1| = |F_2|$, and the error occurs in exactly two rows (consequently in no columns).

*Remark 2.* Let $p$ be any point of $F_2 \setminus F_1$ and let $q$ be any point of $F_1 \setminus F_2$.

(a) From the shape of $F_1$ we have:
   - if $col(p) = col(q)$, then $row(p) > row(q)$;
   - if $row(p) = row(q)$, then $col(p) > col(q)$.
(b) Since no column error occurs: the point $p$ exists if and only if the point $q$ exists, with $col(p) = col(q)$. Similarly, in the rows in which no error occurs: $p$ exists if and only if $q$ exists, with $row(p) = row(q)$.

**Lemma 3.** *Let $i$ be the row where the $-1$ error occurs, and let $j$ be the row where the $+1$ error occurs. Then $i > j$.*

*Proof.* Suppose that $i < j$; then we show that there is an infinite sequence of points in $F_1 \triangle F_2$ starting with a point $p$ of $F_2 \setminus F_1$ in the $i$th row. Since the $-1$ error occurs in the $i$th row, at least a point $p$ of $F_2 \setminus F_1$ exists in this row. By Remark 2 (a) and (b), for any $p$, a point $q$ of $F_1 \setminus F_2$ exists such that $col(p) = col(q)$ and $i = row(p) > row(q)$. Since there is no error in any row with index less than $i$, we conclude with Remark 2 (a) that there exists also a point

$p'$ of $F_2 \setminus F_1$ in $row(q)$, then again a point $q'$ of $F_1 \setminus F_2$, etc., all of which lie in a row with index less than $i$. This leads to an infinite sequence of points in $F_1 \triangle F_2$ which is not possible. $\qquad\square$

## 3.1   Staircases

**Definition 4.** *Let* $A, B \in \mathcal{F}^2$ *with* $A \cap B = \emptyset$. *A staircase* $T = (t_1, \ldots, t_m)$ *according to the columns is a sequence of an even number* $m > 0$ *of distinct points* $t_{2i+1} \in A$ *and* $t_{2i+2} \in B$ *for* $0 \leq i \leq \frac{m}{2} - 1$ *such that*

*(i)* $col(t_{2i+1}) = col(t_{2i+2})$ *and* $row(t_{2i+1}) > row(t_{2i+2})$ *for* $0 \leq i \leq \frac{m}{2} - 1$;
*(ii)* $row(t_{2i}) = row(t_{2i+1})$ *and* $col(t_{2i}) < col(t_{2i+1})$ *for* $1 \leq i \leq \frac{m}{2} - 1$.

The definition of staircase according to the rows is obtained by exchanging the words "row" by "column" and "A" by "B".

A staircase can be interpreted geometrically. It is a rook path of alternating points of $A$ and $B$ with end points $t_1$ and $t_m$. We refer to $(t_{2i}, t_{2i+1})$ as a *horizontal step* and to $(t_{2i+1}, t_{2i+2})$ as a *vertical step* of $T$. Since $A$ and $B$ are finite, every staircase has a finite number of points. A staircase which is not a proper subset of another staircase is called a *maximal staircase*. Below we will show that there is only one single maximal staircase if $A = F_2 \setminus F_1$ and $B = F_1 \setminus F_2$. The points of $A$ will be called *white points*, and the points of $B$ will be called *black points*.

## 4   Technical Lemmas

In the following we will speak about staircases in $F_1 \triangle F_2$, which implicitly means that $A = F_2 \setminus F_1$ and $B = F_1 \setminus F_2$. Now, we are going to show that the symmetric difference of $F_1$ and $F_2$ is a maximal staircase in $F_1 \triangle F_2$.

*Remark 5.* Notice that for every $p \in F_1 \triangle F_2$ there exists a staircase $T = (t_1, \ldots, t_m)$ (possibly constituted by two points) such that $p$ is an element of $T$. This can be easily deduced from Remark 2.

**Lemma 6.** *Any two maximal staircases in* $F_1 \triangle F_2$ *have the same starting point and the same end point.*

*Proof.* Let $T_1 = (t_1, \ldots, t_m)$ and $T_2 = (s_1, \ldots, s_n)$ be any two maximal staircases in $F_1 \triangle F_2$. We are going to show that $t_1 = s_1$ and $t_m = s_n$. Since the $-1$ error occurs in exactly one row (say $i$), it follows that $row(t_1) = row(s_1) = i$. If $col(t_1) \neq col(s_1)$, a black point, say $q$, exists in the $i$th row. By Remark 2 a white point $p$ exists such that $col(p) = col(q)$ and $row(p) > i$. Since there is no error in any row $k > i$ this leads to an infinite sequence of points in $F_1 \triangle F_2$.

Analogously, since the $+1$ error occurs in exactly one row (say $j$), we have $row(t_m) = row(s_m) = j$, and $col(t_m) = col(s_n)$ because otherwise (with Remark 2) there is an infinite sequence of black and white points in $F_1 \triangle F_2$. $\quad\square$

*Remark 7.* By Lemma 6 every maximal staircase starts in a white point $t_1$ and ends in a black point $t_m$. So, there is exactly one white point in $row(t_1)$ and $col(t_1)$ and one black point in $row(t_m)$ and $col(t_m)$. Moreover there is no black or white point outside the rectangle which is made up of the rows between $row(t_1)$ and $row(t_m)$, and the columns between $col(t_1)$ and $row(t_m)$.

**Lemma 8.** *Any two maximal staircases in $F_1 \triangle F_2$ coincide.*

*Proof.* Let $T_1 = (t_1, \ldots, t_m)$ and $T_2 = (s_1, \ldots, s_n)$ be any two maximal staircases according to the columns in $F_1 \triangle F_2$. We are going to show that $m = n$ and $t_i = s_i$ for $i = 1, \ldots, m$. Since $t_1 = s_1$ there is no white point other than $t_1$ in $col(t_1)$ by Remark 7. It follows that exactly one black point lies on this column, that is, $t_2 = s_2$. Analogously, we can conclude that $t_{m-1} = s_{n-1}$. Consider now $T_1 \setminus \{t_1, t_m\}$ and $T_2 \setminus \{s_1, s_n\}$. They are two staircases according to the rows. Proceeding as before, we conclude that $t_3 = s_3$ and $t_{m-2} = s_{n-2}$. We repeat the procedure alternatively on a staircase according to the rows and one according to the columns until an empty set is achieved. So, $t_i = s_i$ for $i = 1, \ldots, m$.  $\square$

The previous lemmas prove the following proposition.

**Proposition 9.** *The points of $F_1 \triangle F_2$ constitute a maximal staircase.*

## 5   Bounds

In this section we give an upper bound for the number of points on any maximal staircase, when we fix $F_1$ but may vary $F_2$. This gives a sharp bound on $|F_1 \triangle F_2|$, since the maximal staircase contains exactly the points of $F_1 \triangle F_2$.

Let $T = (t_1, \ldots, t_m)$ denote this staircase, and let $\mathcal{R} = \{1, \ldots, a\} \times \{1, \ldots, b\}$ be the rectangle containing $F_1$ having non-empty rows and columns. Clearly, there is at most one point $t_1$ of $T$ outside of $\mathcal{R}$, and for this point we have $1 \le col(t_1) \le b$, while all the other points of $T$ are inside of $\mathcal{R}$. Without loss of generality we can assume that all points of $F_2 \subseteq \{0, \ldots, a\} \times \{1, \ldots, b\}$. So, we have $R_1 \in \mathbb{N}_0^a$ and $C_1 \in \mathbb{N}_0^b$.

**Proposition 10.** *Let $R_1 \in \mathbb{N}_0^a$ and $C_1 \in \mathbb{N}_0^b$ uniquely determine $F_1 \in \mathcal{F}^2$. Then,*

$$\max_{F_2 \in \mathcal{F}^2(F_1)} |F_1 \triangle F_2| \le 2 \min(a+1, b).$$

*Proof.* Since $F_1 \triangle F_2$ forms a staircase $T$, it is immediately clear that an upper bound for the symmetric difference can be obtained by counting two times the number of vertical steps in $T$ that, in turn, is less than $b$. But another bound is given by adding 2 (for including $t_1$ and $t_m$) to two times the number of horizontal steps. Since $t_1$ is the only point that can lie outside $\mathcal{R}$, we have $\max_{F_2 \in \mathcal{F}^2(F_1)} |F_1 \triangle F_2| \le \min(2a+2, 2b) = 2 \min(a+1, b)$.  $\square$

**Proposition 11.** *Under the same hypothesis of Proposition 10, let $l$ be the number of pairwise different row sums of $F_1$. Then,*

$$\max_{F_2 \in \mathcal{F}^2(F_1)} |F_1 \triangle F_2| \leq 2l.$$

*Proof.* Clearly, $|T|$ equals two times the number of vertical steps in $T$. Since by definition of staircase, $row(t_{2i}) = row(t_{2i+1})$ and $col(t_{2i+2}) = col(t_{2i+1})$ with $t_{2i}, t_{2i+2} \in F_1 \setminus F_2$ and $t_{2i+1} \in F_2 \setminus F_1$, we know that the number of points of $F_1$ in $row(t_{2i})$ is less than the number of points of $F_1$ in $row(t_{2i+2})$. So the maximal number of vertical steps in any staircase $T$ for $F_1$ equals $l$. □

The next two lemmas provide lower bounds to the symmetric difference of $F_1$ and $F_2$. These are used later on to show that the derived bounds are sharp.

**Lemma 12.** *For every $n \in \mathbb{N}$ there exist $F_1, F_2 \in \mathcal{F}^2$ with $|F_1|=|F_2|=\frac{1}{2}n(n+1)$ such that $F_1 \triangle F_2$ is a staircase with $2n$ points.*

*Proof.* Taking the sets $F_1 = \{(i,j) : 1 \leq i \leq n, 1 \leq j \leq n+1-i\}$ and $F_2 = \{(i,j) : 1 \leq i \leq n-1, 1 \leq j \leq n-i\} \cup \{(i, n+2-i) \mid 2 \leq i \leq n+1\}$ one can easily verify that the desired properties are fulfilled. □

**Lemma 13.** *For every $k, n \in \mathbb{N}$ with $0 < k < n$, there exist $F_1, F_2 \in \mathcal{F}^2$ with $|F_1| = |F_2| = \frac{1}{2}n(n-1) + k$ such that $F_1 \triangle F_2$ is a staircase with $2(n-1)$ points.*

*Proof.* We define (see Figure 1 (b))

$$S_1 = \{(i,j) : 1 \leq i \leq n-2, 1 \leq j \leq n-i-1\}, S_2 = \{(i, n-i) : 1 \leq i \leq k\},$$
$$B_1 = \{(i, n-i+1) : 1 \leq i \leq k\}, \quad B_2 = \{(i, n-i) : k+1 \leq i \leq n-1\},$$
$$W_1 = \{(i+1, n-i+1) : 1 \leq i \leq k\}, W_2 = \{(i+1, n-i) : k+1 \leq i \leq n-1\}.$$

Then, $F_1 = S_1 \cup S_2 \cup B_1 \cup B_2$ and $F_2 = S_1 \cup S_2 \cup W_1 \cup W_2$ are sets with $|F_1| = |F_2| = \frac{1}{2}n(n-1) + k$. It is easy to see that $F_1$ has a triangular shape. The points $F_1 \triangle F_2 = B_1 \cup B_2 \cup W_1 \cup W_2$ form a staircase with $2(n-1)$ points, namely $T = (p_1, q_2, p_2, q_2, \ldots, p_{n-2}, q_{n-2})$ with

$$p_i = \begin{cases} (n+1-i, i) \in W_2 & : & 1 \leq i \leq n-k-1 \\ (n-i+1, i+1) \in W_1 & : & n-k \leq i \leq n-1 \end{cases}$$

and

$$q_i = \begin{cases} (n-i, i) \in B_2 & : & 1 \leq i \leq n-k-1 \\ (n-i, i+1) \in B_1 & : & n-k \leq i \leq n-1. \end{cases}$$ □

The next lemma is used in the following for bounding the number of different consecutive row sums for a given set $F_1$.

**Lemma 14.** *For any $n+j$ integers $r_1 \geq \cdots \geq r_{n+j} \geq 1$ with $n \in \mathbb{N}$, $j \in \mathbb{N}_0$ and*

*(i) $j \geq 1$ and $\sum_{i=1}^{n+j} r_i = \frac{1}{2}n(n+1)$, or*
*(ii) $j = 0$ and $\sum_{i=1}^{n+j} r_i < \frac{1}{2}n(n+1)$*

*there are at most $n-1$ pairwise different $r_i$'s.*

*Proof.* Suppose there are more than $n - 1$ pairwise different $r_i$'s, which means that in $r_1 \geq \cdots \geq r_{n+j}$ we have at least $n - 1$ times a strict inequality. This implies $r_i \geq n - i + r_{n+j}$ for $1 \leq i \leq n - 1$, and $r_i \geq r_{n+j}$ for $n \leq i \leq n+j$. Summation yields the contradiction

$$\sum_{i=1}^{n+j} r_i \geq (n+j)r_{n+j} + n(n-1) - \sum_{i=1}^{n-1} i \geq n + j + n(n-1) - \frac{1}{2}n(n-1)$$

$$= \frac{1}{2}n(n+1) + j. \qquad \square$$

**Lemma 15.** *Let* $F_1 \in \mathcal{F}^2$ *with* $|F_1| = \frac{1}{2}n(n+1)$ *for an* $n \in \mathbb{N}$. *Then, we have* $\max_{F_2 \in \mathcal{F}^2(F_1)} |F_1 \triangle F_2| = 2n$.

*Proof.* By Lemma 12 we have $\max_{F_2 \in \mathcal{F}^2(F_1)} |F_1 \triangle F_2| \geq 2n$. If $F_1$ has $n + j$ non-empty rows, where $j \geq 1$, then we have by Lemma 14 (i) at most $n - 1$ different consecutive row sums. This leads only, by Proposition 11, to a staircase with at most $2(n-1)$ points. If $F_1$ has less than $n + 1$ non-empty rows, then we again conclude (by Proposition 10) that any staircase contains at most $2n$ points. $\square$

**Lemma 16.** *Let* $F_1 \in \mathcal{F}^2$ *with* $\frac{1}{2}n(n-1) < |F_1| < \frac{1}{2}n(n+1)$ *for an* $n \in \mathbb{N}$. *Then,* $\max_{F_2 \in \mathcal{F}^2(F_1)} |F_1 \triangle F_2| = 2(n-1)$.

*Proof.* Because of Lemma 13 we have $\max_{F_2 \in \mathcal{F}(F_1)} |F_1 \triangle F_2| \geq 2(n-1)$ for any $F_1$ with $\frac{1}{2}n(n-1) < |F_1| < \frac{1}{2}n(n+1)$. If $F_1$ has $n + j$ non-empty rows $(j \geq 0)$, then we have, by Lemma 14 (ii), at most $n - 1$ different row sums. Consequently, by Proposition 11, this leads to a staircases with at most $2(n-1)$ points. If $F_1$ has less than $n$ non-empty rows, then we again conclude (by Proposition 10) that any staircase contains at most $2(n-1)$ points. $\square$

Now, we can summarize the results in the following theorem.

**Theorem 17.** *For any two lattice directions and any two sets* $F_1, F_2 \in \mathcal{F}^2$ *with the three properties*

*(i)* $F_1$ *is uniquely determined by its X-rays along the two prescribed directions;*
*(ii)* $|F_1| = |F_2|$;
*(iii)* $||R_1 - R_2||_1 + ||C_1 - C_2||_1 = 2$,

*we have:*

$$|F_1 \triangle F_2| \leq \begin{cases} 2n & : \quad \text{if } |F_1| = \frac{1}{2}n(n+1) \text{ with } n \in \mathbb{N} \\ 2(n-1) & : \quad \text{if } \frac{1}{2}n(n-1) < |F_1| < \frac{1}{2}n(n+1) \text{ with } n \in \mathbb{N}. \end{cases}$$

*These bounds are sharp and imply*

$$|F_1 \triangle F_2| \leq 2\sqrt{2|F_1| + \frac{1}{4}} - 1.$$

*Proof.* The first bound results from Lemma 15 and Lemma 16. The sharpness of the bounds follows from the constructions given in Lemma 12 and Lemma 13. The second bound follows when $n$ is expressed in terms of $|F_1|$. $\square$

# 6    A Generalization to Lattice Sets with Invariant Points

In this section we study the stability problem under the weaker condition that the X-rays do not uniquely determine the set $F_1$, but we have some "invariant" points.

Let $\mathcal{U}(R_1, C_1)$ denote the class containing lattice sets having row and column sum vectors $R_1$ and $C_1$. The class is *normalized* if $R_1$ and $C_1$ are monotone. If $R_1$ and $C_1$ do not determine $F_1$, then more than one set belongs to $\mathcal{U}(R_1, C_1)$. In this context it is meaningful to study the case where $\mathcal{U}(R_1, C_1)$ has some *invariant points* (these are points belonging to every set in $\mathcal{U}(R_1, C_1)$, or to none of these sets).

It is well-known ([4]) that the normalized class $\mathcal{U}(R_1, C_1)$ has invariant points if and only if the lattice sets in $\mathcal{U}(R_1, C_1)$ are of the form illustrated in Figure 2 (a). To be more precise, let $\mathcal{R} = \{1, \dots, a\} \times \{1, \dots, b\}$ be the rectangle containing $F_1$; there exist pairwise disjoint subsets $K_1, \dots, K_h \subseteq \{1, \dots, a\}$ and pairwise disjoint subsets $L_1, \dots, L_h \subseteq \{1, \dots, b\}$ such that $\mathcal{R} \setminus \bigcup_{u=1}^{h} K_u \times L_u$ contains only invariant points. E.g., the black points in the Figure 2 (a) are invariant points belonging to every set in $\mathcal{U}(R_1, C_1)$ (also called 1-invariant points), while the smaller dots indicate invariant points belonging to no set of $\mathcal{U}(R_1, C_1)$ (also called 0-invariant points).

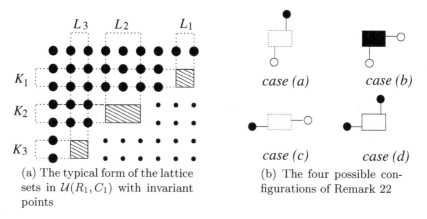

(a) The typical form of the lattice sets in $\mathcal{U}(R_1, C_1)$ with invariant points

(b) The four possible configurations of Remark 22

**Fig. 2.** The illustrations used in Section 6

*Remark 18.* We assume that $\mathcal{U}(R_1, C_1)$ is normalized, $|F_1| = |F_2|$, and the X-ray error is 2, so again we assume that the error occurs in two rows.

Let $p$ be any point of $F_2 \setminus F_1$ and let $q$ be any point of $F_1 \setminus F_2$. Clearly, statement (a) and (b) of Remark 2 hold when $p$ or $q$ is not in $\bigcup_{u=1}^{h} K_u \times L_u$.

*Remark 19.* Consider a so called *non-trivial component* $K_u \times L_u$, $u \in \{1, \dots, h\}$, and suppose that no error occurs on the lines intersecting this component.

- Let $q \in F_1 \setminus F_2$ with $row(q) \in K_u$ and $col(q) \notin L_u$ (implying that $q$ is an 1-invariant point). Then, there exists either one point $p' \in F_2 \setminus F_1$ such that $row(p') \in K_u$ and $col(p') \notin L_u$, or one point $q' \in F_1 \setminus F_2$ such that $col(q') \in L_u$ and $row(q') \notin K_u$. Otherwise we would obtain a contradiction to the assumption about the shape of $\mathcal{U}(R_1, C_1)$, and the assumption that there is no error in the rows and columns intersecting $K_u \times L_u$.
- Similarly, let $p \in F_2 \setminus F_1$ with $row(p) \in K_u$ and $col(p) \notin L_u$ (meaning that $p$ is a 0-invariant point). Then, there exists either one point $q' \in F_1 \setminus F_2$ such that $row(q') \in K_u$ and $col(q') \notin L_u$, or one point $p' \in F_2 \setminus F_1$ such that $col(p') \in L_u$ and $row(p') \notin K_u$.

Suppose now that exactly one error occurs on a line intersecting $K_u \times L_u$. From the assumptions about the shape of $\mathcal{U}(R_1, C_1)$ there follows:

- If the $-1$ error occurs, then there exists $q' \in F_1 \setminus F_2$ such that $col(q') \in L_u$ and $row(q') \notin K_u$.
- If the $+1$ error occurs, then there exists $p' \in F_2 \setminus F_1$ such that $row(p') \in K_u$ and $col(p') \notin L_u$.

From the previous remark there easily follows that, if not both rows $i$ (the row with $-1$ error) and $j$ (the row with $+1$ error) intersect the same non-trivial component, then we have $i > j$ like in Lemma 3.

**Definition 20.** *Let $A, B \in \mathcal{F}^2$ with $A \cap B = \emptyset$. An $(A, B)$-switching component (or a switching component for short) is a sequence of an even number $m > 0$ of distinct points $t_1, \ldots, t_m$ such that $t_{2i+1} \in A$, $t_{2i+2} \in B$, $col(t_{2i+1}) = col(t_{2i+2})$ for $0 \le i \le \frac{m}{2} - 1$, $row(t_{2i+2}) = row(t_{2i+3})$ for $0 \le i \le \frac{m}{2} - 2$, and $row(t_1) = row(t_m)$.*

**Definition 21.** *Let $\mathcal{R}$ be a rectangle of size $a \times b$ containing the disjoint sets $A, B \in \mathcal{F}^2$, and let $K_1, \ldots, K_h$ and $L_1, \ldots, L_h$ be pairwise disjoint subsets of $\{1, \ldots, a\}$ and $\{1, \ldots, b\}$, respectively. An almost-staircase $T = (t_1, \ldots, t_m)$ according to the columns in $\mathcal{R}$ is a sequence of an even number of $m > 0$ distinct points $t_{2i+1} \in A$, $t_{2i+2} \in B$, $0 \le i \le \frac{m}{2} - 1$ such that:*

(i) *For every $i \in \mathbb{N}_0$ with $0 \le i \le \frac{m}{2} - 1$ it holds that $col(t_{2i+1}) = col(t_{2i+2})$ and, if $t_{2i+1} \notin \bigcup_{u=1}^{h} K_u \times L_u$ or $t_{2i+2} \notin \bigcup_{u=1}^{h} K_u \times L_u$, then $row(t_{2i+1}) > row(t_{2i+2})$;*
(ii) *For every $i \in \mathbb{N}_0$ with $0 \le i \le \frac{m}{2} - 2$ it holds that $row(t_{2i+2}) = row(t_{2i+3})$ and, if $t_{2i+2} \notin \bigcup_{u=1}^{h} K_u \times L_u$ or $t_{2i+3} \notin \bigcup_{u=1}^{h} K_u \times L_u$, then $col(t_{2i+2}) < col(t_{2i+3})$;*
(iii) *For every $i \in \mathbb{N}$ with $1 \le i \le m$ we have that $t_i$ is no member of an $(A, B)$-switching component.*

An almost-staircase is a staircase "almost everywhere" except that the properties in (i) and (ii) of Definition 4 are relaxed for points in components $K_u \times L_u$.

*Remark 22.* If for a staircase $T = (t_1, \ldots, t_m)$ and a component $K_u \times L_u$ we have $\{t_1, \ldots, t_m\} \cap (K_u \times L_u) = \{t_i, \ldots, t_j\}$ with $i > 1$ and $j < m$ then, by definition of almost-staircases, the following cases can arise:

- If $t_i \in B$, then $t_{i-1} \in A$, and (a) $col(t_j) = col(t_{j+1})$ implies $t_{j+1} \in B$, whereas (b) $row(t_j) = row(t_{j+1})$ implies $t_{j+1} \in A$;
- If $t_i \in A$, then $t_{i-1} \in B$, and (c) $row(t_j) = row(t_{j+1})$ implies $t_{j+1} \in A$, whereas (d) $col(t_j) = col(t_{j+1})$ implies $t_{j+1} \in B$.

If $i = 1$, then either case (a) or (b) occurs, and if $j = m$, then either case (a) or (d) occurs. Figure 2 (b) illustrates the four configurations: the three kinds of rectangles represent $K_u \times L_u$.

The rectangle of configuration (b) in Figure 2 (b) is colored black because the number of black points inside is greater than the number of white points; rectangle of configuration (d) is colored white because the number of black points inside is smaller than the number of white points, and finally rectangles of configurations (a) and (c) have dotted edges because the number of white and black points inside is the same. Again, a *maximal almost-staircase* is an almost staircase which is no proper subset of another almost-staircase.

**Lemma 23.** *Any two maximal almost-staircases in $F_1 \triangle F_2$ have the same starting point and the same end point.*

*Proof.* We just stress the differences in the proof of Lemma 6 by using the same notations. The case that remains to be considered is the following: $row(t_1) = row(s_1) = i \in K_u \times L_u$, $col(t_1) > col(s_1)$, and a black point $q$ on row $i$ is in $K_u \times L_u$, and this point is not in a switching component (otherwise the error would be too large). Since there is no error on the columns a white point exists such that $col(p) = col(q)$ but we cannot claim that $row(p) > row(q)$. Anyway, points alternate each other such that this sequence visits a black point to the left of the rectangle or a white point to the bottom of the rectangle (leading to an infinite sequence), or it infinitely alternates within $K_u \times L_u$, or forms a switching component with $q$. In all cases, this is a contradiction to the assumptions.   □

**Lemma 24.** *Any two maximal almost-staircases in $F_1 \triangle F_2$ coincide.*

*Proof.* The proof follows like in Lemma 8. Indeed the case to analyze is that of $t_1 = s_1 \in K_u \times L_u$. One can easily show that if there is another white point in $col(t_1)$, then it belongs to the same $K_u \times L_u$, so proving that $t_2 = s_2$. Similarly, one deduces that $t_{m-1} = s_{n-1}$. This allows to apply the procedure used in the proof of Lemma 8.   □

Again, every $p \in F_1 \triangle F_2$ (outside of a switching component) is contained in an almost-staircase which is possibly constituted by two points (see Remark 19 and Remark 22). Because of the shape of $F_1$ and $F_2$ we also have that points in a switching component can only lie in a single component $K_u \times L_u$.

Let $\mathcal{P} := \{p \in F_1 \cup F_2 : p$ lies in an $(F_1, F_2) -$ switching component$\}$ and $T_{\max}$ denote the maximal almost-staircase. Then, we can summarize the results as follows.

**Theorem 25.** *Let $F_i \in \mathcal{F}^2$ be given, with row sum vector $R_i$ and column sum vector $C_i$, where $i = 1, 2$. Suppose the following properties are fulfilled*

*(i) $\mathcal{U}(R_1, C_1)$ has invariant points;*
*(ii) $|F_1| = |F_2|$ and*
*(iii) $\|R_1 - R_2\|_1 + \|C_1 - C_2\|_1 = 2$.*

*Then, $F_1 \triangle F_2 = T_{\max} \cup \mathcal{P}$.*

This theorem can be used to obtain a bound (similar to Section 5) which depends only on $\sum_{u=1}^{h} |K_u \times L_u|$, $R_1$, and on $C_1$.


# 7  Conclusion

In this paper we have answered a question which was left open in [3]. Under the assumption of a small X-ray error and that the original set is uniquely determined by its X-rays, we proved that the reconstruction of lattice sets from two directions is a stable task. This is in contrast to its unstable counterparts, i.e., to the case of reconstruction from more than two directions. Additionally, we have shown that similar arguments can be carried over to the more general case where the original set has invariant points.


# References

[1] A. Alpers, Instability and Stability in Discrete Tomography, *PhD thesis*, Technische Universität München, Shaker Verlag, ISBN 3-8322-2355-X, (2003).

[2] A. Alpers, P. Gritzmann: On stability, error correction and noise compensation in discrete tomography, *in preparation*

[3] A. Alpers, P. Gritzmann, L. Thorens: Stability and instability in discrete tomography, *LNCS 2243;* Digital and Image Geometry, G. Bertrand, A. Imiya, R. Klette (Eds.), (2002) 175-186.

[4] R.A. Brualdi: Matrices of zeros and ones with fixed row and column sum vectors, *Linear Algebra Appl.* 33, (1980) 159-231.

[5] S. Brunetti, A. Daurat: Stability in discrete tomography: Linear programming, additivity and convexity, *LNCS 2886;* Discrete Geometry for Computer Imagery, I. Nyström, G. Sanniti di Baja, S. Svensson (Eds.), (2003) 398-407.

[6] S. Brunetti, A. Daurat: Stability in discrete tomography: Some positive results, *to appear in Discrete Appl. Math.*

[7] R.M. Haber: Term rank of 0,1 matrices, *Rend. Sem. Mat. Univ. Padova* 30, (1960) 24-51.

[8] G.T. Herman, A. Kuba: Discrete tomography: Foundations, algorithms and applications, *Birkhäuser*, (1999)

[9] A. Kuba: Determination of the structure class $\mathcal{A}(\mathcal{R}, \mathcal{S})$ of $(0, 1)$-matrices, *Acta Cybernet.*, 9-2, (1989) 121-132.

[10] G.G. Lorentz: A problem of plane measure, *Amer. J. Math.* 71, (1949) 417-426.

[11] S. Matej, A. Vardi, G.T. Herman, E. Vardi: Binary tomography using Gibbs priors, *Discrete tomography: Foundations, algorithms and applications (chapter 8)*, (1999)

[12] H.J. Ryser: Combinatorial properties of matrices of zeros and ones, *Can. J. Mathematics* 9, (1957) 371-377.

[13] H.J. Ryser: The term rank of a matrix, *Canad. J. Math.* 10, (1958) 57-65.

[14] H.J. Ryser: Matrices of zeros and ones, *Bull. Amer. Math.* 66, (1960) 442-464.

[15] C. Valenti: An experimental study of the stability problem in discrete tomography, *Electron. Notes Discrete Math.* 12, (2003)

# Reconstruction of Decomposable Discrete Sets from Four Projections

Péter Balázs

Department of Informatics, University of Szeged,
Árpád tér 2, H-6720 Szeged, Hungary
pbalazs@inf.u-szeged.hu

**Abstract.** In this paper we introduce the class of decomposable discrete sets and give a polynomial algorithm for reconstructing discrete sets of this class from four projections. It is also shown that the class of decomposable discrete sets is more general than the class $\mathcal{S}'_8$ of $hv$-convex 8- but not 4-connected discrete sets which was studied in [3]. As a consequence we also get that the reconstruction from four projections in $\mathcal{S}'_8$ can be solved in $O(mn)$ time.

**Keywords:** discrete tomography, reconstruction from projections, decomposable discrete set

## 1 Introduction

One of the most frequently studied problems in the area of discrete tomography [14, 15] is the reconstruction of 2-dimensional discrete sets from few (usually up to four) projections. Several theoretical questions are connected with reconstruction such as existence and uniqueness (as a summary see [6, 9, 12]). There are also reconstruction algorithms for different classes of discrete sets (e.g., [4, 5, 7, 8, 11, 16, 17, 19]). However, the reconstruction problem is usually underdetermined and the number of solutions can be very large. Moreover, the reconstruction in certain classes can be NP-hard (see [21]). In order to keep the reconstruction process tractable and to reduce the number of solutions a commonly used technique is to suppose having some a priori information of the set to be reconstructed. The most frequently used properties are connectedness, directedness and some kind of discrete versions of the convexity. In this paper we introduce a new property of discrete sets, namely the decomposability, and study uniqueness and reconstruction problems in the class of discrete sets having this property.

This article is structured as follows. First, the necessary definitions are introduced in Section 2. In Section 3 we define the class of decomposable discrete sets and give a polynomial algorithm for reconstructing sets belonging to this class using four projections. In Section 4 we show that every 8- but not 4-connected set is decomposable and applying the results of Section 3 we get an $O(mn)$ algorithm for the reconstruction problem in this class using four projections. Finally, in Section 5 we conclude our results.

E. Andres et al. (Eds.): DGCI 2005, LNCS 3429, pp. 104–114, 2005.

## 2    Definitions and Notation

Let $\hat{F} = (\hat{f}_{ij})_{m \times n}$ be a binary matrix where $m, n \geq 1$. Let $F$ denote the set of positions $(i, j)$ where $\hat{f}_{ij} = 1$, i.e., $F = \{(i,j) | \hat{f}_{ij} = 1\}$. $F$ is called a *discrete set*, its elements are called *points* or *positions*. The *k-th diagonal/antidiagonal* $(k = 1, \ldots, m + n - 1)$ of $\hat{F}$ are defined by the set $D_k / A_k$, respectively, where

$$D_k = \{(i,j) \in \{1, \ldots, m\} \times \{1, \ldots, n\} \mid i + (n - j) = k\} \ , \tag{1}$$

$$A_k = \{(i,j) \in \{1, \ldots, m\} \times \{1, \ldots, n\} \mid i + j = k + 1\} \ . \tag{2}$$

Let $\mathcal{F}$ denote the class of discrete sets. For any discrete set $F \in \mathcal{F}$ we define the functions $\mathcal{H}$, $\mathcal{V}$, $\mathcal{D}$, and $\mathcal{A}$ as follows.
$\mathcal{H} : \mathcal{F} \longrightarrow \mathbb{N}_0^m$, $\mathcal{H}(F) = H = (h_1, \ldots, h_m)$, where

$$h_i = \sum_{j=1}^{n} \hat{f}_{ij}, \qquad i = 1, \ldots, m \ , \tag{3}$$

$\mathcal{V} : \mathcal{F} \longrightarrow \mathbb{N}_0^n$, $\mathcal{V}(F) = V = (v_1, \ldots, v_n)$, where

$$v_j = \sum_{i=1}^{m} \hat{f}_{ij}, \qquad j = 1, \ldots, n \ , \tag{4}$$

$\mathcal{D} : \mathcal{F} \longrightarrow \mathbb{N}_0^{m+n-1}$, $\mathcal{D}(F) = D = (d_1, \ldots, d_{m+n-1})$, where

$$d_k = \sum_{(i,j) \in D_k} \hat{f}_{ij} = |F \cap D_k|, \qquad k = 1, \ldots, m + n - 1 \ , \tag{5}$$

$\mathcal{A} : \mathcal{F} \longrightarrow \mathbb{N}_0^{m+n-1}$, $\mathcal{A}(F) = A = (a_1, \ldots, a_{m+n-1})$, where

$$a_k = \sum_{(i,j) \in A_k} \hat{f}_{ij} = |F \cap A_k|, \qquad k = 1, \ldots, m + n - 1 \ . \tag{6}$$

The vectors $H, V, D$ and $A$ are called the *row, column, diagonal* and *antidiagonal* projections of $F$, respectively (see Fig. 1). In the following we suppose that $h_i > 0$ and $v_j > 0$ for all $i \in \{1, \ldots, m\}$ and $j \in \{1, \ldots, n\}$. The *cumulated horizontal/vertical/diagonal/antidiagonal* vectors are denoted by $\tilde{H} = (\tilde{h}_1, \ldots, \tilde{h}_m)$, $\tilde{V} = (\tilde{v}_1, \ldots, \tilde{v}_n)$, $\tilde{D} = (\tilde{d}_1, \ldots, \tilde{d}_{m+n-1})$, and $\tilde{A} = (\tilde{a}_1, \ldots, \tilde{a}_{m+n-1})$, respectively, and defined by the following formulas (see Fig. 1)

$$\tilde{h}_i = \sum_{l=1}^{i} h_l, \qquad i = 1, \ldots, m \ , \tag{7}$$

$$\tilde{v}_j = \sum_{l=1}^{j} v_l, \qquad j = 1, \ldots, n \ , \tag{8}$$

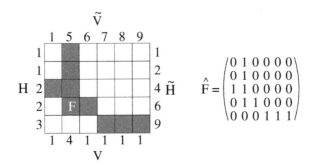

**Fig. 1.** An $hv$-convex 8- but not 4-connected discrete set $F$ and the corresponding binary matrix $\hat{F}$. The elements of $F$ are marked with grey squares. The projections of $F$ are the vectors $H$, $V$, $D = (0,0,0,0,2,2,3,2,0,0)$, and $A = (0,1,2,1,1,1,0,1,1,1)$. The cumulated vectors are $\tilde{H}$, $\tilde{V}$, $\tilde{D} = (0,0,0,0,2,4,7,9,9,9)$, and $\tilde{A} = (0,1,3,4,5,6,6,7,8,9)$

$$\tilde{d}_k = \sum_{l=1}^{k} d_l, \qquad \tilde{a}_k = \sum_{l=1}^{k} a_l, \qquad k = 1,\ldots,m+n-1 . \qquad (9)$$

Given a class $\mathcal{G} \subseteq \mathcal{F}$ of discrete sets, we say that the discrete set $F \in \mathcal{G}$ is *unique in the class* $\mathcal{G}$ (with respect to some projections) if there is no different discrete set $F' \in \mathcal{G}$ with the same projections.

Two points $P = (p_1, p_2)$ and $Q = (q_1, q_2)$ in a discrete set $F$ are said to be *4-adjacent* if $|p_1 - q_1| + |p_2 - q_2| = 1$. The points $P$ and $Q$ are said to be *8-adjacent* if they are 4-adjacent or $|p_1 - q_1| = 1$ and $|p_2 - q_2| = 1$. The sequence of distinct points $P_0, \ldots, P_k$ is a *4/8-path* from point $P_0$ to point $P_k$ in a discrete set $F$ if each point of the sequence is in $F$ and $P_l$ is 4/8-adjacent, respectively, to $P_{l-1}$ for each $l = 1, \ldots, k$. Two points are 4/8-connected in the discrete set $F$ if there is a 4/8-path, respectively, in $F$ between them. A discrete set $F$ is *4/8-connected* if any two points in $F$ are 4/8-connected, respectively, in $F$. The 4-connected set is also called as polyomino. The discrete set $F$ is *horizontally convex/vertically convex* (or shortly, *h-convex/v-convex*) if its rows/columns are 4-connected, respectively. The $h$- and $v$-convex sets are called $hv$-convex (see Fig. 1). In this paper we are going to study the reconstruction problem from four projections in several classes. Given a class $\mathcal{G} \subseteq \mathcal{F}$ the problem can be formulated as follows

4-RECONSTRUCTION($\mathcal{G}$).

**Instance:**    Four non-negative vectors $H \in \mathbb{N}^m$, $V \in \mathbb{N}^n$, $D \in \mathbb{N}_0^{m+n-1}$ and $A \in \mathbb{N}_0^{m+n-1}$.

**Task:**    Construct a discrete set $F \in \mathcal{G}$ with $\mathcal{H}(F) = H$, $\mathcal{V}(F) = V$, $\mathcal{D}(F) = D$ and $\mathcal{A}(F) = A$.

# 3    Reconstruction of Decomposable Discrete Sets

Let $F$ be a discrete set. A maximal 4-connected subset of $F$ is called a *component* of $F$ (e.g., in Fig. 1 there are two components: $\{(5,4),(5,5),(5,6)\}$ and $\{(1,2),(2,2),(3,1),(3,2),(4,2),(4,3)\}$). Clearly, the components of $F$ give a (uniquely determined) partition of $F$. We will use the concept of *smallest containing discrete rectangle of $F$* (SCDR) which corresponds to the notion of strong convex hull of $F$ (see [20]). Throughout this paper we always study discrete sets having the following properties

($\alpha$) the components are uniquely reconstructible from their horizontal and vertical projections in polynomial time, and

($\beta$) the sets of the row/column indices of the components are disjoint, i.e., if $I \times J \subseteq \{1,\ldots,m\} \times \{1,\ldots,n\}$ is the SCDR of a component of the discrete set $F$, then $\bar{I} \times J \cap F = I \times \bar{J} \cap F = \emptyset$ (where $\bar{I} = \{1,\ldots,m\} \setminus I$ and $\bar{J} = \{1,\ldots,n\} \setminus J$).

In fact, to satisfy property ($\alpha$) we need to have some a priori information about the components. For example, NW-directed $hv$-convex discrete sets can be used as components since in this class property ($\alpha$) is fulfilled [10].

The *North West-gluing* (or shortly, NW-gluing) is an operator defined by

$$\mathcal{F}^2 \longrightarrow \mathcal{F} \; : \; C \times D \to F, \text{ where } \hat{F} = \begin{pmatrix} \hat{C} & \mathbf{0} \\ \mathbf{0} & \hat{D} \end{pmatrix} . \tag{10}$$

If $C$ is a single component then we say that $C$ is the *NW-component of $F$*. NE-, SE-, SW-gluings and -components are defined similarly. We say that a discrete set $F$ consisting of $k$ ($k \geq 2$) components is *decomposable* if

(i) $F$ satisfies properties ($\alpha$) and ($\beta$), and

(ii) if $k > 2$ then we get $F$ by gluing a single component to a decomposable discrete set consisting of $k - 1$ components using one of the four gluing operators.

As a straight consequence of the definition we get that every discrete set consisting of three components and satisfying properties ($\alpha$) and ($\beta$) is decomposable. Figure 2 shows some decomposable and undecomposable configurations if the set consists of four components. The class of decomposable discrete sets is denoted by $\mathcal{DEC}$. The following lemma shows an important property of the decomposable discrete sets.

**Lemma 1.** *Let $F \in \mathcal{DEC}$ having more than two components, $C$ be a component of $F$ with the SCDR $I \times J$. Let $F'$ be the discrete set that we get by deleting rows $I \times \{1,\ldots,n\}$ and columns $\{1,\ldots,m\} \times J$ from $F$. Then $F' \in \mathcal{DEC}$.*

*Proof.* See [1].

On the basis of properties ($\alpha$) and ($\beta$) in the reconstruction of a decomposable discrete set it is sufficient to identify the SCDRs of the components. In order to do this we first give a necessary condition.

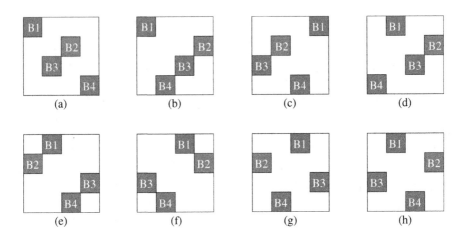

**Fig. 2.** Some decomposable (first row) and undecomposable (second row) configurations of the components. The SCDRs of the components are denoted by B1, B2, B3, and B4

**Theorem 1.** *Let $F \in \mathcal{DEC}$. If $(i,j)$ is the bottom right position of the SCDR of the NW-component of $F$ then $i$ is the smallest integer for which there exists an integer $j$ such that $\widetilde{h}_i = \widetilde{v}_j = \widetilde{a}_{i+j-1}$ and $a_{i+j} = 0$.*

*Proof.* Define a set $E$ as follows

$$E = (\{1, \ldots, i\} \times \{j+1, \ldots, n\}) \cup (\{i+1, \ldots, m\} \times \{1, \ldots, j\}) \ . \qquad (11)$$

If $(i,j)$ is the bottom right position of the SCDR of the NW-component then $F \cap E = \emptyset$ (see Fig. 3), and so

$$\widetilde{h}_i = \sum_{t=1}^{i} h_t = |F \cap \{1, \ldots, i\} \times \{1, \ldots, n\}| = |F \cap \{1, \ldots, i\} \times \{1, \ldots, j\}|$$

$$= |F \cap \{1, \ldots, m\} \times \{1, \ldots, j\}| = \sum_{t=1}^{j} v_t = \widetilde{v}_j \ . \qquad (12)$$

Furthermore, $(F \cap A_k) \cap E \subseteq F \cap E = \emptyset$ for every $k = 1, \ldots, m+n-1$ (see again Fig. 3). Then, recalling that $a_k = |F \cap A_k|$ for $k = 1, \ldots, n+m-1$ we get that

$$\widetilde{a}_{i+j-1} = \sum_{k=1}^{i+j-1} |F \cap A_k| = |F \cap \{1, \ldots, i\} \times \{1, \ldots, j\}| = \widetilde{h}_i = \widetilde{v}_j \ . \qquad (13)$$

Moreover, $A_{i+j} \subseteq E$ (see Fig. 3). Then, $F \cap A_{i+j} \subseteq F \cap E = \emptyset$ and we get that

$$a_{i+j} = |F \cap A_{i+j}| \leq |F \cap E| = 0 \ . \qquad (14)$$

Finally, assume that an integer $i' < i$ exists for which an integer $j'$ exists such that $\widetilde{h}_{i'} = \widetilde{v}_{j'} = \widetilde{a}_{i'+j'-1}$ and $a_{i'+j'} = 0$. Clearly, in this case $j' < j$. Since

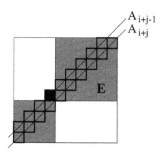

**Fig. 3.** The relations between the sets $A_{i+j-1}$, $A_{i+j}$ and $E$. The position $(i,j)$ is marked with black square. The antidiagonals $A_{i+j-1}$ and $A_{i+j}$ are marked with bold squares. The set $E$ is drawn with grey squares.

$(i,j)$ is the bottom right position of the SCDR of the NW-component and every component is a polyomino we get that the 1st, $\dots$, $(i+j)$-th coordinates of the antidiagonal projection have to be of the form $(0,\dots,0,a_{k_1},\dots,a_{k_2},0,\dots,0)$, where $1 \leq k_1 \leq k_2 < i+j$ and $a_l \neq 0$ for every $k_1 \leq l \leq k_2$. But then $a_{i'+j'} = 0$ only if $i'+j' < k_1$ or $i'+j' > k_2$. If $i'+j' < k_1$ then $\tilde{a}_{i'+j'-1} = 0$. Since the cumulated horizontal sums of a discrete set having nonzero rows are always satisfy the relation

$$0 < \tilde{h}_1 < \tilde{h}_2 < \dots < \tilde{h}_m \qquad (15)$$

we get that $\tilde{a}_{i'+j'-1} < \tilde{h}_{i'}$ which is a contradiction. Otherwise, i.e., if $i'+j' > k_2$ then $\tilde{h}_i > \tilde{h}_{i'}$ (since $i > i'$ and (15) holds) and we get that $\tilde{a}_{i'+j'-1} = \tilde{a}_{i+j-1} = \tilde{h}_i > \tilde{h}_{i'}$ which is, again, a contradiction. $\square$

Similar theorems can be given for NE-, SE-, and SW-components. Before giving a sufficient condition for finding the SCDR of a component of $F$ we introduce some further concepts. Let $F, F' \in \mathcal{F}$ such that $F' \setminus F = \{(p_1, q_1)\}$ and $F \setminus F' = \{(p_2, q_2)\}$. If $(p_1+k, q_1+k) = (p_2, q_2)$ for a $k \in \mathbb{Z} \setminus \{0\}$ then we say that we get $F'$ by applying a *slide* on $F$. Similarly, if $(p_1+k, q_1-k) = (p_2, q_2)$ for a $k \in \mathbb{Z} \setminus \{0\}$ then we say that we get $F'$ by applying an *antislide* on $F$. Clearly, applying slides/antislides on a discrete set, the diagonal/antidiagonal projection does not change, respectively. The following lemma shows an important relation between polyominoes and decomposable discrete sets having the same horizontal, vertical, and antidiagonal projections.

**Lemma 2.** *Let $P$ be a polyomino and $F \in \mathcal{DEC}$ such that $\mathcal{H}(P) = \mathcal{H}(F)$, $\mathcal{V}(P) = \mathcal{V}(F)$, and $\mathcal{A}(P) = \mathcal{A}(F)$. Then, the SCDRs of the components of $F$ are connected to each other with their bottom left and upper right corners.*

*Proof.* See [1] for the proof based on the results of [13].

An analogue of Lemma 2 replacing the antidiagonal projection with the diagonal one can also be proven. In this case the components of the decomposable

discrete set must be connected to each other with their bottom right and upper left corners. With the aid of the following theorem it is possible to test
whether the decomposable discrete set has a NW-component and if so then the
component itself can also be reconstructed.

**Theorem 2.** *Let $F \in \mathcal{DEC}$, $\mathcal{H}(F) = (h_1, \ldots, h_m)$, $\mathcal{V}(F) = (v_1, \ldots, v_n)$, and
$\mathcal{A}(F) = (a_1, \ldots, a_{m+n-1})$. Moreover, let $(i, j)$ be a position satisfying the necessary conditions of Theorem 1. If a polyomino $P$ exists according to the a priori
information which guarantees that property $(\alpha)$ is satisfied such that $\mathcal{H}(P) =
(h_1, \ldots, h_i)$, $\mathcal{V}(P) = (v_1, \ldots, v_j)$, and $\mathcal{A}(P) = (a_1, \ldots, a_{i+j-1})$ and there is no
$P' \in \mathcal{DEC}$ with $\mathcal{H}(P') = \mathcal{H}(P)$, $\mathcal{V}(P') = \mathcal{V}(P)$, and $\mathcal{A}(P') = \mathcal{A}(P)$ such that
the SCDRs of $P'$ are connected to each other with their bottom left and upper
right corners then $P$ is the NW-component of $F$. If no such polyomino exists
then $F$ does not have a NW-component.*

*Proof.* Define a set by $T = \bigcup_{k=1}^{i+j-1} A_k$ and let $Q' = F \cap T$. Since $\mathcal{A}(P) =
(a_1, \ldots, a_{i+j-1})$ we get $Q'$ by applying some (possibly none) antislides on $P$.
Let $Q$ be an arbitrary discrete set with the projections $\mathcal{H}(Q) = (h_1^q, \ldots, h_m^q)$,
$\mathcal{V}(Q) = (v_1^q, \ldots, v_n^q)$, and $\mathcal{A}(Q) = (a_1^q, \ldots, a_{m+n-1}^q)$ that we get by applying
some antislides on $P$. This time we allow that some of the coordinates of $\mathcal{H}(Q)$
and $\mathcal{V}(Q)$ are zero. Clearly, $a_l^q = a_l$ for each $l = 1, \ldots, i + j - 1$ and $Q \subseteq T$.
Moreover, for the horizontal and vertical projections of $Q$ exactly one of the
following cases holds

(i) $\exists\, i' \leq i$ such that $h_{i'}^q \neq h_{i'}$ or $\exists\, j' \leq j$ such that $v_{j'}^q \neq v_{j'}$,
(ii) $h_{i'}^q = h_{i'}$ for each $i' = 1, \ldots, i$ and $v_{j'}^q = v_{j'}$ for each $j' = 1, \ldots, j$.

Assume that Case (i) is true with $h_{i'}^q \neq h_{i'}$ for an $i' \leq i$. Then, there also exists
an $i'' \leq i$ such that $h_{i''}^q > h_{i''}$ or a $j'' \leq j$ such that $v_{j''}^q > v_{j''}$. Clearly, in
this case there is no discrete set $F'$ with the projections $\mathcal{H}(F') = (h_1, \ldots, h_m)$
and $\mathcal{V}(F') = (v_1, \ldots, v_n)$ such that $F' \cap T = Q$. Assuming that Case (i) is
true with $v_{j'}^q \neq v_{j'}$ for a $j' \leq j$ we get the same in a similar way. Therefore
$F$ can have the given projections if and only if for $Q'$ Case (ii) is true which
is possible only if $Q' \subseteq \{1, \ldots, i\} \times \{1, \ldots, j\}$. Since $\mathcal{H}(Q') = (h_1, \ldots, h_i)$ and
$\mathcal{V}(Q') = (v_1, \ldots, v_n)$ it follows that $F$ can have the prescribed projections if and
only if $F \cap E = \emptyset$ where $E$ is defined by (11). Then, $\{1, \ldots, i\} \times \{1, \ldots, j\}$ is the
SCDR of a set $G \subseteq F$ consisting of one or more components of $F$. However, if $G$
consists of several components of $F$ then $G \in \mathcal{DEC}$ based on Lemma 1 and the
SDCRs of the components of $G$ are connected to each other with their bottom
left and upper right corners on the basis of Lemma 2 which is a contradiction.
Consequently, $G$ is a simple polyomino. Since $P$ satisfies the conditions which
guarantees that property $(\alpha)$ holds $G = P$ and so the first part of the theorem is
proven. The second part of the theorem follows from the fact that the position
that satisfies the necessary conditions of Theorem 1 is uniquely determined.    □

Similar theorems can be given for testing the existence of NE-, SE-, and
SW-components. We now can outline an algorithm for reconstructing decomposable discrete sets with given horizontal, vertical, diagonal, and antidiagonal
projections.

**Algorithm 4-DEC**

**Input:** the vectors $H \in \mathbb{N}^m$, $V \in \mathbb{N}^n$, $D \in \mathbb{N}_0^{m+n-1}$, and $A \in \mathbb{N}_0^{m+n-1}$.
**Output:** the uniquely determined decomposable discrete set with projections $H$, $V$, $D$, and $A$ or FAIL (if no such set exists).
1: **repeat**

try to reconstruct a NW-component by identifying its SCDR and using the corresponding elements of the vectors $H$, $V$, and $A$;
**if** not succeed **then** try to reconstruct a NE-component by identifying its SCDR and using the corresponding elements of the vectors $H$, $V$, and $D$;
**if** not succeed **then** try to reconstruct a SE-component by identifying its SCDR and using the corresponding elements of the vectors $H$, $V$, and $A$;
**if** not succeed **then** try to reconstruct a SW-component by identifying its SCDR and using the corresponding elements of the vectors $H$, $V$, and $D$;
**if** not succeed **then break**;
modify $H$, $V$, $D$, and $A$ according to the reconstructed component and in the following omit the reconstructed part of the discrete set;
**until** all the components are reconstructed;
2: **if** the diagonal/antidiagonal projection is not equal to the given vector $D/A$, respectively **then**

{ let $P_1, \ldots, P_l$ denote the polyominoes reconstructed in Step 1;
$i = 0$;
**repeat**

assuming that $P_1, \ldots, P_i$ are components try to decompose further components in order SW, SE, NE, NW similarly as in Step 1;
$i = i + 1$;
**until** all the components are reconstructed **or** $i = l$; }
3: **if** the diagonal/antidiagonal projection is not equal to the given vector $D/A$, respectively **then** FAIL (no solution);

Turning to the analysis of the algorithm we can say the following

**Theorem 3.** *Algorithm 4-DEC solves the problem* 4-RECONSTRUCTION($\mathcal{DEC}$). *Assume that the reconstructed set consists of components $F_1, \ldots, F_k$ and let $C_i$ denote the time complexity of reconstructing the component $F_i$ ($i = 1, \ldots, k$). Then, the worst case time complexity of the algorithm is of $k \cdot \max_{1 \leq i \leq k} C_i$ which is polynomial. The solution is uniquely determined.*

*Proof.* As a straight consequence of the algorithm we get that the reconstructed set is decomposable and has the given projections. Assuming that the $l$-th $(l = 1, \ldots, k)$ component to be reconstructed is a NW-component it takes $O(m + n)$ time to find the (uniquely determined) position which satisfies the necessary conditions of Theorem 1. We do it simply by scanning the vectors $\widetilde{H}$ and $\widetilde{V}$. In order to test whether this position is the bottom right position of the SCDR of the NW-component we try to reconstruct this component based on Theorem 2 which takes $C_l$ time. The same is true if the $l$-th component is a NE-, SE- or SW-component. In the worst case the component is a SW-component, i.e., we try to reconstruct the $l$-th component at most four times and so the reconstruction complexity of Step 1 is $\max_{1 \le i \le k} C_i$ which is polynomial because of property $(\alpha)$. Theorem 2 guarantees the existence of a NW-component only when there is no decomposable discrete set with the same horizontal, vertical, and antidiagonal projections such that the components are connected to each other with their bottom left and upper right corners. Therefore it can occur that we accept the reconstructed polyomino as a NW-component although the decomposable discrete set to be reconstrcuted has no NW-component at all. The same is true for NE-, SE-, and SW-components, too. These situations result that the algorithm cannot reconstruct the decomposable discrete set with the given projections in Step 1. However, it reconstructs some polyominoes $P_1, \ldots, P_l$ such that there exists an $l' \le l$ for which $P_1, \ldots, P_{l'-1}$ are components and $P_{l'}$ is not a component of the decomposable discrete set. Then, the components $F_1, \ldots, F_{l'-1}$ are already reconstructed and the remaining components of the discrete set can be reconstructed in reversed order in Step 2. If $l'$ is known then all the remaining components can be reconstructed in Step 1. Since $l'$ is not known in Step 2 we have to call Step 1 at most $k$ times and so the reconstruction complexity of Step 2 is of $k \cdot \max_{1 \le i \le k} C_i$ in the worst case. The uniqueness of the solution follows from property $(\alpha)$.    □

## 4    Reconstruction of $hv$-Convex 8- but Not 4-Connected Discrete Sets from Four Projections

The class of $hv$-convex 8- but not 4-connected discrete sets (denoted by $\mathcal{S}_8'$) was introduced in [2]. In the same paper the authors gave a reconstruction algorithm in this class using the horizontal and vertical projections. This algorithm has worst case time complexity of $O(mn \cdot \min\{m, n\})$ and the solution is not always uniquely determined. Then, in [3] it is shown that using also the diagonal and antidiagonal projections the algorithm can be speeded up having complexity of $O(mn)$ and in this case uniqueness also holds. In the following we show that this is a consequence of

**Theorem 4.** $\mathcal{S}_8' \subseteq \mathcal{DEC}$.

*Proof.* Let $F \in \mathcal{S}_8'$. Then, clearly, the number of components of $F$ is at least 2. Since $F$ is $hv$-convex the sets of the row/column indices of the components consist of consecutive integers and they are disjoint. Then, property $(\beta)$ is satisfied. Moreover, on the basis of Theorem 5 in [3] and Theorem 3 in [18] the components can be reconstructed uniquely from the horizontal and vertical projections in $O(mn)$ time, i.e., property $(\alpha)$ also holds. Finally, the configuration of the components can follow only two cases (see Theorem 2 in [3]). Namely, the SCDRs of the components are connected to each other with their bottom right and upper left or with their bottom left and upper right positions (see Fig. 4a and 4b, respectively). Clearly, both configurations are decomposable.    □

Then, applying Theorem 3 we get the following

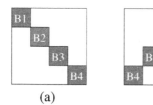

(a)          (b)

**Fig. 4.** The two possible configurations of the components in the class $\mathcal{S}_8'$

**Corollary 1.** *Algorithm 4-DEC solves the problem* 4-RECONSTRUCTION($\mathcal{S}_8'$) *in* $O(mn)$ *time. The reconstructed set is uniquely determined.*

## 5    Conclusions and Further Work

In this paper we have introduced a new class of discrete sets, the class of decomposable discrete sets and we have given a reconstruction algorithm in this class using four projections. It is shown that the algorithm has polynomial time complexity. Then, it is proven that the class of $hv$-convex 8- but not 4-connected sets is a subclass of $\mathcal{DEC}$. As a consequence we got that the problem 4-RECONSTRUCTION($\mathcal{S}_8'$) can be solved in $O(mn)$ time. Since the complexity of our algorithm strongly depends on the fact that the components are uniquely determined by the horizontal and vertical projections it seems to be important to find classes of discrete sets where the reconstruction problem can be solved uniquely.

It is shown that in some cases the discrete set can be decomposed along the diagonal and antidiagonal projections to facilitate the reconstruction. However, in some cases the decomposition into components is impossible. For example, the configuration in Fig. 2e can be decomposed into two parts (one containing $B1$ and $B2$, and the other containing $B3$ and $B4$) by the antidiagonal projection but then, the two parts cannot be further decomposed into components since the diagonal projections of the two parts are not independent. In some unfortunate cases the components cannot be separated at all (see, e.g., Fig. 2g and 2h). If the set is not decomposable then our algorithm simply FAILs without giving a solution. It is an interesting question whether it could be decided in advance if a discrete set is decomposable. Further investigation of the undecomposable configurations is also needed.

Throughout the paper it was assumed that every coordinate of the horizontal and vertical projections is nonzero, i.e., $h_i > 0$ and $v_j > 0$ for all $i \in \{1, \ldots, m\}$ and $j \in \{1, \ldots, n\}$. However, the results can be generalized easily to handle decomposable discrete sets where some of the coordinates of the horizontal or vertical projections are zero. In our work we concentrated on discrete sets consisting of components which satisfy some special properties (namely, properties ($\alpha$) and ($\beta$)). More work has to be done on the field whether assuming weaker properties about the components the reconstruction process remains tractable. Further work in this field can lead us towards designing efficient reconstruction algorithms for important classes like the one of $hv$-convex sets.

# References

1. P. Balázs, Reconstruction of decomposable discrete sets from four projections, *Technical Report at the University of Szeged* (2004)
   http://www.inf.u-szeged.hu/~pbalazs/research/research.html

2. P. Balázs, E. Balogh, A. Kuba, A fast algorithm for reconstructing $hv$-convex 8-connected but not 4-connected discrete sets, *Lecture Notes in Computer Science* **2886** (2003) 388-397.

3. P. Balázs, E. Balogh, A. Kuba, Reconstruction of 8-connected but not 4-connected $hv$-convex discrete sets, *Discrete Applied Mathematics*, accepted.

4. E. Balogh, A. Kuba, Cs. Dévényi, A. Del Lungo, Comparison of algorithms for reconstructing $hv$-convex discrete sets, *Lin. Alg. and Its Appl.* **339** (2001) 23–35.

5. E. Barcucci, A. Del Lungo, M. Nivat, R. Pinzani, Reconstructing convex polyominoes from horizontal and vertical projections, *Theor. Comput. Sci.* **155** (1996) 321–347.

6. R.A. Brualdi, Matrices of zeros and ones with fixed row and column sum vectors, *Lin. Algebra and Its Appl.* **33** (1980) 159–231.

7. S. Brunetti, A. Daurat, Reconstruction of discrete sets from two or more X-rays in any direction, *Proceedings of the seventh International Workshop on Combinatorial Image Analysis* (2000) 241–258.

8. M. Chrobak, Ch. Dürr, Reconstructing $hv$-convex polyominoes from orthogonal projections, *Information Processing Letters* **69(6)** (1999) 283–289.

9. A. Daurat, Convexité dans le plan discret. Application à la tomographie, *Thèse de doctorat de l'Université Paris 7* (2000)
   http://llaic3.u-clermont1.fr/~daurat/these.html

10. A. Del Lungo, Polyominoes defined by two vectors, *Theor. Comput. Sci.* **127** (1994) 187–198.

11. A. Del Lungo, M. Nivat, R. Pinzani, The number of convex polyominoes reconstructible from their orthogonal projections, *Discrete Math.* **157** (1996) 65–78.

12. R.J. Gardner, P. Gritzmann, Uniqueness and complexity in discrete tomography, In [15] (1999) 85–113.

13. L. Hajdu, R. Tijdeman, Algebraic aspects of discrete tomography, *Journal für die reine und angewandte Mathematik* **534** (2001) 119–128.

14. G.T. Herman, A. Kuba (Eds.), Discrete Tomography, Special Issue. *Int. J. Imaging Systems and Techn.* **9** (1998) No. 2/3.

15. G.T. Herman, A. Kuba (Eds.), *Discrete Tomography: Foundations, Algorithms and Applications* (Birkhäuser, Boston, 1999).

16. A. Kuba, The reconstruction of two-directionally connected binary patterns from their two orthogonal projections, *Comp. Vision, Graphics, and Image Proc.* **27** (1984) 249–265.

17. A. Kuba, Reconstruction in different classes of 2D discrete sets, *Lecture Notes on Computer Sciences* **1568** (1999) 153–163.

18. A. Kuba, E. Balogh, Reconstruction of convex 2D discrete sets in polynomial time, *Theor. Comput. Sci.* **283** (2002) 223–242.

19. H.J. Ryser, Combinatorial properties of matrices of zeros and ones, *Canad. J. Math.* **9** (1957) 371–377.

20. P. Soille, From binary to grey scale convex hulls, *Fundamenta Informaticae* **41** (2000) 131–146.

21. G.W. Woeginger, The reconstruction of polyominoes from their orthogonal projections, *Inform. Process. Lett.* **77** (2001) 225–229.

# A Tomographical Characterization of L-Convex Polyominoes

Giusi Castiglione[1], Andrea Frosini[2], Antonio Restivo[1], and Simone Rinaldi[2]

[1] Università di Palermo, Dipartimento di Matematica e Applicazioni,
via Archirafi, 34 - 90123 Palermo, Italy
{giusi, restivo}@math.unipa.it.
[2] Università di Siena, Dipartimento di Scienze Matematiche ed Informatiche,
Pian dei Mantellini, 44 - 53100 Siena, Italy
{frosini, rinaldi}@unisi.it.

**Abstract.** Our main purpose is to characterize the class of L-convex polyominoes introduced in [3] by means of their horizontal and vertical projections. The achieved results allow an answer to one of the most relevant questions in tomography i.e. the uniqueness of discrete sets, with respect to their horizontal and vertical projections. In this paper, by giving a characterization of L-convex polyominoes, we investigate the connection between uniqueness property and unimodality of vectors of horizontal and vertical projections. In the last section we consider the continuum environment; we extend the definition of L-convex set, and we obtain some results analogous to those for the discrete case.

## 1 Definitions and Preliminaries

Let our environment be the integer lattice $\mathbb{Z} \times \mathbb{Z}$. A *discrete set* is a finite subset $S$ of $\mathbb{Z} \times \mathbb{Z}$ considered up to translations.

Usually, a discrete set is represented by a binary matrix or by a set of *cells* (unitary squares), as depicted in Fig. 1. In the sequel, we will use the latter representation, and we number the rows and the columns of the set starting from the upper left corner of the minimum rectangle containing it. We denote by $(i, j)$ the cell in the $i$-th row and $j$-th column of the rectangle.

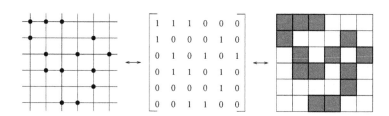

**Fig. 1.** A finite set of $\mathbb{Z}^2$, and its representation in terms of a binary matrix and a set of cells

E. Andres et al. (Eds.): DGCI 2005, LNCS 3429, pp. 115–125, 2005.
© Springer-Verlag Berlin Heidelberg 2005

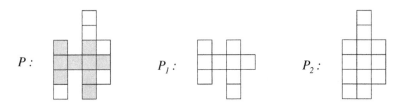

**Fig. 2.** The polyomino $P$ contains the polyomino $P_1$, but does not contain the polyomino $P_2$. The shaded cells of $P$ show the inclusion

In this paper we study a particular class of discrete sets, i.e. the well known class of *polyominoes* (cf.[4]). A polyomino is defined as a finite union of cells whose interior is connected. Given two polyominoes $P$ and $P'$, we say that $P$ is contained in $P'$ if $P \subseteq P'$ with respect to the standard set-inclusion (see Fig. 2).

A polyomino is said to be *h-convex* (resp. *v-convex*) if every its row (resp. column) is connected. A polyomino is said to be *hv-convex*, or simply *convex*, if it is both h-convex and v-convex (see Fig. 3).

For any two cells $A$ and $B$ in a polyomino, a *path* $\Pi_{AB}$, from $A$ to $B$, is a sequence $(i_1, j_1), (i_2, j_2), ..., (i_r, j_r)$ of adjacent disjoint cells, with $A = (i_1, j_1)$, and $B = (i_r, j_r)$. For each $1 \leq k < r$, we say that the two consecutive cells $(i_k, j_k), (i_{k+1}, j_{k+1})$ form:

– an *east* step if $i_{k+1} = i_k + 1$ and $j_{k+1} = j_k$;
– a *north* step if $i_{k+1} = h_i$ and $j_{k+1} = j_k + 1$;
– a *west* step if $i_{k+1} = h_i - 1$ and $j_{k+1} = j_k$;
– a *south* step if $i_{k+1} = h_i$ and $j_{k+1} = j_k - 1$.

Finally, we define a path to be *monotone* if it is entirely made of only two of the four types of steps defined above.

The cells in a convex polyomino satisfy a particular connection property that involves the shape of the paths connecting any pair of them.

**Proposition 1.** *A polyomino $P$ is convex iff every pair of cells is connected by a monotone path.*

The property in Proposition 1, allows us to introduce a particular family of convex polyominoes, called *L-convex* polyominoes, defined and studied in [3].

## 1.1    The Class of L-Convex Polyominoes

Let us consider a polyomino $P$. A path in $P$ has a *change of direction* in the cell $(i_k, j_k)$, for $2 \leq k \leq r - 1$, if

$$i_k \neq i_{k-1} \iff j_{k+1} \neq j_k.$$

In [3] it is proposed a classification of convex polyominoes based on the number of changes of direction in the paths connecting any two cells of a polyomino.

**Fig. 3.** The convex polyomino on the left is not L-convex, while the one on the right is L-convex. For both the polyominoes two cells are highlighted, and a monotone path which connects them and which contains the minimum number of possible changes of direction, is depicted

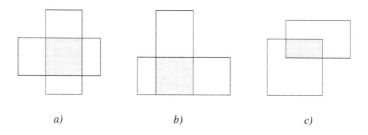

*a)*                          *b)*                          *c)*

**Fig. 4.** In *a)* and *b)* we have crossing intersections among the rectangles, while the intersection in *c)* is not crossing

More precisely, we call *k-convex* a convex polyomino such that every pair of its cells can be connected by a monotone path with at most *k* changes of direction. For $k = 1$ we have the class of *L-convex polyominoes*, i.e. those polyominoes such that every pair of their cells can be connected by a path with at most one change of direction (see Fig.3).

In the same paper it is given a nice characterization of L-convex polyominoes that involves the following notion of maximal rectangle.

A *rectangle*, that we denote by $[x, y]$, with $x, y \in \mathbb{N} \setminus \{0\}$, is a rectangular polyomino whose dimensions are $x$ and $y$ ($x$ rows and $y$ columns). We say $[x, y]$ to be *maximal* in $P$ if

$$\forall [x', y'], \ [x, y] \subseteq [x', y'] \subseteq P \ \Rightarrow \ [x, y] = [x', y'].$$

Two rectangles $[x, y]$ and $[x', y']$ contained in $P$ have a *crossing intersection* if their intersection is a rectangle having as basis the smallest of the two bases, and as height the smallest of the two heights, i.e.

$$[x, y] \cap [x', y'] = [min\{x, x'\}, min\{y, y'\}].$$

Figure 4 shows examples of crossing and non-crossing intersections.

**Theorem 1.** *A convex polyomino is L-convex iff every pair of its maximal rectangles has crossing intersection.*

**Fig. 5.** A L-convex polyomino $P$ obtained by one of the overlappings of the four comparable rectangles $a$, $b$, $c$ and $d$ having crossing intersection

From Theorem 1, it immediately follows that all the maximal rectangles of a L-convex polyomino are distinct. The same result allows to characterize a L-convex polyomino as one of the overlapping of its maximal rectangles.

Since the set of maximal rectangles can be partially ordered as follows:

$$[x_1, y_1] > [x_2, y_2] \text{ if } x_1 > x_2 \text{ and } y_1 < y_2 ,$$

then each finite overlapping of comparable rectangles such that any pair of them has a crossing intersection, determines a L-convex polyomino (see Fig.5 for an example).

## 1.2    Basic Notions of Discrete Tomography

To each discrete set $S$, we can associate two integer vectors $H = (h_1, \ldots, h_m)$ and $V = (v_1, \ldots, v_n)$ such that, for each $1 \leq i \leq m$, $1 \leq j \leq n$, $h_i$ and $v_j$ are the number of cells of $S$ which lie on row $i$ and column $j$, respectively. The vectors $H$ and $V$ are called the horizontal and vertical projections of $S$, respectively. Given two vectors $H$ and $V$, we will denote by $\mathcal{U}(H, V)$ the class of discrete sets having $H$ and $V$ as projections.

A discrete set $S$ is *unique* (with respect to H and V) if $\mathcal{U}(H, V) = \{S\}$. In such a case also $H$ and $V$ are said to be *unique*.

Fundamental problems of discrete tomography concern the retrieval of information about some geometrical aspects (cf. [1], [2], [7]) of discrete sets, from the knowledge of their projections (for a survey cf. [5]).

In general, the horizontal and vertical projections of a discrete set are not sufficient to uniquely determine it (see Fig. 7), as it is known from [9], where Ryser pointed out that a discrete set is unique if and only if it does not contain particular configurations of points called *switching components*. Figure 6 shows the two simplest of them, called *elementary switching components*, and defined as follows: a discrete set $S$ contains the elementary switching component $a$) [resp. $b$)] if there exists two rows $i$ and $i'$, and two columns $j$ and $j'$ such that the cells in positions $(i, j)$, and $(i', j')$ [resp. $(i', j)$, and $(i, j')$] belong to $S$ (represented in the figure by filled squares), while the cells in positions $(i', j)$, and $(i, j')$ [resp. $(i, j)$, and $(i', j')$] do not belong to $S$ (represented in the figure by dotted squares).

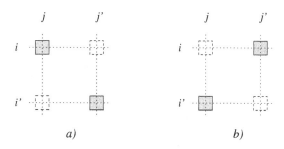

**Fig. 6.** The two elementary switching components. The presence of one of them in a discrete set assures its non-uniqueness

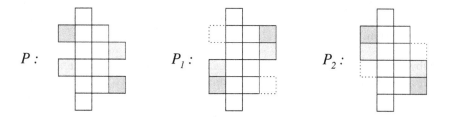

**Fig. 7.** Three polyominoes belonging to the class $\mathcal{U}(H, V)$, with $H = (1, 3, 3, 3, 3, 1)$ and $V = (2, 6, 4, 2)$

Furthermore, Ryser defined an operator, called *interchange* and successively *switching (operator)*, which modifies a discrete set by changing one of its switching component, if it exists, into the other.

In Fig. 7, the discrete sets $P_1$ and $P_2$ are obtained from $P$ by performing the two highlighted switchings.

Clearly, switching does not modify the projections of a discrete set, which consequently reveals to be non-unique (cf. [10]). The reverse of this property is also true, as stated in the following

**Theorem 2.** *(Ryser's Theorem) A discrete set is non-unique (with respect to its horizontal and vertical projections) if and only if it has a switching component.*

## 2 A Characterization Theorem for L-Convex Polyominoes

In this section we furnish a series of results that produce a characterization of L-convex polyominoes in terms of horizontal and vertical projections.

**Lemma 1.** *A L-convex polyomino $P$ is uniquely determined by its horizontal and vertical projections.*

*Proof.* Theorem 2 allows us to achieve the uniqueness of $P$ by proving that it does not contain any switching component. So, let us assume that there exists a switching component involving the two cells $A$ and $B$ of $P$, in positions $(i, j)$ and $(i', j')$, respectively, where $i \neq i'$ and $j \neq j'$. By definition of switching, the two cells in positions $(i, j')$ and $(i', j)$ do not belong to $P$, and consequently a monotone path $\Pi_{AB}$ having at most one change of direction does not exist. This fact contradicts the hypothesis of L-convexity of $P$.                                            □

**Lemma 2.** *Let $j$ and $j'$ be two different columns of a L-convex polyomino $P$, such that $v_j \leq v_{j'}$. For each row $i$ of $P$, if $(i, j) \in P$, then $(i, j') \in P$.*

*Proof.* Let us proceed by contradiction and assume that there exists a row $i'$ of $P$ such that $(i', j) \in P$ and $(i', j') \notin P$. Since $v_j \leq v_{j'}$, there exists a row $i''$ such that $(i'', j) \notin P$ and $(i'', j') \in P$. These four cells form a switching component, a contradiction by Lemma 1.                                            □

Obviously a result similar to that of Lemma 2 holds if $v_j > v_{j'}$, and, furthermore, if we replace the two different columns of $P$ with two of its rows.

We define an integer vector $X = (x_1, \dots, x_k)$ to be *unimodal*, if there exists $0 \leq i \leq k$, such that $x_1 \leq x_2 \leq \dots, \leq x_i$ and $x_i \geq x_{i+1} \geq \dots \geq x_k$.

**Lemma 3.** *If $P$ is a L-convex polyomino then its horizontal and vertical projections are unimodal.*

*Proof.* Let $P$ be a L-convex polyomino belonging to $\mathcal{U}(H, V)$, with $H \in \mathbb{N}^m$ and $V \in \mathbb{N}^n$. By Theorem 1, it follows that each element $h_i$ of $H$ is the basis of a maximal rectangle of $P$. Let us proceed by contradiction and assume $H$ to be non-unimodal, i.e. there exist $1 \leq i < j < k \leq m$ such that $h_j < h_k$ and $h_j < h_i$. The following three cases arise:

$h_i = h_k$: the cells of $P$ lying on row $i$ and row $k$ belong to the same maximal rectangle, so $h_j \geq h_i$, a contradiction;

$h_i < h_k$: the two values $h_i$ and $h_j$ are the bases of two different maximal rectangles. Since each pair of maximal rectangles has crossing intersection, then $h_j \geq h_i$, a contradiction;

$h_i > h_k$: analogous to the previous case.

Since each element $v_j$ of $V$, with $j = 1, \dots, m$, is the height of a maximal rectangle of $P$, a similar reasoning leads to prove that also $V$ is unimodal.    □

The properties stated in Lemmas 2 and 3 directly follow from the definition of L-convexity. A less intuitive result is the characterization of L-convex polyominoes by means of the uniqueness and monotonicity of its projections.

**Theorem 3.** *Let $P \in \mathcal{U}(H, V)$, with $H \in \mathbb{N}^m$ and $V \in \mathbb{N}^n$.*

$$\left.\begin{array}{l} H \text{ and } V \text{ are unimodal} \\ H \text{ and } V \text{ are unique} \end{array}\right\} \Leftrightarrow P \text{ is a L-convex polyomino.}$$

*Proof.* ($\Rightarrow$) We prove by contradiction the $h$-convexity of $P$: let us assume that there exist three cells $(i,j) \in P$, $(i,j') \notin P$ and $(i,j'') \in P$, with $i < i' < i''$. The unimodality of $V$ allows the following three cases:

$v_j \geq v_{j'} \geq v_{j''}$: Lemma 2 applied to columns $j''$ and $j'$, implies that $(i,j') \in P$, which is clearly a contradiction;

$v_j \leq v_{j'} \leq v_{j''}$: Lemma 2 applied to columns $j$ and $j'$, implies that $(i,j') \in P$, a contradiction;

$v_j \leq v_{j'}$ and $v_{j'} \geq v_{j''}$: Lemma 2 applied or to columns $j$ and $j'$, or to columns $j''$ and $j'$ implies that $(i,j') \in P$, again a contradiction.

A similar reasoning leads to the $v$-convexity of $P$.

Finally, for any pair of cells $(i,j)$ and $(i',j')$ belonging to $P$, the uniqueness of $P$ implies that $(i',j) \in P$ or $(i,j') \in P$, so the cells $(i,j)$ and $(i',j')$ can be connected by a path having at most one change of direction. This determines the connectedness and the L-convexity of $P$.

($\Leftarrow$) The result follows from Lemmas 1 and 3.                            $\square$

The following remark is a direct consequence of the proof of Theorem 3:

*Remark 1.* A convex discrete set is unique if and only if it is L-convex.

## 3   Extension to Measurable Plane Sets

In this section, we introduce the concept of L-convex plane set in order to extend to the continuum the uniqueness results stated in Section 2.

In the case of generic measurable plane sets, G.G.Lorentz gave in [8] necessary and sufficient conditions for a pair of projections to be respectively unique, non-unique and consistent. These results were obtained by using analytic transformations of the projection functions. Further studies considered the same problem from a geometrical point of view, with the aim of defining a switching theory which translates in the continuum what was introduced for discrete sets. In particular in [6], the authors introduced the notion of switching components in the continuum, and stated a result similar to Theorem 2. Furthermore, they furnished other nice characterizations of plane sets related to their geometrical properties. In this section we will often rely on these works in order to support our results.

So, let us start by recalling the following standard definitions: a set $S$ of $\mathbb{R}^2$ is called *h-convex* (resp. *v-convex*) if, for each pair of points $(x,y), (u,v) \in S$, with $y = v$ (resp. $x = u$), the horizontal (resp. vertical) line segment which join them is entirely contained in $S$. We call *hv-convex* the plane sets that are both $h$-convex and $v$-convex.

Furthermore, a *step polygon* is a polygonal curve consisting of horizontal and vertical line segments and having no self-intersections. A step polygon joining two distinct points $(x,y), (u,v) \in \mathbb{R}^2$ can be represented as a finite sequence of vertices $(x_0, y_0), (x_1, y_1), ..., (x_k, y_k)$ such that each vertex is connected by a line segment to the next one, $(x_0, y_0) = (x,y)$, and $(x_k, y_k) = (u,v)$. To our purpose, line segments are the continuum counterpart of the four kinds of steps defined

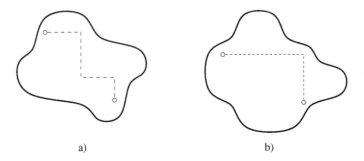

a)                                    b)

**Fig. 8.** a) a hv-convex plane set and a monotone step polygon lying inside it; b) a L-convex plane set, and two of its cells joined with a three vertex monotone step polygon

in Section 1 for the discrete lattice, and so they can be classified as north, south, east and west segments. A step polygon is called *monotone* if it is composed of at most two different kinds of these segments.

Hence, we have the following natural translation of Proposition 1 to *hv*-convex plane sets:

**Proposition 2.** *A plane set S is hv-convex iff every pair of points in S can be joined by a monotone step polygon lying in S.*

Now we can finally define a plane set $S$ to be *L-convex* if each pair of its points can be joined by a monotone step polygon with at most three vertices, and entirely contained in $S$ (see Fig.8).

### 3.1 A Characterization Theorem for L-Convex Plane Sets

In this paragraph, the reader will encounter some basic definitions together with the continuum counterparts of the results stated in Section 2

A function $f(x)$, defined in the interval $[a, b] \subset \mathbb{R}$, is *unimodal* if there exists $\bar{x} \in [a, b]$ (called *mode*) such that $f(x)$ increases from $a$ to $\bar{x}$ and decreases from $\bar{x}$ to $b$.

Let $S \subseteq \mathbb{R}^2$ be a measurable set such that $\lambda_2(S) < \infty$ ($\lambda_2$ being the two dimensional Lebesgue measure), and let $f(x, y)$ be its characteristic function. Using notations and definitions from [6], we call *horizontal projection* of $S$ the function

$$f_x(y) = \int_{-\infty}^{\infty} f(x, y) dx \tag{1}$$

and *vertical projection* of $S$ the function

$$f_y(x) = \int_{-\infty}^{\infty} f(x, y) dy. \tag{2}$$

These functions exist almost everywhere on $\mathbb{R}$ and they are integrable (Fubini's theorem).

In [6], it is introduced a notion of switching components in the continuum which naturally extends the one for discrete sets.

In particular, let $t, u$ be two real numbers. The sets

$$S(t, 0) = \{(x, y) | (x - t, y) \in S\}$$

$$S(0, u) = \{(x, y) | (x, y - u) \in S\}$$

are called horizontal and vertical translation of $S$, respectively.

We say that $S$ admits a *switching component* if there exist four sets $A, B, C, D$ and two real numbers $t$ and $u$ such that $B \cup C \subseteq S$, $A \cup D \cap S = \emptyset$, and such that $B = A_{(t,0)}$, $C = A_{(0,u)}$ and $D = A_{(t,u)}$.

We have that if $S$ admits a switching component, then $S$ is not uniquely determined by its projections, in fact the set

$$S' = (S - (B \cup C)) \cup (A \cup D)$$

is different from $S$, and it has its same horizontal and vertical projections.

The existence of a switching component is also a necessary condition to guarantee the non-uniqueness of the set $S$ (see [6] for a proof), so we have the following result, analogous to Lemma 1:

**Theorem 4.** *A measurable plane set having finite measure is non-uniquely determined by its projections iff it has a switching component.*

Finally, L-convexity of a plane set causes the existence of a mode both in its horizontal and in its vertical projections:

**Lemma 4.** *If a plane set is L-convex, then both its horizontal and its vertical projections are unimodal.*

The proof can be simple inferred from that of Lemma 3. As a consequence we can obtain, for the continuous case, the same characterization result as for discrete sets:

**Theorem 5.** *Let $f_x$ and $f_y$ be projection functions defined in $\mathbb{R}^2$ of a plane set $S$. It holds that*

$$\left.\begin{array}{l} f_x \text{ and } f_y \text{ are unimodal} \\ f_x \text{ and } f_y \text{ are unique} \end{array}\right\} \Leftrightarrow S \text{ is L-convex}.$$

A last remark is needed: in [6], a different and interesting characterization of unique plane sets is provided. Let $S$ be a measurable plane set of finite measure. The rectangle $X \times Y$ is *measurably inscribed* (briefly *m*-inscribed) in $S$ if

$$X \times Y \subseteq S \quad \text{and} \quad \overline{X} \times \overline{Y} \subseteq \overline{S}.$$

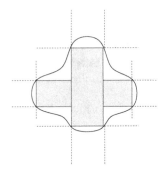

**Fig. 9.** Two $m$-inscribed rectangles inside a L-convex plane set

The set $S$ is *m-inscribable* if it is the union of $m$-inscribed rectangles. We can immediately argue that the presence of $m$-inscribed rectangles inside the set $S$ is similar to the presence of maximal rectangles inside an L-convex polyomino.

This idea is strengthened by the fact that, using Theorem 4, in [6] it is proved the following

**Theorem 6.** *A measurable plane set having finite measure is uniquely determined by its projection functions iff it is m-inscribable.*

We want to observe that the notion of crossing intersection in the continuum environment, leads to the equivalence between L-convex plane sets and $m$-inscribable plane sets. In fact, at the same time, we obtain a nice generalization of Theorem 1 and an uniqueness result.

**Theorem 7.** *Let $S$ be a measurable plane set. It holds that $S$ is L-convex iff $S$ is m-inscribable by rectangles with crossing intersection.*

## 4    Conclusions and Further Work

In this work we have proposed a characterization of L-convex sets in terms of features relevant to discrete tomography. In particular, we observed that each L-convex set is unique with respect to its horizontal and vertical projections, and that both the projections show a unimodal behavior. The characterization is achieved after showing that these two properties are also sufficient to obtain a L-convex set.

Finally, the last section of the paper concerns the natural extension of our main result to the continuum environment.

We would like to point out some open questions: one can ask wether similar tomographical characterizations can be proved when generalizing the notion of L-convexity by taking into consideration two or more directions different from (or possibly strictly including) the horizontal and the vertical ones.

Furthermore, we could consider the extension of the notion of L-convex polyomino to the three dimensional lattice. In fact, it seems not so trivial to keep maintaining the crucial equivalence between the characterizations of L-convexity by means of monotone path and of maximal rectangles in such environment.

# References

1. Barcucci, E., Del Lungo, A., Nivat, M., Pinzani, R.: Reconstructing convex polyominoes from horizontal and vertical projections. Theoret. Comput. Sci. **155** (1996) 321–347
2. Brualdi, R.A.: Matrices of zeros and ones with fixed row and column sum vectors. Lin. Algebra and Its Applications **33**, (1980) 159–231
3. Castiglione, G., Restivo, A.: Reconstruction of L-convex Polyominoes. Electron. Notes in Discrete Math. **12** Elsevier Science (2003)
4. Golomb, S.W.: Polyominoes. Scribner, New York (1965)
5. Herman, G.T., Kuba, A. (eds.): *Discrete Tomography: Foundations, Algorithms and Applications*, Birkhauser Boston, Cambridge, MA (1999)
6. Kuba, A., Volčič, A.: Characterization of measurable plane sets which are reconstructable from their two projections. Inverse Problems **4** (1988) 513–527
7. Kuba, A., Balogh, E.: Reconstruction of convex 2D discrete sets in polynomial time Theoret. Comput. Sci. **283** (2002) 223–242
8. Lorentz, G.G.: A problem of plane measure. Am. J. Math. **71** (1949) 417–426
9. Ryser, H.J.: Combinatorial properties of matrices of zeros and ones. Canad. J. Math. **9** (1957) 371–377
10. Shliferstein, A.R., Chien, Y.T.: Switching components and the ambiguity problem in the reconstruction of pictures from their projections. Pattern Recognition, **10** (1978) 327–340

# Computerized Tomography with Digital Lines and Linear Programming

Fabien Feschet and Yan Gérard

LLAIC - IUT Clermont-Ferrand,
Campus des Cézeaux,
63172 Aubière Cedex - France
{feschet, gerard}@llaic.u-clermont1.fr

**Abstract.** We present a new method of computerized tomography based on linear programming. The approach is based on three main ideas: covering the set of pixels by digital lines, introducing a variable of maximal error in the linear constraints and adding in the objective function an entropy term.

## 1  Introduction

Since the invention of the first X-rays scanner by G.N. Hounsfield and A. McCormack in the beginning of the seventies, computerized tomography has become an industrial, medical and mathematical stake. The principle of this technology is to reconstruct an image from X-rays. This purpose requires to solve a mathematical problem belonging to the family of inverse problems: compute the inverse of the radon transform. A solution is given by the formula of the Fourier slice theorem [1] but this result belongs to the framework of "continuous" mathematics whereas the technology of the measurements of the X-rays provides only discrete data. The computation of the image with the Fourier slice theorem requires to know the X-rays for all the angles going from 0 to 360 degrees, which is of course not possible. The camera can only provide X-rays according to a finite panel of directions. This reason and the fact that the captors of the camera provide digital data with a finite precision make rational to consider the problem in a discrete framework (Fig 1).

The most popular methods of computerized tomography are the filtered back-projection algorithm and its variant called the convolution back-projection algorithm. Since the seventies, the back-projections methods have been chosen in computerized tomography in preference to iterative [2, 3, 4, 5] reconstruction techniques. These algebraic approaches [6] have however recently come to the lead in both frameworks of PET imagery and discrete tomography. Discrete tomography deals with the reconstruction of binary images, namely lattice sets instead of grey-level images (see [7] for details). One of the characteristics of discrete tomography is the small number of directions which makes the Fourier analytical tools inappropriate. In this particular framework, the algebraic approach and especially linear programming has provided a lot of interesting results ([8],

E. Andres et al. (Eds.): DGCI 2005, LNCS 3429, pp. 126–135, 2005.

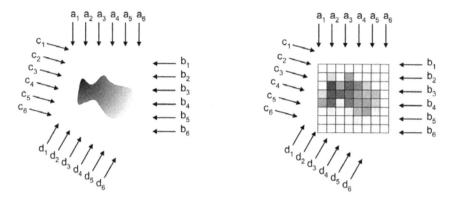

**Fig. 1.** Moving the camera in four positions with a panel of directions of angles 0, 90, 150 and 210 around the object provides four vectors $(a_i)_{1 \leq i \leq I}$, $(b_i)_{1 \leq i \leq I}$, $(c_i)_{1 \leq i \leq I}$ and $(d_i)_{1 \leq i \leq I}$ called X-rays. The purpose of tomography is to obtain from the X-rays a significative grey-level image of the object

[9], [10], [11], [12]). This paper is deeply inspired by these earlier works. Its purpose is a presentation of a new method of computerized tomography based on linear programming. The choice of reconstructing grey levels images instead of lattice sets avoids the main difficulty met by this kind of methods in discrete tomography. Then we can focus on the generation of the linear constraints and of the objective function. They are based in the new method on three elementary ideas:

- We cover the set of the pixels by digital lines (a digital line is a subset of $\mathbb{Z}^2$ characterized by a double-inequality $\mu \leq ax + by < \mu + \delta$ [13]) instead of using Diophantine straight lines or weights equal to the lengths of the intersection between lines and pixels.
- We introduce in the problem a variable of maximal error that we denote $h$. The linear constraints of the linear programming instance come from the definition of $h$: each difference between the given X-rays and the X-rays of the image should belong to the interval between the maximal error and its opposite. The choice of the objective function is made in order to minimize $h$.
- We add in the objective function an entropy factor as done in [10].

Next section 2 is focused on the literature of linear programming and tomography. The purpose of the third section 3 is a detailed description of our new method while section 4 provides experimental results.

## 2    Tomography and Linear Programming

According to [8], the idea to use linear programming in the framework of discrete tomography came to P. Fishburn, P. Schwander, L. Shepp, and R. Vanderbei after a talk of A. Kuba at the first mini-symposium on discrete tomography (DIMACS, 1994, Rutgers University) about his joint work with G. Herman and R. Aharoni. It was the first intrusion of linear programming theory in discrete tomography and probably in the whole domain of tomography. Since ten years, this idea has get on well by providing a lot of interesting results going from algorithms to more theoretical questions as for instance uniqueness [8] or stability of the solutions [11]. The basic principle is to relax the binary constraint on the image for computing with linear programming a grey-level image. The instance is made from the constraint on the grey level to belong to an interval while some partial sums according to lines are equal with the given X-rays [8]. This first formulation searches only exact solutions with the disadvantage that approximations are not supported (if only one data is modified, the feasibility of the instance is not guaranteed). It makes this approach inappropriate to real data. Then P. Gritzmann, S. de Vries, and M. Wiegelmann have relaxed the equalities between the partial sums and the X-rays: whether they consider that the sums are less than the X-rays and they maximize the sum of the grey levels (method Best-Inner-Fit), or they consider that the sums are greater than the X-rays and they minimize the sum of the grey levels (method Best-Outer-Fit) [9]. This approach can deal with real data because any instance is feasible. Next point was the introduction by S. Weber, C. Schnörr, and J. Hornegger of an entropy term in the objective function. This technic requires auxiliary variables with an expensive increase in the number of constraints. The computation is much more expensive in time but it provides results that the human eyes should find more obvious and less noisy than the ones obtained without entropy. It is the reason why we kept this term for reconstructing grey levels images.

Finally, we mention other works trying to use sophisticated tools of convex optimization [5] or Difference Convex programming [14].

## 3    New Method

### 3.1    Digital Lines

To project a continuous image, real lines are used and the projected value is the integral of density along the line of projection. When dealing with discrete images, projections may be modeled with different models. First, it is possible to consider the intersection between the digital grid and rational lines [12, 9]. In this case, a line is a set of integer points nearly always disconnected. The value of the projection is the sum of the values of the points of the intersection. Second, it is possible to consider real lines and to discretize the intersection of the lines with the discrete grid. More precisely, each point of the discrete grid might be modeled as a unit square. Then the intersection of a line with each square is an interval whose length is used as a weight associated to the integer points situated at the center of the square. This is done for instance in [14].

None of the above methods are satisfactory because discrete projections are computed with disconnected sets in the first case, and weights introduce a mix between the discrete space $\mathbb{Z}^n$ and the real space $\mathbb{R}^n$. We propose another strategy which is a purely discrete approach. Our projection operator is based on arithmetical lines.

The $\mu$ parameter controls the translation of any given arithmetical line over the grid whose size is the size of the image to be reconstructed. If the width parameter $\delta$ is greater than $\max(|a|, |b|)$ then the arithmetical lines are connected and hence projections are computed from connected subsets of the image. In the experiments, we have chosen to use digital lines with $\delta = \max(|a|, |b|)$ (these digital lines are called "naive"). The translation parameter $\mu$ has also a deep impact on the projection operator. Indeed, it must be increased or decreased by $\delta$ to ensure that consecutive lines do not intersect. Hence smaller values force the consecutive lines to overlap, which makes the projection operator closer to the real process of acquisition used in computerized tomography. As a by-product, it is possible to deal with image having an imported resolution while keeping limited resolutions for the projections. This is also achieved thanks to the parameters $\delta$ and $\mu$ of arithmetical lines.

## 3.2    Linear Programming

As previously described in the paper, discrete tomography can be algebraically reduced to a problem of the form

$$Ax = b \tag{1}$$

However, the matrix $A$ is rarely invertible. Hence, the previous system cannot be solved algebraically. To build an approximate solution, we introduce the error term: $e = Ax - b$. Ideally $e = 0$ but since this never happens, we try to minimize $e$. Minimization corresponds in fact to the minimization of the norm $\|e\|$ of $e$ for some norm $\|.\|$ which can be the $L_1$, $L_2$ or any $L_p$ norm or the $L_\infty$ norm. $L_2$ norm leads to a mean-square approximation of the solution whereas the $L_\infty$ norm controls only the worse term of the approximation. Hence, we bound $e$ by $h\mathbb{1}$ and $-h\mathbb{1}$, where $\mathbb{1}$ is the constant vector with all components equal to one, and we minimize $h$. It is known [15] that $L_\infty$ norm problem might be solved by linear programming. More precisely, to minimize the $L_\infty$ error, it is sufficient to solve the following problem

$$\begin{cases} \min & h \\ \text{such that} \\ -h\mathbb{1} \leq Ax - b \leq h\mathbb{1} \end{cases} \tag{2}$$

This linear programming problem can be solved by any available linear programming solver. We have chosen the program `soplex` [16].

Approximation of the computerized tomography reconstruction problem with linear programming is an interesting approach but lacks of topological constraints [10].

### 3.3    Entropic Regularization

To solve this problem, regularization constraints must be added to the linear programming approach. In [10], the following strategy was used. Some variables $z_{(j,k)}$ were added for each position $(j,k)$ where $j$ and $k$ are 4-neighbors in the grid containing the image. Then, $z_{(j,k)}$ is forced to be lower than the absolute value of the difference between the values at position $j$ and $k$. Moreover, each density value is searched in the continuous interval $[0,1]$.

We have adopted another strategy to test the influence of an entropic term onto the reconstruction. First if we want to reconstruct images with $g$ grey levels, we have forced all variables in the continuous interval $[0, g - 1]$. Second, let us consider a position $(m,n)$ in the grid. $x_{(m,n)}$ denotes the unknown density value at position $(m,n)$. We introduce two variables, $u_{(m,n)}$ and $v_{(m,n)}$ and we add the constraints : $u_{(m,n)} \leq \min(x_{(m,n)}, x_{(m+1,n)})$ and $v_{(m,n)} \leq \min(x_{(m,n)}, x_{(m,n+1)})$. Obviously, our entropic term is not symmetric whereas the previous one of Weber et al [10] was. This might result in privileging high values.

The final problem to solve becomes:

$$
\begin{cases}
\min \quad h - K \left( \displaystyle\sum_{(m,n)} u_{(m,n)} + v_{(m,n)} \right) \\[2mm]
\text{such that} \\[1mm]
-h\mathbb{1} \leq Ax - b \leq h\mathbb{1} \\[1mm]
u_{(m,n)} \leq \min \left( x_{(m,n)}, x_{(m+1,n)} \right), \quad \forall\, (m,n) \\[1mm]
v_{(m,n)} \leq \min \left( x_{(m,n)}, x_{(m,n+1)} \right), \quad \forall\, (m,n) \\[1mm]
0 \leq x \leq (g-1)\mathbb{1}
\end{cases}
\tag{3}
$$

$K$ is a constant used to control the influence of the topological constraints versus the minimization of $h$.

## 4    Experiments

We present in this section some experiments to illustrate the quality of the reconstruction method. Despite the fact that the method was designed to deal with grey level images, we have applied it successfully to black and white image usually encountered in classical discrete tomography. The method has been tested on two types of simulated images. One image was used to test the influence of the constant $K$ parameter and the other image, closed to real images acquired in practice, was used to study the influence of the number of directions in the reconstruction process.

For each test, the method was the following. Starting with the resolution of the reconstructed image, we have generated the list of constraints using non overlapping naive lines. Projections were computed free of error using the synthetic images. The linear programming problem was solved by `soplex` without any special parameters. The output of `soplex` was converted to an image by rounding, since the variables were forced to be in the interval of allowed grey values.

## 4.1    B and W Tests

We start with a black and white test (see Fig. 2). $K$ was fixed to 0.001. No difference was constated between the two images. When considering real values, the output of `soplex` was either 0 or 255, so no rounding was in fact added to the process. It must be noticed that the reconstruction uses 8 directions: $(1,0)$, $(0,1)$, $(1,1)$, $(1,-1)$, $(1,2)$, $(2,1)$, $(1,-2)$ and $(2,-1)$.

A second test (see Fig. 3) was done on an imperfect black and white image. We have used the same $K$ value and the same projections than for the previous experiments. Due to noise, some grey levels appears inside the objects which is in accordance to what was expected since noise perturbates the projections.

We now illustrate the influence of the constant $K$ on the reconstruction process. The image in Fig. 3 was reconstructed with $K = 0.001$. In Fig. 4, we present

**Fig. 2.** (left) Original Image  (right) Reconstructed Image

**Fig. 3.** (left) Original Noisy Image  (right) Reconstructed Noisy Image

**Fig. 4.** (left) $K = 0.01$  (middle) $K = 0.007$  (right) $K = 0.0001$

the reconstruction obtained for $K = 0.01$, $K = 0.007$ and $K = 0.0001$. When $K$ has a too high value, the topological constraints are stronger than the constraint on $h$. Hence, this forces to have regions with regular grey level even if objects inside the image are not well reconstructed. When $K$ decreases, the regularity of the grey levels appears inside the objects but when $K$ is very small, the influence of the second term becomes neglectable and thus peaks of grey level appear. All reconstructions were done with the given eight directions.

### 4.2    Grey Level Tests

We have also tested the method on a synthetic image which has simply been constructed by adding grey levels inside the objects of the black and white noise free image. We still used 8 directions and $K$ was fixed to 0.001. Results are given in Fig. 5. Images look similar even if they are not. The only difference is in the grey level regularity. The effect of the topological constraints was to regularize variations of grey levels in the reconstructed image. Visually, this effect was not detected first. When zooming both images (see Fig. 6), the effect of the

**Fig. 5.** (left) Synthetic grey level image  (middle) Reconstruction with $K = 0.001$ (right) Reconstruction with $K = 0.0$

**Fig. 6.** (left) Original image  (right) Reconstructed image

topological constraints becomes clearer in the variations of grey levels in the reconstructed images.

In Fig. 5, we provide the result obtained when $K$ is null. Hence, the linear program only corresponds to the minimization of $h$, that is the $L_\infty$ norm. As it can be seen, the solution is not acceptable since the reconstructed grey levels have no meaning at all. Minimizing only the error on the projections does not permit to reconstruct the grey levels inside the image.

### 4.3    Influence of the Number of Directions

We now carry on our analysis with a more realistic synthetic image by the analysis of the influence of the number of directions on the quality of reconstruction, see Fig. 7. We have used at most 16 directions by adding the following directions to the previous ones: $(1, 4)$, $(4, 1)$, $(2, 3)$, $(3, 2)$, $(1, -4)$, $(4, -1)$, $(2, -3)$ and $(3, -2)$. The experiments were performed by taking the first four values of the 8 directions given earlier in the paper, then by taking all previous 8 directions and finally the sixteen directions.

As it can be seen, the influence of the directions is certainly not neglectable. $K$ was fixed to a constant value $K = 0.001$ and this explains the regularity of the grey levels. With four directions, it was nearly impossible to reconstruct a good image. However, a careful analysis of the images shows that the result is not so bad since high and low values of grey levels correspond inside the object of the original position of extremal values. It is clear however, that the information given by the four projections were insufficient. With eight direction, the result is better. It is also better localized. This image can be sufficient in practice but is not perfect. The area which is totally missing is the hole between the two objects. Eight projections is insufficient to separate the objects where it was sufficient to reconstruct nearly well the interior grey values of the objects. When using sixteen directions, separation of the objects appears as well as a perfect localization of grey level extrema. This result is totally coherent with what was expected. More projections can be added but it is clear that visually sixteen are sufficient. Since reducing the number of directions is a challenging problem, it seems that our method does succeed well in this task.

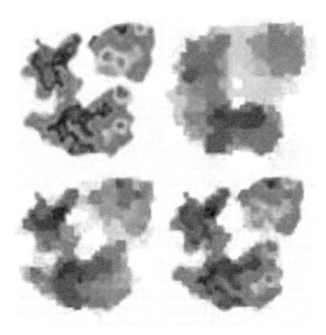

**Fig. 7.** In order, original image, reconstructed with 4, 8 and 16 directions

## 5    Conclusion

We have presented in this paper a promising method for reconstructing grey level images in computerized tomography. Based on linear programming and digital lines, this method has a pure discrete nature. It was successfully tested on synthetic images and seems also to perform well on black and white images. We believe that this method can be a good starting point for constructing efficient and reliable reconstruction method. All experiments were done on $64 \times 64$ images but is virtually not limited. Times of computations approach one hour with `soplex` at this time but the solution is nearly obtained in 20 minutes. Computing time compatible with clinical available time is a challenging problem that must be addressed in future works.

## References

1. Kak, A., Slaney, M.: Principles of Computerized Tomographic Imaging. IEEE Press (1988)
2. Herman, G., Lent, A.: Iterative reconstruction algorithms. Comput. Biol. Med. **6** (1976) 273–274
3. Gordon, R., Herman, G.: Reconstruction of pictures from their projections. Communication of the ACM **14** (1971) 759–768
4. Gordon, R.: A tutorial on ART (Algebraic Reconstruction Techniques). IEEE Transactions on Nuclear Science **NS-21** (1974) 31–43

5. Ben-Tal, A., Margalit, T., Nemirovski, A.: The ordered subsets mirror descent optimization method with applications to tomography. SIAM J. Optimization **12** (2001) 79–108
6. Aharoni, R., Herman, G., Kuba, A.: Binary vectors partially determined by linear equation systems. Discrete Mathematics **171** (1997) 1–16
7. Kuba, A., Herman, G., eds.: Discrete Tomography: Foundations, Algorithms and Applications. Birkhaüser (1999)
8. Fishburn, P., Schwander, P., Shepp, L., Vanderbei, R.: The discrete radon transform and its approximate inversion via linear programming. Discrete Applied Math. **75** (1997) 39–61
9. Gritzmann, P., de Vries, S., Wiegelmann, M.: Approximating binary images from discrete X-rays. SIAM J. Optimization **11** (2000) 522–546
10. Weber, S., Schnörr, C., Hornegger, J.: A linear programming relaxation for binary tomography with smoothness priors. In: Int. Workshop on Combinatorial Image Analysis IWCIA'03. Volume 12 of Electronic Notes in Discrete Math., Elsevier (2003)
11. Brunetti, S., Daurat, A.: Stability in discrete tomography: Linear programming, additivity and convexity. In: 11th Discrete Geometry and Computer Imagery. Volume 2886 of LNCS., Springer-Verlag (2003) 398–408
12. Hajdu, L., Tijdeman, R.: An algorithm for discrete tomography. Linear Algebra and Appl. **339** (2001) 147–169
13. Reveillès, J.P.: Géométrie discrète, calcul en nombres entiers et algorithmique. Thèse d'état, Université ULP - Strasbourg (1991)
14. Weber, S., T. Schüle, C.S., Hornegger, J.: Discrete tomography by convex-concave regularization and d.c. programming. Technical report, Mannheim University (2003)
15. Schrijver, A.: Theory of Linear and Integer Programming. J. Wiley and Sons (1986)
16. Wunderling, R.: Paralleler und Objektorientierter Simplex-Algorithmus. PhD thesis, ZIB TR 96-09, Berlin (1996)

# A Discrete Modulo $N$ Projective Radon Transform for $N \times N$ Images

Andrew Kingston and Imants Svalbe

Centre for X-ray Physics and Imaging,
School of Physics and Materials Engineering,
Monash University, VIC 3800, AUS
{Andrew.Kingston, Imants.Svalbe}@spme.monash.edu.au

**Abstract.** This paper presents a Discrete Radon Transform (DRT) based on congruent mathematics that applies to $N \times N$ arrays where $N \in \mathbb{N}$. This definition incorporates and is a natural extension of the more restricted cases of the finite Radon transform [1] where $N$ must be prime, the discrete periodic Radon transform [2] where $N$ must be a power of 2, and the DRT over $p^n$ [3], where $N$ must be a power of a single prime. The DRT exactly and invertibly maps a 2-D image to a set of 1-D projections of length $N$. Projections are found as the sum of the pixels centred on a parallel set of discrete lines. The image is assumed to be periodic and these lines wrap around the array under modulo $N$ arithmetic. Properties of the continuous Radon transform are preserved in the DRT; a discrete form of the Fourier slice theorem applies, as does the convolution property. A formula is given to find the projection set required to be exactly invertible for arrays with $N$ any composite number, as well as a means to determine the level of redundancy in sampling that is introduced on such composite arrays.

## 1   Introduction

This paper presents a discrete projective transform for $N \times N$ arrays of image data where $N \in \mathbb{N}$. The motivation for this work is to remove the current restrictions on array size $N$ of such transforms to be a power of a prime.

The Radon transform maps a 2-D function $f(x, y)$ to projection space $r(\rho, \theta)$ as the integral of $f$ along the line $y = \tan \theta x + \rho / \cos \theta$. A projection is a set of all parallel line integrals at some angle $\theta$. It was first defined by Johan Radon in 1917, however applications in tomography and image processing were not fully realised until the advent of computers. As the transform is predominantly implemented on discrete data sets, a discrete version of the Radon transform that minimises the need for interpolation is required. The Radon transform has no 1-D analogue, so defining discrete projection sets is not a trivial task and there have been many proposed formalisms.

A general algebraic definition for the discrete form of the Radon transform was presented by Beylkin in 1987 [4]. A specific class of Beylkin's definition, termed the Finite Radon Transform (FRT), was defined by Matus and Flusser in

E. Andres et al. (Eds.): DGCI 2005, LNCS 3429, pp. 136–147, 2005.

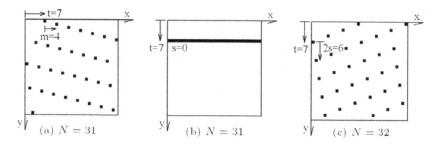

**Fig. 1.** Examples of discrete lines on $N \times N$ grids. Black pixels lie on the centre of the wrapped lines, the values of these pixels are summed to give the value of one element in the transform. (a) Pixels on $x \equiv 4y + 7$ (mod 31) sum to give $R_4(7)$. $N$ is prime here so only one element on each row and column is sampled by a discrete wrapped line. (b) Pixels on $y \equiv 0x + 7$ (mod 31) sum to give $R_0^{\perp}(7)$. (c) Pixels on $y \equiv 6x + t$ (mod 32) sum to give $R_3^{\perp}(7)$, i.e., $s = 3$. $N$ is now composite and the gradient $2s$ has a common factor with $N$, each row and column is no longer uniquely sampled

1993 [1]. This transform is an exact and invertible projective mapping, requiring only additive operations. This contrasts with the projections of the general DRT in [4] which require an algebraic solution for the inversion process. The FRT applies to square arrays of prime size, $p \times p$. A discrete form of both the Fourier slice theorem and convolution property of the continuous transform hold [1]. Thus the transformation and inversion can also be achieved very efficiently via the 2-D discrete Fourier transform of the function. These properties of the FRT make it an attractive tool to transform and interpret discrete data [5, 6].

An image defined on the 2-D array, $I(x, y)$, is mapped to a set of 1-D projections of length $p$. The discrete FRT projections are found as a set of parallel discrete line sums with intercepts $t$. There are $p + 1$ projections in total, $p$ projections $R_m(t)$ for $0 \le m < p$, and one "perpendicular" projection, $R_0^{\perp}(t)$. $R_m(t)$ is defined as the sum of pixel values centred on the discrete lines $x \equiv my + t$ (mod $p$), while $R_0^{\perp}(t)$ is found as the sum of all pixel values centred on the line $y \equiv 0x + t$ (mod $p$). The transform is then defined as

$$
\begin{aligned}
R_m(t) &= \sum_{y=0}^{p-1} I(\langle my + t \rangle_p, y) \quad \text{for } 0 \le m < p, \\
R_0^{\perp}(t) &= \sum_{x=0}^{p-1} I(x, t),
\end{aligned}
\tag{1}
$$

where $\langle x \rangle_\eta$ denotes $x$ (mod $\eta$). An example of these discrete lines for $p = 31$ is presented in Fig. 1a and 1b.

The FRT formalism was extended by Hsung, Lun and Siu in 1996 to apply to square arrays of size $2^n$ for any positive integer $n$ [2]. This version, termed the Discrete Periodic Radon Transform (DPRT), makes the transform more conducive to image processing, which commonly utilises images of size $2^n$ for computational efficiency. The DPRT has projections $R_m(t)$, as defined above, for $0 \le m < N = 2^n$. However, since the array size is not prime, the image size $N$ now has factors other than $N \equiv 0$ (mod $N$). Additional "perpendicular" projections are required as $R_s^\perp(t)$ for $0 \le s < N/2$, which are discrete line sums along $y \equiv 2sx + t \pmod{N}$. An example of these discrete lines for $N = 32$ is presented in Fig. 1c. Including all these discrete projections represents $I(x, y)$ exactly in the transform space. The DPRT is then defined as

$$\begin{aligned} R_m(t) &= \sum_{y=0}^{N-1} I(\langle my + t \rangle_N, y) && \text{for } 0 \le m < N, \\ R_s^\perp(t) &= \sum_{x=0}^{N-1} I(x, \langle 2sx + t \rangle_N) && \text{for } 0 \le s < \tfrac{N}{2}. \end{aligned} \tag{2}$$

The DPRT is a redundant representation of $I(x, y)$ as it has $N(1 + 1/2)$ projections of length $N$. A non-redundant form of the DPRT with orthogonal bases was presented by Lun and Hsung and Shen in 2003 [7], it is termed the Orthogonal DPRT (ODPRT).

The formalism for the DRT applied to arrays of size $p^n \times p^n$ is a natural extension of the DPRT and was developed in [3]. Projections are defined as for the DPRT, with $p$ replacing 2 in (2). This is a redundant transform but requires only $N(1 + 1/p)$ projections of length $N$. A non-redundant form of this DRT, with orthogonal bases, was also presented in [3].

The above DRTs are all restricted to apply to square arrays with dimensions based on a single prime. Should the DRT be required of an array of arbitrary size $M \times N$, the image must be padded with zeroes to a square array with minimum size being the smallest $p^n \ge \sup\{M, N\}$. Padding arrays adds redundant information and unwanted computational complexity. This paper extends the FRT (over $p$), the DPRT (over $2^n$) and the DRT (over $p^n$) to apply to square arrays of any composite size $N = p_1^{n_1} p_2^{n_2} p_3^{n_3} \ldots$, where $p_i$ is prime and $n_i$ is any positive integer giving the prime decomposition of $N$. This definition is referred to in this paper as the DRT over $N$. A DRT for non-square arrays is the subject of ongoing research.

Section 2 establishes the set of projections that are required for an exact and invertible DRT and provides a method to obtain that set. This leads to the definition for the DRT over $N \in \mathbb{N}$ in section 3. A discrete form of the Fourier slice theorem which applies to this DRT and the related convolution property is also explained in this section. The inversion process for exact image reconstruction from composite array projections is presented in section 4.

## 2    Projection Sets for Invertible Mappings

For the FRT, each projection, $m$, is generated by lines with gradient $m$, however $\tan^{-1}(1/m)$ is not necessarily the best way to define the angle of the projection. It is defined here as the angle, $\theta_m$, which provides the smallest translation between adjacent sampled pixels of the discrete line. Due to the assumption that the image is periodic, The sample pattern for each discrete line $x \equiv my + t \pmod{p}$ produces an infinite 2-D lattice with basis vectors $\{m, 1\}$ and $\{0, p\}$ (example depicted in Fig. 2b). Identifying the angle of the shortest vector between elements of this lattice is equivalent to finding the projection angle. The design of efficient algorithms to find the shortest length vectors in lattices has received

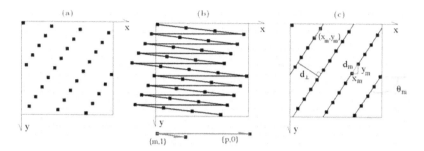

**Fig. 2.** Example of discrete line sums for $p = 29$, $m = 9$ and $t = 0$. Black pixels are those summed as a periodic discrete line (a) $x \equiv 9y + 0 \pmod{29}$. (b) Lattice basis $\{9, 1\}$ and $\{29, 0\}$ generates the sampling pattern of this periodic discrete line. (c) Shortest vector in lattice, $\{x_m, y_m\} = \{-2, 3\}$ at rational projection angle, $\theta_m = \tan^{-1}(y_m/x_m) = 123.7°$

considerable attention. For example, Rote showed that any shortest vector can be determined very efficiently through finding the reduced basis for 2D lattices [8]. This shortest vector is denoted $\{x_m, y_m\}$. The projection angle is then defined as $\theta_m = \tan^{-1}(y_m/x_m)$ (see Fig. 2c). Each projection, $m$, has a unique rational angle. These fractions are irreducible and form a subset of the Farey sequence [9].

The lattice can be generated by $\{0, p\}$ and/or $\{p, 0\}$, along with any other vector in the lattice $\{m, 1\}$, $\{2m, 2\}$, ..., $\{(p-1)m, p-1\}$, reduced modulo $p$. These vectors are equivalent for all parallel lines in a projection, regardless of the translate, $t$. Therefore each vector specifies an entire projection. Let $\Theta_p$ be the set of these vectors that define the projection set for a $p \times p$ array. $\Theta_p$ contains $\{m, 1\}$ (or any equivalent basis vector) for $0 \leq m < p$ and $\{1, 0\}$.

The pixels sampled by each discrete line with $t = 0$ correspond to the $p + 1$ unique cyclic subgroups of order $p$ in the group $\mathbb{Z}_p \times \mathbb{Z}_p$ under addition. Each subgroup contains the identity $(0, 0)$ and, since $p$ is prime, each subgroup contains $p - 1$ unique elements. These elements also uniquely sample each row and column of the image, (see the example in Fig. 1a). Therefore there must be $p + 1$ subgroups to contain all $p^2$ elements, as $(p+1)(p-1)+1 = p^2$. Let $\Upsilon(N)$ represent the number of discrete lines in the projection set for the DRT of an $N \times N$ array. It can be seen $\Upsilon(p) = p + 1 = p(1 + 1/p)$ where $p$ is prime. $\Upsilon(N)$ also gives the number of unique cyclic subgroups of order $N$ in the group $\mathbb{Z}_N \times \mathbb{Z}_N$ under addition.

For the DPRT, the additional "perpendicular" discrete lines $y \equiv 2sx + t$ $\pmod{2^n}$ are necessary, since $N = 2^n$ has factors so each pixel is not necessarily contained in a unique cyclic subgroup. Here $\Theta_{2^n}$ contains $\{m, 1\}$ for $0 \leq m < 2^n$ and $\{1, 2s\}$ for $0 \leq s < 2^{n-1}$. There are $N + N/2$ cyclic subgroups of order $N$ in the group $\mathbb{Z}_N \times \mathbb{Z}_N$ under addition when $N = 2^n$. There are also cyclic subgroups of order less than $N$, however it is unnecessary to include discrete lines resulting from these, as each one is entirely contained within a cyclic subgroup of order

$N$. Therefore $\Upsilon(2^n) = 2^n + 2^{n-1} = 2^n(1 + 1/2)$. Similarly for the DRT over $p^n$, $\Theta_{p^n}$ contains $\{m, 1\}$ for $0 \le m < p^n$ and $\{1, ps\}$ for $0 \le s < p^{n-1}$ with $\Upsilon(p^n) = p^n + p^{n-1} = p^n(1 + 1/p)$.

Define the discrete lines for the DRT over square arrays of composite size as $ax \equiv by + t \pmod{N}$. $\Theta_N$ then contains a number of vectors $\{b, a\}$. How many of these are there to ensure the transform is invertible and how are they found?

A function $f(n)$ is said to be *multiplicative* if $\gcd(m, n) = 1$ implies that $f(mn) = f(m)f(n)$ [10]. We can show that $\Upsilon(N)$ is multiplicative and can then define

$$\Upsilon(N) = N \prod_{p|N}(1 + 1/p). \tag{3}$$

Suppose that $\gcd(m, n) = 1$ and $\Theta_m$ contains $\{b_i, a_i\}_m$ for $0 \le i < \Upsilon(m)$ which gives the complete projection set for an $m \times m$ array and $\Theta_n$ contains $\{d_i, c_i\}_n$ for $0 \le i < \Upsilon(n)$ which gives the complete projection set for an $n \times n$ array. Then $n\Theta_m + m\Theta_n$ gives the projection set for the $mn \times mn$ array which will have $\Upsilon(m)\Upsilon(n)$ discrete lines.

If         $n\{b_1, a_1\}_m + m\{d_1, c_1\}_n \equiv n\{b_2, a_2\}_m + m\{d_2, c_2\}_n \pmod{mn}$
then        $n\{b_1, a_1\}_m \equiv n\{b_2, a_2\}_m$         $\pmod{m}$
and so      $\{b_1, a_1\}_m \equiv \{b_2, a_2\}_m$         $\pmod{m}$,
similarly   $\{d_1, c_1\}_n \equiv \{d_2, c_2\}_n$         $\pmod{n}$.

So all the new discrete lines defined are incongruent $\pmod{mn}$ and form the complete projection set for $mn \times mn$. The projection set for a composite $N$ can be found via the projection sets for all the primes making up $N$. Below is an example demonstrating how to obtain the projection set for $N = 6$ from the sets for the two primes that make up 6, $N = 2$ and $N = 3$:

Given   $\Theta_2 = \{1, 0\}, \{0, 1\}, \{1, 1\}$.
        $\Theta_3 = \{1, 0\}, \{0, 1\}, \{1, 1\}, \{2, 1\}$.

| $+$ | $\{1, 0\}$ | $\{0, 1\}$ | $\{1, 1\}$ | $\{2, 1\}$ $\times 2$ |
|---|---|---|---|---|
| $\{1, 0\}$ | $\{5, 0\}$ | $\{3, 2\}$ | $\{5, 2\}$ | $\{1, 2\}$ |
| $\{0, 1\}$ | $\{2, 3\}$ | $\{0, 5\}$ | $\{2, 5\}$ | $\{4, 5\}$ |
| $\{1, 1\}$ | $\{5, 3\}$ | $\{3, 5\}$ | $\{5, 5\}$ | $\{1, 5\}$ |

$\Theta_6 = 2\Theta_3 + 3\Theta_2$, i.e., (with $\times 3$ below)

This property is the underlying basis for generating the DRT for an arbitrary composite array size $N$.

## 3   DRT Formalism

### 3.1   Definition

Denote each element of the DRT as $R_{b,a}(t)$ which is found as the sum of all pixels in $I(x, y)$ such that $ax \equiv by + t \pmod{N}$. Here for the FRT and DPRT $R_m(t) =$

$R_{m,1}(t)$ and $R_s^\perp(t) = R_{1,2s}(t)$. The DRT over square arrays of composite size, $N \times N$, can be defined as

$$R_{b,a}(t) = \sum_{x=0}^{N-1} \sum_{y=0}^{N-1} I(x,y)\delta\langle ax - by - t\rangle_N \quad \text{for} \quad \{b,a\} \in \Theta_N. \tag{4}$$

where $\delta\langle x\rangle_\eta$ is 1 when $x \equiv 0 \pmod{\eta}$ and 0 otherwise and $\Theta_N$ is as defined in section 2.

## 3.2    Properties

An important property of the continuous RT is the Fourier slice theorem which states that the 1-D Fourier transform (FT) of a continuous projection at angle $\theta$ is equivalent to a central radial slice through the 2-D FT of the original object/function at the angle $\theta^\perp = \theta + \pi/2$ [11]. A discrete form of this Fourier slice theorem can be demonstrated for the DRT. Denote $\widehat{R}_{b,a}(u)$ as the 1-D DFT of $R_{b,a}(t)$ for $\{b,a\} \in \Theta_N$, then

$$\widehat{R}_{b,a}(u) = \sum_{t=0}^{N-1} R_{b,a}(t)e^{-i2\pi ut/N}$$

$$= \sum_{t=0}^{N-1} \sum_{x=0}^{N-1} \sum_{y=0}^{N-1} I(x,y)e^{-i2\pi ut/N}\delta\langle ax - by - t\rangle_N$$

$$= \sum_{x=0}^{N-1} \sum_{y=0}^{N-1} I(x,y)e^{-i2\pi(axu-byu)/N}$$

$$= \widehat{I}(au, -bu), \tag{5}$$

where $\widehat{I}(u,v)$ is the 2-D discrete FT (DFT) of $I(x,y)$. This gives a wrapped discrete line through the origin of Fourier space, $-bu \equiv av \pmod{N}$, which is perpendicular to the discrete line of projection as the product of the gradients is -1.

The DRT of an image can be obtained in $O(N\Upsilon(N)\log N)$ operations by utilising this property. $\widehat{I}(u,v)$, the 2-D DFT of the image $I(x,y)$, can be obtained in $O(N^2 \log N)$ operations. Each of the $\Upsilon(N)$ projections of length $N$ can be obtained as the inverse 1-D DFT of the corresponding discrete slices in $O(N \log N)$ operations.

Another useful property of the continuous RT, which results from the Fourier slice theorem, is the convolution property. A discrete form of this property is also conserved in the DRT. Assume the function $F(x,y)$ on an $N \times N$ array is to be determined from the $N \times N$ function $G(x,y)$ and $L \times M$ function H(x,y), $0 < L, M \leq N$, by the 2D convolution

$$F(x,y) = \sum_{\alpha=0}^{L-1} \sum_{\beta=0}^{M-1} G(\langle x - \alpha\rangle_N, \langle y - \beta\rangle_N)H(\alpha, \beta). \tag{6}$$

This operation can be performed on one projection at a time in the DRT. Denote the DRT over $N$ of $\eta(x, y)$ as $R_{b,a}^{\eta}(t)$. The value of $F(x, y)$ can be found through the DRT of $G(x, y)$ and $H(x, y)$ as

$$R_{b,a}^{F}(t) = \sum_{k=0}^{N-1} R_{b,a}^{G}(\langle t - k \rangle_N) R_{b,a}^{H}(k). \qquad (7)$$

This shows the 2-D convolution of arrays can be performed in DRT space as a set of 1-D circular convolutions of each discrete projection. This reduces the computational complexity of 2-D problems such as filtering or matching, analogous to the applications for the FRT and DPRT outlined in [2].

## 4   DRT Inversion

The inverse transform (or image reconstruction from projections) for the DRT is achieved via a form of back-projection similar to the process of projection (4). To recover the value of the original function at pixel $(i, j)$, the discrete line sums in each of the $\Upsilon(N)$ projections that contain this pixel are summed. This incorporates the value of the pixel $(i, j)$ $\Upsilon(N)$ times and all other pixels at least once (depending on the degree of compositeness of $N$). The next section investigates how to determine the degree of over-representation, and the following section establishes a method to correct for it, allowing the exact value for the pixel $(i, j)$ to be recovered.

### 4.1   Sampling Function, $v_N(x, y)$

Let $v_N(x, y)$ represent the number of times a pixel $(x, y)$ is sampled by all discrete lines in the $N \times N$ DRT that include the origin. This function shows the over-representation of each pixel after back-projection when reconstructing the origin. This over-representation must be accounted for in reconstructing the original function. Since the array is congruent (mod $N$), the over-representation for any pixel $(x, y)$ from the back-projection to reconstruct a pixel $(i, j)$ can be found as $v_N(x - i, y - j)$, so it is sufficient to investigate reconstructing the origin only.

For the FRT case, where $N = p$, a prime, $v_p(x, y) = 1$ for all pixels except the origin which is $\Upsilon(N)$ or $p + 1$. An example of this for $p = 7$ is shown in Fig. 3a. This can be written as $v_p(x, y) = d \prod_{d=p}(1 + 1/p)$ where $d = \gcd(x, y, p)$. For the DPRT case, where $N = 2^n$, it was shown in [2] that $v_{2^n}(x, y) = \gcd(x, y, 2^n)$ for all pixels except the origin which is $\Upsilon(N) = 2^n(1 + 1/2)$. An example of this for $N = 8$ is shown in Fig. 3b. This can be written as $v_{2^n}(x, y) = d \prod_{d=2^n}(1 + 1/2)$ where $d = \gcd(x, y, 2^n)$. Similarly, for the DRT over $p^n$, it was shown in [3] that $v_{p^n}(x, y) = d \prod_{d=p^n}(1 + 1/p)$ where $d = \gcd(x, y, p^n)$. An example of this for $N = 9$ is shown in Fig. 3c.

Suppose that $\gcd(m, n) = 1$ where $\{a_i, b_i\}$ for $0 \le i < v_m(x, y)$ is the set of solutions $ax + by \equiv 0 \pmod{m}$ and $\{c_i, d_i\}$ for $0 \le i < v_n(x, y)$ is the set of solutions $cx + dy \equiv 0 \pmod{n}$. Then $n\{a, b\} + m\{c, d\}$ gives the solution

**Fig. 3.** Examples of the sampling function for (a) FRT over $p$, $v_7(x,y)$ (b) DPRT over $2^n$, $v_8(x,y)$ (c) DRT over $p^n$, $v_9(x,y)$

set (mod $mn$) and there are $v_m(x,y)v_n(x,y)$ discrete lines. The proof that each new discrete line is unique is identical to that for $\Upsilon(N)$ in section 2, so all the new solutions defined are incongruent (mod $mn$) and form the complete set of solutions. Therefore we can say $v_N(x,y)$ is multiplicative and define

$$v_N(x,y) = d \prod_{p\,|\,d\,\nmid\,\frac{N}{d}} (1+1/p) \quad \text{where } d = \gcd(x,y,N). \tag{8}$$

where $p \mid d \nmid \frac{N}{d}$ denotes some prime $p$ that divides $d$ but does not divide $N/d$. Below is an example for finding $v_6(2,4)$ as $v_3(2,4)v_2(2,4)$, a graphical representation of the discrete lines is presented in Fig. 4.

$$\text{Given} \quad \begin{cases} v_2(2,4) = v_2(0,0) = 3 & \text{as } \{1,0\}, \{0,1\}, \{1,1\}. \\ v_3(2,4) = v_3(2,1) = 1 & \text{as } \{2,1\}. \end{cases}$$

$$v_6(2,4) = 3 \text{ found as } \begin{array}{c|ccc} + & \{1,0\} & \{0,1\} & \{1,1\} & \times 3 \\ \hline \{2,1\} & \{1,2\} & \{4,5\} & \{1,5\} \\ \times 2 & & & \end{array}$$

$$\text{so the lines} \quad \begin{cases} 2x \equiv 1y \pmod 6 \\ 5x \equiv 4y \pmod 6 \quad \text{all include pixel } (i,j) \\ 5x \equiv y \pmod 6 \end{cases}$$

From the discrete Fourier slice theorem, given in section 3.2, $v_n$ defined about the origin of the image also gives the over-representation of spatial frequencies in the 2-D discrete Fourier transform, $\widehat{I}(u,v)$ as $v_N(v,-u) = v_N(u,v)$.

## 4.2   Correcting for $v_N(x,y)$ in Back-Projection

For the FRT, $v_p(x,y)$ is $\Upsilon(p) = p+1$ at the origin and 1 at all other $(x,y)$. The sum of all discrete line sums containing a specific pixel $(i,j)$,[i.e., $R_{1,0}(j)$ and $R_{m,1}\langle i - mj \rangle_p$ for $0 \le m < p$] gives the sum of the entire image, $I_{\text{sum}}$, with $I(i,j)$ an additional $p$ times, (see example in Fig. 3b). The over-representation can be corrected for by subtracting the sum of the image, $I_{\text{sum}}$ and dividing the result by $p$. So the image is recovered from its FRT as

$$I(x,y) = \frac{1}{p}\left( \sum_{m=0}^{p-1} R_{m,1}\langle x - ym \rangle_p + R_{1,0}(y) - I_{\text{sum}} \right), \tag{9}$$

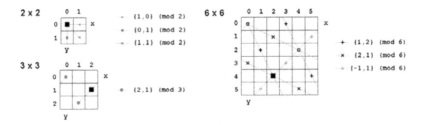

**Fig. 4.** Depiction of discrete lines presented in $v_6(2,4)$ example. Note in $6 \times 6$ case the basis vector $\{2,1\}$ generates the same lattice as $\{4,5\}$ and the basis vector $\{-1,1\}$ generates the same lattice as $\{1,5\}$, those used in this figure are simply the reduced bases

**Fig. 5.** (a) $v_2(x,y)$ (b) All discrete lines in $\Theta_2$ on the $8 \times 8$ array, with intercept 0 (mod 2) yields $v_2(x,y)$ replicated (mod 2) (c) $v_4(x,y)$ (d) All discrete lines in $\Theta_4$ on the $8 \times 8$ array, with intercept 0 (mod 4) yields $v_4(x,y)$ replicated (mod 4)

where $I_{\mathrm{sum}}$ can be obtained as the sum of all discrete line sums from any one projection, i.e., $\sum_{t=0}^{p-1} R_{q,1}(t)$ for any $0 \leq q < p$.

For the DPRT and DRT over $p^n$, $v_p(x,y)$ is $\Upsilon(p^n)$ at the origin and $\gcd(x,y,p^n)$ at all other $(x,y)$. This gives an over-representation at each resolution $p \times p$, $p^2 \times p^2$, ..., $p^{n-1} \times p^{n-1}$. To correct for this, an important property to note is that the sampling achieved by taking the projection set for a $p^k \times p^k$ array, $\Theta_{p^k}$, for all discrete line sums $t$ (mod $p^k$), (i.e., $\sum_{j=0}^{p^{n-k}-1} R_{b,a}(t+jp^k)$ for all $\{b,a\}$ in $\Theta_{p^k}$) yields the sampling pattern for $p^k$, i.e., $v_{p^k}(x,y)$ replicated (mod $p^k$) in the $x$ and $y$ directions. An example of this for $p^k = 2$ and $p^k = 4$ within an $8 \times 8$ array is depicted in Fig. 5b and 5d.

This can be used to correct for $v_{p^n}(x,y)$ at resolution $p^k \times p^k$. $v_{p^n}(x,y)$ can therefore be corrected through a multi-resolutional process; each step, $\xi_i$, corrects for the over-representation described above at resolution $N/p^i \times N/p^i$. The inversion process is given as

$$I(x,y) = \xi_0(x,y) - \sum_{i=1}^{n-1} \frac{(p-1)}{p^i} \xi_i(x,y) - \frac{p}{N^2} I_{\mathrm{sum}}, \qquad (10)$$

where $\xi_i = \dfrac{1}{N} \left[ \begin{array}{l} \sum_{m=0}^{\frac{N}{p^i}-1} \sum_{j=0}^{p^i-1} R_{m,1}\left( \langle x - my \rangle_{\frac{N}{p^i}} + j\frac{N}{p^i} \right) \\ + \sum_{s=0}^{\frac{N}{p^{i+1}}-1} \sum_{j=0}^{p^i-1} R_{1,s}\left( \langle y - psx \rangle_{\frac{N}{p^i}} + j\frac{N}{p^i} \right) \end{array} \right].$

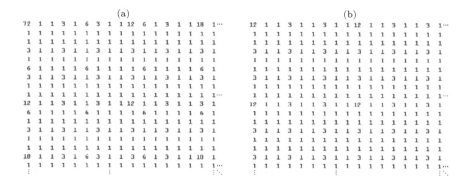

**Fig. 6.** Top left corner of (a) $v_{45}(x,y)$ (b) All discrete lines in $\Theta_9$ on the $45 \times 45$ array, with intercept 0 (mod 9) yields $v_9(x,y)$ replicated (mod 9)

The inversion process for composite $N$ is a natural extension of that for $N = p^n$. It is also undertaken by correcting for the over-representation at each resolution. The resolutions requiring correction correspond to all the factors, $F$, of $N$ $(F|N)$. For the example in Fig. 6a, where $N = 3^2 5 = 45$, the oversampling must be corrected for at scales $15 \times 15, 9 \times 9, 5 \times 5$ and $3 \times 3$. Subtracting a certain fraction of $I_{\text{sum}}$ corrects at resolution $1 \times 1$ leaving only a multiple $(N)$ of $I(x,y)$.

The sampling achieved by taking the projection set for some $F|N$ for all projections $t$ (mod $F$) yields the sampling pattern $v_F(x,y)$ repeated (mod $F$) in the $x$ and $y$ directions. An example of this is depicted in Fig. 6b for $N = 45$ and $F = 9$. This sampling pattern can be used to correct the over-representation from back-projection at a resolution of $9 \times 9$. This technique is used to correct the back-projection at each resolution.

Back projection for pixel $(x,y)$ is defined as $\left( \sum_{\{a,b\} \in P_N} R_{a,b} \langle ay - bx \rangle_N \right) / N$. Correction for the over-representation at each resolution $F \times F$, for $F|P$, is found as $A_F \sum_{\{b,a\} \in \Theta_F} \sum_{j=0}^{N/F-1} R_{b,a} \langle ay - bx + jF \rangle_N$ where the scaling factors, $A_F$, required are found as

$$A_F = \frac{F}{N} \prod_{p | \frac{N}{F}} (1 - p) \prod_{p | \frac{N}{F} \nmid F} \frac{p}{(1 - p)}. \tag{11}$$

where $p \mid \frac{N}{F} \nmid F$ denotes some prime, $p$, that divides $N/F$ but does not divide $F$. Therefore the entire inversion process can be written as

$$I(x,y) = \frac{1}{N} \left[ \begin{array}{l} \sum_{\{a,b\} \in P_N} R_{b,a} \langle ay - bx \rangle_N \\ - \sum_{F|N} A_F \sum_{\{b,a\} \in \Theta_F} \sum_{j=0}^{N/F-1} R_{b,a} \langle ax - by + jF \rangle_N \\ - A_1 I_{\text{sum}} \end{array} \right]. \tag{12}$$

(a)                (b)                (c)                (d)

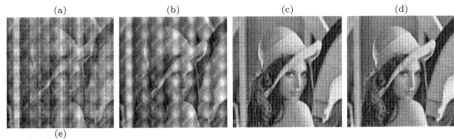

(e)

**Fig. 7.** Depiction of the correction made by each resolution of the reconstruction of the $\Upsilon(175) = 240 \times 175$ DRT of a 175 Lena image. (a) Uncorrected back-projection. (b) Corrected at resolution $35 \times 35$. (c) Corrected at resolution $25 \times 25$. (d) Corrected at resolution $7 \times 7$. (e) Correcting at resolution $5 \times 5$ and subtracting $A_1 I_{\mathrm{sum}}$ completes all required corrections and exactly reproduces the original image

The result of the correction process at each resolution for the inversion of the $\Upsilon(175) \times 175$ DRT of a $175 \times 175$ Lena image is depicted in Fig. 4.2.

The reconstruction can also be performed via Fourier space by taking the 1-D DFT of each projection as $O(N\Upsilon(N)\log N)$ and mapping it onto the 2-D DFT of the image, using the discrete Fourier slice theorem given in section 3.2. This will over-represent some spatial frequencies according to $v_N(u,v)$. Dividing the value at each spatial frequency by $v_N(u,v)$ and applying the inverse 2-D DFT to this data in $O(N^2\log N)$ operations recovers the original image.

## 5    Conclusion

A DRT based on modulo arithmetic which applies to $N \times N$ arrays for $N \in \mathbb{N}$ has been presented. It projects the 2-D image into a set of $N \prod_{p|N}(1 + 1/p)$ projections of length $N$. It is a redundant transform as with the DPRT. The multi-resolutional nature of the transform may prove useful in image analysis, particularly for textures and patterns. Research into this aspect and the development of an orthogonal DRT over $N$ and $M \times N$ are the subject of ongoing work. An investigation into the distribution of the discrete angles required for $N \times N$ as compared to the $p \times p$ set and the uniformly distributed angle set for the continuous RT is also the subject of ongoing research.

Important properties of the continuous Radon transform are preserved in this discrete formalism. A discrete form of the Fourier slice theorem and the convolution property hold for this DRT over $N$. These properties allow the DRT to be obtained in $O(N^2 \log N \prod_{p|N}(1 + 1/p))$ operations and make it a useful image processing tool, reducing 2-D problems to a set of 1-D problems.

## Acknowledgements

Thanks to Fred Ninio, School of Physics and Materials Engineering, Monash University, for discussions on Group theory applied to give the DRT. AK is a Monash University postgraduate student in receipt of an Australian Postgraduate Award scholarship, provided through the Australian Government.

## References

1. Matus, F., Flusser, J.: Image representation via a finite Radon transform. IEEE Transactions on Pattern Analysis & Machine Intelligence **15** (1993) 996–1006
2. Hsung, T., Lun, D., Siu, W.: The discrete periodic Radon transform. IEEE Transactions on Signal Processing **44** (1996) 2651–2657
3. Kingston, A.: Orthogonal discrete Radon transform over $p^n$. Signal Processing (submitted November 2003)
4. Beylkin, G.: Discrete radon transform. IEEE Transactions on Acoustics, Speech, & Signal Processing **35** (1987) 162–172
5. Svalbe, I., van der Spek, D.: Reconstruction of tomographic images using analog projections and the digital Radon transform. Linear Algebra and Its Applications **339** (2001) 125–145
6. Kingston, A., Svalbe, I.: Adaptive discrete Radon transforms for grayscale images. In: Electronic Notes in Discrete Mathematics. Volume 12., Elsevier (2003)
7. Lun, D., Hsung, T., Shen, T.: Orthogonal discrete periodic Radon transform. Part I: theory and realization. Signal Processing **83** (2003) 941–955
8. Rote, G.: Finding the shortest vector in a two dimensional lattice modulo m. Theoretical Computer Science **172** (1997) 303–308
9. Svalbe, I., Kingston, A.: Farey sequences and discrete Radon transform projection angles. In: Electronic Notes in Discrete Mathematics. Volume 12., Elsevier (2003)
10. Hardy, G., Wright, E.: An introduction to the theory of numbers. 5 edn. Clarendon Press, Oxford (1979)
11. Kak, A., Slaney, M.: Principles of Computerized Tomographic Imaging. IEEE Press (1988)

# Two Remarks on Reconstructing Binary Vectors from Their Absorbed Projections

Attila Kuba[1] and Gerhard J. Woeginger[2]

[1] Department of Image Processing and Computer Graphics,
University of Szeged, H-6720, Szeged Árpád tér 2., Hungary
kuba@inf.u-szeged.hu
[2] Department of Mathematics and Computer Science,
TU Eindhoven, P.O. Box 513, 5600 MB Eindhoven, The Netherlands
kgwoegi@win.tue.nl

**Abstract.** We prove two small results on the reconstruction of binary matrices from their absorbed projections: (1) If the absorption constant is the positive root of $x^2 + x - 1 = 0$, then every row is uniquely determined by its left and right projections. (2) If the absorption constant is the root of $x^4 - x^3 - x^2 - x + 1 = 0$ with $0 < x < 1$, then in general a row is not uniquely determined by its left and right projections.

## 1 Introduction

The reconstruction of binary matrices from their row and column sums is a basic problem in discrete tomography, and work on this problem goes back to a seminal paper of Ryser [5] from 1957. Recently, Kuba and Nivat [4] introduced a new discrete tomography model that they call the *emission discrete tomography* model, EDT for short. In this EDT model, the whole space is filled with some homogeneous partially absorbing material, and the function to be reconstructed represents an object emitting radioactive rays into the surrounding space. Hence, the measurements in EDT are *absorbed* projections that depend on both, the emitting object and the absorption.

Formally, a *picture* is a binary $m \times n$ matrix $A = (a_{i,j})_{m \times n}$. The entries $a_{i,j}$ are from $\{0, 1\}$, and they are sometimes called *pixels*. Measurements of the picture are taken by sending rays through the material; since the material is partially absorbing, these rays become gradually weaker as they proceed through the material. In [2, 3, 4], the absorption behavior is characterized by the *absorption constant*

$$\beta \; = \; \frac{1}{2}(-1 + \sqrt{5}), \tag{1}$$

where $\beta \approx 0.618$ is the positive root of the equation

$$\beta^2 + \beta \; = \; 1. \tag{2}$$

E. Andres et al. (Eds.): DGCI 2005, LNCS 3429, pp. 148–152, 2005.

Then the *absorbed left projection* $\text{ALP}_i$ of row $i$ is defined as

$$\text{ALP}_i = \sum_{j=1}^{n} a_{i,j} \cdot \beta^j \tag{3}$$

and the *absorbed right projection* $\text{ARP}_i$ of row $i$ is defined as

$$\text{ARP}_i = \sum_{j=1}^{n} a_{i,j} \cdot \beta^{n+1-j}. \tag{4}$$

Intuitively speaking, in the left projection the ray enters the row from the left hand side. As the ray passes through the pixels at the beginning of the row, it is still strong, and provides precise information. Then the ray becomes gradually absorbed, and the factors $\beta^j$ in (3) encode this absorption. In the right projection, the ray enters the row symmetrically from the right hand side, and this leads to the symmetric formula given in (4). Absorbed upward and downward projections of columns are defined analogously.

The papers by Kuba and Nivat [4], Balogh et al. [2], and Kuba et al. [3] discuss a variety of algorithmical and combinatorial questions around reconstruction problems with absorbed projections. Noteworthy, all these papers only study the cases with absorbed *left* projections of rows (but without right projections) and with absorbed *upward* projections of columns (but without downward projections). Barcucci et al. [1] proved that the left and right absorbed projections determine uniquely the binary matrix if the absorbtion coefficient is $\mu = \log((1 + \sqrt{5})/2)$ and they gave an algorithm for the reconstruction.

In this note, we will show that the corresponding problems where the left and right absorbed row projections are known simultaneously are rigid and highly constrained. Our combinatorial proof is different from the proof given by Barcucci et al. [1] in the sense that it does not use the concenpt of switching component.

**Theorem 1.** *In the EDT model with absorption constant $\beta = \frac{1}{2}(-1 + \sqrt{5})$, the left and the right absorbed projections $\text{ALP}_i$ and $\text{ARP}_i$ determine the row $r_i$ uniquely.*

Up to now, we have only considered the basic EDT model with absorption constant $\beta = \frac{1}{2}(-1 + \sqrt{5})$ (which arguably is the simplest non-trivial EDT model). Interestingly, the statement in Theorem 1 does *not* generalize to arbitrary values of the absorption constant:

**Theorem 2.** *Consider the EDT model where the absorption constant*

$$\gamma = \frac{1}{4}\left(\sqrt{13} + 1 - \sqrt{2\sqrt{13} - 2}\right) \approx 0.581 \tag{5}$$

*is the unique real root of $\gamma^4 - \gamma^3 - \gamma^2 - \gamma + 1 = 0$ that satisfies $0 < \gamma < 1$. Then the left and the right absorbed projections $\text{ALP}_i$ and $\text{ARP}_i$ in general do not determine a row $r_i$ uniquely.*

## 2   Proof of the Unique Reconstruction Result

In this section we prove Theorem 1. Throughout, let $\beta = \frac{1}{2}(-1+\sqrt{5})$ be defined as in (2), and let $\alpha = \frac{1}{2}(1+\sqrt{5})$ denote the positive root of $\alpha^2 = \alpha + 1$. Note that $\alpha = 1/\beta$. Furthermore, we define the Fibonacci numbers as usual by

$$F_{-3} = -1 \qquad F_{-2} = 1 \qquad F_{-1} = 0 \qquad F_0 = 1 \qquad F_1 = 1 \qquad (6)$$

and by $F_n = F_{n-1} + F_{n-2}$ for all $n \geq 2$.

**Proposition 1.** *The following statements on the Fibonacci numbers are well-known:*

*(i) $\alpha^i = F_{i-1}\alpha + F_{i-2}$ for all $i \geq 0$.*
*(ii) $\beta^i = (-1)^{i-1}F_{i-1}\beta + (-1)^{i-2}F_{i-2}$ for all $i \geq 0$.*
*(iii) $F_{i+2} - 1 = F_i + F_{i-1} + \cdots + F_1 + F_0$ for all $i \geq 0$.*

*Proof.* By induction on $i$.                                                    □

**Proposition 2.** *Let $x_i \in \{-1, 0, 1\}$ for $i \geq 0$. Then the following statements hold true:*

*(i) If $\sum_{i=0}^{\ell} F_{2i} \cdot x_{2i} = 0$, then $x_{2i} = 0$ for $i = 0, \ldots, \ell$.*
*(ii) If $\sum_{i=0}^{\ell} F_{2i+1} \cdot x_{2i+1} = 0$, then $x_{2i+1} = 0$ for $i = 0, \ldots, \ell$.*

*Proof.* Proof of statement (i). Suppose for the sake of contradiction that $x_{2i} \neq 0$ for some $i$, and consider the largest index $k$ with $x_{2k} \neq 0$. Without loss of generality, we may assume that $x_{2k} = -1$ (and otherwise, we multiply all $x_{2k}$ by $-1$). Then $\sum_{i=1}^{k-1} F_{2i} \cdot x_{2i} - F_{2k} = 0$. This leads to

$$F_{2k} = \sum_{i=0}^{k-1} F_{2i} \cdot x_{2i} \leq \sum_{i=0}^{k-1} F_{2i}$$

$$= \sum_{i=0}^{k-1} F_{2i-2} + F_{2i-1} = \sum_{i=-2}^{2k-3} F_i < 1 + \sum_{i=0}^{2k-2} F_i.$$

This blatantly contradicts statement (iii) in Proposition 1, and proves statement (i). Statement (ii) can be handled similarly: Let $k$ be the largest index $k$ with $x_{2k+1} \neq 0$, and assume that $x_{2k+1} = -1$. Then

$$F_{2k+1} = \sum_{i=0}^{k-1} F_{2i+1} \cdot x_{2i+1} \leq \sum_{i=0}^{k-1} F_{2i+1}$$

$$= \sum_{i=0}^{k-1} F_{2i-1} + F_{2i} = \sum_{i=-1}^{2k-2} F_i = \sum_{i=0}^{2k-2} F_i.$$

This contradicts Proposition 1.(iii) and completes the proof of statement (ii).   □

Now consider two rows $r_1 = \langle a_1, \ldots, a_n \rangle$ and $r_2 = \langle b_1, \ldots, b_n \rangle$ with pixels $a_i, b_i \in \{0, 1\}$ in some binary picture. We assume that the left projections of these two rows are identical, i.e.,

$$\sum_{i=1}^{n} a_i \, \beta^i = \sum_{i=1}^{n} b_i \, \beta^i, \tag{7}$$

and that also their right projections are identical, i.e.,

$$\sum_{i=1}^{n} a_i \, \beta^{n+1-i} = \sum_{i=1}^{n} b_i \, \beta^{n+1-i}. \tag{8}$$

For $i = 1, \ldots, n$ we define $c_i = a_i - b_i$ with $c_i \in \{-1, 0, +1\}$. Furthermore, we introduce the four auxiliary quantities $A = \sum_{i=1}^{n} F_{i-2} \, c_i$, $B = \sum_{i=1}^{n} F_{i-3} \, c_i$, $C = \sum_{i=1}^{n} (-1)^{i-2} F_{i-2} \, c_i$, and $D = \sum_{i=1}^{n} (-1)^{i-3} F_{i-3} \, c_i$.

Now Equation (7) can be written as $\sum_{i=1}^{n} \beta^i c_i = 0$. By using statements (i) and (ii) in Proposition 1, this leads to

$$0 = \sum_{i=1}^{n} \beta^i c_i = \sum_{i=1}^{n} ((-1)^{i-1} F_{i-1}\beta + (-1)^{i-2} F_{i-2}) c_i = \beta(D - C) + C. \tag{9}$$

Similarly, Equation (8) can be written as $\sum_{i=1}^{n} \alpha^i c_i = 0$ and yields

$$0 = \sum_{i=1}^{n} \alpha^i c_i = \sum_{i=1}^{n} (F_{i-1}\alpha + F_{i-2}) c_i = \alpha(A + B) + A. \tag{10}$$

Since $\alpha$ and $\beta$ are irrational numbers, whereas $D - C$, $C$, $A + B$, $A$ are integers, the Equations (9) and (10) imply $D - C = 0$, $C = 0$, $A + B = 0$, and $A = 0$. Hence, $A = B = C = D = 0$ holds. From this we derive

$$0 = A + C = \sum_{i=1}^{n} (1 + (-1)^{i-2}) \, F_{i-2} \, c_i = 2 \sum_{i=1}^{\lfloor n/2 \rfloor} F_{2i-2} c_{2i}. \tag{11}$$

Equation (11) together with Proposition 2.(i) now yields that all $c_i$ with an even index are 0. Furthermore, we derive for the odd indices that $A - C = 0$ and therefore

$$0 = \sum_{i=1}^{n} (1 - (-1)^{i-2}) \, F_{i-2} \, c_i = 2 \sum_{i=1}^{\lceil n/2 \rceil} F_{2i-3} c_{2i-1} = 2 \sum_{i=2}^{\lceil n/2 \rceil} F_{2i-3} c_{2i-1}. \tag{12}$$

Here we used in the final step that $F_{-1} = 0$. Equation (12) together with Proposition 2.(ii) now yields that all $c_i$ with an odd index $i \geq 3$ are 0. What about $c_1$? The value of $B = \sum_{i=1}^{n} F_{i-3} \, c_i$ boils down to $B = F_{-2} c_1 = c_1$, and together with $B = 0$ this implies that also $c_1 = 0$. Summarizing, $c_i = 0$ must hold for all $i = 1, \ldots, n$. This implies $a_i = b_i$ for all $i = 1, \ldots, n$. Hence, the two rows $r_1$ and $r_2$ must be identical, and any row is uniquely determined by its left and right absorbed projections. This completes the proof of Theorem 1.

*Remark 1.* About absorption values for which any binary matrix is uniquely determined by its absorbed row sums see Section 6 in [3].

# 3   Proof of the Non-unique Reconstruction Result

In this section we prove Theorem 2. Hence, consider an absorption constant $\gamma$ with $0 < \gamma < 1$ that is a root of $1 + \gamma^4 = \gamma + \gamma^2 + \gamma^3$ as defined in (5). Consider a $2 \times 10$ matrix with two rows $r_1 = \langle 1,0,0,0,1,1,0,0,0,1 \rangle$ and $r_2 = \langle 0,1,1,1,0,0,1,1,1,0 \rangle$. Then

$$\begin{aligned}
\mathrm{ALP}_1 &= \gamma + \gamma^5 + \gamma^6 + \gamma^{10} = (\gamma + \gamma^6)(1 + \gamma^4) \\
&= (\gamma + \gamma^6)(\gamma + \gamma^2 + \gamma^3) \\
&= \gamma^2 + \gamma^3 + \gamma^4 + \gamma^7 + \gamma^8 + \gamma^9 = \mathrm{ALP}_2.
\end{aligned}$$

Since rows $r_1$ and $r_2$ both are left-right symmetric, this chain of equations also yields $\mathrm{ARP}_1 = \mathrm{ARP}_2$. Summarizing, these two rows both have the same absorbed left projection and both have the same absorbed right projection. Generally, it is not possible to uniquely reconstruct a row from its two projections. This proves Theorem 2.

*Remark 2.* Of course, the left and right absorbed projections determine the binary vector uniquely if its size is small enough, that is, if $n \leq 4$. The explanation is that a binary vector is non-uniquely determined by its left and right absorbed projections if and only if the equation $1 + \gamma^4 = \gamma + \gamma^2 + \gamma^3$ can be applied in the description of its value. This equation requires at least 5 binary digits.

## Acknowledgment

This work was supported by the NSF Grant DMS 0306215.

## References

1. Barcucci, E., Frosini, A., Rinaldi, S.: Reconstruction of discrete sets from two absorbed projections: an algorithm. Electronic Notes on Discrete Mathematics **12** (2003)
2. Balogh, E., Kuba, A., Del Lungo, A., Nivat, M.: Reconstruction of binary matrices from absorbed projections. In Proceedings of the 10th International Conference on Discrete Geometry for Computer Imagery (DGCI'2002), Springer LNCS **2301** (2002) 392–403
3. Kuba, A., Nagy, A., Balogh, E.: Reconstuction of hv-convex binary matrices from their absorbed projections. Discrete Applied Mathematics **139** (2004) 137–148
4. Kuba, A., Nivat, M.: Reconstruction of discrete sets with absorption. Linear Algebra and Applications **339** (2001) 171–194
5. Ryser, H.J.: Combinatorial properties of matrices of zeros and ones. Canadian Journal of Mathematics **9** (1957) 371–377

# How to Obtain a Lattice Basis from a Discrete Projected Space

Nicolas Normand, Myriam Servières, and JeanPierre Guédon

IRCCyN/IVC, École polytechnique, University of Nantes,
Rue Christian Pauc, 44306 Nantes

**Abstract.** Euclidean spaces of dimension $n$ are characterized in discrete spaces by the choice of lattices. The goal of this paper is to provide a simple algorithm finding a lattice onto subspaces of lower dimensions onto which these discrete spaces are projected. This first obtained by depicting a tile in a space of dimension $n - 1$ when starting from an hypercubic grid in dimension $n$. Iterating this process across dimensions gives the final result.

## 1 Introduction

Regular lattices constitute the cornerstone for the building of discrete geometry tools and also for bases of classical continuous space of functions. Of course, these regular lattices can be defined without any outside reference. However, the definition of the resulting lattice is not obvious when a problem is designed in a discrete space and when it is mandatory to go back and forth from this first space to an other discrete space using a given discrete transform. When the same problem is entirely defined in a discrete manner, the lattice identification problem can become even harder. This paradigm was used to construct a cryptographic/signature scheme using the NTRU lattice [1].In this case, the lattice identification leads to an NP-problem.

Lattice identifications have also been investigated by Conway and Sloane for sphere packing which results are mainly employed for vector quantization in the multimedia coding area [2]. In Sect. 2, the starting point to review the literature will lie into the continuous/discrete correspondence firstly established by Shannon and generalized by Unser-Aldroubi. This work allows to start with a basis through a tensorial product, to sample the continuous space to give a lattice onto which a Riesz functional basis will be defined.

The aim of this paper is then to demonstrate how to obtain an unitary tile with a regular lattice on the projection hyperplane with discrete projection directions. This will be performed in a general manner in Sect. 3. The fact that we restrict the projection operator to discrete projection is based on the attempt to get a simple way to obtain a lattice onto the hyperplane. The difficulty is then to extract the lattice from regular projection grids, *i.e.* not to oversample the hyperplane grid with unused points nor to undersample a grid from which the lattice would not be obtained.

E. Andres et al. (Eds.): DGCI 2005, LNCS 3429, pp. 153–160, 2005.
© Springer-Verlag Berlin Heidelberg 2005

In other words, each point of the discrete projection plane must have a pre-decessor in the initial space and each point of the initial lattice is projected onto an existing point.

## 2   Related Work

Starting with the construction of orthonormal bases that give regular tiling leads to the Gram-Schmidt orthonormalisation procedure that can be found in almost any algebra textbook [3]. Because of the normalization, the resulting basis can be easily discretised and replicated to give a regular tiling.

Following this path (starting from the continuous point of view and discretizing afterward) any $n$ dimensional continuous space will give regular tiling from this unconstrained orthonormal continuous grid. As a matter of fact, Unser and Aldroubi have generalized the Shannon-Whittaker-Kotelnikov Sampling Theorem starting with this $n$-dimensional orthogonal basis then lattice [4, 5]. The initial purpose of this theorem is to use other functional bases than the Riesz bases $\{(\operatorname{sinc}(\mathrm{kx}), k \in \mathbb{Z})\}$.

This theorem is described in Fig. 1

The first step corresponds to the orthogonal projection of the $L_2$ function $f(x)$ onto a closed subspace of functions generated by a Riesz basis $\{\eta(kx), k \in \mathbb{Z}\}$. In other words, these functions already need to be defined from a tiling (the first versions of the theorem corresponds to cardinal functions as the sinc for Shannon or cardinal splines for Unser-Aldroubi).

The second step also uses the same tiling but explicitly since it just picks up the values onto the tiling and throw out the rest of the continuous functions.

The third step re-generates this previous continuous function from three different informations:

1. the dual functional basis $\mathring{\eta}$ (defined onto the tile)
2. the sample $f_\eta(k)$ (defined onto the tile)
3. the tile which allows to perform the discrete convolution that lies onto the box 3.

There are two major points with this great theorem:

1. The tiling is the subsequent material that links continuous and discrete words. The strength of this theorem is to allow to work only into a discrete word $f_\eta(k)$ and go back into a continuous word only when mandatory.

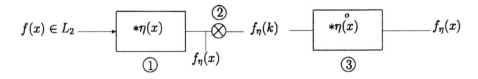

**Fig. 1.** Representation of the Unser-Aldroubi theorem

2. The only tilling known to allow the conditions of the theorem are obtained by tensorial products over higher dimensions. In other words, the Riesz basis structure in one dimension $\{\eta(k-x), k \in \mathbb{Z}\}$ can not be used in two dimensions as $\{\eta(\sqrt{((k-l)^2+(l-y)^2)}), (k,l) \in \mathbb{Z}^2\}$ but the only known extension is $\{\eta(k-l).\eta(l-y), (k,l) \in \mathbb{Z}^2\}$. In this latter extension, Fig. 1 where $x$ is a $n$-dimensional vector still holds.

As a consequence using specific discrete operator (in our case a projector operator) between step 2 and step 3 must be done with a correspondence between grids not to loose information and the benefits of the theorem. This correspondance has been applied for the continuous and discrete Radon transforms defined into spline spaces leading to new filtered backprojection algorithms [7, 8].

# 3   Obtaining a Tile in a $m$-Dimensional Space from a $n$-Dimensional Space

## 3.1   From $n$-Dimension to $(n-1)$-Dimension

The initial space $\mathcal{L}$ is a lattice in a Euclidean space. This $n$-dimensional discrete space can be seen as the regular sampling of a continuous $n$-dimensional space structured by a hypercubic grid $\{i_1, \ldots, i_n\}$. Each point $(b_1, \ldots, b_n) \in \mathbb{Z}^n$ in the discrete space corresponds to the point $b_1 \times i_1 + \ldots + b_n \times i_n$ in the continuous space (each lattice point is described by a $n$-dimensional vector $b = \{b_1, \ldots, b_n\}$ relative to the lattice basis $\{i_1, \ldots, i_n\}$). The continuous space is translation invariant according to any vector $i_m$ or any integer combination of vectors $i_1, \ldots, i_n$.

By projecting the initial $n$-dimensional space along a line direction, we create a $(n-1)$-dimensional hyperplane. It can be easily seen (Fig. 2) that if the line direction is discrete (an integer combination of $i_1, \ldots, i_n$), then the hyperplane has a regular discrete structure: it is also a lattice. It is translation-invariant along any integer combination of $i'_1, \ldots, i'_n$, where each $i'_m$ is the projection of $i_m$ on the hyperplane. However, the set $\{i'_1, \ldots, i'_n\}$ is not a base (its dimension is $n-1$) and a $n-1$-vector subset does not generally define a tile.

The purpose is to extract a $n-1$-vector basis from $\{i'_1, \ldots, i'_n\}$ that defines a lattice basis for the projected hyperplane. Equivalently, it will lead to a discrete $(n-1) \times n$ projection matrix.

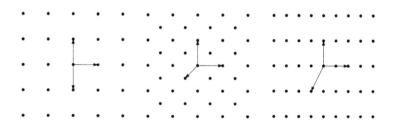

**Fig. 2.** Some examples of 3D projected grids with projection directions $(0, 1, 1)$, $(1, 1, 2)$ and $(1, 2, 2)$

The proposed method will conceptually use the set of vectors $\{i'_1, \ldots, i'_n\}$, obtained by projecting $\{i_1, \ldots, i_n\}$ along the line direction $(v_1, \ldots, v_n)$ onto the hyperplane. The relationship that links these vectors together is given by the projection direction:

$$v_1 \times i'_1 + \ldots + v_n \times i'_n = 0 \ . \tag{1}$$

In the following, we will assume that the subset $\{i'_1, \ldots, i'_{n-1}\}$ is a vector basis (i.e. that $i'_n$ is a linear combination of these vectors). Hence, $i'_1, \ldots, i'_{n-1}$ generates the continuous projected hyperplane. This is always true except when the last component of the projection direction, $v_n$, is zero. In this particular case, (1) creates a linear dependence between the vectors $i'_1, \ldots, i'_{n-1}$. But since the projection direction is not the null vector, there is at least one non-zero $v_m$ component. The previous assumption can be ensured by permuting the vectors twice: before applying the method in order to put $i'_m$ at the end and after applying the method to put the vectors back in the initial order.

The method iteratively creates a set of lattice basis $(j_1, \ldots, j_k)$ with increasing dimension $k$. Each intermediate basis $(j_1, \ldots, j_k)$ generates the part of $\mathcal{L}$ contained in the space spanned by $(i'_1, \ldots, i'_k)$. At step $k$, the $k-1$ previously found vectors build a tile i.e. a parallelepiped which vertices correspond to projected discrete points and which interior does not contain any discrete points.

**First Basis Vector.** The first vector $j_1$ is chosen following the $i'_1$ direction from the origin point $O$. The end point of $j_1$ is the visible point in $i'_1$ direction (the closest point from $O$ in this direction). Since the dimension of the projection matrix is $n-1$ for a line projection, there are exactly two linearly independant ways to follow this direction, either with $i'_1$ or with a linear combination of $i'_2, \ldots, i'_n$ according to (1):

$$v_1 \times i'_1 = -v_2 \times i'_2 - \ldots - v_n \times i'_n \ . \tag{2}$$

The second term of this equation can be written as an integral linear combination with a division by the greatest common divisor of its coefficients:

$$\frac{v_1}{\gcd(v_2, \ldots, v_n)} \times i'_1 = -\frac{v_2 \times i'_2 + \ldots + v_n \times i'_n}{\gcd(v_2, \ldots, v_n)} \ . \tag{3}$$

All the linear combinations of $i'_1$ and $v_1 i'_1 / \gcd(v_2, \ldots, v_n)$ are obviously collinear to $i'_1$. The visible point from the origin $O$ in this direction is given by the minimal linear combination with integer coefficients. Let's remark that $\frac{\gcd(v_2, \ldots, v_n)}{\gcd(v_1, \ldots, v_n)}$ and $\frac{v_1}{\gcd(v_1, \ldots, v_n)}$ are integers and relatively prime. Following the Bézout's theorem, there exist $\alpha_1, \beta_1 \in \mathbb{Z}$ such that:

$$\alpha_1 \times \frac{\gcd(v_2, \ldots, v_n)}{\gcd(v_1, \ldots, v_n)} + \beta_1 \times \frac{v_1}{\gcd(v_1, \ldots, v_n)} = 1$$

$$\alpha_1 + \beta_1 \frac{v_1}{\gcd(v_2, \ldots, v_n)} = \frac{\gcd(v_1, \ldots, v_n)}{\gcd(v_2, \ldots, v_n)} \ . \tag{4}$$

The minimal integral combination of $i'_1$ and $\frac{v_1 i'_1}{\gcd(v_1,\ldots,v_n)}$ is then chosen as the first vector of the projected lattice basis:

$$j_1 = \frac{\gcd(v_1, \ldots, v_n)}{\gcd(v_2, \ldots, v_n)} i'_1 .$$

From $\alpha_1$ and $\beta_1$ (obtained with the extended Euclidean algorithm) we can find two points that project onto $j_1$:

$$A_1 = \left( \alpha_1, -\beta_1 \frac{v_2}{\gcd(v_2, \ldots, v_n)}, \ldots, -\beta_1 \frac{v_n}{\gcd(v_2, \ldots, v_n)} \right) ,$$

$$B_1 = \left( \alpha_1 + \beta_1 \frac{v_1}{\gcd(v_2, \ldots, v_n)}, \underbrace{0, \ldots, 0}_{n-1} \right) . \tag{5}$$

Conversely to $B_1$, $A_1$ always has integer components and thus belongs to $\mathcal{L}$.

**The $k^{th}$ Basis Vector.** Let assume that vectors $j_1$ to $j_{k-1}$ have already been found. The lattice spanned by $(j_1, \ldots, j_{k-1})$ is the subset of $\mathcal{L}$ restricted to the continuous subspace generated by $(i'_1, \ldots, i'_{k-1})$.

The vector $i'_k$ introduces a new dimension because $i'_1, \ldots, i'_k$ are linearly independent. $j_k$ must have the smallest non null component in this new direction. There are two independent ways to move along $i'_k$, following $i'_k$ itself or a linear combination of $i'_{k+1}, \ldots, i'_n$ according to (1):

$$\frac{v_1 i'_1 + \ldots + v_k i'_k}{\gcd(v_{k+1}, \ldots, v_n)} = -\frac{v_{k+1} i'_{k+1} + \ldots + v_n i'_n}{\gcd(v_{k+1}, \ldots, v_n)} . \tag{6}$$

The closeness of the hyperplanes to the origin is measured by the projection onto $i'_k$ and gives respectively 1 and $\frac{v_k}{\gcd(v_{k+1}, \ldots, v_n)}$. The minimum integral combination is given by $\alpha_k$ and $\beta_k$:

$$\alpha_k + \beta_k \frac{v_k}{\gcd(v_{k+1}, \ldots, v_n)} = \frac{\gcd(v_k, \ldots, v_n)}{\gcd(v_{k+1}, \ldots, v_n)} . \tag{7}$$

The new basis vector $j_k$ is directly derived:

$$j_k = \alpha_k i'_k + \beta_k \frac{v_1 i'_1 + \ldots + v_k i'_k}{\gcd(v_{k+1}, \ldots, v_n)} . \tag{8}$$

Two antecedents of $j_k$ can be obtained by:

$$A_k = \left( \underbrace{0, \ldots, 0}_{k-1}, \alpha_k, -\beta_k \frac{v_{k+1}}{\gcd(v_{k+1}..v_n)}, \ldots, -\beta_k \frac{v_n}{\gcd(v_{k+1}..v_n)} \right) ,$$

$$B_k = \left( \frac{\beta_k v_1}{\gcd(v_{k+1}..v_n)}, \ldots, \frac{\beta_k v_{k-1}}{\gcd(v_{k+1}..v_n)}, \alpha_k + \frac{\beta_k v_k}{\gcd(v_{k+1}..v_n)}, \underbrace{0, \ldots, 0}_{n-k} \right) . \tag{9}$$

**Projection Matrix.** Any point $M(a_1, \ldots, a_n)$ in $\mathcal{L}$ is projected on the hyperplane to a point $p(M) = M'$ with integer coordinates $(b_1, \ldots, b_{n-1})$ in the $(j_1, \ldots, j_{n-1})$ basis. The preimage of $M'$ is the set of points in $\mathcal{L}$ that are aligned with $M$ relative to the projection direction $V$. These points can be reached from the known point $b_1 A_1 + \ldots + b_{n-1} A_{n-1}$:

$$p^{-1}(M') = \{m | p(m) = M'\} = \left\{ b_1 A_1 + \ldots + b_{n-1} A_n + k \frac{V}{\gcd(v_1..v_n)}, k \in \mathbb{Z} \right\} .$$
(10)

$(A_1, \ldots, A_{n-1}, V / \gcd(v_1..v_n))$ is a basis of the initial lattice $\mathcal{L}$. An extra row corresponding to the direction of the projection is added:

$$A = \left[ A_1 | \ldots | A_{n-1} | \frac{V}{\gcd(v_1..v_n)} \right]$$

$$= \begin{bmatrix} \alpha_1 & 0 & \cdots & 0 & \frac{v_1}{\gcd(v_1..v_n)} \\ -\beta_1 \frac{v_2}{\gcd(v_2..v_n)} & \alpha_2 & \ddots & \vdots & \vdots \\ -\beta_1 \frac{v_3}{\gcd(v_3..v_n)} & -\beta_2 \frac{v_3}{\gcd(v_3..v_n)} & \ddots & 0 & \vdots \\ \vdots & \vdots & & \alpha_{n-1} & \frac{v_{n-1}}{\gcd(v_1..v_n)} \\ -\beta_1 \frac{v_n}{\gcd(v_2..v_n)} & -\beta_2 \frac{v_n}{\gcd(v_3..v_n)} & \cdots & -\beta_{n1} \frac{v_n}{\gcd(v_{n-1},v_n)} & \frac{v_n}{\gcd(v_1..v_n)} \end{bmatrix} . \quad (11)$$

$$P = [Id_{n-1} | 0] . A^{-1} . \quad (12)$$

Figure 3 pictures the results of our algorithm for a simple 3D to 2D projection of angle direction $(6, 10, 15)$. The projection matrix is given by:

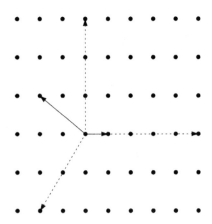

**Fig. 3.** The result of the projection of a 3D lattice with angle direction $(6, 10, 15)$. The three dashed lines vectors represent the projection of the initial 3D basis vectors whereas the two plain lines vectors are the result of the computed lattice

$$P = \begin{bmatrix} 5 & 6 & -6 \\ 0 & 3 & -2 \end{bmatrix} ,$$

and

$$A = \begin{bmatrix} -1 & 0 & 6 \\ -2 & 1 & 10 \\ -3 & 1 & 15 \end{bmatrix} .$$

## 3.2    From $(n-1)$-Dimension to $m$-Dimension

To go from a $n$-dimensional space to a $m$-dimensional space $(m \geq 1)$, there are $(n-m)$ projection directions $V$:

$$V = \begin{bmatrix} v_{1,1} & \cdots & v_{1,n} \\ \vdots & & \vdots \\ v_{n-m,1} & \cdots & v_{n-m,n} \end{bmatrix} . \tag{13}$$

The described method is followed along the first direction:

$$V_1 = \begin{bmatrix} v_{1,1} & \cdots & v_{1,n} \end{bmatrix} , \tag{14}$$

and gives the transformation of the basis $(i_1, \ldots, i_n)$ to $(j_1, \ldots, j_{n-1})$. Each $V_i, i \in \{2, \ldots, (n-m)\}$ is projected onto $(j_1, \ldots, j_{n-1})$. They give the projection directions in the $(n-1)$-dimensional space with the $(j_1, \ldots, j_{n-1})$ basis. The process is iterated to reach a $m$-dimensional space.

# 4    The Projection Matrix

We will define the projection matrix from a n-dimensional space onto a m-dimensional space on the regular grid defined before.

To obtain the final projection matrix, the projection directions can be followed in different order. One projection direction is chosen, the other projection directions are projected following this first direction and the projection matrix following this direction is derivated. In the projected direction, an other direction is chosen and the same process is iterated. Finally the projection matrix from the $n$-dimensional space to the $m$-dimensional space is obtained by putting together the intermediate matrix.

To go from a n-dimensional space to an m-dimensional space $(m \geq 1)$, there are $(n-m)$ projection directions $V_k$.

To obtain the projection matrix from a $n$-dimensional to an $m$-dimensional space we will calculate the projection matrix step by step by lowering the dimension. The final $nD$ to $mD$ projection matrix is composed by products of all the projection matrices describing hyperplane lattices. This matrix being the product of integer matrices has only integer coefficients.

## 5    Conclusion

Tiling a discrete projected space from higher dimensional Euclidean spaces was the subject of this paper. We first demonstrated how to obtain a tile in a $(n-1)$-dimensional space from the Euclidean hypercubic grid of dimension $n$. The generalization across dimensions is quite straightforward. The obtained results are directly useful for discrete Radon transforms.

## References

1. Jeffrey Hoffstein, Jill Pipher, Joseph H. Silverman: NTRU A Ring-Based Public Key Cryptosystem. in Algorithmic Number Theory (ANTS III), Portland, OR, June 1998, J.P. Buhler (ed.), Lecture Notes in Computer Science 1423, Springer-Verlag, Berlin, 1998, 267-288.
2. John H. Conway and N.J.A. Sloan: Sphere Packings, Lattices and Groups. Springer(1998)
3. Harris, J.W., Stocker, H.: Handbook of Mathematics and Computational Science. Springer(1998)
4. Aldroubi, A., Unser, M: Sampling procedures in function spaces and asymptotic equivalence with Shannonś sampling theory. Numer. Funct. Anal. and Optimiz. **15**(1994)1–21
5. Unser, M., Aldroubi, A., Eden, M.: Polynomial Spline Signal Approximations: Filter Desing and Assymptotic Equivalence with Shannonś Sampling Theorem. IEEE Transaction on Information theory **38**(1992)95–103
6. Guédon, JP., Bizais, Y.: Separable and radial bases for medical image processing. SPIE lmage Processing **1898**(1993)652–661
7. Guédon, JP., Bizais, Y.: Bandlimited and Haar Filtered Back-Projection Reconstuction. IEEE Transaction on Medical Imaging **13**(1994)430–440
8. Guédon, J.P., Unser, M., Bizais, Y.: Pixel Intensity Distribution Models for Filtered Back-Projection. Conference Record of the 1991 IEEE Nuclear Science Symposium and Medical Imaging Conference

# Local Characterization of a Maximum Set of Digital $(26, 6)$-Surfaces[*]

Jose C. Ciria[1], Angel de Miguel[1], Eladio Domínguez[1], Angel R. Francés[1], and Antonio Quintero[2]

[1] Dpt. de Informática e Ingeniería de Sistemas, Facultad de Ciencias, Universidad de Zaragoza, E-50009 – Zaragoza, Spain
{jcciria, admiguel, afrances}@unizar.es
[2] Dpt. de Geometría y Topología, Facultad de Matemáticas, Universidad de Sevilla, Apto. 1160, E-41080 – Sevilla, Spain
quintero@us.es

**Abstract.** This paper provides a local characterization for a set of digital surfaces $S^U$ defined in [6] by mean of continuous analogues. For this, we firstly identify the set of admissible plates for any surface $S \in S^U$ (i.e., the intersection $S \cap C$ of $S$ with a unit cube $C$ of $\mathbb{Z}^3$). Then, the characterization is given in terms of a graph representing the intersection of plates. In addition, we establish a further condition that detects the digital surfaces in $S^U$ which are strongly separating objects.

The family $S^U$ consists of all objects which are a digital surface in some homogeneous $(26,6)$-connected digital space in the sense of [3]. Moreover, the subset of strongly separating surfaces of $S^U$ contains the family of simplicity 26-surfaces and other surfaces in literature as well.

## 1 Introduction

One of the most interesting problems in Digital Topology is, probably, to obtain a general notion of digital surface, defined as a "thin" set of voxels (or, equivalently, points of $\mathbb{Z}^3$), that naturally extends to higher dimensions and such that these digital surfaces have properties similar to those held by topological surfaces. In this paper we restrict our interest to such a notion for the usual $(26,6)$-adjacency in image processing. Morgenthaler and Rosenfeld gave in [10] the first definition of digital surface for this adjacency pair. Later on, Bertrand and Malgouyres [4] introduced the family of *strong 26-surfaces* and showed that it strictly contains the set of Morgenthaler's surfaces. More recently, Couprie and Bertrand [7] have proved that the strong 26-surfaces are still contained in a larger set of surfaces called *simplicity 26-surfaces*. Despite of these contributions the most general definition of digital surface for the $(26,6)$-adjacency is not yet clear. For example, in [9] it is suggested that the $(26,6)$-connected digital object

---

[*] This work has been partially supported by the projects BFM2001-3195-C03-01 and BFM2001-3195-C03-02 (MCYT Spain).

E. Andres et al. (Eds.): DGCI 2005, LNCS 3429, pp. 161–171, 2005.

shown Fig. 3(a) "should normally be interpreted as a simple surface", however it is not a simplicity 26-surface. Thus, a new definition of digital surface is needed to include this kind of objects.

Within the framework for Digital Topology proposed in [3] we have recently found in [6] a homogeneous $(26, 6)$-connected digital space $E^U$ whose set of digital surfaces $S^U$ is the largest in that class of digital spaces. In particular, the Jordan–Brouwer and Index Theorems, proved in [3] and [2] for digital manifolds of arbitrary dimensions, automatically hold for these surfaces. Moreover the set $S^U$ strictly contains the family of simplicity 26-surfaces and the object in Fig. 3(a) as well.

In spite of these nice properties, the digital surfaces in $S^U$ were defined in [6] by constructing a continuous analogue, and thus it might not be considered a completely digital notion. In this paper we give a purely digital characterization of the family $S^U$ by the use of plates and graphs extending the Kong and Roscoe method in [8] (see Theorem 3).

The digital objects characterizing the surfaces in $S^U$ are introduced in Section 2. In Section 3 we recall the basic elements of the framework in [3], and the definition of $E^U$, needed to obtained this characterization in Section 4.

Despite of all surfaces in $S^U$ are *Jordan* objects, some of them might be considered as pathological examples since the deletion of one of their points might yield an object that still separates its complement into two 6-components. In Section 5, we state a characterization of the subset of non-pathological surfaces of $S^U$, which still strictly contains the family of simplicity 26-surfaces [6].

As a conclusion, the results in [6] and this work lead to a good and general enough approach to a notion of digital surface since:

1. it is suitable for image processing as it is given in terms of a graph;
2. these digital surfaces have properties similar to those of topological surfaces;
3. and, in a certain sense, they constitute a maximum family of digital surfaces that, as far as we know, contains any set of surfaces defined on $\mathbb{Z}^3$ using the graph-based approach in the literature.

## 2   A Set of Jordan Objects

In this section we introduce a family of objects in the discrete space $\mathbb{Z}^3$ that satisfies a Jordan property for the usual $(26, 6)$-adjacency. These objects are made of small surface pieces, called *plates*, which are adequately glued to each other in the way defined by an *assembly graph*. In order to introduce these notions, we firstly recall some basic definitions of the graph–theoretical approach to Digital Topology.

Two voxels $\sigma = (\sigma_1, \sigma_2, \sigma_3), \tau = (\tau_1, \tau_2, \tau_3) \in \mathbb{Z}^3$ are said to be 6-, 18- or 26-adjacent if $\max\{|\sigma_i - \tau_i|; 1 \leq i \leq 3\} \leq 1$ and they differ in, at most, one, two or three of their coordinates, respectively. Moreover, we say that two 18-adjacent voxels are *strictly 18-adjacent* if they are not 6-adjacent. A *unit cube* of $\mathbb{Z}^3$ is any subset $C$ of eight mutually 26-adjacent voxels. Similarly, a *unit square* of $\mathbb{Z}^3$ is a subset of four mutually 18-adjacent voxels.

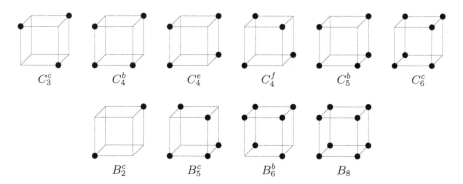

**Fig. 1.** Some patterns that may appear in a digital object $O \subseteq \mathbb{Z}^3$. Each picture represents a unit cube $C$ of $\mathbb{Z}^3$, and the black dots are the set of voxels in $O \cap C$. The row above contains the six non-square plates that may appear in a digital object. The patterns in the row below cannot appear in a 26-presurface

For each subset $A \subseteq \mathbb{Z}^3$, the transitive closure of the $n$-adjacency, $n \in \{6, 18, 26\}$, defines an equivalence relation whose classes are called the $n$-components of $A$. Moreover, $A$ is said to be $n$-connected if it has only one $n$-component.

**Definition 1.** *Let $O \subseteq \mathbb{Z}^3$ be a digital object. A subset $P \subseteq O$ is said to be a plate in $O$ if either $P$ is a unit square of $\mathbb{Z}^3$ or $P = O \cap C$, where $C$ is a unit cube of $\mathbb{Z}^3$, and $P$ correspond (up to rotations and symmetries) to one of the patterns in the set $\mathfrak{C} = \{C_3^c, C_4^b, C_4^e, C_4^f, C_5^b, C_6^c\}$; see Fig. 1 and Remark 3 below.*
    *For any voxel $\sigma \in O$ we denote by $\mathbf{P}(O, \sigma)$ the set of all plates in $O$ containing $\sigma$, and $\mathbf{P}(O)$ the set of all plates in $O$.*

From the set $\mathbf{P}(O)$ of plates in a given object $O$ we construct a bipartite graph $G(O)$, termed the *assembly graph of $O$*, whose nodes are the elements of the set $\mathbf{P}(O) \cup \mathbf{E}(O)$, where each $q \in \mathbf{E}(O) = \cup_{\sigma \in O} \mathbf{E}(O, \sigma)$ is associated to a pair of plates $p_1, p_2$ in $O$ such that $p_1 \cap p_2$ contains at least two voxels (notice that $p_1 \cap p_2$ consists of three voxels at most.) More precisely, the sets $\mathbf{E}(O, \sigma)$ and the edges of $G(O)$ are chosen according to the following criteria. Let $p_1, p_2 \in \mathbf{P}(O)$ be two plates, then:

1. If $p_1 \cap p_2 = \{\tau_1, \tau_2\}$ we consider a new node $q \in \mathbf{E}(O, \tau_1) \cap \mathbf{E}(O, \tau_2)$ and two edges $\langle p_i, q \rangle$, $i = 1, 2$, in $G(O)$.
2. If $p_1 \cap p_2 = \{\tau_0, \tau_1, \tau_2\}$ then we consider two new nodes $q_1$ and $q_2$ and four edges $\langle p_i, q_j \rangle$, $i, j \in \{1, 2\}$, in $G(O)$. Moreover, assume $\tau_0$ is the only voxel in $p_1 \cap p_2$ which is 6-adjacent to both $\tau_1$ and $\tau_2$, then we set $q_j \in \mathbf{E}(O, \tau_0) \cap \mathbf{E}(O, \tau_j)$, $j \in \{1, 2\}$.
3. For each $\sigma \in O$, the set $\mathbf{E}(O, \sigma)$ consists exclusively of the elements $q$ and $q_j$ introduced above for all intersections $p_1 \cap p_2$ containing $\sigma$.

For each voxel $\sigma \in O$ we define the *assembly graph of $O$ around $\sigma$*, $G(O, \sigma)$, as the subgraph of $G(O)$ induced by the nodes in $\mathbf{P}(O, \sigma) \cup \mathbf{E}(O, \sigma)$.

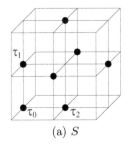

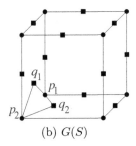

(a) $S$                          (b) $G(S)$

**Fig. 2.** A 26-presurface $S$ and its assembly graph $G(S)$. Each dot in (b) represents one of the eight unit cubes shown in (a), which are actually plates of $S$. Squares in (b) are the points that, together with plates, define the assembly graph of $S$; each one represents the intersection between two plates

**Definition 2.** *A digital object $S \subseteq \mathbb{Z}^3$ is said to be a* 26-presurface *if the following conditions hold for each voxel $\sigma \in S$.*

1. *For each unit cube $C$ of $\mathbb{Z}^3$ the intersection $S \cap C$ does not corresponds (up to rotations and symmetries) to any pattern in $\mathfrak{B} = \{B_2^c, B_5^c, B_6^b, B_8\}$; see Fig. 1.*
2. *If $\tau \in S$ is strictly 18-adjacent to $\sigma$ and no other voxel in $S$ is 6-adjacent to both $\sigma$ and $\tau$ then $\mathbf{P}(S, \sigma) \cap \mathbf{P}(S, \tau) \neq \emptyset$.*
3. *If $\tau \in S$ is 6-adjacent to $\sigma$ then $\mathbf{P}(S, \sigma) \cap \mathbf{P}(S, \tau)$ consists exactly of two plates.*
4. *$\mathbf{P}(S, \sigma) \neq \emptyset$ and $G(S, \sigma)$ is a cycle.*

*Example 1.* **Figure 2** depicts a 26-presurface $S$, made of eight plates, and its assembly graph $G(S)$. Notice that the voxel $\tau_0 \in S$ belongs to exactly two plates $p_1, p_2 \in \mathbf{P}(O)$ for which $p_1 \cap p_2 = \{\tau_0, \tau_1, \tau_2\}$. Hence, $\mathbf{E}(O, \tau_0) = \{q_1, q_2\}$ and the assembly graph of $S$ around $\tau_0$, $G(S, \tau_0)$, is the cycle defined by the vertices $p_1, p_2, q_1$ and $q_2$ in Fig. 2(b).

The main goal in this paper is to show that the assembly graph endows each 26-presurface with the combinatorial structure of a surface. More precisely, we will show in Theorem 3 that 26-presurfaces characterize the digital surfaces of a certain digital space $E^U = (R^3, f^U)$ defined in [6] within the approach to Digital Topology in [3]. In order to state and prove Theorem 3 we recall in the next section the basic elements of this framework and the definition of $E^U$.

*Remark 1.* The assembly graph of an object $O \subseteq \mathbb{Z}^3$ can be derived from a simpler graph whose nodes are the plates in $O$ and two of them are adjacent if their intersection contains at least two voxels. Moreover, the five conditions defining 26-presurfaces can be rewritten in terms of this simpler graph.

## 3    A Framework for Digital Topology

In [3] we propose a framework for Digital Topology in which a notion of digital surface and, more generally, of digital manifold naturally arises. In this approach a *digital space* is a pair $(K, f)$, where $K$ is a polyhedral complex, called *device model*, which represents the spatial layout of voxels and $f$ is a *lighting function* from which we associate to each digital image an Euclidean polyhedron called its continuous analogue.

In this paper we will only consider digital spaces of the form $(R^3, f)$, where the device model $R^3$, termed the *standard cubical decomposition* of the Euclidean space $R^3$, is the polyhedral complex determined by the collection of unit cubes in $R^3$ whose edges are parallel to the coordinate axes and whose centers are in the set $\mathbb{Z}^3$. Each 3-cell in $R^3$ is representing a voxel, and so the digital object displayed in a digital image is a subset of the set $\mathrm{cell}_3(R^3)$ of 3-cells in $R^3$; while the other lower dimensional cells in $R^3$ (actually, $k$-cubes, $0 \le k < 3$) are used to describe how the voxels could be linked to each other.

*Remark 2.* Each $k$-cell $\sigma \in R^3$ can be associated to its center $c(\sigma)$ which is a point in the set $\mathcal{Z}^3$, where $\mathcal{Z} = \frac{1}{2}\mathbb{Z} = \{x \in \mathbb{R} \,;\, x = z/2, z \in \mathbb{Z}\}$. If $\dim \sigma = 3$ then $c(\sigma) \in \mathbb{Z}^3$, so that every digital object $O$ in $R^3$ can be naturally identified with a subset of the discrete space $\mathbb{Z}^3$. Henceforth we shall use this identification without further comment. Notice also that if $C$ is a unit cube (square) of $\mathbb{Z}^3$ then its center is the point $c(\alpha)$, where $\alpha \in R^3$ is a 0-cell (1-cell, respectively). Thus, if $P = O \cap C$ is a plate in $O$ we call $\alpha$ the *center of $P$*.

As it is usual, given two cells $\gamma, \sigma \in R^3$ we write $\gamma \le \sigma$ if $\gamma$ is a face of $\sigma$, and $\gamma < \sigma$ if in addition $\gamma \ne \sigma$. The interior of a cell $\sigma$ is the set $\mathring{\sigma} = \sigma - \partial\sigma$, where $\partial\sigma = \cup\{\gamma \,;\, \gamma < \sigma\}$ stands for the boundary of $\sigma$. We refer to [11] for further notions on polyhedral topology.

To recall the notion of lighting function we need the following definitions. Given a cell $\alpha \in R^3$ and a digital object $O \subseteq \mathrm{cell}_3(R^3)$ the *star of $\alpha$ in $O$* is the set $\mathrm{st}_3(\alpha; O) = \{\sigma \in O \,;\, \alpha \le \sigma\}$ of 3-cells (voxels) in $O$ having $\alpha$ as a face. Similarly, the *extended star of $\alpha$ in $O$* is the set $\mathrm{st}_3^*(\alpha; O) = \{\sigma \in O \,;\, \alpha \cap \sigma \ne \emptyset\}$. Finally, the *support* of $O$ is the set $\mathrm{supp}(O)$ of cells of $R^3$ (not necessarily voxels) that are the intersection of 3-cells in $O$; that is, $\alpha \in \mathrm{supp}(O)$ if and only if $\alpha = \cap\{\sigma \,;\, \sigma \in \mathrm{st}_3(\alpha; O)\}$. To ease the writing, we use the following notation: $\mathrm{st}_3(\alpha; R^3) = \mathrm{st}_3(\alpha; \mathrm{cell}_3(R^3))$ and $\mathrm{st}_3^*(\alpha; R^3) = \mathrm{st}_3^*(\alpha; \mathrm{cell}_3(R^3))$. Finally, we write $\mathcal{P}(A)$ for the family of all subsets of a given set $A$.

A *lighting function* on the device model $R^3$ is a map $f : \mathcal{P}(\mathrm{cell}_3(R^3)) \times R^3 \to \{0, 1\}$ satisfying the following five axioms for all $O \in \mathcal{P}(\mathrm{cell}_3(R^3))$ and $\alpha \in R^3$:

(1) *object axiom*: if $\alpha \in O$ then $f(O, \alpha) = 1$;
(2) *support axiom*: if $\alpha \notin \mathrm{supp}(O)$ then $f(O, \alpha) = 0$;
(3) *weak monotone axiom*: $f(O, \alpha) \le f(\mathrm{cell}_3(R^3), \alpha)$;
(4) *weak local axiom*: $f(O, \alpha) = f(\mathrm{st}_3^*(\alpha; O), \alpha)$; and,

(5) *complement connectivity axiom*: if $O' \subseteq O \subseteq \text{cell}_3(R^3)$ and $\alpha \in R^3$ are such that $\text{st}_3(\alpha; O) = \text{st}_3(\alpha; O')$, $f(O', \alpha) = 0$ and $f(O, \alpha) = 1$, then the set $\alpha(O', O) = \cup\{\mathring{\omega}; \omega < \alpha, f(O', \omega) = 0, f(O, \omega) = 1\} \subseteq \partial\alpha$ is non-empty and connected.

If $f(O, \alpha) = 1$ we say that $f$ *lights* the cell $\alpha$ for the object $O$, otherwise $f$ *vanishes* on $\alpha$ for $O$.

A digital space $(R^3, f)$ is said to be *homogeneous* if for any spatial motion $\varphi : R^3 \to R^3$ preserving $Z^3$ the equality $f(\varphi(O), \varphi(\alpha)) = f(O, \alpha)$ holds for all cells $\alpha \in R^3$ and digital objects $O \subseteq \text{cell}_3(R^3)$; see [6]. We refer to [3] for a definition of digital spaces on more general device models.

Given a lighting function $f$, we associate to each digital object $O \subseteq \text{cell}_3(R^3)$ a continuous analogue $|\mathcal{A}_O|$ which intends to be a "continuous interpretation" of $O$ in the digital space $(R^3, f)$. Namely, $|\mathcal{A}_O|$ is the underlying polyhedron of the simplicial complex $\mathcal{A}_O$, whose $k$-simplexes are $\langle c(\alpha_0), c(\alpha_1), \ldots, c(\alpha_k) \rangle$ where $\alpha_0 < \alpha_1 < \cdots < \alpha_k$ are cells in $R^3$ such that $f(O, \alpha_i) = 1$, $0 \le i \le k$. The complex $\mathcal{A}_O$ is called the *simplicial analogue* of $O$. Notice that the center $c(\sigma)$ of a 3-cell $\sigma \in \text{cell}_3(R^3)$ is a 0-simplex of $\mathcal{A}_O^f$ if and only if $\sigma \in O$.

For the sake of simplicity we will write $\mathcal{A}_{R^3}$ instead of $\mathcal{A}_{\text{cell}_3(R^3)}$.

Continuous analogues allow us, in a natural way, to introduce digital notions in terms the corresponding continuous ones. For example, we will say that an object $O$ is *connected* in a digital space $(R^3, f)$ if its continuous analogue $|\mathcal{A}_O|$ is a connected polyhedron. And, in the same way, the complement $\text{cell}_3(R^3) - O$ of $O$ is said to be connected if $|\mathcal{A}_{R^3}| - |\mathcal{A}_O|$ is connected. Moreover, we call $C \subseteq \text{cell}_3(R^3)$ a component of $O$ ($\text{cell}_3(R^3) - O$) if it consists of all the voxels $\sigma$ whose centers $c(\sigma)$ belong to a component of $|\mathcal{A}_O|$ ($|\mathcal{A}_{R^3}| - |\mathcal{A}_O|$, respectively). See Section 4 in [3] for more details on these notions of connectedness defined in a much more general context.

For certain digital spaces $(R^3, f)$, these notions of connectedness can be related with the usual ones given on $Z^3$ by mean of adjacency pairs. More precisely, given an adjacency pair $(k, \overline{k})$ on $Z^3$ we say that the digital space $(R^3, f)$ is $(k, \overline{k})$-*connected* if the two following properties hold for any digital object $O \subseteq \text{cell}_3(R^3)$:

1. $C$ is a component of $O$ if and only if it is a $k$-component of $O$; and,
2. $C$ is a component of the complement $\text{cell}_3(R^3) - O$ of $O$ if and only if it is a $\overline{k}$-component.

Several examples of (homogeneous) $(26, 6)$-connected digital spaces $(R^3, f)$ can be found in [1, 3, 5] as well as examples of $(k, \overline{k})$-connected spaces for $k, \overline{k} \in \{6, 18, 26\}$.

## 3.1    A Maximum Set of Digital Surfaces

Similarly to the previous definition of connectedness, we use continuous analogues to introduce a notion of digital surface. Namely, an object $S$ in a digital space $(R^3, f)$ is called a *digital surface* if $|\mathcal{A}_S|$ is a combinatorial surface without

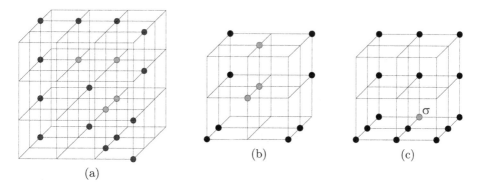

**Fig. 3.** Some surfaces in $(R^3, f^U)$ which are not simplicity 26-surfaces. Both black and grey dots are voxels in the surface. In (a) and (b) the pattern (i.e., the intersection of the surface with a unit cube $C$ of $\mathbb{Z}^3$) consisting of the grey dots is not allowed in a simplicity 26-surface. In (c), the configuration made of the four square plates containing $\sigma$ cannot appear in a simplicity 26-surface

boundary; that is, if for each vertex $v \in \mathcal{A}_S$ its link $\mathrm{lk}(v; \mathcal{A}_S) = \{A \in \mathcal{A}_S ; v, A < B \in \mathcal{A}_S \text{ and } v \notin A\}$ is a 1-sphere.

This notion of digital surface is closely related to the families of 26-connected digital surfaces quoted in the Introduction. Actually, each of these families agrees or is contained in the set of digital surfaces of a particular homogeneous (26, 6)-connected digital space; see [1, 3, 5]. Moreover, in [6] we give a homogeneous (26, 6)-connected digital space $E^U = (R^3, f^U)$ whose set of digital surfaces is the largest in this class of digital spaces. More precisely, we prove

**Theorem 1 (Th. 4 in [6]).** *Any digital surface $S$ in an arbitrary homogeneous (26, 6)-connected digital space $(R^3, f)$ is also a digital surface in $E^U$.*

The lighting function $f^U : \mathcal{P}(\mathrm{cell}_3(R^3)) \times R^3 \to \{0, 1\}$ is defined as follows. Given a digital object $O \subseteq \mathrm{cell}_3(R^3)$ and a cell $\delta \in R^3$, $f^U(O, \delta) = 1$ if and only if one of the following conditions holds:

1. $\dim \delta = 0$ and the intersection $O \cap C$, where $\delta$ is the center of the unit cube $C$ of $\mathbb{Z}^3$, corresponds (up to rotations and symmetries) with one of the patterns in the set $\mathfrak{B} \cup \mathfrak{C}$, with $\mathfrak{B} = \{B_2^c, B_5^c, B_6^b, B_8\}$ and $\mathfrak{C} = \{C_3^c, C_4^b, C_4^e, C_4^f, C_5^b, C_6^c\}$; see Fig. 1.
2. $\dim \delta > 1$ and $\delta \in \mathrm{supp}(O)$.
3. $\dim \delta = 1$ and either $\mathrm{st}_3(\delta; R^3) \subseteq O$ (i.e., $\delta$ is the center of a square plate in $O$) or $\mathrm{st}_3(\delta; O) = \{\sigma, \tau\}$, with $\delta = \sigma \cap \tau$, and $f^U(O, \alpha_1) = f^U(O, \alpha_2)$ for the vertices $\alpha_1, \alpha_2$ of $\delta$.

*Remark 3.* From the definition of the lighting function $f^U$ it is not difficult to check that digital surfaces in $E^U$ can contain each of the patterns in $\mathfrak{C}$, while those in $\mathfrak{B}$ are forbidden for a surface (see Lemma 3 below).

It is not difficult to check that each of the objects in Fig. 3 is (a piece of) a digital surface in $E^U$ and, however, none of them is a simplicity 26-surface. Therefore, the family of simplicity 26-surfaces is strictly contained in the set of digital surfaces of $E^U$. At this point, it is worth mentioning that Malandain et al. [9] suggest to consider the object depicted in Fig. 3(a) as a surface.

We get the following separation theorem for digital surfaces in $E^U$ as a corollary of the Jordan–Brouwer Theorem in [3] for almost arbitrary digital spaces.

**Theorem 2.** *Each 26-connected digital surface $S$ in $E^U$ separates its complement $\mathrm{cell}_3(R^3) - S$ into two 6-components.*

In addition the digital surfaces in $E^U$ also satisfy the general Index Theorem in [2] which provides us with a criterion to determining the components given by Theorem 2.

## 4   Local Characterization of Digital Surfaces in $E^U$

Despite its good properties, digital surfaces in the space $E^U$ might not be considered completely digital since their definition relies on the construction of a continuous analogue. Our goal in this section is to characterize these surfaces as the set of 26-presurfaces. This way, we get a procedure to determine whether a digital object $S$ is a digital surface in $E^U$ which does not depend on the lighting function $f^U$. Moreover, such procedure is local since the five conditions defining 26-surfaces can be checked in the 26-neighbourhood of each voxel $\sigma \in S$. More precisely, we will prove

**Theorem 3.** *A digital object $S \subseteq \mathbb{Z}^3$ is a 26-presurface if and only if $S$ is a digital surface in the space $(R^3, f^U)$.*

Next lemmas, which we will use later in the proof of Theorem 3, state some elementary properties of the lighting function $f^U$ in relation to the conditions defining 26-presurfaces. We do not include their proofs (which are not too hard anyway) in order to keep the paper within the required length.

**Lemma 1.** *Let $O \subseteq \mathbb{Z}^3$ be a digital object satisfying condition (1) in Definition 2. The two following properties hold for any cell $\delta \in R^3$.*

a) *If $\dim \delta = 0$ then $f^U(O, \delta) = 1$ if and only if $\delta$ is the center of a plate in $O$.*
b) *If $\dim \delta = 1$ and $\delta$ is the center of a plate in $O$ then $f^U(O, \delta) = 1$ and $f^U(O, \alpha_i) = 0$ for the two vertices $\alpha_1, \alpha_2 < \delta$.*

**Lemma 2.** *Let $O \subseteq \mathbb{Z}^3$ be a digital object satisfying conditions (1)-(3) in Definition 2. If a cell $\beta \in R^3$, with $\dim \beta \leq 2$ and $f^U(O, \beta) = 1$, is not the center of a plate in $O$ then $\mathrm{st}_3(\beta; O) = \{\sigma_1, \sigma_2\}$ and the set $A = \{\alpha < \beta ; f^U(O, \alpha) = 1\}$ consists of exactly two cells, which are actually the centers of plates in $O$. Moreover, if $\dim \beta = 1$ then $f^U(O, \gamma) = 0$ for any 2-cell $\gamma > \beta$.*

**Lemma 3.** *Let $O \subseteq \mathbb{Z}^3$ be a digital object. If there exists a unit cube $C$ of $\mathbb{Z}^3$ such that $O \cap C \in \mathfrak{B} = \{B_2^c, B_5^c, B_6^b, B_8\}$ then $O$ is not a digital surface in $(R^3, f^U)$.*

The lemmas above are also used in the proof of the following result which is a crucial step in the proof of Theorem 3.

**Proposition 1.** *Let $O \subseteq \mathbb{Z}^3$ be a digital object satisfying conditions (1)-(3) in Definition 2. Then, for each $\sigma \in O$, there exists a simplicial isomorphism $\varphi_\sigma : G(O, \sigma) \to \mathrm{lk}(c(\sigma); \mathcal{A}_O)$.*

*Proof (Sketch).* Firstly notice that $\mathrm{lk}(c(\sigma); \mathcal{A}_O)$ is 1-dimensional since no tetrahedra can appear in the simplicial analogue of any object satisfying condition (1) in Def. 2. Furthermore, by Lemma 1 there is a bijection $p \mapsto \varphi_\sigma(p)$ between $\mathbf{P}(O, \sigma)$ and the set $X$ of centers of cells $c(\alpha) \in L = \mathrm{lk}(c(\sigma), \mathcal{A}_O)$ such that $\alpha$ is also the center of a plate in $O$. Moreover, Lemmas 1 and 2 and the definition of $f^U$ allow us to describe $L$ as a bipartite graph generated by vertices in $X$ on one side and the set $Y$ of those $c(\alpha) \in L$ such that $\alpha$ is not the center of a plate on the other. Moreover, by using the same lemmas one can show that there is a bijection between $\mathbf{E}(O, \sigma)$ and $Y$. Therefore one gets a bijection between the vertex sets of $L$ and $G(O, \sigma)$. Finally, Lemmas 1 and 2 are used one more time to show that this bijection extends to the required simplicial isomorphism.

*Proof (of Theorem 3).* Assume $S \subseteq \mathbb{Z}^3$ is a 26-presurface. It will suffice to check that the link $L = \mathrm{lk}(c(\delta); \mathcal{A}_S)$ is a 1-sphere for each cell $\delta \in R^3$ such that $f^U(S, \delta) = 1$. For any voxel $\delta \in S$ Proposition 1 yields that $L$ can be identified with the assembly graph $G(S, \delta)$ around $\delta$, and hence $L$ is a 1-sphere by Condition (4) in Definition 2.

If $\dim \delta = 2$ then $\mathrm{st}_3(\delta; S) = \{\sigma_1, \sigma_2\}$ by Lemma 2 and, moreover, exactly two faces $\alpha_1, \alpha_2 < \delta$ are lighted by $f^U$. Therefore, $L$ coincides with the cycle defined by the centers $\{c(\alpha_1), c(\alpha_2), c(\sigma_1), c(\sigma_2)\}$.

If $\delta$ is an edge of $R^3$ which is the center of a square plate in $S$ then $L$ is obviously a 1-sphere by Lemma 1 and the definition of $f^U$. Otherwise, if $\delta$ is not the center of a plate, the result follows by Lemma 2 as in the case $\dim \delta = 2$.

Finally, if $\delta$ is a vertex then it is the center of a plate $p$ by Lemma 1. Moreover the plate $p$ belongs to the set $\mathfrak{C}$ in Definition 1. Assume $p$ does not corresponds to pattern $C_4^f$ in Fig. 1. If $\sigma_1, \sigma_2 \in p$ are strictly 18-adjacent voxels, we derive from the fact that $c(\delta) \in \mathrm{lk}(c(\sigma_i); \mathcal{A}_S)$, $i = 1, 2$, which has been proved to be a 1-sphere, that $f^U(S, \sigma_1 \cap \sigma_2) = 1$. Therefore $L$ is necessarily a 1-sphere by the definition of $f^U$.

If $p = \{\sigma_1, \sigma_2, \sigma_3, \sigma_4\}$ corresponds to $C_4^f$ we have some choices to make. For this we set $L_i = \mathrm{lk}(c(\sigma_i); \mathcal{A}_S)$, $1 \leq i \leq 4$. As $c(\delta) \in L_1$ is a 1-sphere, it contains exactly two of the centers $c_i^1 = c(\sigma_1 \cap \sigma_i)$, $i = 2, 3, 4$. Assume they are $c_2^1$ and $c_3^1$. Similarly, by looking at $L_2$ we have that either $c_3^2 = c(\sigma_2 \cap \sigma_3)$ or $c_4^2 = c(\sigma_2 \cap \sigma_4)$ belongs to $L_2$. But $c_3^2$ is ruled out by the fact that $L_4$ is a cycle since otherwise we would get a one-point union of three edges in $L_i$ for some $1 \leq i \leq 3$. Hence $c_4^2 \in L_2$, and the only choice for $L_4$ is to have $c_3^4 = c(\sigma_3 \cap \sigma_4) \in L_4$ since

$c_1^4 = c(\sigma_1 \cap \sigma_4)$ would imply that $L_4$ contains a one-point union of three edges. Therefore $\mathcal{A}_S$ is a combinatorial surface and hence $S$ is a digital surface.

Conversely, assume that $S$ is a digital surface. It will be enough to check conditions (1)-(3) in Definition 2 for $S$ since, under the assumption of these properties, we get (4) as an immediate consequence of Proposition 1.

Condition (1) follows immediately from Lemma 3 above. To check condition (2), let $\sigma, \tau \in S$ be two strictly 18-adjacent 3-cells and assume that they are not 6-connected by a third voxel in $S$. For the edge $\beta = \langle \alpha_1, \alpha_2 \rangle = \sigma \cap \tau$ we consider the two possible cases:

Case $f^U(S, \beta) = 0$. The definition of $f^U$ shows that $f^U(S, \alpha) = 1$ for a vertex $\alpha < \beta$ ($f^U$ vanishes on the other one necessarily). Then by condition (1), already proved, and Lemma 1 it follows that $\alpha$ is the center of a plate in $\mathbf{P}(S, \sigma) \cap \mathbf{P}(S, \tau)$.

Case $f^U(S, \beta) = 1$. Then it can be readily checked that $f^U(S, \alpha_i) = 1$ for the two vertices $\alpha_1, \alpha_2$ of $\beta$, since $S$ is a digital surface in $E^U$. Therefore the vertices $\alpha_i$ are centers of plates in $\mathbf{P}(S, \sigma) \cap \mathbf{P}(S, \tau)$ by Lemma 1.

Finally we prove condition (3). For this let $\sigma, \tau \in S$ be two 6-adjacent voxels $\gamma = \sigma \cap \tau$. Then $f^U(S, \gamma) = 1$ by definition of $f^U$. As $\mathrm{lk}(c(\gamma); \mathcal{A}_S)$ is a 1-sphere there exist exactly two faces $\alpha_1, \alpha_2 < \gamma$ with $f^U(S, \alpha_i) = 1$. If $\dim \alpha_i = 0$ then $\alpha_i$ is the center of a plate in $\mathbf{P}(S, \sigma) \cap \mathbf{P}(S, \tau)$ by Lemma 1. Similarly, if $\dim \alpha_i = 1$ then the definition of $f^U$ yields that $\mathrm{st}_3(\alpha_i; S) = \mathrm{st}_3(\alpha_i; R^3)$ since it contains the two 6-adjacent voxels $\sigma$ and $\tau$, and hence $\alpha_i$ is the center of a square plate.

## 5    Final Results

As a consequence of the characterization above and Theorem 2 we get that each 26-connected 26-presurface $S$ separates its complement $\mathbb{Z}^3 - S$ into two 6-components; that is, $S$ is a *Jordan* object. However, we find in this set examples (as the 26-presurfaces $S_1$ and $S_2$ shown in Fig. 2(a) and 3(c), respectively) which might be considered as pathological surfaces. More precisely, one readily checks that the voxel $\tau_0 \in S_1$ in Fig. 2(a) is not 6-adjacent to both 6-components of $\mathbb{Z}^3 - S_1$; that is, $S_1$ is not a *strongly separating* object as defined in [4, 7]. Hence the deletion of $\tau_0$ from $S_1$ yields an object which is still Jordan (actually, $S_1 - \{\tau_0\}$ is a 26-presurface). We next state without proof a a local characterization for the set of strongly separating 26-presurfaces.

**Theorem 4.** Let $S \subseteq \mathbb{Z}^3$ a 26-connected 26-presurface. Then $S$ is strongly separating if and only if $S$ is 6-thin; that is, for each $\sigma \in S$, the complement of $S$ in the 26-neighbourhood of $\sigma$, $\mathrm{st}_3^*(\sigma; R^3) - S$, contains exactly two 6-components which are 6-adjacent to $S$.

The proof of this result makes use of the difference $\mathrm{lk}(c(\sigma); \mathcal{A}_{R^3}) - \mathrm{lk}(c(\sigma); \mathcal{A}_S)$ of links in the continuous analogue to prove that the 6-components of $\mathrm{st}_3^*(\sigma; R^3) - S$ which are 6-adjacent to $\sigma$ characterize the 6-components of $\mathbb{Z}^3 - S$, as in the proof of the digital Index Theorem given in [2].

This result suggests the following

**Definition 3.** *A strongly separating 26-surface* is any 26-presurface which is also 6-thin.

Finally, we point out that the set of strongly separating 26-surfaces also contains strictly the set of simplicity 26-surfaces since each one of them is a strongly separating object [7] as well as the 26-presurfaces pictured in Fig. 3(a) and (b).

# References

1. R. Ayala, E. Domínguez, A.R. Francés, A. Quintero. Digital Lighting Functions. *Lecture Notes in Computer Science.* **1347** (1997) 139–150.
2. R. Ayala, E. Domínguez, A.R. Francés, A. Quintero. A Digital Index Theorem. *Int. J. Patter Recog. Art. Intell.* **15**(7) (2001) 1–22.
3. R. Ayala, E. Domínguez, A.R. Francés, A. Quintero. Weak Lighting Functions and Strong 26-surfaces. *Theoretical Computer Science.* **283** (2002) 29–66.
4. G. Bertrand, R. Malgouyres. Some Topological Properties of Surfaces in $\mathbb{Z}^3$. *Jour. of Mathematical Imaging and Vision.* **11** (1999) 207–221.
5. J.C. Ciria, E. Domínguez, A.R. Francés. Separation Theorems for Simplicity 26-surfaces. *Lecture Notes in Computer Science.* **2301** (2002) 45–56.
6. J.C. Ciria, A. De Miguel, E. Domínguez, A.R. Francés, A. Quintero. A maximum set of $(26,6)$-connected digital surfaces. *Lecture Notes in Computer Science.* **3322** (2004) 291–306.
7. M. Couprie, G. Bertrand. Simplicity Surfaces: a new definition of surfaces in $\mathbb{Z}^3$. *SPIE Vision Geometry V.* **3454** (1998) 40–51.
8. T.Y. Kong, A.W. Roscoe. Continuous Analogs of Axiomatized Digital Surfaces. *Comput. Vision Graph. Image Process.* **29** (1985) 60–86.
9. G. Malandain, G. Bertrand, N. Ayache. Topological Segmentation of Discrete Surfaces. *Int. Jour. of Computer Vision.* **10:2** (1993) 183–197.
10. D.G. Morgenthaler, A. Rosenfeld. Surfaces in three–dimensional Digital Images. *Inform. Control.* **51** (1981) 227–247.
11. C.P. Rourke, and B.J. Sanderson. *Introduction to Piecewise-Linear Topology.* Ergebnisse der Math. **69**, Springer 1972.

# Algorithms for the Topological Watershed

Michel Couprie, Laurent Najman, and Gilles Bertrand

Laboratoire A2SI, Groupe ESIEE,
BP99, 93162 Noisy-le-Grand Cedex France
IGM, Unité Mixte de Recherche CNRS-UMLV-ESIEE UMR 8049
{m.couprie, l.najman, g.bertrand}@esiee.fr

**Abstract.** The watershed transformation is an efficient tool for segmenting grayscale images. An original approach to the watershed [1, 4] consists in modifying the original image by lowering some points until stability while preserving some topological properties, namely, the connectivity of each lower cross-section. Such a transformation (and its result) is called a topological watershed. In this paper, we propose quasi-linear algorithms for computing topological watersheds. These algorithms are proved to give correct results with respect to the definitions, and their time complexity is analyzed.

## 1   Introduction

The watershed transformation was introduced as a tool for segmenting grayscale images by S. Beucher and C. Lantuéjoul [2] in the late 70's, and is now used as a fundamental step in many powerful segmentation procedures. The most popular presentation of the watershed is based on a flooding paradigm. Let us consider a grayscale image as a topographical relief: the gray level of a pixel becomes the altitude of a point, the basins and valleys of the relief correspond to the dark areas, whereas the mountains and crest lines correspond to the light areas. Let us imagine the surface of this relief being immersed in still water, with holes pierced in local minima. Water fills up basins starting at these local minima, and dams are built at points where waters coming from different basins would meet. As a result, the surface is partitioned into regions or basins which are separated by dams, called watershed lines. Efficient watershed algorithms based on immersion simulation were proposed by L. Vincent, P. Soille [12] and F. Meyer [6, 3] in the early 90's.

A different approach to watersheds, originally proposed by G. Bertrand and M. Couprie [4], is developed in [1]. In this approach, we consider a transformation called topological watershed, which modifies a map (*e.g.* a grayscale image) while preserving some topological properties, namely, the connectivity of each lower cross-section. It is proved in [1] that, among other properties, topological watersheds satisfy a "constrast preservation" property which is, in general, not satisfied by the most popular watershed algorithms [8].

In this paper, we study algorithms to compute topological watersheds. These algorithms are proved [5] to give correct results with respect to the definition, and their time complexity is analyzed.

E. Andres et al. (Eds.): DGCI 2005, LNCS 3429, pp. 172–182, 2005.

## 2    Topological Notions for Weighted Graphs

Let $E$ be a finite set, we denote by $\mathcal{P}(E)$ the set of all subsets of $E$. Throughout this paper, $\Gamma$ will denote a binary relation on $E$ (thus, $\Gamma \subseteq E \times E$), which is reflexive (for all $p$ in $E$, $(p, p) \in \Gamma$) and symmetric (for all $p, q$ in $E$, $(q, p) \in \Gamma$ whenever $(p, q) \in \Gamma$). We say that the pair $(E, \Gamma)$ is a *graph*, each element of $E$ is called a *vertex* or a *point*. We will also denote by $\Gamma$ the map from $E$ into $\mathcal{P}(E)$ such that, for any $p$ in $E$, $\Gamma(p) = \{q \in E; (p, q) \in \Gamma\}$. For any point $p$, the set $\Gamma(p)$ is called the *neighborhood of $p$*. If $q \in \Gamma(p)$ then we say that $p$ and $q$ are *adjacent* or that $q$ is a *neighbor of $p$*. If $X \subseteq E$ and $q$ is adjacent to $p$ for some $p \in X$, we say that $q$ *is adjacent to $X$*.

For applications to digital image processing, assume that $E$ is a finite subset of $\mathbb{Z}^n$ ($n = 2, 3$), where $\mathbb{Z}$ denotes the set of integers. A subset $X$ of $E$ represents the "object", its complementary $\overline{X} = E \backslash X$ represents the "background", and $\Gamma$ corresponds to an adjacency relation between points of $E$. In $\mathbb{Z}^2$, $\Gamma$ may be one of the usual adjacency relations, for example the 4-adjacency or the 8-adjacency in the square grid. In the sequel, the 4-adjacency is assumed.

Let $(E, \Gamma)$ be a graph, let $X \subseteq E$, and let $p_0, p_k \in X$. A *path from $p_0$ to $p_k$ in $X$* is an ordered family $(p_0, p_1, \ldots, p_k)$ of points of $X$ such that $p_{i+1} \in \Gamma(p_i)$, with $i = 0 \ldots k - 1$. Let $p, q \in X$, we say that *$p$ and $q$ are linked for $X$* if there exists a path from $p$ to $q$ in $X$. We say that *$X$ is connected* if any $p$ and $q$ in $X$ are linked for $X$. We say that a subset $Y$ of $E$ is a *(connected) component of $X$* if $Y \subseteq X$, $Y$ is connected, and $Y$ is maximal for these two properties. In the sequel of the article, we will assume that $E$ is connected.

We are interested in transformations that preserve the number of connected components of the background. For this purpose, we introduce the notion of W-simple point (where Wstands for watershed) in a graph. Intuitively, a point of $X$ is W-simple if it may be removed from $X$ while preserving the number of connected components of $\overline{X}$.

**Definition 1.** Let $X \subseteq E$, let $p \in X$. We say that:
- $p$ is a *border point (for $X$)* if $p$ is adjacent to $\overline{X}$;
- $p$ is an *inner point (for $X$)* if $p$ is not a border point for $X$;
- $p$ is *separating (for $X$)* if $p$ is adjacent to at least two components of $\overline{X}$;
- $p$ is *W-simple (for $X$)* if $p$ is adjacent to exactly one component of $\overline{X}$.

Notice that a point which is not W-simple is either an inner point or a separating point. In Fig. 2, the points of the set $X$ are represented by "1"s. The points which are W-simple are circled. It may be easily seen that one cannot locally decide whether a point is W-simple or not. Consider the points $p$ and $q$ in the third row: their neighborhoods are alike, yet $p$ is W-simple (it is adjacent to exactly one connected component of $\overline{X}$), and $q$ is not, since it is adjacent to two different connected components of $\overline{X}$.

Now, we extend this notion to a weighted graph $(E, \Gamma, F)$, where $F$ is a map from $E$ to $\mathbb{Z}$. A weighted graph is a model for a digital grayscale image; for any point $p \in E$, the value $F(p)$ represents the gray level of $p$. We denote by $\mathcal{F}(E)$ the set composed of all maps from $E$ to $\mathbb{Z}$.

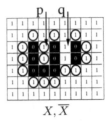

$$X, \overline{X}$$

**Fig. 1.** A set $X$ (the 1's) and its complement $\overline{X}$ (the 0's). All the W-simple points are circled

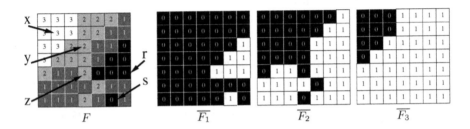

**Fig. 2.** A grayscale image $F$ and its lower sections $\overline{F_1}$, $\overline{F_2}$ and $\overline{F_3}$ (in white). The points $x, r, s$ are inner points, $y$ is W-destructible (with lowest value 1), and $z$ is separating

**Definition 2.** Let $F \in \mathcal{F}(E)$, let $k \in \mathbb{Z}$.
We denote by $F_k$ the set $\{p \in E; F(p) \geq k\}$; $F_k$ is called an *upper section of $F$*, and its complementary $\overline{F_k}$ is called a *lower section of $F$*.
A component $c$ of $\overline{F_k}$ is called a *(regional) minimum for $F$* if $c \cap \overline{F_{k-1}} = \emptyset$.
We denote by $\Gamma^-(p, F)$ the *set of lower neighbors of the point $p$ for the map $F$*, that is, $\Gamma^-(p, F) = \{q \in \Gamma(p); F(q) < F(p)\}$. When no confusion may occur, we write $\Gamma^-(p)$ instead of $\Gamma^-(p, F)$.

Fig. 2 shows a grayscale image $F$ and three lower sections of $F$: $\overline{F_2}$ is made of two components (in white), and $\overline{F_3}$ is made of one component. The set $\overline{F_1}$ is made of two components which are both minima of $F$.

**Definition 3.** Let $F \in \mathcal{F}(E)$, let $p \in E$, let $k = F(p)$.
We say that $p$ is a *border point (for $F$)* if $p$ is a border point for $F_k$.
We say that $p$ is an *inner point (for $F$)* if $p$ is an inner point for $F_k$.
We say that $p$ is *separating (for $F$)* if $p$ is separating for $F_k$.
The point $p$ is *W-destructible (for $F$)* if $p$ is W-simple for $F_k$. Let $v \in \mathbb{Z}$, $v < F(p)$, the point $p$ is *W-destructible with lowest value $v$ (for $F$)* if for any $h$ such that $v < h \leq F(p)$, $p$ is W-simple for $F_h$, and if $p$ is not W-simple for $F_v$.

In other words, the point $p$ is W-destructible for $F$ if and only if $p$ is a border point for $F$ (i.e., $\Gamma^-(p) \neq \emptyset$) and all the points in $\Gamma^-(p)$ belong to the same connected component of $\overline{F_k}$, with $k = F(p)$. In Fig. 2, the points $x, r, s$ are inner points, $y$ is W-destructible (with lowest value 1), and $z$ is separating.

Let $F \in \mathcal{F}(E)$, let $p \in E$, let $v \in \mathbb{Z}$ such that $v < F(p)$, we denote by $[F \backslash p \downarrow v]$ the element of $\mathcal{F}(E)$ such that $[F \backslash p \downarrow v](p) = v$ and $[F \backslash p \downarrow v](q) = F(q)$ for all $q \in E \backslash \{p\}$. Informally, it means that the only difference between the map $F$ and the map $[F \backslash p \downarrow v]$, is that the point $p$ has been lowered down to the value $v$. We also write $[F \backslash p] = [F \backslash p \downarrow v]$ when $v = F(p) - 1$.

If we consider $F' = [F \backslash p \downarrow v]$, it may be easily seen that for any $h$ in $\mathbb{Z}$, the number of connected components of $\overline{F'_h}$ equals the number of connected components of $\overline{F_h}$. That is to say, the value of a W-destructible point may be lowered by one or down to its lowest value without changing the number of connected components of any lower section of $F$.

**Definition 4.** Let $F \in \mathcal{F}(E)$. We say that $G \in \mathcal{F}(E)$ is *a W-thinning of $F$* if
i) G = F, or if
ii) there exists a map $H$ which is a W-thinning of $F$ and there exists a W-destructible point $p$ for $H$ such that $G = [H \backslash p]$.
We say that $G$ is a *(topological) watershed* of $F$ if $G$ is a W-thinning of $F$ and if there is no W-destructible point for $G$.

Let $F \in \mathcal{F}(E)$, let $p \in E$, let $v \in \mathbb{Z}$. It may be easily seen that, if $p$ is W-destructible with lowest value $v$, then $[F \backslash p \downarrow v]$ is a W-thinning of $F$ and $p$ is not W-destructible for $[F \backslash p \downarrow v]$ ; and that the converse is also true.

In other words, one can obtain a W-thinning of a map $F$ by iteratively selecting a W-destructible point and lowering it by one, or directly down to its lowest value. If this process is repeated until stability, one obtains a topological watershed of $F$. Notice that the choice of the W-destructible point is not necessarily unique at each step, thus, in general, there may exist several topological watersheds for the same map.

In Fig. 3, we present an image 3a and a topological watershed 3b of 3a. Note that in 3b, the minima of 3a have been spread and are now separated from each other by a "thin line"; nevertheless, their number and values have been preserved. Fig. 3c shows a W-thinning of 3a which is not a topological watershed of 3a (there are still some W-destructible points).

Let us consider a point $p \in E$ which is not W-destructible for $F \in \mathcal{F}(E)$. Three cases may be distinguished. From Def. 3, such a point is either an inner point or a separating point for $F$. Furthermore, if $p$ is an inner point, then either $p$ belongs to a minimum of $F$ or not.

On the other hand, if $p$ is W-destructible for $F$, then $p$ is not W-destructible for $[F \backslash p \downarrow v]$ where $v$ is the lowest value of $p$. Again, we can distinguish the same three possibilities for the status of $p$ with respect to $[F \backslash p \downarrow v]$. The following definition formalizes these observations (S stands for separating, M for minimum and P for plateau).

**Definition 5.** Let $F \in \mathcal{F}(E)$, let $p \in E$, $p$ not W-destructible for $F$.
We say that $p$ is an *S-point (for $F$)* if $p$ is separating for $F$.
We say that $p$ is an *M-point (for $F$)* if $p$ belongs to a minimum of $F$.
We say that $p$ is an *P-point (for $F$)* if $p$ is an inner point for $F$ which does not belong to a minimum of $F$.

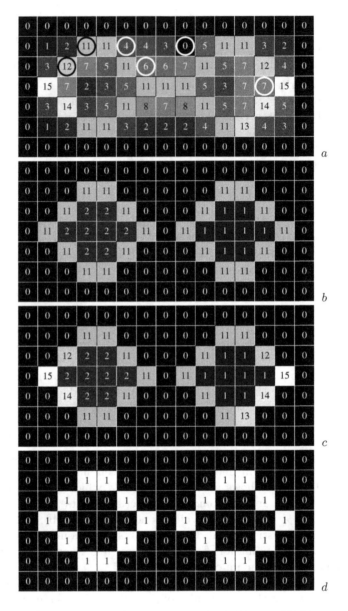

**Fig. 3.** *a*: original image; *b*: a topological watershed of *a*; *c*: a W-thinning of *a* which is also an M-watershed of *a* (see Sec. 3); *d*: a W-crest of *a* (see Sec. 3). In *a*, we have circled six points which have different types (see Def. 5). From left to right: $\tilde{\text{S}}$-point (12), S-point (11), $\tilde{\text{M}}$-point (4), $\tilde{\text{P}}$-point (6), M-point (0), P-point (7).

Let $q$ be a point which is W-destructible for $F$, let $v$ be its lowest value. We say that $q$ is an $\tilde{S}$-*point (for F)* if $q$ is an S-point for $[F \setminus q \downarrow v]$.

We say that $q$ is an $\tilde{M}$-point (for $F$) if $q$ is an M-point for $[F \setminus q \downarrow v]$.
We say that $q$ is a $\tilde{P}$-point (for $F$) if $q$ is a P-point for $[F \setminus q \downarrow v]$.

In Fig. 3a, we have circled six points which are representative of all the possible types.

The components of the lower sections of a map may be organized, thanks to the inclusion relation, in a tree structure that we call *component tree* (see the bibliography of [7] for a list of references). In [5], we propose and prove a characterization of the W-destructible points, which may be checked locally and efficiently implemented thanks the component tree. In [7], we introduce a new algorithm, using Tarjan's union-find procedure [10], to build the component tree of any weighted graph in quasi-linear time, that is, in $O(n \times \alpha(n))$ where $n$ is the size of the graph (number of vertices + number of arcs) and $\alpha(n)$ is a function which grows extremely slowly with $n$ (we have $\alpha(10^{80}) \approx 4$). For a precise definition of $\alpha$, see [10].

Furthermore, for applications to digital image processing, where each point has a fixed (and small) number of neighbors, we can consider that testing whether a point is W-destructible and computing its lowest value can be done in constant time thanks to the component tree.

## 3    M-Thinning and Binary Watershed Algorithm

The outline of a topological watershed algorithm is the following:

**Repeat Until Stability**

>  Select a W-destructible point $p$, using a certain criterion
>  Lower the value of $p$

It can be seen that, even if a W-destructible point is lowered down to its lowest value, it may again become W-destructible in further steps of the W-thinning process, due to the lowering of some of its neighbors. For example, the point at level 6 circled in white in Fig. 3a is W-destructible with lowest value 4. If we lower this point down to 4, we will have to lower it again, after the lowering of its neighbor at level 4 down to 3 or 0.

In order to ensure a linear complexity, we must avoid multiple selections of the same point during the execution of the algorithm. The following properties provide selection criteria which guarantee that a point lowered once will never be W-destructible again during the W-thinning process.

The first criterion concerns points which may be lowered down to the value of a neighbor which belongs to a minimum (*i.e.*, $\tilde{M}$-points). If an $\tilde{M}$-point is lowered down to its lowest value, then we say that the point is *M-lowered*. The aim of theorem 1 is to show that, if $\tilde{M}$-points are sequentially selected and M-lowered, and if we continue this process until stability, giving a result $G$, then it is not possible that a W-thinning of $G$ contains any $\tilde{M}$-point. Since, obviously, a point which has been M-lowered will never be considered again in a W-thinning algorithm, we obtain an "M-thinning algorithm" which considers each point at most once, and produces a result in which the minima cannot be extended by further W-thinning.

**Definition 6.** Let $F, G \in \mathcal{F}(E)$, we say that $G$ is an *M-thinning of F* if $G = F$ or if $G$ can be obtained from $F$ by sequentially M-lowering some M̃-points. We say that $G$ is an *M-watershed of F* if $G$ is a M-thinning of $F$ and has no M̃-point.

**Theorem 1.** *Let $F \in \mathcal{F}(E)$, let $G$ be an M-watershed of $F$. Any W-thinning of $G$ has exactly the same minima as $G$.*

See [5] for a proof. Let $F$ be a map and let $G$ be a topological watershed of $F$, the set of points which do not belong to any regional minimum of $G$ is called a *W-crest* of $F$ (see Fig. 3d). A W-crest of $F$ corresponds to a "binary watershed" of $F$. A corollary of this theorem is that the set of points which do not belong to any minimum of an M-watershed of $F$ is always a W-crest of $F$. Thus, we can compute a W-crest (or binary watershed) by only lowering M̃-points. In Fig. 3c, we see an M-watershed of 3a.

In the following algorithm, we introduce a priority function $\mu$ which is used to select the next M̃-point. The priority function $\mu$ associates to each point $p$ a positive integer $\mu(p)$, called the priority of $p$. This function is used for the management of a priority queue, a data structure which allows one to perform, on a set of points, an arbitrary sequence of the two following operations ($L$ denotes a priority queue and $p$ a point):

**AddPrioQueue**$(L, p, \mu(p))$: store $p$ with the priority $\mu(p)$ into the queue $L$;
**ExtractPrioQueue**$(L)$: remove and return a point which has the minimal priority value among those stored in $L$ (if several points fulfill this condition, an arbitrary choice is made).

The choice and the interest of the priority function will be discussed afterwards, but notice that whatever the chosen priority function (for example a constant function), the output of the procedure will always be an M-watershed of the input.

Given a map $F$ and a point $p$, the procedure call **M-destructible**$(F, p)$ (resp. **W-destructible**$(F, p)$) returns in constant time (see end of Sec. 2) the lowest value for $p$ if $p$ is an M-point (resp. a W-destructible point), or $\infty$ otherwise.

**Procedure M-watershed (Input** $F, \mu$ ; **Output** $F$)
```
01.       L ← EmptyPrioQueue
02.       For All p ∈ E such that M-destructible(F, p) ≠ ∞ Do
03.            AddPrioQueue(L, p, μ(p)) ; mark p
04.       While L ≠ EmptyPrioQueue Do
05.            p ← ExtractPrioQueue(L) ; unmark p
06.            If M-destructible(F, p) ≠ ∞ Then
07.                 F(p) ← M-destructible(F, p)
08.                 For All q ∈ Γ(p), q ≠ p, q not marked Do
09.                      If M-destructible(F, q) ≠ ∞ Then
10.                           AddPrioQueue(L, q, μ(q)) ; mark q
```

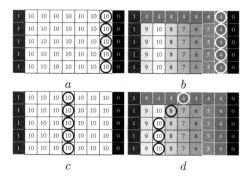

**Fig. 4.** $a, b$: two images with a possible W-crest (circled) computed thanks to procedure **M-watershed** with a constant priority function. $c, d$: the same two images with the W-crest (circled) computed thanks to procedure **M-watershed** with the lexicographic priority function

The following property is a consequence of results proved in [5] and of the fact that, obviously, each point is selected at most once by this algorithm.

**Property 2.** *Whatever the chosen priority function, the output of Procedure* **M-watershed** *is an M-watershed of the input.*
*Let $n$ and $m$ denote respectively the number of vertices and the number of arcs in the graph $(E, \Gamma)$. The time complexity of Procedure* **M-watershed** *is in $O(n + m) + k$, where $k$ is the overall complexity for the management of the priority queue.*

We introduced the priority function and the priority queue in order to take into account some geometrical criteria. For example, with a constant priority function, plateaux or even domes located between basins may be thinned in different ways, depending on the arbitrary choices that are allowed by the calls to **ExtractPrioQueue** with this particular priority function (line 05). See Fig. 4 $a, b$ for a possible result of procedure **M-watershed** with a constant priority function.

In order to "guide" the watershed set towards the highest locations of the domes and the "center" of the plateaux, we choose a lexicographic priority function $\mu$ described below.

Let $F \in \mathcal{F}(E)$, let $d$ be a distance on $E$ (e.g., the Euclidean distance), let $p \in E$. We denote by $D(p)$ be the minimal distance between $p$ and any point $q$ strictly lower than $p$, that is, $D(p) = \min\{d(p, q); F(q) < F(p)\}$.

It is easy to build a function $\mu$ such that, for any $p, q$ in $E$:

– if $F(p) < F(q)$ then $\mu(p) > \mu(q)$;
– if $F(p) = F(q)$ and $D(p) \leq D(q)$ then $\mu(p) \geq \mu(q)$.

See Fig. 4 $c, d$ for the result of **M-watershed** with such a priority function.

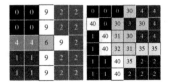

**Fig. 5.** Examples of W-destructible points in an MS-watershed which are neither $\tilde{\text{M}}$-points nor $\tilde{\text{S}}$-points: the point at 6 in the image on the left, the points at 31 and 32 in the image on the right

The values of such a priority function may be pre-computed by a linear-time algorithm, see for example algorithm 4.5 of [9] which is called *lower completion* algorithm. The efficient management of priority queues is a well studied problem, and efficient solutions exist (see e.g. [11]). Furthermore, in most current situations of image analysis, where the number of possible values for the priority function is limited and the number of neighbors of a point is a small constant, specific linear algorithms can be used, avoiding the use of a priority queue. An example of such a linear algorithm is given in the next section, with algorithm **TopologicalWatershed**.

## 4    Watershed Algorithm

After iteratively lowering $\tilde{\text{M}}$-points until stability, we have to process the other W-destructible points in order to get a topological watershed. Let $F \in \mathcal{F}(E)$, let us call an *MS-watershed of F* a map obtained from $F$ by iteratively lowering $\tilde{\text{M}}$-points and $\tilde{\text{S}}$-points until stability. We could think that all $\tilde{\text{P}}$-points will be eventually changed to $\tilde{\text{M}}$-points and then M-lowered in such a process, as it is the case for images like Fig. 3a. But the examples of Fig. 5 show that it is not always the case, in other words, an MS-watershed of $F$ is not always a topological watershed of $F$. Furthermore, there may exist thick regions made of $\tilde{\text{P}}$-points in an MS-watershed, and although $\tilde{\text{M}}$-points and $\tilde{\text{S}}$-points may be lowered directly down to their lowest possible value, we have no such guarantee for the $\tilde{\text{P}}$-points (see theorem 5 of [5]).

Thus, we must propose a criterion for the selection of the remaining W-destructible points, in order to avoid multiple selections of the same point. The idea is to give the greatest priority to a W-destructible point which may be lowered down to the lowest possible value. We prove that an algorithm which uses this strategy never selects the same point twice. A priority queue could be used, as in the previous section, to select W-destructible points in the appropriate order. Here, we propose a specific linear watershed algorithm which may be used when the grayscale range is small. Let $F \in \mathcal{F}(E)$, let $k_{\min} = \min\{F(p); p \in E\}$ and $k_{\max} = \max\{F(p); p \in E\}$.

**Procedure TopologicalWatershed (Input** $F$ **; Output** $F$)
```
01.        For k From kmin To kmax Do Lk ← ∅
02.        For All p ∈ E Do
03.            i ← W-Destructible(F, p)
04.            If i ≠ ∞ Then
05.                Li ← Li ∪ {p} ; K(p) ← i
06.        For k = kmin To kmax Do
07.            While ∃p ∈ Lk Do
08.                Lk = Lk \ {p}
09.                If K(p) = k Then
10.                    F(p) ← k
11.                    For All q ∈ Γ(p), k < F(q) Do
12.                        i ← W-Destructible(F, q)
13.                        If i = ∞ Then K(q) ← ∞
14.                        Else If K(q) ≠ i Then
15.                            Li ← Li ∪ {q} ; K(q) ← i
```

We have the following guarantees:

**Property 3.** *In algorithm* **TopologicalWatershed,**
*i) at the end of the execution, $F$ is a topological watershed of the input map;*
*ii) let $n$ and $m$ denote respectively the number of vertices and the number of arcs in the graph $(E, \Gamma)$. If $k_{max} - k_{min} \leq n$, then the time complexity of the algorithm is in $O(n + m)$.*

As discussed in the previous section, this algorithm provides topological guarantees but does not care about geometrical criteria. If we want to take such criteria into account, we can use first the procedure **M-watershed** with the priority function described at the end of section 3, and then the procedure **TopologicalWatershed**.

## 5   Watershed from Markers

In many applications, instead of finding a separation between the minima of the input function, we need to separate the components of a given set of points called the *marker*. Let us illustrate how to reach this goal using the topological watershed, following a classical approach based on reconstruction.

Fig. 6 illustrates the outline of the whole procedure. Fig. 6$a$ shows the input data, function $F$ and marker $M$. Fig. 6$b$: a function $G$ is generated, such that $G(x) = k_{min}$ for all $x \in M$, and $G(x) = k_{max}$ for all $x \notin M$. The function $F' = \min(F, G)$ is computed. Fig. 6$c$: we compute the morphological geodesic reconstruction $G'$ of $G$ over $F'$. See [13] for a description of this operator, which can be efficiently implemented thanks to the component tree. Notice that, by construction, each minimum of $G'$ contains a component of $M$. Fig. 6$d$: finally, an M-watershed $W$ of $G'$ is extracted, hence, a W-crest $C$.

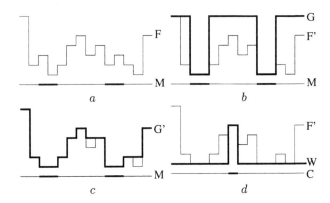

**Fig. 6.** Outline of a topological watershed-from-markers procedure

# References

1. G. Bertrand, "On topological watersheds", *Journal of Mathematical Imaging and Vision*, Vol. 22, pp. 217-230, 2005.
2. S. Beucher, Ch. Lantuéjoul, "Use of watersheds in contour detection", *Proc. Int. Workshop on Image Processing, Real-Time Edge and Motion Detection/Estimation*, Rennes, France, 1979.
3. S. Beucher, F. Meyer, "The morphological approach to segmentation: the watershed transformation", *Mathematical Morphology in Image Processing*, Chap. 12, pp. 433-481, Dougherty Ed., Marcel Dekker, 1993.
4. M. Couprie, G. Bertrand, "Topological grayscale watershed transformation", *Proc. SPIE Vision Geometry VI*, Vol. 3168, pp. 136-146, 1997.
5. M. Couprie, L. Najman, G. Bertrand, "Quasi-linear algorithms for the topological watershed", *Journal of Mathematical Imaging and Vision*, Vol. 22, pp. 231-249, 2005.
6. F. Meyer, "Un algorithme optimal de ligne de partage des eaux", *Proc. 8th Conf. Reconnaissance des Formes et Intelligence Artificielle*, Vol. 2, pp. 847-859, AFCET Ed., Lyon, 1991.
7. L. Najman, M. Couprie, "Quasi-linear algorithm for the component tree", *Proc. SPIE Vision Geometry XII*, Vol. 5300, pp. 98-107, 2004.
8. L. Najman, M. Couprie, G. Bertrand, "Watersheds, extension maps, and the emergence paradigm", report IGM2004-04 of the Institut Gaspard Monge (University of Marne-la-Vallée), to appear in *Discrete Applied Mathematics*, 2004.
9. J. Roerdink, A. Meijster, "The watershed transform: definitions, algorithms and parallelization strategies", *Fundamenta Informaticae*, Vol. 41, pp. 187-228, 2000.
10. R.E. Tarjan, "Disjoint sets" *Data Structures and Network Algorithms*, Chap. 2, pp. 23-31, SIAM, 1978.
11. M. Thorup, "On RAM priority queues", *7th ACM-SIAM Symposium on Discrete Algorithms*, pp. 59-67, 1996.
12. L. Vincent, P. Soille, "Watersheds in digital spaces: an efficient algorithm based on immersion simulations", *IEEE Trans. on PAMI*, Vol. 13, No. 6, pp. 583-598, 1991.
13. L. Vincent, "Morphological Grayscale Reconstruction in Image Analysis: Application and Efficient Algorithms", *IEEE Trans. on PAMI*, Vol. 2, No. 2, pp. 176-201, 1993.

# The Class of Simple Cube-Curves Whose MLPs Cannot Have Vertices at Grid Points

Fajie Li and Reinhard Klette

CITR, University of Auckland, Tamaki Campus,
Building 731, Auckland, New Zealand

**Abstract.** We consider simple cube-curves in the orthogonal 3D grid of cells. The union of all cells contained in such a curve (also called the tube of this curve) is a polyhedrally bounded set. The curve's length is defined to be that of the minimum-length polygonal curve (MLP) fully contained and complete in the tube of the curve. So far only one general algorithm called rubber-band algorithm was known for the approximative calculation of such a MLP. There is an open problem which is related to the design of algorithms for calculation a 3D MLP of a cube-curve: Is there a simple cube-curve such that none of the vertices of its 3D MLP is a grid vertex? This paper constructs an example of such a simple cube-curve. We also characterize this class of cube-curves.

## 1 Introduction

The analysis of cube-curves is related to 3D image data analysis. A cube-curve is, for example, the result of a digitization process which maps a curve-like object into a union $S$ of face-connected closed cubes. The length of a simple cube-curve in 3D Euclidean space is based on the calculation of the minimal length polygonal curve (MLP) in a polyhedrally bounded compact set [3, 4].

The computation of the length of a simple cube-curve in 3D Euclidean space was a subject in [5]. But the method may fail for specific curves. [1] presents an algorithm (rubber-band algorithm) for computing the approximating MLP in S with measured time complexity in $O(n)$, where $n$ is the number of grid cubes of the given cube-curve.

The difficulty of the computation of the MLP in 3D may be illustrated by the fact that the Euclidean shortest path problem (i.e., find a shortest obstacle-avoiding path from source point to target point, for a given finite collection of polyhedral obstacles in 3D space and a given source and a target point) is known to be NP-complete [7]. However, there are some algorithms solving the approximate Euclidean shortest path problem in 3D with polynomial-time, see [8]. The Rubber-band algorithm is not yet proved to be always convergent to the correct 3D-MLP.

Recently, [6] developed of an algorithm for calculation of the correct MLP (with proof) for a special class cube-curves. The main idea is to discompose the cube-curve into some arcs by finding some "end angles" (see Definition 4 below).

E. Andres et al. (Eds.): DGCI 2005, LNCS 3429, pp. 183–194, 2005.

There is an open problem (see [2–page 406]) which is related to designing algorithms for the calculation of the 3D MLP of a cube-curve: It there a simple cube-curve such that none of the vertices of its 3D MLP is a grid vertex? This paper constructs an example of such a simple cube-curve, and generalizes this by characterizing the class of all of those cube-curves. Furthermore it is true that these cube-curves do not have any end angle; and this means that we cannot use the MLP algorithm proposed in [6] which is provable correct. This is the basic importance of the given result: we show the existence of cube-curves which require further algorithmic studies.

Following [1], a grid point $(i, j, k) \in \mathbb{Z}^3$ is assumed to be the center point of a *grid cube* with *faces* parallel to the coordinate planes, with *edges* of length 1, and *vertices* as its corners. *Cells* are either cubes, faces, edges, or vertices. The intersection of two cells is either empty or a joint *side* of both cells. A *cube-curve* is an alternating sequence $g = (f_0, c_0, f_1, c_1, \ldots, f_n, c_n)$ of faces $f_i$ and cubes $c_i$, for $0 \leq i \leq n$, such that faces $f_i$ and $f_{i+1}$ are sides of cube $c_i$, for $0 \leq i \leq n$ and $f_{n+1} = f_0$. It is *simple* iff $n \geq 4$ and for any two cubes $c_i, c_k \in g$ with $|i - k| \geq 2$ (mod $n + 1$), if $c_i \bigcap c_k \neq \phi$ then either $|i - k| = 2$ (mod $n + 1$) and $c_i \bigcap c_k$ is an edge, or $|i - k| \geq 3$ (mod $n + 1$) and $c_i \bigcap c_k$ is a vertex.

A *tube* **g** is the union of all cubes contained in a cube-curve $g$. A tube is a compact set in $\mathbb{R}^3$, its frontier defines a polyhedron, and it is homeomorphic with a torus in case of a simple cube-curve. A curve in $\mathbb{R}^3$ is *complete* in **g** iff it has a nonempty intersection with every cube contained in $g$. Following [3, 4], we define:

**Definition 1.** *A minimum-length polygon (MLP) of a simple cube-curve $g$ is a shortest simple curve $P$ which is contained and complete in tube **g**. The length of a simple cube-curve $g$ is defined to be the length $l(P)$ of an MLP $P$ of $g$.*

It turns out that such a shortest simple curve $P$ is always a polygonal curve, and it is uniquely defined if the cube-curve is not only contained in a single layer of cubes of the 3D grid (see [3, 4]). If it is contained in one layer, then the MLP is uniquely defined up to a translation orthogonal to that layer. We speak about *the* MLP of a simple cube-curve.

A *critical edge* of a cube-curve $g$ is such a grid edge which is incident with exactly three different cubes contained in $g$. Figure 1 shows all the critical edges of a simple cube-curve.

**Definition 2.** *If $e$ is a critical edge of $g$ and $l$ is a straight line such that $e \subset l$, then $l$ is called a* critical line *of $e$ in $g$ or* critical line *for short.*

**Definition 3.** *Let $e$ be a critical edge of $g$. Let $P_1$ and $P_2$ be the two end points of $e$. If one of coordinates of $P_1$ is less than that of $P_2$, then $P_1$ is called the* first *end point of $e$ in $g$. Otherwise $P_1$ is called the* second *end point of $e$ in $g$.*

**Definition 4.** *Assume a simple cube-curve $g$ and a triple of consecutive critical edges $e_1$, $e_2$, and $e_3$ such that $e_i \perp e_j$, for all $i, j = 1, 2, 3$ with $i \neq j$. If $e_2$ is parallel to the x-axis (y-axis, or z-axis) implies the x-coordinates (y-coordinates,* *or z-coordinates) of two vertices (i.e., end points) of $e_1$ and $e_3$ are equal, then we say that $e_1$, $e_2$ and $e_3$ form an end angle, and g has an end angle, denoted by $\angle(e_1, e_2, e_3)$; otherwise we say that $e_1$, $e_2$ and $e_3$ form a middle angle, and g has a middle angle.*

Figure 1 shows a simple cube-curve which has 5 end angles $\angle(e_{21}, e_0, e_1)$, $\angle(e_4, e_5, e_6)$, $\angle(e_6, e_7, e_8)$, $\angle(e_{14}, e_{15}, e_{16}))$, $\angle(e_{16}, e_{17}, e_{18})$, and many middle angles (e.g., $\angle(e_0, e_1, e_2)$, $\angle(e_1, e_2, e_3)$, or $\angle(e_2, e_3, e_4)$).

**Definition 5.** *A simple cube-curve g is called* first class *iff each critical edge of g contains exactly one vertex of the MLP of g.*

We can simply detect a simple cube-curve is first class or not by running rubber band algorithm: the curve is first class iff option $(O_1)$ (see [1]) does not occur.

This paper focuses on first-class simple cube-curves because the general simple cube-curves require further studies.

**Definition 6.** *Let $S \subseteq \mathbb{R}^3$. The set $\{(x, y, 0) : \exists z(z \in \mathbb{R} \wedge (x, y, z) \in S)\}$ is the xy-projection of S, or projection of S for short. Analogously we define the yz- or xz-projection of S.*

**Definition 7.** *If $e_1$, $e_2$, ..., $e_m$ are consecutive critical edges of a cube-curve g and $e_0 \perp e_1$, $e_m \perp e_{m+1}$, and $e_i \parallel e_{i+1}$, where i equals 1, 2, ..., and $m - 1$, $m \geq 2$, then $\{e_1, e_2, ..., e_m\}$ is a set of maximal parallel critical edges of g, and critical edge $e_0$ or $e_{m+1}$ is called adjacent to this set.*

Figure 1 shows a simple cube-curve which has 2 maximal parallel critical edge sets: $\{e_{11}, e_{12}\}$ and $\{e_{18}, e_{19}, e_{20}, e_{21}\}$. The two adjacent critical edges of

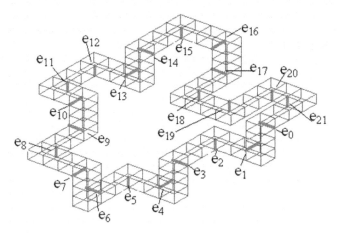

**Fig. 1.** Example of a first-class simple cube-curve which has middle and end angles

$\{e_{11}, e_{12}\}$ are $e_{10}$ and $e_{13}$, they are on two different grid planes. The two adjacent critical edges of $\{e_{18}, e_{19}, e_{20}, e_{21}\}$ are $e_{17}$ and $e_0$, they are on two different grid planes as well.

The paper is organized as follows: Section 2 describes theoretical fundamentals for constructing our example. Section 3 presents the example. Section 4 gives the conclusions.

## 2    Basics

We provide mathematical fundamentals used for constructing a simple cube-curve such that none of the vertices of its 3D MLP is a grid vertex. We start with citing a basic theorem from [1]:

**Theorem 1.** *Let $g$ be a simple cube-curve. Critical edges are the only possible locations of vertices of the MLP of $g$.*

Let $d_e(p, q)$ be the Euclidean distance between points $p$ and $q$.

Let $e_0, e_1, e_2, \ldots, e_m$ and $e_{m+1}$ be $m+2$ consecutive critical edges in a simple cube-curve, and let $l_0, l_1, l_2, \ldots, l_m$ and $l_{m+1}$ be the corresponding critical lines. We express a point $p_i(t_i) = (x_i + k_{x_i} t_i, y_i + k_{y_i} t_i, z_i + k_{z_i} t_i)$ on $l_i$ in general form, with $t_i \in \mathbb{R}$, where $i$ equals $0, 1, \ldots,$ or $m + 1$.

In the following, $p(t_i)$ will be denoted by $p_i$ for short, where $i$ equals $0, 1, \ldots,$ or $m + 1$.

**Lemma 1.** *If $e_1 \perp e_2$, then $\frac{\partial d_e(p_1, p_2)}{\partial t_2}$ can be written as $(t_2 - \alpha)\beta$, where $\beta > 0$, and $\beta$ is a function of $t_1$ and $t_2$, $\alpha$ is $0$ if $e_1$ and the first end point of $e_2$ are on the same grid plane, and $\alpha$ is $1$ otherwise.*

*Proof.* Without loss of generality, we can assume that $e_2$ is parallel to $z$-axis. In this case, the parallel projection (denoted by $g'(e_1, e_2)$) of all of $g$'s cubes, contained between $e_1$ and $e_2$, is illustrated in Figure 2, where $AB$ is the projective image of $e_1$, and $C$ is that of one of the end points of $e_2$.
*Case 1.* $e_1$ and the first end point of $e_2$ are on the same grid plane. Let the two end points of $e_2$ be $(a, b, c)$ and $(a, b, c + 1)$. Then the two end points of $e_1$ are

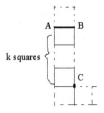

**Fig. 2.** Illustration of the proof of Lemma 1

$(a-1, b+k, c)$ and $(a, b+k, c)$. Then the coordinates of $p_1$ and $p_2$ are $(a-1+t_1, b+k, c)$ and $(a, b, c+t_2)$ respectively, and $d_e(p_1, p_2) = \sqrt{(t_1-1)^2 + k^2 + t_2{}^2}$.

Therefore $\frac{\partial d_e(p_1, p_2)}{\partial t_2} = \frac{t_2}{\sqrt{(t_1-1)^2+k^2+t_2{}^2}}$. Let $\alpha = 0$ and $\beta = \frac{1}{\sqrt{(t_1-1)^2+k^2+t_2{}^2}}$.

This proves the lemma for Case 1.

*Case 2.* $e_1$ and the first end point of $e_2$ are on different grid planes (i.e., $e_1$ and the second end point of $e_2$ are on the same grid plane). Let the two end points of $e_2$ be $(a, b, c)$ and $(a, b, c+1)$. Then the two end points of $e_1$ are $(a-1, b+k, c+1)$ and $(a, b+k, c+1)$. Then the coordinates of $p_1$ and $p_2$ are $(a-1+t_1, b+k, c+1)$ and $(a, b, c+t_2)$ respectively, and $d_e(p_1, p_2) = \sqrt{(t_1-1)^2 + k^2 + (t_2-1)^2}$.

Therefore $\frac{\partial d_e(p_1, p_2)}{\partial t_2} = \frac{t_2-1}{\sqrt{(t_1-1)^2+k^2+(t_2-1)^2}}$. Let $\alpha = 1$ and

$\beta = \frac{1}{\sqrt{(t_1-1)^2+k^2+(t_2-1)^2}}$. This proves the lemma for Case 2.    □

**Lemma 2.** *If $e_1 \parallel e_2$, then $\frac{\partial d_e(p_1, p_2)}{\partial t_2}$ can be written as $(t_2 - t_1)\beta$, where $\beta > 0$, and $\beta$ is a function of $t_1$ and $t_2$*

*Proof.* Without loss of generality, we can assume that $e_2$ is parallel to $z$-axis. In this case, the parallel projection (denoted by $g'(e_1, e_2)$) of all of $g$'s cubes contained between $e_1$ and $e_2$ is illustrated in Figure 3, where $A$ is the projective image of one of the end points of $e_1$, and $B$ is that of one of the end points of $e_2$.

*Case 1.* $e_1$ and $e_2$ are on the same grid plane. Let the two end points of $e_2$ be $(a, b, c)$ and $(a, b, c+1)$. Then the two end points of $e_1$ are $(a, b+k, c)$ and $(a, b+k, c+1)$. Then the coordinates of $p_1$ and $p_2$ are $(a, b+k, c+t_1)$ and $(a, b, c+t_2)$ respectively, and $d_e(p_1, p_2) = \sqrt{(t_2-t_1)^2 + k^2}$.

Therefore $\frac{\partial d_e(p_1, p_2)}{\partial t_2} = \frac{t_2-t_1}{\sqrt{(t_2-t_1)^2+k^2}}$. Let $\beta = \frac{1}{\sqrt{(t_2-t_1)^2+k^2}}$. This proves the lemma for Case 1.

*Case 2.* $e_1$ and $e_2$ are on different grid planes. Let the two end points of $e_2$ be $(a, b, c)$ and $(a, b, c+1)$. Then the two end points of $e_1$ are $(a-1, b+k, c)$ and

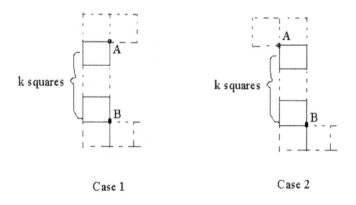

Case 1                    Case 2

**Fig. 3.** Illustration of the proof of Lemma 2

$(a - 1, b + k, c + 1)$. Then the coordinates of $p_1$ and $p_2$ are $(a - 1, b + k, c + t_1)$ and $(a, b, c + t_2)$ respectively, and $d_e(p_1, p_2) = \sqrt{(t_2 - t_1)^2 + k^2 + 1}$.

Therefore $\frac{\partial d_e(p_1, p_2)}{\partial t_2} = \frac{t_2 - t_1}{\sqrt{(t_2 - t_1)^2 + k^2 + 1}}$. Let $\beta = \frac{1}{\sqrt{(t_2 - t_1)^2 + k^2 + 1}}$. This proves the lemma for Case 2.                                                   □

This Lemma will be used when we prove Lemma 6 later.

Let $d_i = d_e(p_{i-1}, p_i) + d_e(p_i, p_{i+1})$, where $i$ equals 1, 2, ..., or $m$.

**Theorem 2.** *If $e_i \perp e_j$, where $i, j = 1, 2, 3$ and $i \neq j$, then $e_1$, $e_2$ and $e_3$ form an end angle iff the equation $\frac{\partial(d_e(p_1, p_2) + d_e(p_2, p_3))}{\partial t_2} = 0$ has a unique root 0 or 1.*

*Proof.* Without loss of generality, we can assume that $e_2$ is parallel to $z$-axis.

(A) If $e_1$, $e_2$ and $e_3$ form an end angle, then by Definition 4, the $z$-coordinates of two end points of $e_1$ and $e_3$ are equal.

*Case A1.* $e_1$, $e_3$ and the first end point of $e_2$ are on the same grid plane. By Lemma 1, $\frac{\partial(d_e(p_1, p_2))}{\partial t_2} = (t_2 - \alpha_1)\beta_1$, where $\alpha_1 = 0$ and $\beta_1 > 0$, and $\frac{\partial(d_e(p_2, p_3))}{\partial t_2} = (t_2 - \alpha_2)\beta_2$, where $\alpha_2 = 0$ and $\beta_2 > 0$. So we have $\frac{\partial(d_e(p_1, p_2) + d_e(p_2, p_3))}{\partial t_2} = t_2(\beta_1 + \beta_2)$. Therefore the equation $\frac{\partial(d_e(p_1, p_2) + d_e(p_2, p_3))}{\partial t_2} = 0$ has a unique root $t_2 = 0$.

*Case A2.* $e_1$, $e_3$ and the second end point of $e_2$ are on the same grid plane. By Lemma 1, $\frac{\partial(d_e(p_1, p_2))}{\partial t_2} = (t_2 - \alpha_1)\beta_1$, where $\alpha_1 = 1$ and $\beta_1 > 0$, and $\frac{\partial(d_e(p_2, p_3))}{\partial t_2} = (t_2 - \alpha_2)\beta_2$, where $\alpha_2 = 1$ and $\beta_2 > 0$. So we have $\frac{\partial(d_e(p_1, p_2) + d_e(p_2, p_3))}{\partial t_2} = (t_2 - 1)(\beta_1 + \beta_2)$. Therefore, equation $\frac{\partial(d_e(p_1, p_2) + d_e(p_2, p_3))}{\partial t_2} = 0$ has a unique root $t_2 = 1$.

(B) Conversely, if equation $\frac{\partial(d_e(p_1, p_2) + d_e(p_2, p_3))}{\partial t_2} = 0$ has a unique root 0 or 1, then $e_1$, $e_2$ and $e_3$ form an end angle. Otherwise, $e_1$, $e_2$ and $e_3$ form a middle angle. By Definition 4, the $z$-coordinates of two end points of $e_1$ are not equal to $z$-coordinates of two end points of $e_3$ (Note: Without loss of generality, we can assume that $e_2 \parallel z$-axis.). So $e_1$ and $e_3$ are not on the same grid plane.

*Case B1.* $e_1$ and the first end point of $e_2$ are on the same grid plane, while $e_3$ and the second end point of $e_2$ are on the same grid plane. By Lemma 1, $\frac{\partial(d_e(p_1, p_2))}{\partial t_2} = (t_2 - \alpha_1)\beta_1$, where $\alpha_1 = 0$ and $\beta_1 > 0$, while $\frac{\partial(d_e(p_2, p_3))}{\partial t_2} = (t_2 - \alpha_2)\beta_2$, where $\alpha_2 = 1$ and $\beta_2 > 0$. So we have $\frac{\partial(d_e(p_1, p_2) + d_e(p_2, p_3))}{\partial t_2} = t_2\beta_1 + (t_2 - 1)\beta_2$. Therefore $t_2 = 0$ or 1 is not a root of the equation $\frac{\partial(d_e(p_1, p_2) + d_e(p_2, p_3))}{\partial t_2} = 0$. This is a contradiction.

*Case B2.* $e_1$ and the second end point of $e_2$ are on the same grid plane, while $e_3$ and the first end point of $e_2$ are on the same grid plane. By Lemma 1, $\frac{\partial(d_e(p_1, p_2))}{\partial t_2} = (t_2 - \alpha_1)\beta_1$, where $\alpha_1 = 1$ and $\beta_1 > 0$, while $\frac{\partial(d_e(p_2, p_3))}{\partial t_2} = (t_2 - \alpha_2)\beta_2$, where $\alpha_2 = 0$ and $\beta_2 > 0$. So we have $\frac{\partial(d_e(p_1, p_2) + d_e(p_2, p_3))}{\partial t_2} = (t_2 - 1)\beta_1 + t_2\beta_2$. Therefore, $t_2 = 0$ or 1 is not a root of the equation $\frac{\partial(d_e(p_1, p_2) + d_e(p_2, p_3))}{\partial t_2} = 0$. This is a contradiction as well.                                                   □

**Theorem 3.** *If $e_i \perp e_j$, where $i, j = 1, 2, 3$ and $i \neq j$, then $e_1$, $e_2$ and $e_3$ form a middle angle iff the equation $\frac{\partial(d_e(p_1, p_2) + d_e(p_2, p_3))}{\partial t_2} = 0$ has a root $t_{2_0}$ such that $0 < t_{2_0} < 1$.*

*Proof.* If $e_1$, $e_2$ and $e_3$ form a middle angle, then by Definition 4, $e_1$, $e_2$ and $e_3$ do not form an end angle. By Theorem 2, 0 or 1 is not a root of the equation $\frac{\partial(d_e(p_1, p_2) + d_e(p_2, p_3))}{\partial t_2} = 0$. By Lemma 1, $\frac{\partial(d_e(p_1, p_2) + d_e(p_2, p_3))}{\partial t_2} = (t_2 - \alpha_1)\beta_1 + (t_2 - \alpha_2)\beta_2$, where $\alpha_1, \alpha_2$ are 0 or 1, $\beta_1 > 0$ is a function of $t_1$ and $t_2$, and $\beta_2 > 0$ is a function of $t_2$ and $t_3$. So $\alpha_1 \neq \alpha_2$. (i.e., $\alpha_1 = 0$ and $\alpha_2 = 1$ or $\alpha_1 = 1$ and $\alpha_2 = 0$). Therefore the equation $\frac{\partial(d_e(p_1, p_2) + d_e(p_2, p_3))}{\partial t_2} = 0$ has a root $t_{2_0}$ such that $0 < t_{2_0} < 1$.

Conversely, if the equation $\frac{\partial(d_e(p_1, p_2) + d_e(p_2, p_3))}{\partial t_2} = 0$ has a root $t_{2_0}$ such that $0 < t_{2_0} < 1$, then by Theorem 2, $e_1$, $e_2$ and $e_3$ do not form an end angle. By Definition 4, $e_1$, $e_2$ and $e_3$ do form a middle angle. □

Assume that $e_0 \perp e_1$, $e_2 \perp e_3$, and $e_1 \parallel e_2$. Assume that $p(t_{i_0})$ is a vertex of the MLP of $g$, where $i$ equals 1 or 2. Then we have

**Lemma 3.** *If $e_0$, $e_3$ and the first end point of $e_1$ are on the same grid plane, and $t_{i_0}$ is a root of $\frac{\partial d_i}{\partial t_i} = 0$, then $t_{i_0} = 0$, where $i$ equals 1 or 2.*

*Proof.* From $p_0(t_0)p_1(0) \perp e_1$ it follows that

$$d_e(p_0(t_0)p_1(0)) = \min\{d_e(p_0(t_0), p_1(t_1)) : t_1 \in [0, 1]\}$$

(see Figure 4). Analogously, we have $d_e(p_2(0)p_3(t_3)) = \min\{d_e(p_2(t_2), p_3(t_3)) : t_2 \in [0, 1]\}$ and $d_e(p_1(0)p_2(0)) = \min\{d_e(p_1(t_1), p_2(t_2)) : t_1, t_2 \in [0, 1]\}$. Therefore we have

$$\min\{d_e(p_0(t_0), p_1(t_1)) + d_e(p_1(t_1), p_2(t_2)) + d_e(p_2(t_2), p_3(t_3)) : t_1, t_2 \in [0, 1]\}$$
$$\geq d_e(p_0(t_0), p_1(0)) + d_e(p_1(0), p_2(0)) + d_e(p_2(0), p_3(t_3)) \qquad □$$

Assume that we have $e_0 \perp e_1$, $e_m \perp e_{m+1}$, and $e_i \parallel e_{i+1}$, (i.e., the set $\{e_1, e_2, \ldots, e_m\}$ is a set of maximal parallel critical edges of $g$, and $e_0$ or $e_{m+1}$ is an adjacent critical edge of this set). Furthermore, let $p(t_{i_0})$ be a vertex of the MLP of $g$, where $i = 1, 2, \ldots, m - 1$. Analogously, we have the following two lemmas:

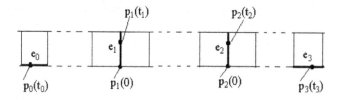

**Fig. 4.** Illustration of the proof of Lemma 3

**Lemma 4.** *If $e_0$, $e_{m+1}$ and the first point of $e_1$ are on the same grid plane, and $t_{i_0}$ is a root of $\frac{\partial d_i}{\partial t_i} = 0$, then $t_{i_0} = 0$, where $i = 1, 2, \ldots, m$.*

**Lemma 5.** *If $e_0$, $e_{m+1}$ and the second end point of $e_1$ are on the same grid plane, and $t_{i_0}$ is a root of $\frac{\partial d_i}{\partial t_i} = 0$, then $t_{i_0} = 1$, where $i = 1, 2, \ldots, m$.*

**Lemma 6.** *If $e_0$ and $e_{m+1}$ are on different grid planes, and $t_{i_0}$ is a root of $\frac{\partial d_i}{\partial t_i} = 0$, where $i = 1, 2, \ldots, m$. Then $0 < t_1 < t_2 < \ldots < t_m < 1$.*

*Proof.* Assume that $e_0$ and the first end point of $e_1$ are on the same grid plane, and $e_{m+1}$ and the second end point of $e_1$ are on the same grid plane. Then by Lemmas 1 and 2, $\frac{\partial d_i}{\partial t_i}$, where $i = 1, 2, \ldots, m$, have the following forms: $\frac{\partial d_1}{\partial t_1} = t_1 b_{1_1} + (t_1 - t_2)b_{1_2}$, $\frac{\partial d_2}{\partial t_2} = (t_2 - t_1)b_{2_1} + (t_2 - t_3)b_{2_2}$, $\frac{\partial d_3}{\partial t_3} = (t_3 - t_2)b_{3_1} + (t_3 - t_4)b_{3_2}$, $\ldots$, $\frac{\partial d_{m-1}}{\partial t_{m-1}} = (t_{m-1} - t_{m-2})b_{m-1_1} + (t_{m-1} - t_m)b_{m-1_2}$, and $\frac{\partial d_m}{\partial t_m} = (t_m - t_{m-1})b_{m_1} + (t_m - 1)b_{m_2}$, where $b_{i_1} > 0$, and $b_{i_1}$ is a function of $t_i$ and $t_{i-1}$, and $b_{i_2} > 0$, and $b_{i_2}$ is a function of $t_i$ and $t_{i+1}$, $i = 1, 2, \ldots, m$.

If $t_{1_0} < 0$, then by $\frac{\partial d_1}{\partial t_1} = 0$, we have $t_{1_0} b_{1_1} + (t_{1_0} - t_{2_0})b_{1_2} = 0$. Since $b_{1_1} > 0$ and $b_{1_2} > 0$, so we have $t_{1_0} - t_{2_0} > 0$, (i.e., $t_{1_0} > t_{2_0}$). Analogously, by $\frac{\partial d_2}{\partial t_2} = 0$, so $(t_{2_0} - t_{1_0})b_{2_1} + (t_{2_0} - t_{3_0})b_{2_2} = 0$. Then we have $t_{2_0} > t_{3_0}$. Analogously, we have $t_{3_0} > t_{4_0}, \ldots, t_{m-1_0} > t_{m_0}$. Therefore, by $\frac{\partial d_m}{\partial t_m} = (t_m - t_{m-1})b_{m_1} + (t_m - 1)b_{m_2}$, we have $t_{m_0} - 1 > 0$. So we have $0 > t_{1_0} > t_{2_0} > t_{3_0} > \ldots > t_{m_0} > 1$. This is a contradiction.

If $t_{1_0} = 0$, then by $\frac{\partial d_1}{\partial t_1} = 0$ we have $t_{2_0} = 0$. Analogously, by $\frac{\partial d_2}{\partial t_2} = 0$ we have $t_{3_0} = 0$. Analogously, we have $t_{4_0} = 0, \ldots, t_{m_0} = 0$. But, by $\frac{\partial d_m}{\partial t_m} = (t_m - t_{m-1})b_{m_1} + (t_m - 1)b_{m_2}$, we have $\frac{\partial d_m}{\partial t_m} = (t_m - 1)b_{m_2} = -b_{m_2} < 0$. This is in contradiction to $\frac{\partial d_m}{\partial t_m} = 0$.

If $t_{1_0} \geq 1$, then by $\frac{\partial d_1}{\partial t_1} = 0$, we have $t_{1_0} b_{1_1} + (t_{1_0} - t_{2_0})b_{1_2} = 0$. Due to $b_{1_1} > 0$ and $b_{1_2} > 0$ we have $t_{1_0} - t_{2_0} < 0$, (i.e., $t_{1_0} < t_{2_0}$). Analogously, by $\frac{\partial d_2}{\partial t_2} = 0$ it follows that $(t_{2_0} - t_{1_0})b_{2_1} + (t_{2_0} - t_{3_0})b_{2_2} = 0$. Then we have $t_{2_0} < t_{3_0}$. Analogously, we have $t_{3_0} < t_{4_0}, \ldots, t_{m-1_0} < t_{m_0}$. Therefore, by $\frac{\partial d_m}{\partial t_m} = (t_m - t_{m-1})b_{m_1} + (t_m - 1)b_{m_2}$, we have $t_{m_0} - 1 < 0$. So we have $1 \leq t_{1_0} < t_{2_0} < t_{3_0} < \ldots < t_{m_0} < 1$. This is a contradiction.     □

Let $t_{i_0}$ be a root of $\frac{\partial d_i}{\partial t_i} = 0$, where $i = 1, 2, \ldots, m$. We apply Lemmas 4, 5 and 6 and obtain

**Theorem 4.** *$e_0$ and $e_{m+1}$ are on different grid plane iff $0 < t_{1_0} < t_{2_0} < \ldots < t_{m_0} < 1$.*

## 3   An Example

We provide one example to show that there is a simple cube-curve such that none of the vertices of its 3D MLP is a grid vertex. See Table 1, which lists the coordinates of the critical edges $e_0, e_1, \ldots, e_{19}$ of $g$. Let $v(t_0), v(t_1), \ldots, v(t_{19})$ be

**Table 1.** Coordinates of endpoints of critical edges in Figure 5

| Critical edge | $x_{i1}$ | $y_{i1}$ | $z_{i1}$ | $x_{i2}$ | $y_{i2}$ | $z_{i2}$ |
|---|---|---|---|---|---|---|
| $e_0$ | -1 | 4 | 7 | -1 | 4 | 8 |
| $e_1$ | 1 | 4 | 7 | 1 | 5 | 7 |
| $e_2$ | 2 | 4 | 5 | 2 | 5 | 5 |
| $e_3$ | 4 | 5 | 4 | 4 | 5 | 5 |
| $e_4$ | 4 | 7 | 4 | 5 | 7 | 4 |
| $e_5$ | 5 | 7 | 2 | 5 | 8 | 2 |
| $e_6$ | 7 | 7 | 2 | 7 | 8 | 2 |
| $e_7$ | 7 | 8 | 4 | 8 | 8 | 4 |
| $e_8$ | 8 | 10 | 4 | 8 | 10 | 5 |
| $e_9$ | 10 | 10 | 4 | 10 | 10 | 5 |
| $e_{10}$ | 10 | 8 | 5 | 11 | 8 | 5 |
| $e_{11}$ | 11 | 7 | 7 | 11 | 8 | 7 |
| $e_{12}$ | 12 | 7 | 7 | 12 | 7 | 8 |
| $e_{13}$ | 12 | 5 | 7 | 12 | 5 | 8 |
| $e_{14}$ | 10 | 4 | 8 | 10 | 5 | 8 |
| $e_{15}$ | 9 | 4 | 10 | 10 | 4 | 10 |
| $e_{16}$ | 9 | 0 | 10 | 10 | 0 | 10 |
| $e_{17}$ | 9 | 0 | 8 | 10 | 0 | 8 |
| $e_{18}$ | 9 | 1 | 7 | 9 | 1 | 8 |
| $e_{19}$ | -1 | 2 | 7 | -1 | 2 | 8 |

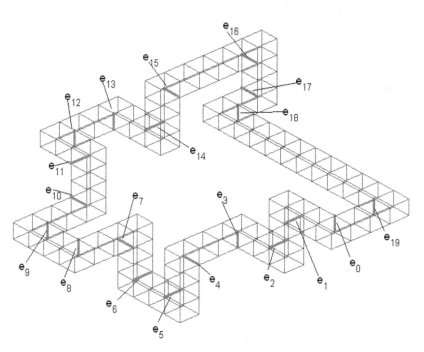

**Fig. 5.** A simple cube-curve such that none of the vertices of its 3D MLP is a grid vertex

the vertex of the MLP of $g$ such that $v(t_i)$ is on $e_i$ and $t_i$ is in $[0, 1]$, where $i$ = 0, 1, 2, ..., 19. By Appendix we can see that there is not any end angle in $g$. In fact, there are 6 middle angles: $\angle(e_2, e_3, e_4))$, $\angle(e_3, e_4, e_5))$, $\angle(e_6, e_7, e_8))$, $\angle(e_9, e_{10}, e_{11}))$, $\angle(e_{10}, e_{11}, e_{12}))$, and $\angle(e_{13}, e_{14}, e_{15}))$. By Theorem 3, we have $t_3, t_4, t_7, t_{10}, t_{11}$ and $t_{14}$ are in $(0, 1)$. By Figure 5 we can see that $e_1 \parallel e_2$ and $e_0$ and $e_3$ are on different grid planes. By Theorem 4, we have $t_1$ and $t_2$ are in $(0, 1)$.

Analogously, we have $t_5$ and $t_6$ are in $(0, 1)$; $t_8$ and $t_9$ are in $(0, 1)$; $t_{12}$ and $t_{13}$ are in $(0, 1)$; $t_{15}, t_{16}$ and $t_{17}$ are in $(0, 1)$; and $t_{18}, t_{19}$ and $t_0$ are in $(0, 1)$. Therefore, each $t_i$ is in $(0, 1)$, where $i = 0, 1, ..., 19$. So $g$ is a simple cube-curve such that none of the vertices of its 3D MLP is a grid vertex.

## 4    Conclusions

We have constructed a non-trivial simple cube-curve such that none of the vertices of its 3D MLP is a grid vertex. Indeed, by Theorems 2 and 4, and Lemmas 5 and  6, we can come to the conclusion that given a simple first class cube-curve $g$, none of the vertices of its 3D MLP is a grid point iff $g$ has not any end angle and for every set of maximal parallel edges of $g$, its two adjacent critical edges are not on the same grid plane.

It follows that the (provable correct) MLP algorithm proposed in [6] cannot be applied to this curve, because it requires at least one end angle for decomposing the curve into arcs. Of course, the rubber-band algorithm is applicable, and will produce a result (i.e., a polygonal curve). However, in this case we are still unable to show whether this result is the MLP of the given cube-curve or not.

**Acknowledgements.** The reviewers' comments have been very helpful for revising an earlier version of this paper.

## Appendix: List of $\frac{\partial d_i}{\partial t_i}$ ($i = 0, 1, ..., 19$)

We compute $\frac{\partial d_i}{\partial t_i}$ ($i = 0, 1, ..., 19$) for $g$ as shown in Figure 5.

$$d_{t_0} = \frac{t_0}{\sqrt{t_0^2 + t_1^2 + 4}} + \frac{t_0 - t_{19}}{\sqrt{(t_0 - t_{19})^2 + 4}} \tag{1}$$

$$d_{t_1} = \frac{t_1}{\sqrt{t_0^2 + t_1^2 + 4}} + \frac{t_1 - t_2}{\sqrt{(t_1 - t_2)^2 + 5}} \tag{2}$$

$$d_{t_2} = \frac{t_2 - t_1}{\sqrt{(t_2 - t_1)^2 + 5}} + \frac{t_2 - 1}{\sqrt{(t_2 - 1)^2 + (t_3 - 1)^2 + 4}} \tag{3}$$

$$d_{t_3} = \frac{t_3 - 1}{\sqrt{(t_2 - 1)^2 + (t_3 - 1)^2 + 4}} + \frac{t_3}{\sqrt{t_3^2 + t_4^2 + 4}} \tag{4}$$

$$d_{t_4} = \frac{t_4}{\sqrt{t_3^2 + t_4^2 + 4}} + \frac{t_4 - 1}{\sqrt{(t_4 - 1)^2 + t_5^2 + 4}} \tag{5}$$

$$d_{t_5} = \frac{t_5}{\sqrt{(t_4 - 1)^2 + t_5^2 + 4}} + \frac{t_5 - t_6}{\sqrt{(t_5 - t_6)^2 + 4}} \tag{6}$$

$$d_{t_6} = \frac{t_6 - t_5}{\sqrt{(t_6 - t_5)^2 + 4}} + \frac{t_6 - 1}{\sqrt{(t_6 - 1)^2 + t_7^2 + 4}} \tag{7}$$

$$d_{t_7} = \frac{t_7}{\sqrt{(t_6 - 1)^2 + t_7^2 + 4}} + \frac{t_7 - 1}{\sqrt{(t_7 - 1)^2 + t_8^2 + 4}} \tag{8}$$

$$d_{t_8} = \frac{t_8}{\sqrt{(t_7 - 1)^2 + t_8^2 + 4}} + \frac{t_8 - t_9}{\sqrt{(t_8 - t_9)^2 + 4}} \tag{9}$$

$$d_{t_9} = \frac{t_9 - t_8}{\sqrt{(t_9 - t_8)^2 + 4}} + \frac{t_9 - 1}{\sqrt{(t_9 - 1)^2 + t_{10}^2 + 4}} \tag{10}$$

$$d_{t_{10}} = \frac{t_{10}}{\sqrt{(t_9 - 1)^2 + t_{10}^2 + 4}} + \frac{t_{10} - 1}{\sqrt{(t_{10} - 1)^2 + (t_{11} - 1)^2 + 4}} \tag{11}$$

$$d_{t_{11}} = \frac{t_{11} - 1}{\sqrt{(t_{11} - 1)^2 + (t_{10} - 1)^2 + 4}} + \frac{t_{11}}{\sqrt{t_{11}^2 + t_{12}^2 + 1}} \tag{12}$$

$$d_{t_{12}} = \frac{t_{12}}{\sqrt{t_{11}^2 + t_{12}^2 + 1}} + \frac{t_{12} - t_{13}}{\sqrt{(t_{12} - t_{13})^2 + 4}} \tag{13}$$

$$d_{t_{13}} = \frac{t_{13} - t_{12}}{\sqrt{(t_{13} - t_{12})^2 + 4}} + \frac{t_{13} - 1}{\sqrt{(t_{13} - 1)^2 + (t_{14} - 1)^2 + 4}} \tag{14}$$

$$d_{t_{14}} = \frac{t_{14} - 1}{\sqrt{(t_{13} - 1)^2 + (t_{14} - 1)^2 + 4}} + \frac{t_{14}}{\sqrt{t_{14}^2 + (t_{15} - 1)^2 + 4}} \tag{15}$$

$$d_{t_{15}} = \frac{t_{15} - 1}{\sqrt{t_{14}^2 + (t_{15} - 1)^2 + 4}} + \frac{t_{15} - t_{16}}{\sqrt{(t_{15} - t_{16})^2 + 16}} \tag{16}$$

$$d_{t_{16}} = \frac{t_{16} - t_{15}}{\sqrt{(t_{16} - t_{15})^2 + 16}} + \frac{t_{16} - t_{17}}{\sqrt{(t_{16} - t_{17})^2 + 4}} \tag{17}$$

$$d_{t_{17}} = \frac{t_{17} - t_{16}}{\sqrt{(t_{17} - t_{16})^2 + 4}} + \frac{t_{17}}{\sqrt{t_{17}^2 + (t_{18} - 1)^2 + 1}} \tag{18}$$

$$d_{t_{18}} = \frac{t_{18} - 1}{\sqrt{t_{17}^2 + (t_{18} - 1)^2 + 1}} + \frac{t_{18} - t_{19}}{\sqrt{(t_{18} - t_{19})^2 + 101}} \tag{19}$$

$$d_{t_{19}} = \frac{t_{19} - t_{18}}{\sqrt{(t_{19} - t_{18})^2 + 101}} + \frac{t_{19} - t_0}{\sqrt{(t_{19} - t_0)^2 + 4}} \tag{20}$$

# References

1. T. Bülow and R. Klette. Digital curves in 3D space and a linear-time length estimation algorithm. *IEEE Trans. Pattern Analysis Machine Intelligence*, **24**:962–970, 2002.
2. R. Klette and A. Rosenfeld. Digital Geometry: Geometric Methods for Digital Picture Analysis. Morgan Kaufmann, San Francisco, 2004., 2004.
3. F. Sloboda, B. Zaťko, and R. Klette. On the topology of grid continua. SPIE *Vision Geometry VII*, **3454**:52–63, 1998.
4. F. Sloboda, B. Zaťko, and J. Stoer. On approximation of planar one-dimensional grid continua. In R. Klette, A. Rosenfeld, and F. Sloboda, editors, *Advances in Digital and Computational Geometry*, pages 113–160. Springer, Singapore, 1998.
5. A. Jonas and N. Kiryati. Length estimation in 3-D using cube quantization, *J. Math. Imaging and Vision*, **8**: 215–238, 1998.
6. F. Li and R. Klette. Minimum-length polygon of a simple cube-curve in 3D space. In Proceedings IWCIA2004, LNCS3322 (to appear).
7. J. Canny and J.H. Reif. New lower bound techniques for robot motion planning problems. Proc. *IEEE Conf. Foundations Computer Science*, pages 49–60, 1987.
8. J. Choi, J. Sellen, and C.-K. Yap. Approximate Euclidean shortest path in 3-space. Proc. *ACM Conf. Computational Geometry*, ACM Press, pages 41–48, 1994.

# Computation of Homology Groups and Generators

Samuel Peltier[1], Sylvie Alayrangues[2], Laurent Fuchs[1],
and Jacques-Olivier Lachaud[2]

[1] SIC (FRE 2731 CNRS), Université de Poitiers,
Boulevard Marie et Pierre Curie, 86962 Futuroscope Chasseneuil Cedex, France
{peltier, fuchs}@sic.univ-poitiers.fr
[2] LaBRI, Université Bordeaux 1,
351 cours de la Libération, 33405 Talence Cedex, France
{alayrang, lachaud}@labri.fr

**Abstract.** Topological invariants are extremely useful in many applications related to digital imaging and geometric modelling, and homology is a classical one. We present an algorithm that computes the whole homology of an object of arbitrary dimension: Betti numbers, torsion coefficients and generators. Results on classical shapes in algebraic topology are presented and discussed.

## 1 Introduction

In digital image analysis, shape invariants are useful for classification, indexation, or, more recently, shape description [ACZ04]. They can be used in object simplification and object thinning. In solid modeling, shape invariants ensure the consistency of constructive operations. Computing topological invariants of objects has thus a significant impact in these domains. The fundamental group is an invariant that carries most of the topological information about an object. It has been studied by many authors [Kon89, Box99, Mal01, ADFQ03] in the image analysis field. But the comparison of such groups is highly related to undecidable problems [Mal01]. Many authors have proposed algorithms to compute the Euler characteristic (some of them summarized in [KR89]), but it is a simpler and less expressive topological invariant. Other approaches compute the Betti numbers [DE95] of embedded objects.

We focus here on homology groups, which are known to be computable in finite dimensions, and which have a good topological characterization power at least in low dimensions. We not only compute these groups but also their generators, to delineate the topological holes on the shapes. For instance, the generators of the homology group of dimension 1 are connectivity lines of the shape: cutting along such lines does not divide the shape into two parts. The contributions of this work are: (i) we report recent works in computational group theory and bring these results to the imagery community, (ii) we combine these works to classical results in homology theory to compute the homology groups

E. Andres et al. (Eds.): DGCI 2005, LNCS 3429, pp. 195–205, 2005.

(Betti numbers and torsion) and their generators, (iii) we effectively implement these algorithms with numerous optimizations.

In the first part of the paper we recall classical definitions in homology theory. We choose here simplicial homology since it is widely used in geometric modeling and is straightforwardly applicable to digital objects. We then present related works. After that, we present our approach for computing homology groups: Smith Normal Form (SNF) of the boundary homomorphisms, modified SNF to compute generators, integer computations performed with a modulo. Lastly, we show some experiments and list some perspectives to this work.

## 2  Simplicial Homology

**Semi-simplicial Set.** Shapes are classically modeled with a cellular subdivision. Several combinatorial structures may represent such a subdivision. We choose here semi-simplicial sets, which can represent indifferently manifold or non-manifold objects. This structure is a subclass of simplicial sets, a structure studied in algebraic topology [May67, Cur71].

**Definition 1.** *[May67] A semi-simplicial set $S = (K, (d_i^q))$ is a graded family of sets $K = (K^q)_{q \in \mathbb{N}}$ together with maps $d_i^q : K^q \to K^{q-1}$ for $i = 0, \dots, q$, which satisfy the following identity: $\forall \sigma \in K^q, d_j^{q-1}(d_i^q(\sigma)) = d_{i-1}^{q-1}(d_j^q(\sigma))$ if $j < i$.*

The elements of $K^q$ are called $q-simplices$. The $d_i^q$ are called *boundary operators* (the subscripts $q$ will generally be dropped later for clarity). Simplices are glued together consistently with these operators (see Fig 1a-b for two examples).

Semi-simplicial sets are clearly adapted to the constructive operations of solid modeling [LL95]. They are also well suited to digital imagery [DG03]. To determine a semi-simplicial set that represents a given digital object, the first

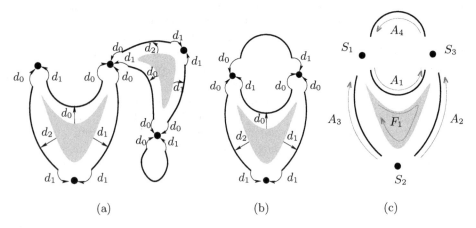

|  (a)  |  (b)  |  (c)  |

**Fig. 1.** On $(a)$ and $(b)$, examples of semi-simplicial sets. On $(c)$, positive orientation of the simplices of $(b)$

step is to construct a simplicial analog. One method is proposed in [GDR03] (see Section 3). The second step is to number the vertices of the simplicial analog; the boundary maps follow directly [LE93].

We can now introduce homology groups in an intuitive way. All objects are assumed to be finite. Note that the homology theory is applicable on most combinatorial structures.

**Chain, Boundary Homomorphism, Chain Complex.** In a first step, we define group structures on semi-simplicial sets. A $p$-chain in $K^p$ is a linear combination of $p$-simplices with integer coefficients. More formally, any $p$-chain is written uniquely as a finite sum $\sum_{i=1}^{n_p} \alpha_i^p \sigma_i^p$, where $n_p$ is the cardinal of $K^p = \{\sigma_1^p, \cdots, \sigma_{n_p}^p\}$, and for all $i$, $\alpha_i^p$ is an integer. The addition over $p$-chains is defined simply by adding coefficients simplex by simplex. The resulting groups are denoted by $C_p$. For all $p$, $K^p$ forms a basis for $C_p$ (see [Mun84] p.28).

A $p$-chain is a purely formal construction. The coefficients $\alpha_i$ have generally not a geometric interpretation, except for the coefficients 1 and $-1$. In this case $1 \cdot \sigma$ means that we consider the simplex $\sigma$ with its orientation and $-1 \cdot \sigma$ means that we consider the simplex $\sigma$ with its opposite orientation. This is consistent with the fact that simplices can be equipped with two orientations, one considered positive and the other negative. Fig. 1($c$) displays the positive orientations of the simplices of Fig. 1($b$). A formal definition of simplex orientation is available in classical algebraic topology books [Mun84, Hat02].

In a second step, we relate chain groups of successive dimension with homomorphisms called boundary operators.

**Definition 2.** *For all $p > 0$, the boundary of a $p$-simplex $\sigma^p$, denoted by $\partial_p(\sigma^p)$, is the $(p-1)$-chain $\sum_{i=0}^{p}(-1)^i d_i(\sigma)$. 0-simplices have an empty boundary. The boundary is extended as an homomorphism from $C_p$ to $C_{p-1}$, meaning for any $p$-chain $c = \sum_{i=1}^{n_p} \alpha_i^p \sigma_i^p$, its boundary $\partial_p(c)$ is equal to $\sum_{i=1}^{n_p} \alpha_i^p \partial_p(\sigma_i^p)$.*

Usually, when no confusion may arise, we simply write $\partial(c)$ for the boundary of a $p$-chain $c$. For example, on Fig. 1c, we have $\partial(F) = A_1 - A_2 + A_3$ and we can verify that $\partial(\partial(F)) = \partial(A_1 - A_2 + A_3) = 0$.

We have just constructed a sequence of chain groups $C_p$ together with homomorphisms $\partial_p$, $C_n \xrightarrow{\partial_n} C_{n-1} \xrightarrow{\partial_{n-1}} \cdots \xrightarrow{\partial_1} C_0 \xrightarrow{\partial_0} 0$. One can check that $\partial_{p-1}(\partial_p(c)) = 0$ for all $p$-chains $c$. This sequence is called a *free chain complex*.

**Cycle, Boundary, Hole.** The homology groups of a combinatorial object are derived from specific subgroups of the chains of a free chain complex.

The $p$-chains whose boundary is empty are called $p$-*cycles*. For example, on Fig. 1c, the 1-chains $A_1 - A_2 + A_3$ and $A_1 + A_4$ are 1-cycles. The set of $p$-cycles is a subgroup of $C_p$, denoted by $Z_p$.

Some $p$-chains are the boundary of a $(p + 1)$-chain. They are called $p$-*boundaries*. For example, on Fig. 1c, the 1-chain $A_1 - A_2 + A_3$ is the boundary of the 2-chain $F$. The set of $p$-boundaries form a subgroup of $C_p$, denoted by $B_p$. Since $\forall c \in C_p, \partial_{p-1}(\partial_p(c)) = 0$, we have $B_p \subset Z_p \subset C_p$.

A *p-dimensional hole* is a $p$-cycle which is not a $p$-boundary. For example, on Fig. 1c, the 1-cycle $z_1 = A_1 + A_4$ is not a boundary. We define an equivalence relation in the group of $p$-cycles as follows: two $p$-cycles $s$ and $t$ are in the same equivalence class iff there exist a chain $c$ with $s = t + \partial_{p+1}c$. They are then said to be *homologous*. In particular, when $s = \partial_{p+1}c$ then $s$ is homologous to 0. The set of cycles is then partitioned by the homology relation, according to the hole they surround. Two cycles in the same equivalence class surround the same hole. The set of $p$-boundaries is the 0-equivalence class. For example, the cycle $z_2 = A_2 - A_3 + A_4$ is in the $z_1$ equivalence class because $z_1 = z_2 + \partial_2(F)$.

**Homology Groups, Weak Boundary.** In any dimension $p$, the *homology group* $H_p$ is defined as the group of the equivalent classes for the homology relation. It is exactly the quotient group of the $p$-cycles by the $p$-boundaries, $H_p = Z_p/B_p$. Homology groups are known to be topological invariants, meaning homeomorphic shapes have isomorphic homology groups.

For all $p$, there exists a finite number of elements of $H_p$ from which we can deduce all $H_p$ elements, thus $H_p$ is called finitely generated. So, the group $H_p$ verifies the fundamental theorem of finitely generated abelian groups [Mun84], and $H_p$ is isomorphic to a direct sum:

$$\underbrace{\mathbb{Z} \oplus \cdots \oplus \mathbb{Z}}_{\beta_p} \oplus \mathbb{Z}/t_1^p\mathbb{Z} \oplus \cdots \oplus \mathbb{Z}/t_n^p\mathbb{Z}.$$

We denote by $\beta_p$ the number of apparitions of $\mathbb{Z}$ in this direct sum: it is the number of elements of $H_p$ with infinite order and is called the *$p$-th Betti number*. The numbers $t_1^p, \ldots, t_n^p$ are called the *torsion coefficients* of $H_p$. To each group $\mathbb{Z}$ of $H_p$ is associated a set of $p$-dimensional homologous cycles: they surround the same $p$-dimensional topological hole and are not the boundary of any $p+1$-chain. It is the same for each group $\mathbb{Z}/t_i^p\mathbb{Z}$: the associated homologous cycles are not the boundary of any $p + 1$-chain. However, when taken $t_i^p$ times, they become the boundary of some $p+1$-chain. An example is the 1-cycle $A_2$ on Fig. 2, which becomes a boundary only when taken two times: $2A_2 = \partial(F_1 + F_2)$.

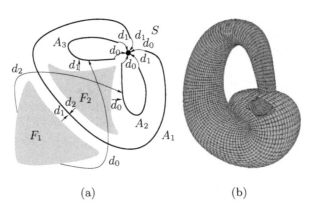

(a)                                    (b)

**Fig. 2.** Klein bottle. (a) Semi-simplicial and (b) geometric representation

# 3   Related Works

Kaczynski *et al.* [KMS98] proposed to compute a chain complex homology with a sequence of reductions. The idea is to derive a new object with less cells while preserving homology at each step of the transformation. To reduce the number of cells, one chooses two cells $a$ and $b$, such that $\partial b = \lambda a + r$ and $\lambda$ is invertible. These cells are then suppressed and the boundary homomorphism is updated. To ensure invertible coefficients, Kaczynski *et al.* choose them in a field. In this case, it can be verified that the reduction algorithm stops on a smallest chain complex with same homology where each cell is a cycle representing the homology class.

González-Díaz and Real [GDR03] recently proposed an algorithm to compute cohomology information on digital objects that are subsets of the 3D body-centered cubic grid. They first construct a simplicial complex with identical topology. After that the cohomology is obtained by the construction of a chain contraction in two passes: (i) a thinning that reduces the size of the data by simplicial collapses, (ii) an incremental algebraic thinning that progressively extracts the equivalence classes of the cohomology groups. A further computation provides the cohomology ring of the digital object, which appears to carry complementary topological information. All coefficients are in $\mathbb{Z}/2\mathbb{Z}$ (also a field).

The preceding approaches are interesting when dealing with embedded objects in 2D or 3D. Homology over a field is then enough to characterize shapes, since objects have no torsion. On the contrary, we choose a more generic approach, valid for arbitrary dimension and shapes. In the following section, we address the problem of computing the whole homology over the coefficient domain $\mathbb{Z}$ (a ring, not a field).

# 4   Computation of Homology Groups and Generators

In this section, we show how to compute the homology groups $H_p$ from the boundary homomorphisms. First, the Betti number and torsion coefficients are deduced from the classical Smith Normal Form (SNF) of $\partial_p$ and $\partial_{p+1}$. Then, we briefly explain why their SNF must be slightly modified to compute a set of generators of $H_p$. We finally discuss about implementation problems linked to that class of methods. We thus propose a new algorithm which benefits from improvements proposed by Dumas *et al.* [DSV01] and computes the Betti numbers, the torsion coefficients and a set of "moduli generators".

## 4.1   Homology Groups via Smith Normal Form

Information on homology groups may be deduced from matrix representations of boundary homomorphisms. A natural basis of the group of $p$-chains of a chain complex is the one made of all its $p$-simplices, i.e. $K^p$. In the following, the matrix $E_{p+1}$, called *$p$-th incidence matrix*, represents the homomorphism $\partial_{p+1}$ relatively to the canonical bases $K^p$ (rows) and $K^{p+1}$ (columns). Each column in $E_{p+1}$ is the boundary of one $p+1$-simplex, decomposed on the base of $p$-simplices.

There exists bases in which any homomorphism has a very specific matrix form, the so-called *Smith Normal Form (SNF)*. It is a matrix full of $0's$ except for an upper left square submatrix which is diagonal with increasing coefficients: $diag(\lambda_1, \ldots, \lambda_l)$ such that each $\lambda_i$ is greater than 1 and divides each $\lambda_j$ for $j > i$. The $(\lambda_i)$ are called *invariant factors* of the homomorphism. Let $D_{p+1}$ be the SNF of $\partial_{p+1}$ with associated bases $(e_k^{p+1})$ and $(f_k^p)$:

$$
D_{p+1} = 
\begin{array}{cccccc}
 & e_1^{p+1} \cdots e_{\gamma_p}^{p+1} & e_{\gamma_p+1}^{p+1} \cdots e_{n_{p+1}}^{p+1} & \\
\end{array}
\left[
\begin{array}{cc|cc}
\lambda_1^p & 0 & & \\
 & \ddots & & 0 \\
0 & \lambda_{\gamma_p}^p & & \\
\hline
 & 0 & & 0 \\
\end{array}
\right]
\begin{array}{c}
f_1^p \\
\vdots \\
f_{\gamma_p}^p \\
f_{\gamma_p+1}^p \\
\vdots \\
f_{n_p}^p
\end{array}
$$

With these notations, it may be proved that:

1. $(e_{\gamma_p+1}^{p+1}, \cdots, e_{n_{p+1}}^{p+1})$ is a basis of $Z_{p+1}$,
2. $(\lambda_1^p f_1^p, \ldots, \lambda_{\gamma_p}^p f_{\gamma_p}^p)$ is a basis of $B_p$,
3. $(f_1^p, \ldots, f_{\gamma_p}^p)$ is a basis of a group $W_p$, known as the group of *weak boundaries* $(W_p = \{c_p \in C_p / \exists \lambda \in \mathbb{Z}^*, \lambda c_p \in B_p\})$.

Moreover, the group $H_p$ is isomorphic to the direct sum $Z_p/W_p \oplus W_p/B_p$ where $Z_p/W_p$ is a free group and $W_p/B_p$ is a torsion group. The torsion coefficients of $H_p$ are exactly the invariant factors of $\partial_{p+1}$ strictly greater than 1 (given by $D_{p+1}$). Furthermore, the Betti number of $H_p$ is equal to $\text{rank}(Z_p) - \text{rank}(W_p)$. They are read respectively on $D_p$ and $D_{p+1}$ with $\text{rank}(Z_p) = n_p - \gamma_{p-1}$ and $\text{rank}(W_p) = \gamma_p$.

However all the generators of the homology groups cannot be deduced from the bases of the SNF. More precisely, we cannot determine the set of cycles which are not weak boundaries. To do it effectively, two successive boundary homomorphisms $\partial_{p-1}$ and $\partial_p$ must respectively share the same upper and lower bases (i.e. $(f_1^p, \cdots, f_{n_p}^p) = (e_1^p, \cdots, e_{n_p}^p)$). This is obviously not the case since $D_{p-1}D_p \neq 0$ (recall that $\partial_{p-1}\partial_p = 0$ in a free chain complex).

## 4.2    Generators with Modified SNF

Cairn [Cai61] proved that it is possible to simultaneously choose bases for each group of $p$-chains such that the matrix $N_p$ representing each boundary operator relatively to these bases is in a normal form quite similar to SNF. Moreover he explains how to deduce a set of generators of the homology group $H_p$ directly from the matrix $N_{p+1}$. $N_{p+1}$ is shown on Tab. 1. The number of invariant factors of $\partial_{p+1}$ is $\gamma_p$ and $\rho_p$ of them are strictly greater than 1.

The set $\{b_1^p, \cdots, b_{\beta_p}^p\}$ generates the free part of $H^p$: they are $p$-cycles when read as a column in $N_p$ and they have no boundary antecedent when read as a

**Table 1.** Modified SNF of boundary homomorphism $\partial_{p+1}$

| | $(p+1) - Cycles$ | | $Weak\ Boundaries$ | $Antecedents$ | |
|---|---|---|---|---|---|
| | $a_1^{p+1} \cdots a_{\gamma_{p+1}}^{p+1}$ | $b_1^{p+1} \cdots b_{\beta_{p+1}}^{p+1}$ | $c_1^{p+1} \cdots \quad c_{\rho_p}^{p+1}$ | $c_{\rho_p+1}^{p+1} \cdots c_{\gamma_p}^{p+1}$ | |
| $a_1^p$ $\vdots$ $a_{\rho_p}^p$ | 0 | 0 | $\lambda_{\gamma_p}^p \qquad 0$ $\ddots$ $0 \quad \lambda_{\gamma_p-\rho_p+1}^p$ | 0 | Weak Boundaries |
| $a_{\rho_p+1}^p$ $\vdots$ $a_{\gamma_p}^p$ | 0 | 0 | 0 | $\begin{matrix}1 & & 0\\ & \ddots & \\ 0 & & 1\end{matrix}$ | |
| $b_1^p$ $\vdots$ $b_{\beta_p}^p$ | 0 | 0 | 0 | 0 | Cycles but not Weak Boundaries |
| $c_1^p$ $\vdots$ $c_{\gamma_{p-1}}^p$ | 0 | 0 | 0 | 0 | |

row in $N_{p+1}$. The set $\{a_1^p, \cdots, a_{\gamma_p}^p\}$ generates the torsion part of $H^p$: they are $p$-cycles when read as a column in $N_p$ and they must be multiplied by the $\lambda_i^p$ to have a boundary antecedent when read as a row in $N_{p+1}$.

Agoston [Ago76] proposed an algorithm to compute all matrices $N_p$ and keep tracks of changes of bases. The idea is to compute successively all matrices $N_p$ from 0 to the maximal index of the desired homology groups. Each homomorphism is successively expressed in four pairs of bases as in Tab. 2.

**Table 2.** Expression of the homomorphisms

| Step | Bases []\\[] and Matrix of $\partial_p$ | Bases []\\[] and Matrix of $\partial_{p+1}$ |
|---|---|---|
| 0. input from iteration $p$ | $[(V_{p-1}U_{p-1}^{-1})^{-1}K^{p-1}]\backslash[U_pK^p]$ (mSNF) $N_p = V_{p-1}U_{p-1}^{-1}E_pU_p$ | |
| 1. Incidence matrix of $\partial_{p+1}$ | | $[K^p]\backslash[K^{p+1}]$ (incidence) $E_{p+1}$ |
| 2. Left-multiply $E_{p+1}$ by $U_p^{-1}$ | | $[(U_p^{-1})^{-1}K^p]\backslash[K^{p+1}]$ $E'_{p+1} = U_p^{-1}E_{p+1}$ |
| 3. Compute the mSNF $N_{p+1}$ of $\partial_{p+1}$ from $E'_{p+1}$ | | $[(V_pU_p^{-1})^{-1}K^p]\backslash[U_{p+1}K^{p+1}]$ (mSNF) $N_{p+1} = V_pU_p^{-1}E_{p+1}U_{p+1}$ |
| 4. Right-multiply $N_p$ by $V_p^{-1}$ | $[(V_{p-1}U_{p-1}^{-1})^{-1}K^{p-1}]\backslash[U_pV_p^{-1}K^p]$ $N_pV_p^{-1}$ (same as $N_p$) | |

At the end of the whole computation, all the matrices $N^p$ represent the homomorphisms $\partial_p$ relatively to bases $\Gamma^p$ such that $\Gamma^0 = V_0^{-1}K^0$, $\Gamma^1 = U_1V_1^{-1}K^1$, $\ldots$, $\Gamma^{n-1} = U_{n-1}V_{n-1}^{-1}K^{n-1}$, $\Gamma^n = U_nK^n$.

### 4.3    Optimizations for Effective Computation

Algorithms for computing the SNF or the presented modified version are well known (e.g. see [Ago76, Mun84]). But major difficulties arise when trying to program them effectively. These problems are mainly linked to the high computational cost of the algorithms and to the appearance of very big integers during the process. The algorithm is namely valid as long as integer computations have an arbitrary precision. With standard 32 or 64 bits integers, the algorithm is no more accurate. This problem arises even in small chain complexes. Hafner *et al.* have exhibited a $10 \times 10$ incidence matrix, with no value greater than 10, that induces huge *intermediate* integer numbers in SNF computation.

Deterministic and stochastic algorithms have been proposed to tackle these difficulties. The best known deterministic algorithm has been proposed by Storjohann [Sto96]. Stochastic algorithms have for example been proposed by Giesbrecht *et al.* [Gie95]. They are generally more efficient than deterministic ones on sparse matrices, but are quite equivalent on dense matrices. They are however restricted to the SNF computation and do not extract generators.

As far as we know, only Agoston [Ago76] proposed an algorithm to compute all homology information (including generators), but its implementation does not address the difficulties mentioned above. We propose here an adaptation of a Gaussian elimination algorithm developed by Dumas *et al.* [DSV01], which was originally only dedicated to the computation of the SNF of unrestricted simplicial complex. We combine this work to the work of Agoston to compute all homology information of semi-simplicial sets: Betti number and torsion coefficients of all homology groups, sets of "moduli generators". The main steps of the algorithm are described below. All operations made on the incidence matrix implies changes of bases that are stored in suitable matrices.

1. *(Prepare matrix for Dumas's algorithm.)* The rows of the incidence matrix are ordered by increasing pivot,
2. *(Same as Dumas.)* The matrix is put in echelon form with as many pivots at 1 as possible by
   - first pass: only elementary row operations are applied,
   - second pass: all rows are reduced according to their gcd.
   - the matrix is now in triangular form: deduce submatrix determinant (which is also the product of the invariant factors).
   - All further integer operations are made modulo twice this determinant. It has indeed been proved (for example by Storjohann) that such a computation using an appropriate modulo preserves the homology information.
3. *(Different from Dumas.)* Elementary rows and columns operations are performed to compute the modified SNF on the submatrix with non-zero rows. Changes of bases are traced. Agoston's algorithm is used to compute the generators, which are "moduli generators" in the sense they have been partly computed with a modulo.

# 5    Experimentations

We validate our approach on shapes classically encountered when testing topological invariants. For each shape, Betti numbers and torsion coefficients are
extracted from the modified SNF. The generators are read in the matrices $\Gamma^i$.
With this information, we are able to delineate each hole of the complex. It
should be noted that we only present the generators for surfaces because the
nature of 2-cycles on volumes is not well captured by 2D pictures.

Fig. 3 and Fig. 4 shows the shapes and the corresponding generators. Only the
generators of the homology group $H_1$ are displayed since the others are trivial.
For the torus, we have two cycles, one for each 1-dimensional hole ($H_1(K) \cong
\mathbb{Z} \oplus \mathbb{Z}$). According to the topological nature of the Moebius strip (homotopic
to a circle), we found only one cycle ($H_1(K) \cong \mathbb{Z}$). For the Klein bottle, two
cycles are found, one for the free part of the homology and one for the torsion
part ($H_1(K) \cong \mathbb{Z} \oplus \mathbb{Z}/2\mathbb{Z}$).

Torus                    Moebius band                    Klein bottle

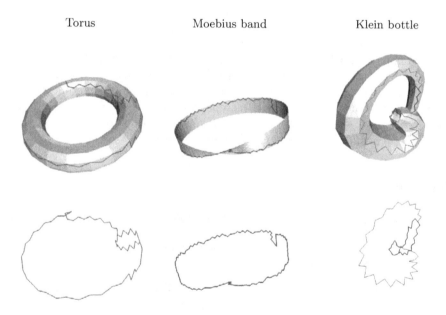

**Fig. 3.** Examples of homology generators on some classical surfaces

We have computed these generators using the previously described method
with moduli. We observe that the "moduli" generators are homologous to those
that would have been computed with arbitrary precision integer. On Fig. 3 each
objects has approximately 2000 triangles.

We guess that this property can be justified in a strict mathematical way
but as far we know there is no indication to invalid or to confirm this property.
Usual mathematical approaches are not really interested by the effective rep-

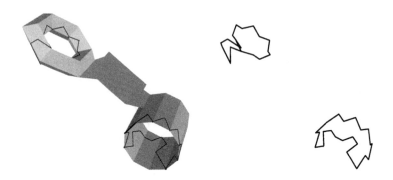

**Fig. 4.** Example of homology generators on a non-manifold complex

resentation of the generators, which explains the lack of theoretical results on "moduli" generators.

To conclude, we have presented and implemented a technique to compute the whole homology of arbitrary finite shapes. We have addressed the problem of extracting generators of the homology groups with a modulo. For future works, we would like to exhibit the theoretical link between generator modulo and $\mathbb{Z}$-generators. We want also to study the simploidal homology for discrete objects. Cubes are indeed simploids and simploidal representations of discrete objects are thus more compact than simplicial ones.

# References

[ACZ04]   M. Allili, D. Corriveau, and D. Ziou. Morse homology desriptor for shape characterization. In *Proc. ICPR 2004*, 2004.
[ADFQ03]  R. Ayala, E. Dominguez, A.R. Francès, and A. Quintero. Homotopy in digital spaces. *DAMATH: Discrete Appl. Math. and Combin. Oper. Research and Comput. Science*, 125, 2003.
[Ago76]   M. K. Agoston. *Algebraic Topology, a first course*. Marcel Dekker Ed., 1976.
[Box99]   L. Boxer. A classical construction for the digital fundamental group. *J. Math. Imaging Vision*, 10:51–62, 1999.
[Cai61]   S. S. Cairns. *Introductory Topology*. Ronald Press Company, 1961.
[Cur71]   E. Curtis. Simplicial homotopy theory. *Adv. Math.*, 6:107–209, 1971.
[DE95]    C. J. A. Delfinado and H. Edelsbrunner. An incremental algorithm for betti numbers of simplicial complexes on the 3-sphere. *Comput. Aided Geom. Design*, 12(7):771–784, 1995.
[DG03]    P. Desbarats and S. Gueorguieva. Topological mainframe for numerical representations of objects. In *Proc. Int. Conf. Comput. Science and its Appl.*, volume 2668 of *LNCS*, pages 498–507. Springer, 2003.

[DSV01]   J. G. Dumas, B.D. Saunders, and G. Villard. On efficient sparse integer matrix smith normal form computations. *Journal of Symbolic Computation*, 2001.

[GDR03]   R. González-Díaz and P. Real. Toward digital cohomology. In G. Sanniti di Baja, S. Svensson, and I. Nyström, editors, *Proc. DGCI'2003*, volume 2886 of *LNCS*, pages 92–101. Springer, 2003.

[Gie95]   M. Giesbrecht. Nearly optimal-algorithms for canonical matrix-forms. *SIAM J. COMPUT.*, 24(5):948–969, OCT 1995.

[Hat02]   A. Hatcher. *Algebraic Topology*. Cambridge University Press, 2002. disponible sur http://www.math.cornell.edu/~hatcher/AT/ATpage.html.

[KMS98]   T. Kaczynski, M. Mrozek, and M. Slusarek. Homology computation by reduction of chain complexes. *Computers & Math. Appl.*, 34(4):59–70, 1998.

[Kon89]   T. Y. Kong. A digital fundamental group. *Computers and Graphics*, 13(2):159–166, 1989.

[KR89]   T. Y. Kong and A. Rosenfeld. Digital Topology: Introduction and Survey. *Comput. Vision, Graphics, and Image Processing*, 48(3):357–393, 1989.

[LE93]   P. Lienhardt and H. Elter. Different combinatorial models based on the map concept for the representation of different types of cellular complexes. In *Proceedings of IFIP TC 5/WG II Work. Conf. on Geom. Modeling in Comp.Graphics*. Springer, 1993.

[LL95]   V. Lang and P. Lienhardt. Geometric modeling with simplicial sets. In *Proc. of Pacific Graphics'95*, Seoul, Korea, 1995.

[Mal01]   R. Malgouyres. Computing the fundamental group in digital spaces. *IJPRAI*, 15(7):1075–1088, 2001.

[May67]   J. P. May. *Simplicial Objects in Algebraic Topology*. Van Nostrand, 1967.

[Mun84]   J. R. Munkres. *Elements of algebraic topology*. Perseus Books, 1984.

[Sto96]   A. Storjohann. Near optimal algorithms for computing Smith normal forms of integer matrices. In Y. N. Lakshman, editor, *Proceedings of the 1996 International Symposium on Symbolic and Algebraic Computation*, pages 267–274. ACM, 1996.

# Inclusion Relationships and Homotopy Issues in Shape Interpolation for Binary Images

Javier Vidal[1,2], Jose Crespo[1], and Victor Maojo[1]

[1] Artificial Intelligence Laboratory, Facultad de Informática,
Universidad Politécnica de Madrid,
28660 Boadilla del Monte (Madrid), Spain
jvidal@infomed.dia.fi.upm.es
{jcrespo, vmaojo}@fi.upm.es
[2] Computer Science Department, Universidad de Concepción, Chile
jvidal@udec.cl
http://www.udec.cl/~jvidal

**Abstract.** Some image processing and analysis applications require performing image interpolation. This paper focuses on interpolation techniques that treat the shapes and the structures of binary images. A summary of some interpolation methods is presented, and their behavior concerning inclusion relationships and homotopy issues is studied. Then, this work discusses an inclusion relationship property that is used in a technique of ours based on median sets that has been recently proposed. The paper shows that such a property can improve shape interpolation results in a relatively easy manner. Several experimental results are provided.

**Keywords:** image processing, interpolation, shape interpolation, mathematical morphology, median set.

## 1   Introduction

Image processing applications normally deal with discrete data organized in slices. In some situations, it is useful, or even required, to perform some type of image interpolation [1]. The objective of image interpolation methods is generally to compute new interpolated slices between those originating from the original data.

Until recently, almost all image interpolation methods have used traditional numerical analysis techniques and linear signal processing approaches. In the last decade, several methods that deal directly with image shapes have been proposed [2, 3, 4, 5, 6, 7, 8, 9], in particular mathematical morphology [10, 11, 12, 13] related approaches.

This paper focuses on the behavior of image interpolation methods concerning inclusion relationships and homotopy issues in binary images. First, we study the behavior of some methods in several situations. We will be concerned, especially, in what happens when nested grains and pores are present in an image.

E. Andres et al. (Eds.): DGCI 2005, LNCS 3429, pp. 206–215, 2005.

Then, we discuss a simple inclusion property satisfied by a technique of ours that has been recently proposed [14]. We will see that this property can improve the experimental results obtained in situations where there exists an inclusion of structures or shapes. There are several aspects in this technique of relative complexity (such as, for example, how connected components are matched), but we will focus in how the application of the inclusion property can be beneficial in these situations.

This paper is organized as follows. Section 2 provides a summary of some interpolation methods. In Section 3, we study the behavior of the methods presented in Section 2 in some experimental situation; afterwards, we discuss the inclusion property of our technique and its application to several experimental cases. Conclusions are commented in Section 4.

## 2    A Summary of Some Shape Interpolation Methods

This section presents a summary of the interpolation methods focusing mainly on morphological ones, which adequately consider images shapes.

1. *The methods based on the Hausdorff distance.* The usage of Hausdorff distance for image interpolation was introduced in [9, 15]. A general definition of the Hausdorff distance between two binary sets $X$ and $Y$ is:

$$\rho(X, Y) = \inf\{\lambda : X \subseteq \delta_\lambda(Y); \ Y \subseteq \delta_\lambda(X)\} \tag{1}$$

where $\delta_\lambda$ denotes the dilation by a compact disk of radius $\lambda$ centered at the origin.

The most basic Hausdorff geodesic between $X$ and $Y$ is obtained by:

$$Z_\alpha(X, Y) = \delta_{\alpha\rho}(X) \cap \delta_{(1-\alpha)\rho}(Y), \alpha \in [0, 1] \tag{2}$$

where the coefficient $\alpha$ is incremented from 0 to 1 in order to generate the different sets $Z$.

A possible problem of (2) is that the interpolated set $Z_\alpha$ depends on the distance between the input sets and, in general, its size is bigger than $X$ and $Y$. Other related methods based on Hausdorff distance intend to correct this issue. For example, an additional Hausdorff geodesic corresponding to the cross-dilation $Z'_\alpha = \delta_{\alpha Y}((1-\alpha)X)$ can be used to intersect (2). Another variant, the reduced Hausdorff distance method, defines $X_a$ and $Y_b$ as the translated initial sets $X$ and $Y$ using vectors $a$ and $b$, respectively. This way, the initial sets are "aligned" before applying (2). Finally, the incorporation of a mask $G$ (to be intersected with (2)) was proposed in [6, 16]. The mask $G$ is obtained as $G = \varepsilon_\lambda(\mathrm{CH}(\delta_\lambda(X \cup Y)))$, where CH computes the convex hull of a set [12].

2. *Median set-based interpolation.* The median set [2, 4] is related to the morphological notion of skeleton by influence zone (SKIZ).

Let us consider two sets $X$ and $Y$, such that $X \cap Y \neq \phi$. The median set can be defined as:

$$M(X,Y) = \bigcup_{\lambda \geq 0} \{\delta_\lambda(X \cap Y) \cap \varepsilon_\lambda(X \cup Y)\} \tag{3}$$

where $\delta_\lambda$ and $\varepsilon_\lambda$ represents the dilation and erosion with a disk of radius $\lambda$. Initially this method was restricted to intersected sets. Subsequently, the notion of median set to non-intersected sets using affine transformations (translation, rotation and scaling) is described in [6, 17]. The interpolation sequence is obtained computing iteratively (3).

3. *Interpolation function.* The method above can also be implemented using two relative distances to generate the interpolated set [8]. This technique is also applied to intersected sets.

Let $X$ and $Y$ be the input sets, such that $X \cap Y \neq \phi$. The interpolation functions corresponds to the distance from $X \cap Y$ in $X$ ($\mathrm{int}^X_{X \cap Y}$) and to the distance from $X \cap Y$ in $Y$ ($\mathrm{int}^Y_{X \cap Y}$):

$$\mathrm{int}^X_{X \cap Y}(x) = \begin{cases} 0; & x \in X^C \\ d(x); & x \in \frac{X}{(X \cap Y)} \\ 1; & x \in (X \cap Y) \end{cases} \qquad \mathrm{int}^Y_{X \cap Y}(x) = \begin{cases} 0; & x \in Y^C \\ d(x); & x \in \frac{Y}{(X \cap Y)} \\ 1; & x \in (X \cap Y) \end{cases} \tag{4}$$

where $d$ is the relative distance from $x$ to $X \cap Y$ and from $x$ to $X^C$ (in the left case) or to $Y^C$ (in the right case). The value of $d$ is between 0 and 1. The interpolated set $Z$ at distance $\alpha$ from $X$ and $(1-\alpha)$ from $Y$ is the union of two sets:

$$Z_\alpha = \{x : \mathrm{int}^X_{X \cap Y}(x) \leq \alpha\} \cup \{y : \mathrm{int}^Y_{X \cap Y}(y) \leq (1-\alpha)\}; \quad \alpha \in [0,1] \tag{5}$$

The interpolation function can be applied to non-intersected sets using affine transformations [6].

This summary is not exhaustive, and other methods were also studied during the early stages of our work. For example we can mention the methods described in [5, 18, 19, 20, 7].

# 3   Inclusion Relationships and Homotopy Issues

We think that the consideration of inclusion relationships and homotopy aspects is important to adequately interpolate image shapes.

In this section, we will first study the behavior of the methods presented in Section 2 concerning homotopy issues with binary images. Then, we will see that many of these situations can be better treated with a technique of ours that implements a simple inclusion property.

## 3.1 Behavior of Interpolation Methods

In the following, we will discuss some results obtained with the interpolation methods described in Sec. 2.

Figure 1 displays and interpolation computed using the Hausdorff distance method with convex mask (computed as described in Sec.2). The two input slices, Fig. 1(a) and Fig. 1(k), have the same homotopy. The intermediate slices, with $\alpha$-values from 0.1 to 0.9, are the interpolated ones. We can clearly see that the homotopy changes in the sequence of images. The hole disappears in Figs. 1(c)-(i). We can also mention that the interpolated shapes do not sometimes correspond satisfactorily to the shapes observed in the input slices.

We use in this paper homotopy trees [12] to easily visualize the homotopy of an image. Two binary images are homotopic if their respective homotopy trees are the same[1].

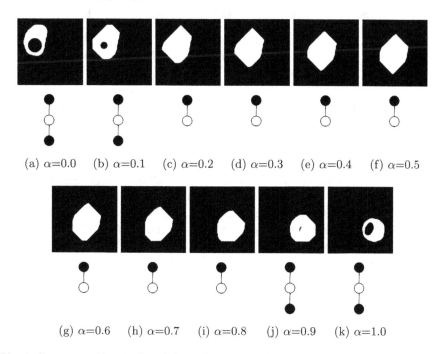

(a) $\alpha=0.0$  (b) $\alpha=0.1$  (c) $\alpha=0.2$  (d) $\alpha=0.3$  (e) $\alpha=0.4$  (f) $\alpha=0.5$

(g) $\alpha=0.6$  (h) $\alpha=0.7$  (i) $\alpha=0.8$  (j) $\alpha=0.9$  (k) $\alpha=1.0$

**Fig. 1.** Sequence of interpolated slices obtained using interpolation based on Hausdorff distance with convex hull and their homotopy trees

A result computed using the median set method (with affine transformation) is shown in Fig. 2. The input slices are displayed in Fig. 2(a) and in Fig. 2(j). In

---

[1] Nodes corresponding to the background and holes appear in black, and nodes corresponding to grains are displayed in white. The upper nodes of the trees correspond to the outer background.

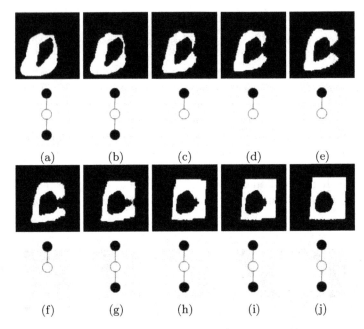

**Fig. 2.** Sequence of interpolated slices generated by median set-based interpolation method and their homotopy trees

this example, the homotopy of the input slices is also the same. However, we can see clearly that the homotopy of the interpolated slices of Figs. 2(c)-(f) differs from that of the input slices: the ring-shaped connected component "breaks", and it does not surround a hole inside.

Figure 3 shows a result computed using the interpolation function method. In this case the homotopy changes several times in the interpolated image sequence, as can be clearly observed in the associated homotopy trees.

## 3.2    The Inclusion Property and Experimental Results

We have proposed [14] an interpolation technique that satisfies the inclusion property that is stated next. It is not our intention to treat here the details of the algorithm, but to emphasize that this property can improve shape interpolation results in a relatively easy manner.

**Inclusion Property:** Suppose $A_1$ and $B_1$ are two sets of the input slice 1, such as $A_1 \subset B_1$, and suppose $C_2$ and $D_2$ are two sets of input slice 2, such as $C_2 \subset D_2$. If there is a correspondence between these two pairs (i.e., we want to interpolate $A_1$ with $C_2$, and $B_1$ with $D_2$), then the following condition should be satisfied:

$$\text{Interpolate}(A_1 \setminus B_1, C_2 \setminus D_2) = \text{Interpolate}(A_1, C_2) \setminus \text{Interpolate}(B_1, D_2) \quad (6)$$

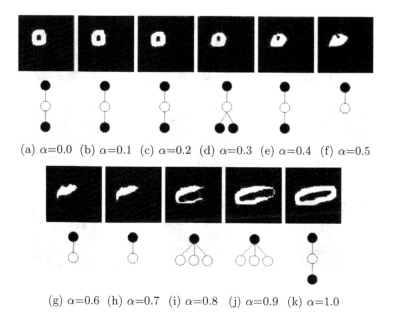

(a) $\alpha$=0.0  (b) $\alpha$=0.1  (c) $\alpha$=0.2  (d) $\alpha$=0.3  (e) $\alpha$=0.4  (f) $\alpha$=0.5

(g) $\alpha$=0.6  (h) $\alpha$=0.7  (i) $\alpha$=0.8  (j) $\alpha$=0.9  (k) $\alpha$=1.0

**Fig. 3.** Sequence of interpolated slices generated by the interpolation function method and their homotopy trees. (Note that there are two pores in image (d) and three grains in images (i) and (j).)

where "Interpolate" denotes our interpolation technique, and "\" symbolizes the set difference. The arguments of "Interpolate" are the two input images, and the result will be an interpolated slice.

The inclusion property is illustrated in Fig. 4. If (6) is satisfied, then image in Fig. 4(h) is equal to the image in Fig. 4(b) minus the image in Fig. 4(d).

Our interpolation technique is based on median sets [4]. Moreover, this technique processes recursively the connected components (CCs) of the slices applying (6) to the inner holes and inner grains of the images. The pseudo-code of the algorithm that implements this technique is shown in the Appendix, and a complete description can be found in [14]. The recursive application of (6) is performed by the last sentence of the while loop in the pseudo-code.

We will discuss next the performance of our technique over the images previously used in Sec. 3.1, and with a more complex case as well.

Figure 5 and Fig. 6 display the application of our technique to the cases visualized in Fig. 2 and Fig. 3, respectively. In both cases, the input slices contain a ring-like CC (a grain with an hole inside). Our technique first interpolates the filled grains. We suppose that the correspondence between them have been established using a matching procedure, and that the CCs are aligned (see [14] for details).

Then, in the next recursive iteration, pores are independently interpolated. The overall interpolation result is the result of interpolated grain filled minus the result of the interpolated pore. This is clearly what (6) establishes. We can

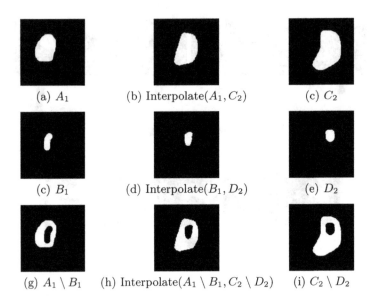

(a) $A_1$          (b) Interpolate$(A_1, C_2)$          (c) $C_2$

(c) $B_1$          (d) Interpolate$(B_1, D_2)$          (e) $D_2$

(g) $A_1 \setminus B_1$     (h) Interpolate$(A_1 \setminus B_1, C_2 \setminus D_2)$     (i) $C_2 \setminus D_2$

**Fig. 4.** An example of the inclusion property

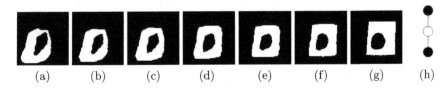

(a)     (b)     (c)     (d)     (e)     (f)     (g)     (h)

**Fig. 5.** Interpolated slices using our technique (compare with Fig. 2)

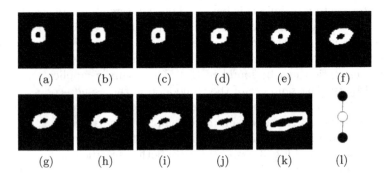

(a)     (b)     (c)     (d)     (e)     (f)

(g)     (h)     (i)     (j)     (k)     (l)

**Fig. 6.** Interpolated slices using our technique (compare with Fig. 3)

observe that all intermediate interpolated slices are a grain with an hole inside (the same homotopy as that of the input slices), and that these results compare favorably with those obtained in Fig. 2 and in Fig. 3.

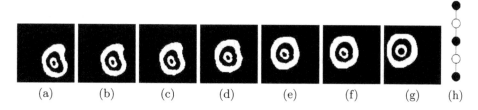

**Fig. 7.** Interpolation of nested CCs

Finally, we will present the application of expression (6) of the inclusion property to a more complex situation. In Fig. 7, we deal with nested CCs. In this case, our technique applies recursively (6) to the inner structures. Again, we can say our technique obtains satisfactory results.

## 4    Conclusions

In this paper we have studied homotopy related issues that arise in binary image interpolation. We have commented the behavior of some interpolation methods in some situations, and we have observed that problems can sometimes arise. We have discussed a technique based on median set that applies a relatively simple inclusion property, which can significantly improve the experimental results. Nevertheless, all homotopy issues are not completely solved. A complete treatment of them would require to consider how the connected components are matched and aligned, and the relative position of the nested shapes.

**Acknowledgements.** This work has been supported in part by the University of Concepción (Chile), by the MECESUP programme of Chilean Education Ministry, and by the Artificial Intelligence Laboratory, School of Computer Science at "Universidad Politécnica de Madrid".

## References

[1] Meijering, E.: A chronology of interpolation: From ancient astronomy to modern signal and image processing. Proceedings of the IEEE **90** (2002) 319–342
[2] Soille, P.: Spatial distributions from contour lines: an efficient methodology based on distance transformations. Journal of Visual Communication and Image Representation **2** (1991) 138–150
[3] Herman, G.T., Zheng, J., Bucholtz, C.A.: Shape-based interpolation. IEEE Computer Graphics and Application **12** (1992) 69–79
[4] Beucher, S.: Interpolations d'ensembles, de partitions et de fonctions. Technical Report N-18/94/MM, Centre de Morphologie Mathmatique (1994)
[5] Guo, J.F., Cai, Y.L., Wang, Y.P.: Morphology-based interpolation for 3D medical image reconstruction. Computarized Medical Imaging and Graphics **19** (1995) 267–279

[6] Iwanowski, M.: Application of Mathematical Morphology to Image Interpolation. PhD thesis, School of Mines of Paris - Warsaw University of Technology (2000)

[7] Lee, T.Y., Wang, W.H.: Morphology-based three-dimensional interpolation. IEEE Transactions on Medical Imaging **19** (2000) 711–721

[8] Meyer, F.: Interpolations. Technical Report N-16/94/MM, Centre de Morphologie Mathmatique (1994)

[9] Serra, J.: Interpolations et distances of Hausdorff. Technical Report N-15/94/MM, Centre de Morphologie Mathmatique (1994)

[10] Serra, J.: Mathematical Morphology. Volume I. London: Academic Press (1982)

[11] Serra, J., ed.: Mathematical Morphology. Volume II: Theoretical advances. London: Academic Press (1988)

[12] Soille, P.: Morphological Image Analysis: Principles and Applications. 2nd edn. Springer-Verlag (2003)

[13] Dougherty, E.R., Lotufo, R.A.: Hands-on Morphological Image Processing. SPIE Press, Bellingham, WA (2003)

[14] Vidal, J., Crespo, J., Maojo, V.: Recursive interpolation technique for binary images based on morphological median sets. In: *accepted in* International Symposium on Mathematical Morphology (ISMM'05), Paris, France. (2005)

[15] Serra, J.: Hausdorff distances and interpolations. In Heijmans, H.J., Roerdink, J.B., eds.: Mathematical Morphology and its Applications to Images and Signal Processing, Dordrecht, The Netherlands. Kluwer Academics Publishers (1998)

[16] Iwanowski, M.: Morphological binary interpolation with convex mask. In: Proceedings International Conference on Computer Vision and Graphics, Zakopane, Poland. (2002)

[17] Iwanowski, M., Serra, J.: The morphologycal-affine object deformation. In John Goutsias, L.V., Bloomberg, D.S., eds.: International Symposium on Mathematical Morphology, Palo Alto, CA. Kluwer Academics Publishers (2000) 445

[18] Migeon, B., Charreyron, R., Deforge, P., Marché, P.: Improvement of morphology-based interpolation. In: Proceedings of the 20th Annual International Conference of the IEEE Engineering in Medicine and Biology Society. (1998) 585–587

[19] Chatzis, V., Pitas, I.: Interpolation of 3d binary images based on morphological skeletonizations. In: Proceedings IEEE International Conference on Multimedia Computing Systems, Florence, Italy. Volume II. (1999) 939–943

[20] Chatzis, V., Pitas, I.: Interpolation of 3D binary images based on morphological skeletonizations. IEEE Transaction on Medical Imaging **19** (2000) 699–710

# Appendix: Interpolation Algorithm Pseudo-Code

Our interpolation technique [14] combines the use of median sets with the recursive application of (6), i.e., the inclusion property.

The algorithm consists of three main steps: (A) the separation of outer filled CCs within the input slices, (B) the matching between the fill CCs of input slices, and (C) the interpolation step that includes the recursive application of (6). The CCs matching step, step (B), determines the correspondence between the CCs (grains or pores) of the two input slices at each recursive iteration.

$INTERPOLATOR(Current\_S_1 : Slice, Current\_S_2 : Slice) : Slice$

$\equiv \lceil$

    // (A) Separation of outer CCs
    $OS_1 = Extract\_Outer\_CC(Current\_S_1);$
    $OS_1 = Extract\_Outer\_CC(Current\_S_2);$
    // (B) Matching
    $P = MATCHING(OS_1, OS_2);$
    // P is a vector of pairs of matched CCs and their MSPs
    $S = \emptyset;$   // init result binary image (set notation)
    // (C) Interpolation
    <u>while</u> $P \neq \emptyset$ <u>do</u>
        $(FS_1, MSP_1, FS_2, MSP_2) = Extract\_Pair\_CCs(P);$
        // $(FS_1, FS_2)$ is a pair of matched filled CCs
        // Computation of median set MS
        $MS = MSF(FS_1, MSP_1, FS_2, MSP_2);$
        // MSF : median set function using MSPs
        // Holes are stored as CCs
        $HS_1 = (FS_1 \cap Current\_S_1)^C \setminus [FS_1]^C;$
        $HS_2 = (FS_2 \cap Current\_S_2)^C \setminus [FS_2]^C;$
        // Result image is updated; inner structures
        // (holes and grains) are treated recursively
        <u>if</u> $((HS_1 = \emptyset) \wedge (HS_2 = \emptyset))$
          <u>then</u> $S = S \cup MS;$
          <u>else</u> $S = S \cup [MS \setminus INTERPOLATOR(HS_1, HS_2)];$ // Recursive call
        <u>fi</u>
    <u>od</u>
    $Return(S);$

$\rfloor$

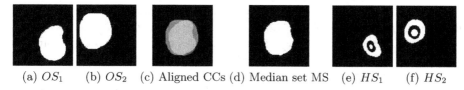

(a) $OS_1$     (b) $OS_2$    (c) Aligned CCs (d) Median set MS    (e) $HS_1$     (f) $HS_2$

**Fig. 8.** Algorithm execution example

In the following, we will briefly comment the execution of the algorithm applied to the input slices of Fig. 7. Input slices are those in Fig. 7(a) and Fig. 7(g). In the first iteration, step (A) extracts the outer filled grains (Fig. 8(a) and Fig. 8(b)) of the input slices. The matching step, step (B), matches these extracted CCs (Fig. 8(c) displays the CCs superimposed and aligned as part of the matching process; see details in [14]). Step (C), the main interpolation step, first computes the median set of the aligned filled CCs (Fig. 8(d)). Then, the inner grains and pores are extracted and complemented (Fig. 8(e) and Fig. 8(f)), and the recursive call is performed. Finally, the interpolated slice after all recursive iterations (in this case, 4 iterations) can be found in Fig. 7(d).

# Discrete Bisector Function and Euclidean Skeleton

Michel Couprie[1,2] and Rita Zrour[1,2]

[1] Laboratoire A2SI, Groupe ESIEE, BP99, 93162 Noisy-le-Grand Cedex France
[2] IGM, Unité Mixte de Recherche CNRS-UMLV-ESIEE UMR 8049
m.couprie@esiee.fr
zrour@llaic3.u-clermont1.fr

**Abstract.** We propose a new definition and an exact algorithm for the discrete bisector function, which is an important tool for analyzing and filtering Euclidean skeletons. We also introduce a new thinning method which produces homotopic discrete Euclidean skeletons. Unlike previouly proposed approaches, this method is still valid in 3D.

## 1 Introduction

The notion of skeleton plays a major role in shape analysis. It has been introduced by Blum [5] in 1961 and is the subject of an abundant literature, which deals with both metrical and topological aspects (see e.g. [1, 2, 6, 12, 13, 15, 16, 19, 21, 25, 26]). In this paper, we focus on skeletons in the discrete grid $\mathbb{Z}^2$ or $\mathbb{Z}^3$, which are centered in the shape with respect to the Euclidean distance, and which have the same topology as the original shape.

Introduced by Talbot and Vincent [25] and generalizing a notion proposed by Meyer [18], the bisector function can play an important role in analyzing and filtering skeletons [1, 2, 19]. Informally, the bisector function associates to each object point $x$ the maximal angle formed by $x$ (as the vertex) and the points of the background which are nearest to $x$. Until now, the algorithms proposed to compute the bisector function in $\mathbb{Z}^2$ were based on the use of vectors produced by distance transformation algorithms (e.g., [7]). To each object point, with such algorithms, only one vector indicates the location of a closest background point, and some other points at the same distance may be ignored. In fact, the aim of previous approaches was to compute an approximation of the bisector function as defined in a continuous framework.

This paper contains two original contributions. First, we propose a new definition and an exact algorithm to compute a discrete bisector function. This algorithm was inspired by the methods recently introduced [6, 21] to compute the exact Euclidean medial axis, and is also based on a pre-computed look-up table. Second, we introduce a new thinning method which produces homotopic discrete Euclidean skeletons. Unlike previously proposed approaches, this method is still valid in 3D.

E. Andres et al. (Eds.): DGCI 2005, LNCS 3429, pp. 216–227, 2005.

## 2    Basic Notions

In this section, we recall some basic metrical and topological notions for binary images [8, 15]. For simplicity, we limit this presentation to the 2D case.

We denote by $\mathbb{Z}$ the set of integers, by $\mathbb{N}$ the set of nonnegative integers, and by $\mathbb{N}^*$ the set of strictly positive integers. We denote by $E$ the discrete plane $\mathbb{Z}^2$. A point $x$ in $E$ is defined by $(x_1, x_2)$ with $x_i$ in $\mathbb{Z}$. Let $x, y \in E$, we denote by $d^2(x, y)$ the square of the Euclidean distance between $x$ and $y$, that is, $d^2(x, y) = (x_1 - y_1)^2 + (x_2 - y_2)^2$. Let $Y \subset E$, we denote by $d^2(x, Y)$ the square of the Euclidean distance between $x$ and the set $Y$, that is, $d^2(x, Y) = \min\{d^2(x, y); y \in Y\}$. Let $X \subset E$ (the "object"), we denote by $D_X^2$ the map from $E$ to $\mathbb{N}$ which associates, to each point $x$ of $E$, the value $D_X^2(x) = d^2(x, \overline{X})$, where $\overline{X}$ denotes the complementary of $X$ (the "background"). The map $D_X^2$ is called the *(squared Euclidean) distance map* of $X$. Let $x \in E, r \in \mathbb{N}^*$, we denote by $B_r(x)$ the *ball of (squared) radius $r$ centered on $x$*, defined by $B_r(x) = \{y \in E, d^2(x, y) < r\}$. Notice that, for any point $x$ in $X$, the value $D_X^2(x)$ is precisely the radius of a ball centered on $x$ and included in $X$, which is not included in any other ball centered on $x$ and included in $X$.

Let us recall the notion of medial axis (see also [20, 25]). Let $X \subseteq E, x \in X$, $r \in \mathbb{N}^*$. A ball $B_r(x) \subseteq X$ is *maximal for $X$* if it is not strictly included in any other ball included in $X$. The *medial axis of $X$*, denoted by $\mathrm{MA}(X)$, is the set of the centers of all the maximal balls for $X$ (see Fig. 1d, see also Fig. 2).

Efficient algorithms have been proposed to compute exact squared Euclidean distance maps [22, 23], and also to extract the exact Euclidean medial axis of a shape, from an exact squared Euclidean distance map and using pre-computed look-up tables [6, 21].

In discrete spaces, it is well known that the topology of the medial axis is generally not the same as the topology of the original object. In particular, if $X$ is connected, $\mathrm{MA}(X)$ is generally not connected (see Fig. 1d). Let us now introduce some topological notions in $\mathbb{Z}^2$ that we use in the sequel.

We consider the two neighborhood relations $\Gamma_4$ and $\Gamma_8$ defined by, for each point $x \in E$: $\Gamma_4(x) = \{y \in E; |y_1 - x_1| + |y_2 - x_2| \leq 1\}$, $\Gamma_8(x) = \{y \in E; \max(|y_1 - x_1|, |y_2 - x_2|) \leq 1\}$. In the following, we will denote by $n$ a number such that $n = 4$ or $n = 8$. We define $\Gamma_n^*(x) = \Gamma_n(x) \setminus \{x\}$. The point $y \in E$ is *n-adjacent* to $x \in E$ if $y \in \Gamma_n^*(x)$. An *n-path* is a sequence of points $x_0 \ldots x_k$ with $x_i$ $n$-adjacent to $x_{i-1}$ for $i = 1 \ldots k$.

Let $X$ be a non-empty subset of $E$. We say that two points $x, y$ of $X$ are *n-connected in $X$* if there is an $n$-path in $X$ between these two points. This defines an equivalence relation. The equivalence classes for this relation are the *n-connected components* of $X$, or *n-components* in short. The set $X$ is said to be *n-connected* if it consists of exactly one $n$-component. The set composed of all $n$-components of $X$ which are $n$-adjacent to a point $x$ is denoted by $C_n[x, X]$.

In order to have a correspondence between the topology of $X$ and the topology of $\overline{X}$, we have to consider two different kinds of adjacency for $X$ and $\overline{X}$ [15]: if we use the $n$-adjacency for $X$, we must use the $\overline{n}$-adjacency for $\overline{X}$, with $(n, \overline{n}) = (8, 4)$ or $(4, 8)$. In the sequel, we assume that the adjacency pair $(n, \overline{n}) = (8, 4)$

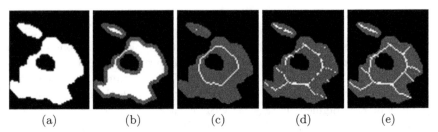

(a)          (b)          (c)          (d)          (e)

**Fig. 1.**  (a): a set $X$ (in white); (b): an homotopic thinning of $X$; (c): ultimate homotopic skeleton of $X$; (d): medial axis of $X$; (e): ultimate homotopic skeleton of $X$ constrained by the medial axis of $X$. In (b,c,d,e) the original set $X$ appears in dark gray for comparison

has been chosen and we do not write the subscripts $n, \overline{n}$ unless necessary; but the results also hold for $(n, \overline{n}) = (4, 8)$.

Informally, a *simple point* $p$ of a discrete object $X$ is a point which is "inessential" to the topology of $X$. In other words, we can remove the point $p$ from $X$ without "changing the topology of $X$". The notion of simple point is fundamental to the definition of topology-preserving transformations in discrete spaces. We now give a definition and a local characterization of simple points in $E = \mathbb{Z}^2$. For the 3D case, see [3].

The point $x \in X$ is *simple (for $X$)* if each $n$-component of $X$ contains exactly one $n$-component of $X \setminus \{x\}$ and if each $\overline{n}$-component of $\overline{X} \cup \{x\}$ contains exactly one $\overline{n}$-component of $\overline{X}$. Let $X \subseteq E$ and $x \in E$, the two *connectivity numbers* are defined as follows ($\#X$ stands for the cardinality of $X$):
$$T(x, X) = \#C_n[x, \Gamma_8^*(x) \cap X]; \quad \overline{T}(x, X) = \#C_{\overline{n}}[x, \Gamma_8^*(x) \cap \overline{X}].$$
The following property allows us to locally characterize simple points [15, 3], hence to implement efficiently topology preserving operators:
$$x \in E \text{ is simple for } X \subseteq E \Leftrightarrow T(x, X) = 1 \text{ and } \overline{T}(x, X) = 1.$$

Let $X$ be any finite subset of $E$. The subset $Y$ of $E$ is an *homotopic thinning of $X$* if $Y = X$ or if $Y$ may be obtained from $X$ by iterative deletion of simple points. We say that $Y$ is an *ultimate homotopic skeleton of $X$* if $Y$ is an homotopic thinning of $X$ and if there is no simple point for $Y$.

Let $C$ be a subset of $X$. We say that $Y$ is an *ultimate homotopic skeleton of $X$ constrained by $C$* if $C \subseteq Y$, if $Y$ is an homotopic thinning of $X$ and if there is no simple point for $Y$ in $Y \setminus C$ (see e.g. [12, 26]). The set $C$ is called the *constraint set* relative to this skeleton.

## 3   The Bisector Function: New Definition and Exact Algorithm

Let $X$ be a non-empty subset of $E$, and let $x \in X$. The *downstream of $x$ in $X$*, denoted by $\mathrm{Ds}(x, X)$ or by $\mathrm{Ds}(x)$ when no confusion may occur, is the set of points $y$ of $\overline{X}$ which are at minimal distance from $x$; more precisely, $\mathrm{Ds}(x, X) = \{y \in \overline{X}, \forall z \in \overline{X}, d^2(y, x) \leq d^2(z, x)\}$. For example in Fig. 2, we have $\mathrm{Ds}(x) =$

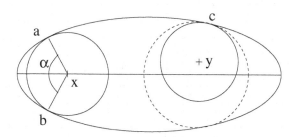

**Fig. 2.** A set $X$ (full ellipsis, represented by its border) and its medial axis (horizontal line), a point $x$ and its downstream $\{a, b\}$, a point $y$ and its downstream $\{c\}$. Notice that $y$ does not belong to the medial axis, since no ball centered on $y$ and included in $X$ is maximal for $X$

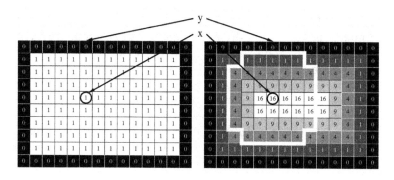

**Fig. 3.** A set $X$ and its distance map $D_X^2$. The point $x$ belongs to the medial axis of $X$ (it is the center of a maximal ball for $X$, delineated in white). The downstream of $x$ is the singleton $\{y\}$

$\{a, b\}$ and $\mathrm{Ds}(y) = \{c\}$. The bisector angle of a point $x$ in $X$ can be defined informally, in the continuous framework, as the maximal unsigned angle formed by $x$ (as the vertex) and any two points in $\mathrm{Ds}(x)$. In Fig. 2, the bisector angle of $x$ is $\alpha$, and the bisector angle of $y$ is 0.

In the continuous framework, a point $x$ which belongs to the medial axis is characterized by the fact that $\mathrm{Ds}(x)$ contains strictly more than one point, in other words, its bisector angle is non-zero. In the discrete case such a characterization is not possible, due to configurations like Fig. 3, which are very common. At least, we want such medial axis points to have a non-zero bisector angle; this motivates the introduction of the following notion.

Let $X \subset E$, and let $x \in X$. The *extended downstream of $x$ in $X$*, denoted by $\mathrm{EDs}(x, X)$ or by $\mathrm{EDs}(x)$ when no confusion may occur, is the union of the sets $\mathrm{Ds}(y, X)$, for all $y$ in $\Gamma_4(x)$.

Now we can propose our definition of the discrete bisector function.

Let $X \subset E$, and let $x \in X$. The *bisector angle of $x$ in $X$*, denoted by $\theta_X(x)$, is the maximal unsigned angle between the vectors $\overrightarrow{xy}, \overrightarrow{xz}$, for all $y, z$ in $EDs(x)$. The *bisector function of $X$*, denoted by $\theta_X$, is the function which associates to each point $x$ of $X$, its bisector angle in $X$.

We now propose a method to exactly and efficiently compute the bisector function, with the help of a squared Euclidean distance map. The following property, which may be easily established, is the key of the method.

**Property 1.** *Let $X \subset E$, let $x \in X$, and let $y \in E$. The point $y$ belongs to the downstream of $x$ if and only if $d^2(x, \overline{X}) = d^2(x, y)$ and $d^2(y, \overline{X}) = 0$.*

Now, observe that for any given point $x \in X$, the value $R = d^2(x, \overline{X}) = D_X^2(x)$ can be read in a pre-computed squared Euclidean distance map. The positions of the points $y$ such that $R = d^2(x, y)$ can be found by solving the diophantine equation $(x_1 - y_1)^2 + (x_2 - y_2)^2 = R$, or more simply $\alpha^2 + \beta^2 = R$. This amounts to compute the different decompositions of a given integer into a sum of two squares (or three squares in 3D), a problem which has been studied extensively (see [14]).

Furthermore, these decompositions can also be pre-computed and stored in a look-up table. The program which computes this look-up table is very simple (see Annex). Notice that the same look-up table was used by Fontoura-Costa et al. to compute an exact Euclidean dilation[17]. Once all decompositions are known, one must take into account the different symmetries of the space (their number is 8 for $\mathbb{Z}^2$, 48 for $\mathbb{Z}^3$) and check those positions $y$ which satisfy the second condition of prop. 1: $d^2(y, \overline{X}) = 0$, again using the distance map.

The following algorithm summarizes the computation of the extended downstream of a given point $x \in X$.

**Procedure ExtendedDownstream (Input $D_X^2$, $x$, Output $EDs$)**
01. $EDs \leftarrow \emptyset$
02. **ForEach** $v \in \Gamma_4(x)$
03.    $T \leftarrow \{(z_1, z_2) \in \mathbb{Z}^2; z_1 \geq z_2 \geq 0; z_1^2 + z_2^2 = D_X^2(v)\}$ (from LUT)
04.    $T' \leftarrow \{(z_1, z_2) \in \mathbb{Z}^2; (|z_1|, |z_2|)$ is a permutation of an element of $T\}$
05.    **ForEach** $z \in T'$ **Do**
06.        **If** $D_X^2(v + z) = 0$ **Then** $EDs \leftarrow EDs \cup \{v + z\}$

The last step to obtain the bisector angle consists in the computation of the maximum unsigned angle between all the pairs of vectors $\{\overrightarrow{xy}, \overrightarrow{xz}\}$ for all $y, z$ in $EDs(x)$. If we denote by $k$ the number of points in $EDs(x)$, the number of such pairs is quadratic with respect to $k$, more precisely, it is equal to $k(k-1)/2$. By normalizing all these vectors, we can easily see that the problem of finding a maximum angle reduces to the problem of finding a maximum diameter of a convex polygon in 2D. This last problem has been solved in 1978 by Shamos [24], who provided a simple linear-time algorithm (that is, in $O(k)$). In 3D, the problem is more complicated but some efficient algorithms (in $O(n \log n)$ or less) have been proposed for the maximal diameter of a set of points, see e.g. [4]. However, in practice, the mean cardinal of the extended downstream for a given shape is

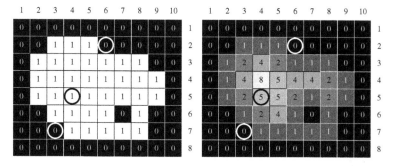

**Fig. 4.** Left: a set $X$ (in white); Right: the squared Euclidean map of $X$. The point $x = (4,5)$ is circled in black, the points $y_1 = (6,2)$ and $y_2 = (3,7)$ are circled in white

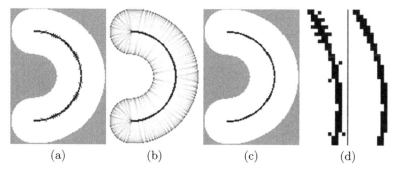

(a)                (b)                (c)                (d)

**Fig. 5.** (a): a set $X$ and its medial axis (in black); (b): the bisector function $\theta_X$ (dark colors correspond to wide angles); (c): filtered medial axis, based on the values of $\theta_X$; (d): detail of the non-filtered and filtered medial axis

usually quite small. Thus, the straightforward algorithm which considers all the pairs of points in the extended downstream is the best choice in most cases.

We illustrate this procedure using the example of Fig. 4. We consider the point $x = (4,5)$, which is circled in black in the figure. Thus, $\Gamma_4(x) = \{x, x_1, x_2, x_3, x_4\}$ with $x_1 = (5,5)$, $x_2 = (4,4)$ $x_3 = (3,5)$, $x_4 = (4,6)$. Let us begin with $x = (4,5)$, and $D_X^2(x) = 5$. From the Look-up table (see Fig. 9), we see that the only decomposition of 5 into a sum of two squares is $2^2 + 1^2$. Applying the 8 symmetries of $\mathbb{Z}^2$, we find the offsets $\{(2,1), (-2,1), (-2,-1), (2,-1), (1,2), (-1,2), (-1,-2), (1,-2)\}$, hence the points $\{(6,6), (2,6), (2,4), (6,4), (5,7), (3,7), (3,3), (5,3)\}$. Among these points, only $(2,6)$ and $(3,7)$ are in the downstream of $x$ since $D_X^2[(2,6)] = D_X^2[(3,7)] = 0$. Carrying on with the points $x_1, x_2, x_3, x_4$, we find that $\mathrm{EDs}(x) = \{(2,6), (3,7), (7,6), (2,2), (6,2)\}$. Let $y_1 = (6,2)$ and $y_2 = (3,7)$, it may be easily checked that the maximum unsigned angle corresponds to the couple of vectors $\{\vec{xy_1}, \vec{xy_2}\}$ and is close to 3.02 rad.

In Fig. 5, we show a set $X$ together with its medial axis (a) and the bisector function $\theta_X$ (b). We illustrate the use of this function to eliminate spurious points of the medial axis: in (c), we show the points of the medial axis (in black)

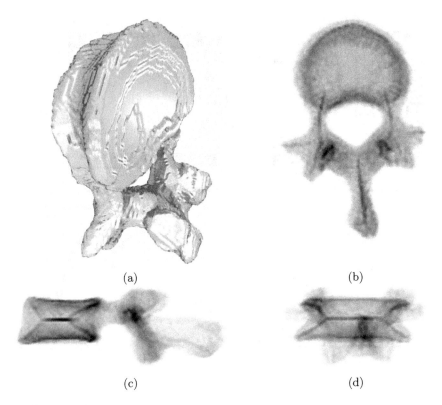

(a)                                    (b)

(c)                                    (d)

**Fig. 6.** (a): a view of a subset $X$ of $\mathbb{Z}^3$ (vertebra), generated thanks to a topologically sound "Marching Cubes-like" algorithm [11]; (b,c,d): the bisector function, illustrated in an "X-ray" manner: the gray level of a point corresponds to the average of the bisector angles on a straight line parallel to one of the three axes

which have a bisector angle greater than 0.7 rad. A zoomed detail of both axes is shown in (d). Notice that only the bisector angles of the medial axis points need to be computed for this application.

The definition and exact computation of this discrete bisector function can be straightforwardly extended to $\mathbb{Z}^3$. To conclude this section, we present in Fig. 6 an illustration of our bisector function of a three-dimensional object (a vertebra). As a matter of fact, the number of decompositions of an integer $k$ grows is a "reasonable way" when $k$ increases, see [10] for more details. The size of the original image of Fig. 6 is $122 \times 144 \times 53 = 931104$ points, among which 103302 are object points. On a standard PC, computing the bisector function on all object points for this image takes 2.9 s, and 0.5 s for only medial axis points. The look-up table used for processing this image occupies 82750 bytes of memory.

# 4   New Distance-Guided Homotopic Thinning Algorithm

The skeletonization methods which are based on homotopic thinnings, in the sense of section 2, provide a formal guarantee that the skeleton and the original object have the same topology. The simplest such method consists in computing an ultimate homotopic skeleton of the object $X$ constrained by the medial axis of $X$, that is, removing iteratively simple points from $X$ which do not belong to $\mathrm{MA}(X)$, taking the distance map as a priority function in order to select first the points which are closest to the background. This can be done using the following procedure, with $P = D_X$ and $Y = \mathrm{MA}(X)$.

**Procedure UltimateSkeleton (Input** $X, P, Y$, **Output** $Z$)
01. $Z \leftarrow X$
02. $Q \leftarrow \{(P(x), x);$ where $x$ is any point of $X \setminus Y\}$
03. **While** $Q \neq \emptyset$ **Do**
04.       choose $(p, x)$ in $Q$ such that $p$ is minimal
05.       **If** $x$ is simple for $Z$ **Then**
06.          $Z \leftarrow Z \setminus \{x\}$
07.          $Q \leftarrow Q \cup \{(P(y), y);$ where $y \in \Gamma(x) \cap (Z \setminus Y)\}$

The drawback of this method has been well analyzed in [25]. Roughly speaking the method does not guarantee that points of the skeleton outside the medial axis are "well centered" in the object; more precisely, such a point may have a null or quasi-null bisector angle.

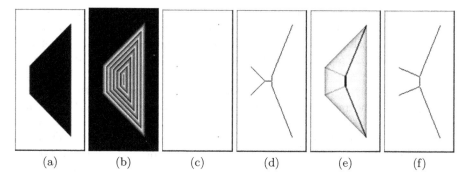

|     |     |     |     |     |     |
| --- | --- | --- | --- | --- | --- |
| (a) | (b) | (c) | (d) | (e) | (f) |

**Fig. 7.** (a): a set $X$ (in black); (b): isolines of the Euclidean map $D_X$; (c): a constraint set $Y$ (four points located at the "corners" of the shape); (d): homotopic skeleton of $X$ guided by $D_X^2$ and constrained by $Y$; (e): the bisector function of $X$; (f): the result of our algorithm EuclideanSkeleton

Let us consider the object $X$ in Fig. 7a, and the constraint set $Y$ in Fig. 7c, which idealizes a filtered medial axis of $X$. The result of **UltimateSkeleton**($X$, $D_X, Y$) is depicted in Fig. 7d. The isolines of the distance map $D_X$ and the bisector function of $X$ are depicted in Fig. 7b,e respectively, for comparison.

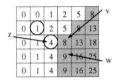

**Fig. 8.** A detail of the preceding example

To understand what happens, let us concentrate on a detail of the above example, depicted in Fig. 8. The numbers correspond to the distance map values. The circled point with value 1 is one of the points belonging to the constraint set $Y$. Suppose that all the points with a value below 8 have been processed by the homotopic thinning algorithm. At this step, the points in gray are still in $X$, as well as the two circled points (the point at 1 because it belongs to $Y$, and the one at 4 because it is not a simple point). All other points are not in $X$. Obviously, the point $v$ at 8 adjacent to $z$ at 4 will be selected before its neighbor $w$ at 9, and since it will be a simple point at this stage, it will be deleted. Such a behaviour will be reproduced at later stages, generating a diagonal branch of the skeleton, and is in contradiction with a property of the skeleton in the continuous framework: informally, skeleton branches follow lines of steepest slope of the Euclidean distance map. Let us compute the slopes of the segments $zv$ and $zw$ in our example: $(\sqrt{8} - \sqrt{4})/1 \approx 0.83$, and $(\sqrt{9} - \sqrt{4})/\sqrt{2} \approx 0.71$. Thus, the point $v$ should be kept in the skeleton and the point at $w$ deleted according to this criterion.

In [25], Talbot and Vincent propose the following strategy to cope with this problem. During the thinning process guided by the distance map, having detected a point $x$ which belongs to the skeleton (either a point of the constraint set or a non-simple point), the neighbor of $x$ which corresponds to the steepest ascending slope is dynamically added to the constraint set. Although this method gives satisfactory results in 2D, it cannot be extended to the 3D case.

We propose another strategy which gives equivalent results in 2D and which also applies to the 3D case. The idea is to define a priority function which takes into account both the distance map and an auxiliary function defined in the neighborhood of each dynamically detected skeleton point. Let $x$ be such a point, then to any neighbor $y$ of $x$ which is still in $X$ and not in the constraint set $Y$, we associate the value $p_y = D_X(x) + (D_X(y) - D_X(x))/d(x,y)$, with $D_X(x) = \sqrt{D_X^2(x)}$ and $d(x,y) = \sqrt{d^2(x,y)}$. The new priority function, for the point $y$, is defined as $\min(p_y, D_X(y))$. We see that $(D_X(y) - D_X(x))/d(x,y)$ is the slope of $xy$, thus the neighbors of $x$ will be examined in increasing order of slope, since the value $p_y$ is always less or equal to the corresponding distance value $D_X(y)$ (for all $x, y$ in $\mathbb{Z}^2$ or $\mathbb{Z}^3$ with $x \neq y$, we have $d(x,y) \geq 1$).

For example, in the above situation, we have $D_X(v) = \sqrt{8} \approx 2.83$, $D_X(w) = 3$, $p_v = \sqrt{4} + (\sqrt{8} - \sqrt{4})/1 = \sqrt{8}$ and $p_w = \sqrt{4} + (\sqrt{9} - \sqrt{4})/\sqrt{2} \approx 2.71$; thus the point $w$ will be selected before $v$ with this strategy. Our algorithm is given below.

**Procedure EuclideanSkeleton (Input** $X, D_X, Y,$ **Output** $Z$)
01. $Z \leftarrow X$
02. $Q \leftarrow \{(D_X(x), x)$ where $x$ is any point of $Z \setminus Y\}$
03. $R \leftarrow \{(p_x, x)$ where $x$ is any point of $Z \setminus Y$ adjacent to $Y$,
04.         and where $p_x = \min\{D_X(z) + (D_X(x) - D_X(z))/d(x,z), z \in Y\}\ \}$
05. **While** $Q \neq \emptyset$ **Or** $R \neq \emptyset$ **Do**
06.         choose $(p, x)$ in $Q \cup R$ such that $p$ is minimal
07.         **If** $x \in Z \setminus Y$ **Then**
08.             **If** $x$ is simple for $Z$ **Then**
09.                 $Z \leftarrow Z \setminus \{x\}$
10.             **Else**
11.                 $Y \leftarrow Y \cup \{x\}$
12.                 $R \leftarrow R \cup \{(p_y, y),$ where $y \in \Gamma(x) \cap (Z \setminus Y)$
13.                     and where $p_y = D_X(x) + (D_X(y) - D_X(x))/d(x,y)\}$

In Fig. 7f, we see the result of this algorithm applied to the preceding example. Compare the shape of this skeleton with the distance map and with the bisector function of $X$ depicted in 7b,e respectively. The complexity of this algorithm depends on the data structure used to represent the sets $Q$ and $R$. To be more precise, this data structure must allow to perform the choice in line 6 efficiently, and also the insertions in lines 11 and 12. Using for example balanced binary trees [9], the overall complexity of the algorithm is in $O(n \log n)$, where $n$ is the number of image points.

## 5   Conclusion

We introduced a new definition and an exact algorithm for the discrete bisector function, and proposed a new thinning algorithm which produces homotopic discrete Euclidean skeletons. Both constitute significant improvements with respect to previous approaches, and apply to the 2D and 3D cases.

In an extended version of this paper [10], we will analyse more deeply the difference between the proposed approach and the previous ones, we will present more results and applications (especially with "real world" data and in 3D), and we will analyze the performance of these algorithms both in terms of time and memory space.

## References

1. D. Attali, J.O. Lachaud: "Delaunay Conforming Iso-surface, Skeleton Extraction and Noise Removal", *Computational Geometry: Theory and Applications*, Vol. 19, pp. 175-189, 2001.
2. D. Attali, A. Montanvert: "Modelling noise for a better simplification of skeletons", *Procs. International Conference on Image Processing*, Vol. 3, pp. 13-16, 1996.
3. G. Bertrand: "Simple points, topological numbers and geodesic neighborhoods in cubic grids", *Pattern Recognition Letters*, Vol. 15, pp. 1003-1011, 1994.

4. S.N. Bespamyatnikh, "An efficient algorithm for the three-dimensional diameter problem", *Procs. ACM-SIAM symp. on discrete algorithms*, pp. 137-146, 1998.

5. H. Blum, "An associative machine for dealing with the visual field and some of its biological implications", *Biological prototypes and synthetic systems*, Vol. 1, pp. 244-260, 1961.

6. G. Borgefors, I. Ragnemalm, G. Sanniti di Baja, "The Euclidean distance transform: finding the local maxima and reconstructing the shape", *Procs. of the 7th Scand. Conf. on image analysis*, Vol. 2, pp. 974-981, 1991.

7. P.E. Danielsson: "Euclidean distance mapping", *Computer Graphics and Image Processing*, 14, pp. 227-248, 1980.

8. J.M. Chassery, A. Montanvert: *Géométrie discrète*, Hermès, 1991.

9. T.H. Cormen, C.E. Leiserson, R.L. Rivest: *Introduction to algorithms*, MIT Press, 1990.

10. M. Couprie, R. Zrour: "Discrete bisector function and Euclidean skeleton in 2D and 3D", report IGM2004-12 of the Institut Gaspard Monge (University of Marne-la-Vallée),   http://www-igm.univ-mlv.fr/LabInfo/rapportsInternes/2004/12.pdf, 2004.

11. X. Daragon, M. Couprie, G. Bertrand: "Discrete frontiers", Springer (Ed.), *Discrete Geometry for Computer Imagery*, Springer LNCS, Vol. 2886, pp. 236-245, 2003.

12. E.R. Davies, A.P.N. Plummer: "Thinning algorithms: a critique and a new methodology", *Pattern Recognition*, Vol. 14, pp. 53-63, 1981.

13. Y. Ge, J.M. Fitzpatrick: "On the generation of skeletons from discrete Euclidean distance maps", *IEEE Trans. on Pattern Analysis and Machine Intelligence*, Vol. 18, No. 11, pp. 1055-1066, 1996.

14. G.H. Hardy, E.M. Wright: *An introduction to the theory of numbers*, Oxford University Press, 1938.

15. T. Yung Kong, A. Rosenfeld: "Digital topology: introduction and survey", *Computer Vision, Graphics and Image Processing*, Vol. 48, pp. 357-393, 1989.

16. L. Lam, S-W. Lee, C.Y. Suen: "Thinning methodologies - a comprehensive survey", *IEEE PAMI*, Vol. 14, No. 9, pp. 869-885, 1992.

17. M. Luppe, L. da Fontoura Costa, V. Obac Roda: "Parallel implementation of exact dilations and multi-scale skeletonization", *Real-Time Imaging*, Vol. 9, pp. 163-169, 2003.

18. F. Meyer, *Cytologie quantitative et morphologie mathématique*, PhD thesis, École des mines de Paris, 1979.

19. G. Malandain, S. Fernández-Vidal, "Euclidean Skeletons", *Image and vision computing*, Vol. 16, pp. 317-327, 1998.

20. A. Rosenfeld, A.C. Kak: *Digital Image processing*, Academic Press, 1982.

21. E. Rémy, E. Thiel: "Exact Medial Axis with Euclidean Distance", to appear in *Image and Vision Computing*, 2004.

22. T. Saito, J.I. Toriwaki: "New algorithms for Euclidean distance transformation of an $n$-dimensional digitized picture with applications", *Pattern Recognition*, Vol. 27, pp. 1551-1565, 1994.

23. T. Hirata: "A unified linear-time algorithm for computing distance maps", *Information Processing Letters*, Vol. 58(3), pp. 129-133, 1996.

24. M.I. Shamos: *Computational geometry*, PhD thesis, Yale University, 1978.

25. H. Talbot, L. Vincent: "Euclidean skeletons and conditional bisectors", *Proceedings of VCIP'92*, SPIE, Vol. 1818, pp. 862-876, 1992.

26. L. Vincent: "Efficient Computation of Various Types of Skeletons", *Proceedings of Medical Imaging V, SPIE*, Vol. 1445, pp. 297-311, 1991.

# Annex: Building the Look-Up Table

**Procedure Build2dLUT (Input** $N$, **Output** LUT**)**
01. $n \leftarrow \lceil \sqrt{N} \rceil$; **For** $i$ **From** $0$ **To** $N$ **Do** LUT[i]$\leftarrow \emptyset$
02. **For** $x$ **From** $0$ **To** $n$ **Do**
02.       **For** $y$ **From** $0$ **To** $x$ **Do**
03.             $i \leftarrow x^2 + y^2$
04.             **If** $i \leq N$ **Then**
05.                   LUT[i] $\leftarrow$ LUT[i] $\cup\{(x,y)\}$

| | | | | |
|---|---|---|---|---|
| 0: $(0,0)$ | 8: $(2,2)$ | 17: $(4,1)$ | 29: $(5,2)$ | 40: $(6,2)$ |
| 1: $(1,0)$ | 9: $(3,0)$ | 18: $(3,3)$ | 32: $(4,4)$ | 41: $(5,4)$ |
| 2: $(1,1)$ | 10: $(3,1)$ | 20: $(4,2)$ | 34: $(5,3)$ | 45: $(6,3)$ |
| 4: $(2,0)$ | 13: $(3,2)$ | 25: $(4,3),(5,0)$ | 36: $(6,0)$ | 49: $(7,0)$ |
| 5: $(2,1)$ | 16: $(4,0)$ | 26: $(5,1)$ | 37: $(6,1)$ | 50: $(5,5),(7,1)$ |

**Fig. 9.** Look-up table (2D case): numbers 0 to 50

**Procedure Build3dLUT (Input** $N$, **Output** LUT**)**
01. $n \leftarrow \lceil \sqrt{N} \rceil$; **For** $i$ **From** $0$ **To** $N$ **Do** LUT[i]$\leftarrow \emptyset$
02. **For** $x$ **From** $0$ **To** $n$ **Do**
02.       **For** $y$ **From** $0$ **To** $x$ **Do**
03.             **For** $z$ **From** $0$ **To** $y$ **Do**
04.                   $i \leftarrow x^2 + y^2 + z^2$
05.                   **If** $i \leq N$ **Then**
06.                         LUT[i] $\leftarrow$ LUT[i] $\cup\{(x,y,z)\}$

| | | | |
|---|---|---|---|
| 0: $(0,0,0)$ | 9: $(2,2,1),(3,0,0)$ | 18: $(3,3,0),(4,1,1)$ | 27: $(3,3,3),(5,1,1)$ |
| 1: $(1,0,0)$ | 10: $(3,1,0)$ | 19: $(3,3,1)$ | 29: $(4,3,2),(5,2,0)$ |
| 2: $(1,1,0)$ | 11: $(3,1,1)$ | 20: $(4,2,0)$ | 30: $(5,2,1)$ |
| 3: $(1,1,1)$ | 12: $(2,2,2)$ | 21: $(4,2,1)$ | 32: $(4,4,0)$ |
| 4: $(2,0,0)$ | 13: $(3,2,0)$ | 22: $(3,3,2)$ | 33: $(4,4,1),(5,2,2)$ |
| 5: $(2,1,0)$ | 14: $(3,2,1)$ | 24: $(4,2,2)$ | 34: $(4,3,3),(5,3,0)$ |
| 6: $(2,1,1)$ | 16: $(4,0,0)$ | 25: $(4,3,0),(5,0,0)$ | 35: $(5,3,1)$ |
| 8: $(2,2,0)$ | 17: $(3,2,2),(4,1,0)$ | 26: $(4,3,1),(5,1,0)$ | 36: $(4,4,2),(6,0,0)$ |

**Fig. 10.** Look-up table (3D case): numbers 0 to 36

# Pixel Queue Algorithm for Geodesic Distance Transforms

Leena Ikonen

Lappeenranta University of Technology,
Department of Information Technology,
PO Box 20, 53851 Lappeenranta, Finland
leena.ikonen@lut.fi

**Abstract.** Geodesic distance transforms are usually computed with sequential mask operations, which may have to be iterated several times to get a globally optimal distance map. This article presents an efficient propagation algorithm based on a best-first pixel queue for computing the Distance Transform on Curved Space (DTOCS), applicable also for other geodesic distance transforms. It eliminates repetitions of local distance calculations, and performs in near-linear time.

## 1  Introduction

Distance transformations were among the first operations developed for digital images. Sequential local transformation algorithms for binary images were presented already in the 1960s [8], and similar chamfering techniques have been used successfully in 2D, 3D and even higher dimensions, see e.g. [2], [3], [1]. By modifying the definitions local distances, the chamfering can be applied to gray-level distance transforms as well. The Distance Transform on Curved Space (DTOCS) and its locally Euclidean modification Weighted DTOCS (WDTOCS), which compute distances to nearest feature along a surface represented as a gray-level height map, have been implented as mask operations [12].

Instead of propagating local distances in a predefined scanning order, the distance transformation can begin from the set of feature pixels, and propagate to points further away in the calculation area. A recursive propagation algorithm was presented in [7], where the distance value propagates from the previously processed neighbor. If the new value is accepted into the distance map, i.e. it is smaller than the previous distance value of the same pixel, the procedure is repeated recursively for each neighbor. The efficiency of the recursive propagation is highly dependent on the order in which the neighbors are processed. An unwise or unlucky choice of propagation order causes numerous repetitions of distance calculations, as shorter paths are found later on in the transformation. The ordered propagation algorithm, also presented in [7], eliminates some of the repetitions. First the boundary of the feature set, and then neighbors of already processed pixels, are placed in a queue, from which they are then taken to be processed in order. Similar pixel queue algorithms are also presented in [9] and [14].

E. Andres et al. (Eds.): DGCI 2005, LNCS 3429, pp. 228–239, 2005.

The recursive and ordered propagation, and pixel queue algorithms, can be seen as applications of graph search, where each pixel represents a vertex, and edges exist between neighbor pixels. Local distances can be defined as weights of connecting edges. The recursive propagation proceeds as a depth-first-search, and first-in-first-out pixel queue algorithms are applications of breadth-first-search. This article presents a best-first-search algorithm for computing gray-level distance transforms based on a priority queue, which is implemented efficiently as a minimum heap. A distance transform algorithm utilizing the priority queue idea was presented in [13]. Bucket sorting is used to find the pixel with the smallest current distance. The algorithm is applicable only for integer distances, as a separate storing bucket is needed for each distance value. Our heap based priority queue works for any distance values, including the real valued modifications of the DTOCS. Experiments demonstrate that convergence of the sequential transformation as well as the ordered propagation algorithm is highly dependent on the image size and complexity, whereas the near-linear pixel queue algorithm slows down only slightly with increasing surface variance.

## 2   Distance Transforms

The Distance Transform on Curved Space (DTOCS) calculates distances along a gray-level surface, when gray-levels are understood as height values. Local distances are defined using gray-level differences. The basic DTOCS simply adds the gray-level difference to the chessboard distance in the horizontal plane, i.e. the distance between neighbor pixels is:

$$d(p_i, p_{i-1}) = |\mathcal{G}(p_i) - \mathcal{G}(p_{i-1})| + 1 \tag{1}$$

where $\mathcal{G}(p)$ denotes the gray-value of pixel $p$, and $p_{i-1}$ and $p_i$ are subsequent pixels on a path. The locally Euclidean Weighted DTOCS (WDTOCS) is calculated from the height difference and the horizontal distance using Pythagoras:

$$d(p_i, p_{i-1}) = \begin{cases} \sqrt{|\mathcal{G}(p_i) - \mathcal{G}(p_{i-1})|^2 + 1} \,, & p_{i-1} \in N_4(p_i) \\ \sqrt{|\mathcal{G}(p_i) - \mathcal{G}(p_{i-1})|^2 + 2} \,, & p_{i-1} \in N_8(p_i) \setminus N_4(p_i) \end{cases} \tag{2}$$

The diagonal neighbors of pixel $p$ are denoted by $N_8(p) \setminus N_4(p)$, where $N_8(p)$ consists of all pixel neighbors in a square grid, and $N_4(p)$ of square neighbors. More accurate global distances can be achieved by introducing weights, which are proven to be optimal for binary distance transforms, to local distances in the horizontal plane. The Optimal DTOCS is defined in [6] as

$$d(p_i, p_{i-1}) = \begin{cases} \sqrt{|\mathcal{G}(p_i) - \mathcal{G}(p_{i-1})|^2 + a_{opt}^2} \,, & p_{i-1} \in N_4(p_i) \\ \sqrt{|\mathcal{G}(p_i) - \mathcal{G}(p_{i-1})|^2 + b_{opt}^2} \,, & p_{i-1} \in N_8(p_i) \setminus N_4(p_i) \end{cases} \tag{3}$$

where $a_{opt} = (\sqrt{2\sqrt{2} - 2} + 1)/2 \approx 0.95509$ and $b_{opt} = \sqrt{2} + (\sqrt{2\sqrt{2} - 2} - 1)/2 \approx 1.36930$ as derived in [2] by minimizing the maximum difference from the Euclidean distance that can occur between points on the binary image plane.

# 3    Pixel Queue Transformation Algorithm

Pixel queue algorithms are simple to implement for binary distance transforms. With equal step lengths the distances propagate smoothly from the feature set outwards, and the distance corresponds to the number of steps. As step lengths vary in the DTOCS transformations, several short steps along a smooth area of the image can create a shorter path than just one or a few steps along an area with high variance. Distances can not propagate as pixel fronts moving outwards from the feature set, or a path with a few long steps might be found instead of a shorter path consisting of many short steps. Both recursive and ordered propagation algorithms can compute the correct global distances also in the DTOCS setting, if neighbors of updated pixels are processed whether or not they have been processed before. However, this is very inefficient, as numerous repetitions of local distance calculations are needed. The new efficient **pixel queue algorithm** utilizes a priority queue implemented as a minimum heap:

1. Define binary image $\mathcal{F}(x) = 0$ for each pixel $x$ in feature set, and $\mathcal{F}(x) = max$ for each $x$ in calculation area.
2. Put feature pixels (or boundary) to *priority queue* Q.
3. While Q not empty
      $p = dequeue(Q)$, $\mathcal{F}_q(p)$ was the smallest distance in Q.
      If $\mathcal{F}_q(p) > \mathcal{F}(p)$ (obsolete value), continue from step 3.
      $\mathcal{F}(p)$ becomes $\mathcal{F}^*(p)$ (value is final).
      For neighbors $x$ of $p$ with $\mathcal{F}(x) > \mathcal{F}^*(p)$
            Compute local distance $d(p,x)$ from original image $\mathcal{G}$
            If $\mathcal{F}^*(p) + d(p,x) < \mathcal{F}(x)$
                  Set $\mathcal{F}(x) = \mathcal{F}(p) + d(p,x)$
                  $enqueue(x)$
            end if
      end for
   end while

The initialization of the queue can be implemented in two different ways without affecting the result. Only feature boundary pixels need to be enqueued in the initial step, but enqueueing all feature pixels yields the same result. Processing non-boundary feature pixels does not cause any changes in the distance image, and hence no further enqueueings of neighbor pixels. The application determines which approach is more efficient, e.g., if distances from the background into a small object are calculated, the external boundary of the object should be used rather than enqueueing the whole background.

The best-first approach eliminates repetition of local distance calculations. Using the priority queue ensures that the propagation always proceeds from a point, which already has its final distance value. As local distances, which by definition are non-negative, are added to distance values taken from the queue, the currently smallest distance can never decrease further. So once a pixel is dequeued, it will not be enqueued again. However, as step lengths vary, a distance value that has propagated from a point with a final optimal value, may still be replaced with a smaller one. Small local distances can create new shorter paths. This will cause the same pixel to be enqueued repeatedly, first with a larger distance value and then with smaller ones, before the first instance has been dequeued. Once the final distance value is dequeued, other instances of the pixel in the queue become obsolete, and could be removed. However, it is easier to just discard them when they are dequeued in the normal priority queue order. Not processing neighbors $x$ of point $p$, which already have a distance value smaller or equal to $\mathcal{F}(p)$, eliminates a significant amount of local distance calculations, including the reverse directions of previously calculated distances, i.e., if $d(p_i, p_j)$ is calculated during the transformation, $d(p_j, p_i)$ will never be needed.

The local distances are treated similarly as in the pixel queue transformation in [9]. The current pixel is considered the source point, and new distance values are assigned to all neighbors, for which the path via the source point is the shortest found so far. The recursive and ordered propagation algorithms in [7] as well as the sequential transforms view the current point as the destination with each neighbor as a possible source. Local distances from all neighbors within mask must be calculated to obtain one new distance value. The "greedy" approach of calculating distances forward from a source point was tested also for the sequential algorithm, but the effect on convergence was insignificant.

## 4    Complexity Analysis

The forward and reverse pass of a sequential local transformation can be done in linear time, as there is a constant number of operations per pixel. The problem with the complexity analysis is that the number of passes needed varies a lot depending on the size and the complexity of the image surface. Smooth and simple images can usually be transformed in just a few iterations, but it is possible to construct example images, which require one iteration for each pixel on the path with the most pixels. Typical values for test images in our previous works have been about 10-15 two-pass-iterations, which for an image of size $128 \times 128$ is in the ballpark of $\log n = 14$, which would make the whole algorithm about $\mathcal{O}(n \log n)$. However, with larger images and more complex surfaces, the number of iterations needed increases. The Experiments section will present $512 \times 512$ example images converging in about 70 iterations, which is clearly more than $\log n = 18$.

The priority queue transformation propagates local distances from each pixel only when it is dequeued with its final distance value. This means that each local distance in the image is computed only once, or some not at all, if neighbor pixels can be discarded due to already smaller distance values. Sequential algorithms recalculate each local distance at each iteration, which can be very costly, especially in transformations requiring heavier floating point calculations, like the WDTOCS. Updating the priority queue adds a factor to the computation time, as each enqueueing and dequeueing takes $\mathcal{O}(\log n_q)$ time, where $n_q$ is the number of pixels in the queue. The value $n_q$ varies through the transformation representing the boundary of the area, where distances are already calculated.

An upper limit on the complexity can be estimated using the fact that at each step after dequeueing one pixel, at most 7 pixels can be enqueued. The path through the current point must come from somewhere, so at least one neighbor must already have its final value. At each step one pixel value becomes final, so the number of efficient steps is $n - n_f$, where $n_f$ is the number of feature pixels. Even with the extra enqueueings, and dequeueings of obsolete pixels, the number of steps is in $\mathcal{O}(n)$, which makes the complexity of the whole algorithm $\mathcal{O}(n \log n_q)$, or worst case complexity $\mathcal{O}(n \log n)$. The theoretically maximal queue length, about $6n$, is a gross overestimate, as distances propagate locally as pixel fronts, which means that in practise only about half the neighbors of a pixel are enqueued with new distance values. Also after the $n - n_f$ efficient steps leaving one final distance value, the queue should be empty, and certainly not at its maximum length. Experimental results will provide a more realistic estimate on the number of queue operations and the average length of the queue.

## 5    Experiments

The priority queue algorithm was tested on gray-level images with varying surface complexity to compare with the sequential local transformation, and also with the ordered propagation algorithm implemented with a first-in-first-out pixel queue, like in [9]. The distance images were compared to make sure they were identical - and at first they were not. The sequential implementation calculated distances only at points, where the whole mask fit on the image, so errors appeared in areas, where the shortest path from the feature passed via edge pixels. Instead of modifying the mask at the edges, the border effects were corrected by adding an extra row or column to each edge before the mask transformation, copying the edge values to the corresponding extra row or column. With this correction the distance images were identical for the DTOCS, and within calculation accuracy tolerance for the WDTOCS. The pixel queue algorithms propagate distances to existing neighbors, so distances near edges are calculated correctly without tricks.

The performance of the algorithms was compared using the images seen in Fig. 1. The Mercury height map, Fig. 1 a), and the Lena image, Fig. 1 b), represent highly varying surfaces. The Lena image is obviously not an actual height map, but is used similarly in these tests. The Ball image, Fig. 1 c), is constructed as a digitization of the sphere function, i.e. the highest gray-value in the center corresponds to the radius of the sphere. The fourth test image, Flat, consists of a constant gray-value representing the smoothest surface possible. Testpoint grids were created (see example on the Ball image, Fig. 1 c), and distances from one testpoint to everywhere else in the image were calculated. The grids contained 244 points, and averages calculated from these 244 independent runs are visualized in figures 2 - 6. The sequential algorithm was faster only for the integer DTOCS on the Flat images. The larger and more complex the surfaces were, the more clearly the pixel queue algorithm outperformed the sequential transformation, and also the ordered propagation. The ordered propagation was

(a) Mercury                  (b) Lena                  (c) Ball

**Fig. 1.** Test images used. An example of a test point grid is shown on the Ball image

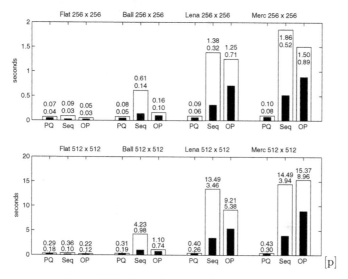

**Fig. 2.** Average run times of DTOCS (black bar) and WDTOCS (white bar) using Priority Queue, Sequential and Ordered Propagation algorithms

slightly faster than the sequential algorithm in most cases, as despite numerous repeated pixel enqueuings, processing all pixels several times in the sequential transformations is more costly. For very smooth surfaces where distances proceed evenly as pixel fronts, the ordered propagation is faster than the priority queue, as first-in-first-out queue operations take constant time.

The run times (Fig. 2), and the number of local distance calculations (Fig. 3) are proportional to the number of iterations in the sequential algorithms, and the number of iterations needed grows with the size and the complexity of the image (Fig. 4). The pixel queue transformation eliminates a lot of computation by calculating only those local distances, which are needed. If each local distance was calculated exactly once, the $256 * 256$ images would require 260610

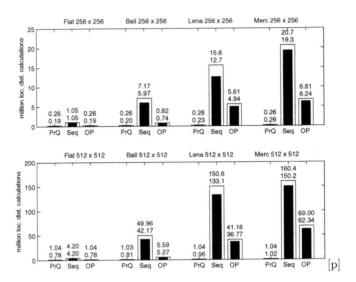

**Fig. 3.** Average number of local distance calculations needed in DTOCS (black bar) and WDTOCS (white bar) using Priority Queue, Sequential and Ordered Propagation algorithms

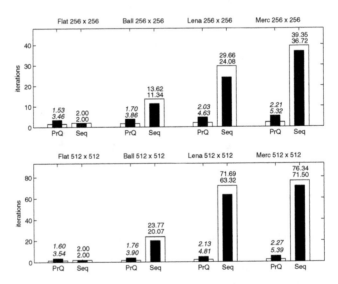

**Fig. 4.** Average number of iterations needed in sequential DTOCS (black bar) and WDTOCS (white bar). The number of iterations indicated for the pixel queue algorithm is a comparison number calculated from the run times

local distances, and the $512 * 512$ images 1045506 ($rows * (columns - 1)$) horizontal, $columns * (rows - 1)$ vertical, and $2 * (rows - 1) * (columns - 1)$ diagonal distances). Each iteration of the sequential transformation calculates each of these local distances twice, once in both directions. Some local distance calculations could have been eliminated from the first iteration by scanning the image to the feature pixel without calculating distances, saving about half an iteration.

The only source for repetition in the priority pixel queue algorithm is the calculation accuracy of floating point distance transforms. A distance value may be considered new, and consequently the pixel enqueued, even if it is smaller than the previous value only because of computation accuracy. Despite adding a threshold to the comparisons (the new value must be 0.001 smaller to be accepted as new), a few pixels ended up being enqueued repeatedly in the complex surfaces, e.g., the number of enqueueings minus the number of obsolete pixels found from the queue was at most 262190 for the $512 * 512$ Mercury surface of 262144 pixels. In the WDTOCS transformations of the smooth images, and of course in all the DTOCS transformations, the number of enqueueings minus the number of obsolete pixels equals the number of pixels.

The running times of C-implementations of the algorithms on a Linux computer with an AMD Athlon 1.678 GHz processor indicate that particularly for the floating point WDTOCS distances the pixel queue algorithm is superiorior. The speed of the priority queue operations, enqueue and dequeue, is not affected by the choice of floating point versus integer distances, so the relative cost of repeating the local distance calculations in numerous iterations is higher when using floating point values. In addition, the WDTOCS typically requires a few more iterations, causing even more repetitions. For example for the Mercury height map of size $512 * 512$ the speedup of the pixel queue transform compared to the sequential transform is $3.94/0.30 \approx 13$ for the integer DTOCS and $14.49/0.43 \approx 34$ for the floating point WDTOCS. The Optimal DTOCS was not tested here, as one integer and one floating point distance transform were enough to demonstrate the efficiency of the pixel queue algorithm. The advantage would be even more clear in the case of the Optimal DTOCS, which requires an additional multiplication operation to calculate each local distance.

The number of iterations marked for the pixel queue algorithm in Fig. 4 is calculated as the number of sequential iterations that could have been performed in the time consumed by the pixel queue algorithm. As the running time for one iteration should be constant for a certain image size and local distance definition, the comparison number can be used to estimate how much the performance of the pixel queue algorithm depends on the complexity of the image surface. The value ranged in the DTOCS tests of $512 * 512$ images from 3.54 (Flat image) to 5.39 (complex Mercury surface), while the number of iterations of the sequential DTOCS ranged from 2 to 71.50. This means that the running time of the pixel queue algorithm is much better predictable. One larger image, the Mercury $768 * 768$ surface, was tested to provide experimental basis to the claim of near-linear complexity. The average runtimes were 0.66 and 1.01 seconds for the priority queue DTOCS and WDTOCS, and 9.52 and 41.72 seconds for the sequential DTOCS and WDTOCS. Compared to the $256 * 256$ images, the corresponding $512*512$ images took about 4 times longer to transform with the priority pixel queue algorithm, and the fact that the $768*768$ Mercury image took about 9 times longer than the $256*256$ image suggests a continuing linear trend.

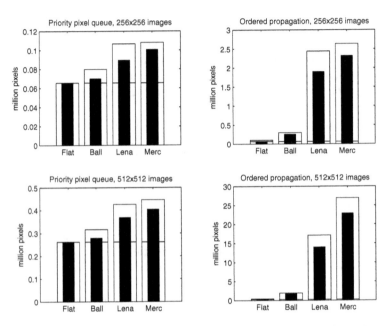

**Fig. 5.** Number of pixel enqueuings in DTOCS (black bar) and WDTOCS (white bar) for the priority queue (left) and the ordered propagation (right). The horizontal line on each bar indicates the number of pixels in the image, so the section of the bar above the line shows how many pixels get enqueued repeatedly. Notice the different scales

More statistics on the pixel queue transformation are shown in Fig. 5 and Fig. 6. The number of enqueued pixels, i.e. the number of enqueue and dequeue operations, is somewhat larger than the number of pixels. The more complex the surface, the more pixels get enqueued repeatedly when new shorter paths are found. The number of pixel enqueuings in the ordered propagation algorithm is in a larger magnitude, and also grows very rapidly with the size and complexity of the image (see Fig. 5). The average and maximum queue lengths (Fig. 6) are calculated from the average and maximum queue lengths recorded at each run. The largest average and the largest maximum queue length for each test image is indicated as lines on top of the bars. The average queue lengths for the $768*768$ Mercury surface not included in the graphs were 5295 for the DTOCS and 6073 for the WDTOCS, and the longest queue encountered contained 14069 pixels $\ll n = 768*768 = 589824$. In general, the queue lengths seem to grow

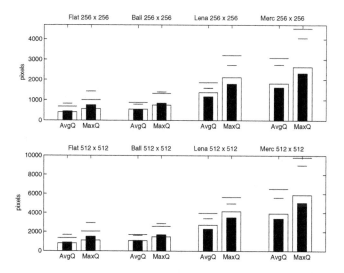

**Fig. 6.** Average and maximum queue lengths in DTOCS (black bar) and WDTOCS (white bar). The horizontal lines above the bars indicate the maximum values, i.e. the largest average queue length and largest maximum queue length

sublinearly with the size of the image. As queue lengths are in a clearly smaller magnitude than the number of pixels, the algorithm is in practise linear.

## 6    Discussion

The DTOCS algorithms have been presented as geodesic distance transforms without proper explanation on how and why they may be called geodesic. The DTOCS distances resemble geographical geodesic distances. Discrete paths follow the gray-level surface like the shortest path between two cities follow the surface of the geoid. In image analysis the term geodesic distance refers to a situation where paths linking image pixels are constrained to remain within a subset of the image plane [11]. In the DTOCS setting paths can cross any areas of the image, but path lengths can become huge. The DTOCS can be used in the same manner as constrained distance transforms, marking constraint pixels with values differing so much from the rest of the image plane that the shortest paths will never cross those pixels. In such a situation the distances propagate similarly as in a geodesic, i.e. constrained, distance transform. The pixel queue algorithm could be used to calculate both types of transforms, as well as gray-level distance transforms calculating minimal cost paths, e.g. the geodesic time transform with distances defined as the sum of gray-values along the path [10].

The presented pixel queue algorithm was demonstrated to be efficient, outperforming the sequential algorithm in almost all test cases. The running times do grow a bit with increased surface complexity, but not nearly is much as the running times of the sequential transformation. The complexity of the algorithm is $\mathcal{O}(n \log n_q)$, but as $n_q \ll n$ it performs in near-linear time. The number of local distance calculations is minimized, i.e. each local distance in the image is computed at most once, which is a clear benefit compared to the iterated sequential transforms, particuarly if the local distances require heavier floating point computations.

Previous DTOCS experiments have been made on quite small images. The experiments here demonstrate the expected effect of increasing the image size, i.e. the number of iterations needed for convergence becomes quite unpredictable. The sequential transformation may still be useful, but in applications with high resolution images, the pixel queue algorithm is more efficient. Another benefit of the pixel queue approach is that distances are calculated exactly where they are needed. If, for example, an image of an object on a background is transformed, the sequential transformation calculates unnecessary distances on the background. The pixel queue algorithm naturally proceeds from the border into the object. Also, as distance values are known to be final once they are dequeued, a real time application could utilize some values before the whole transformation is done. If the feature set is disconnected, the distance values propagated from each feature will be mixed in the priority queue, but distance values near each feature will be calculated early in the transformation. When the propagating fronts meet, the transformation is final. This idea could be utilized for developing a tesselation method.

Another approach, which ensures that obtained distance values are immediately final is presented in [4]. The parallel implementation is based on the fact that in binary distance transforms each pixel with distance value $N$ must have a neigbor with distance value $N - a$ or $N - b$, where $a$ and $b$ are the local distances to square and diagonal neighbors. Pixels with a 0-valued neighbor are updated first, and then pixels with a neighbor of each possible successively increasing distance value. Thus, the distance values propagate similarly as in the pixel queue transformation presented here.

Pixel queue algorithms can be implemented also in higher dimensions. For binary voxel images in 3D, as well as for binary images in 2D, where distances propagate as smooth fronts, ordered propagation with a first-in-first-out queue would probably work as well or even better than the priority queue approach. However, if the voxels have values other than 0 and 1, and path lengths are defined using voxel values on the path resulting in varying local distances, the priority queue algorithm could be useful. Larger neighborhoods, for example $5*5$ in 2D or $5*5*5$ in 3D, could be introduced to the pixel queue algorithm, but in the DTOCS setting larger neighborhoods need to be used with care, as they can result in illegal paths across very narrow obstacles.

The pixel queue algorithm could easily be modified to record the path of the shortest distance, by storing the direction from which the path propagated to each pixel. However, only the first found path would be recorded even though there are usually several equally short paths. The Route DTOCS algorithm for finding the route between two points [6] or point sets [5] requires two distance maps, one for each end-point set. The route consists of points on any optimal path, and a distinct path can be extracted using backtracking. In shortest route applications large complex images with long paths are typical, so the priority pixel queue algorithm improves the method significantly.

# References

1. G. Borgefors. Distance Transformations in Arbitrary Dimensions. *Computer Vision, Graphics, and Image Processing*, 27:321–345, 1984.
2. G. Borgefors. Distance Transformations in Digital Images. *Computer Vision, Graphics, and Image Processing*, 34:344–371, 1986.
3. G. Borgefors. On digital distance transforms in three dimensions. *Computer Vision and Image Understanding*, 64(3):368–376, 1996.
4. G. Borgefors, T. Hartmann, and S. L. Tamimoto. Parallell distance transforms on pyramid machines: theory and implementation. *Signal Processing*, 21(1):61–86, 1990.
5. L. Ikonen and P. Toivanen. Shortest routes between sets on gray-level surfaces. In *Patter recognition and Image Analysis (PRIA)*, pages 244–247, St. Petersburg, Russia, October 2004.
6. L. Ikonen and P. Toivanen. Shortest routes on varying height surfaces using gray-level distance transforms. *Image and Vision Computing*, 23(2):133–141, February 2005.
7. J. Piper and E. Granum. Computing Distance Transformations in Convex and Non-convex Domains. *Pattern Recognition*, 20(6):599–615, 1987.
8. A. Rosenfeld and J. L. Pfaltz. Sequential Operations in Digital Picture Processing. *Journal of the Association for Computing Machinery*, 13(4):471–494, October 1966.
9. J. Silvela and J. Portillo. Breadth-first search and its application to image processing problems. *IEEE Transactions on Image Processing*, 10(8):1194–1199, 2001.
10. P. Soille. Generalized geodesy via geodesic time. *Pattern Recognition Letters*, 15(12):1235–1240, 1994.
11. P. Soille. *Morphological Image Processing: Principles and Applications*. Springer-Verlag, 2 edition, 2003 and 2004.
12. P. Toivanen. New geodesic distance transforms for gray-scale images. *Pattern Recognition Letters*, 17:437–450, 1996.
13. Ben J. H. Verwer, Piet W Verbeek, and Simon T. Dekker. An efficient uniform cost algorithm applied to distance transforms. *IEEE Trans. on Pattern Analysis and Machine Intelligence*, 11(4):425–429, April 1989.
14. L. Vincent. New trends in morphological algorithms. In *Proc. SPIE/SPSE*, volume 1451, pages 158–170, 1991.

# Analysis and Comparative Evaluation of Discrete Tangent Estimators

Jacques-Olivier Lachaud, Anne Vialard, and François de Vieilleville

LaBRI, Univ. Bordeaux 1, 351 cours de la Libération,
33405 Talence Cedex, France
{lachaud, vialard, devieill}@labri.fr

**Abstract.** This paper presents a comparative evaluation of tangent estimators based on digital line recognition on digital curves. The comparison is carried out with a comprehensive set of criteria: accuracy on smooth or polygonal shapes, behaviour on convex/concave parts, computation time, isotropy, asymptotic convergence. We further propose a new estimator mixing the qualities of existing ones and outperforming them on most mentioned points.

## 1 Introduction

In this paper, we address the problem of tangent estimation along contours of digitized 2D objects. Tangent estimation has many applications in discrete geometry. For instance, the length of a digital contour is accurately estimated from tangents by integration [3, 9]. Derivating the orientation of the tangent provides an estimation of the curvature [9, 11, 12]. The previous geometric parameters are used in classical pattern recognition applications. They also define the internal energies of discrete deformable models [7]. When rendering 3D digitized objects, the normal vector field can be estimated from tangent directions along slice contours [8, 9].

When trying to estimate geometric properties of digitized objects, we face the issue that infinitely many shapes have the same digitization: there is no good approximation since there is no reference shape. Other hypotheses are thus required. The common assumption is that the original continuous object has some "natural" properties such as: compactness (not a fractal), bounded curvature, sometimes piecewise linear geometry (i.e. polygon). Therefore we restrict the class of shapes we are interested in. Discrete boundaries will come from the digitization of continuous shapes composed of polygonal parts and of smooth parts with bounded curvature.

Many tangent estimators are based on a fixed-size window of curve points around the point of interest [1, 9, 10, 12]. However these methods cannot converge asymptotically to the value on the continuous shape because the computation scale is not adapted to the local shape geometry. This is why we take into account in this comparative analysis only estimators based on digital straight segment extraction which use an adaptative window size [6, 8, 11].

E. Andres et al. (Eds.): DGCI 2005, LNCS 3429, pp. 240–251, 2005.

In Section 2, we recall the existing definitions of discrete tangents and compare qualitatively their advantages and drawbacks. We then propose in Section 3 a new tangent estimator, called $\lambda$-*MST*, that takes the best out of the existing ones. This estimator is based on the set of maximal digital straight segments going through the point of interest. We prove it has two interesting properties: it identifies convex and concave parts of the shape and behaves accordingly, its computational complexity is equivalent to the other existing estimators both locally and globally for the whole curve. Section 4 is devoted to an experimental comparative evaluation of the tangent estimators. We have checked the following points: tangent estimation on smooth and straight parts of the shape, sharp corner recognition, isotropy, mean and maximal asymptotical error with different shapes. The $\lambda$-MST appears to have the best behaviour in most practical cases.

# 2     Estimating Tangent with Digital Straight Segments

We restrict our study to the geometry of 4-connected digital curves. Indeed, a digital object is a set of pixels and its boundary when seen as a collection of pointels and linels is a 4-connected curve. Besides this work may easily be adapted to 8-connected curves since it relies on DSS recognition. We introduce some notations to get homogeneous definitions of existing tangent estimators based on digital straight lines. In the remaining of the paper, the digital curve is denoted by $C$. Its points $(C_k)$ are assumed to be indexed from 0 to $N - 1$. A set of successive points of $C$ ordered increasingly from index $i$ to $j$ will be conveniently denoted by $C_{i,j}$.

## 2.1     Standard Line, Digital Straight Segment, Maximal Segments

**Definition 1.** *The set of points $(x, y)$ of the digital plane verifying $\mu \leq ax - by < \mu + |a| + |b|$, with $a$, $b$ and $\mu$ integer numbers, is called the* standard line *with slope $a/b$ and shift $\mu$.*

The *standard lines* are the 4-connected discrete lines. As we will see later, all discrete tangents are defined as particular connected subset of standard lines included in 4-connected digital curves.

Since the tangent is a local property of the curve, we can always assume that we look at a restricted part of $C$, where the indices are totally ordered (the curve can be re-indexed differently so that its indices are totally ordered on the subpart of interest). The following definition is thus valid.

**Definition 2.** *We say that a set of successive points $C_{i,j}$ of the digital curve $C$ is a* digital straight segment (DSS) *iff there exists a standard line $(a, b, \mu)$ containing them. The predicate $C_{i,j}$ is a DSS is denoted by $S(i, j)$. When $S(i, j)$, we denote by $D(i, j)$ the characteristics associated with the digital straight segment [4]: the characteristics $(a, b, \mu)$ of the standard line containing all the points $C_{i,j}$, the end points $C_i$ and $C_j$, the principal upper and lower leaning points $U_m$, $U_M$, $L_m$, $L_M$.*

The first index $j$, $i \leq j$, such that $S(i,j)$ and $\neg S(i,j+1)$ is called the *front* of $i$. The map associating any $i$ to its front is denoted by $F$. Symmetrically, the first index $i$ such that $S(i,j)$ and $\neg S(i-1,j)$ is called the *back* of $j$ and the corresponding mapping is denoted by $B$.

The definition of maximal segments will be central for estimating tangents. They form the longest possible DSS in the curve. They are used for polygonizing the curve into the minimum number of segments [6].

**Definition 3.** *Any set of points $C_{i,j}$ is called a* maximal segment *iff any of the following equivalent characterizations holds: (1) $S(i,j)$ and $\neg S(i,j+1)$ and $\neg S(i-1,j)$, (2) $B(j) = i$ and $F(i) = j$, (3) $\exists k, i = B(k)$ and $j = F(B(k))$, (4) $\exists k', i = B(F(k'))$ and $j = F(k')$.*

## 2.2 Discrete Tangents

Based on local DSS recognition, several tangent estimators at a digital curve point have been proposed. Their quality is to adapt the computation window to the local shape of the curve. Exact tangent estimation for digitizations of straight lines can thus be achieved. They all try to make the right balance between longest and most centered DSS around the point of interest.

**Definition 4.** *The following DSS may be defined around any point $C_k$ of the digital curve $C$. They correspond to the notion of discrete tangent (see Fig. 1).*

- *The DSS $C_{k-l,k+l}$ with $S(k-l,k+l)$ and $\neg S(k-l-1,k+l+1)$ is called the* symmetric tangent (ST) *at $C_k$ [8].*
- *The maximal segment with biggest indices that includes the symmetric tangent at $C_k$ is called the* Feschet-Tougne tangent (FTT) *at $C_k$ [6].*
- *The* extended tangent (ET) *at $C_k$ includes the symmetric tangent $C_{k-l,k+l}$ but may be extended in the two following cases: (i) if $S(k-l,k+l+1) \wedge \neg S(k-l-1,k+l)$ then it is extended forward as the maximal segment $C_{k-l,F(k-l)}$, (ii) if $S(k-l-1,k+l) \wedge \neg S(k-l,k+l+1)$ then it is extended backward as the maximal segment $C_{B(k+l),k+l}$.*
- *The* forward half-tangent *at $C_k$ is the DSS $C_{k,F(k)}$ and the* backward half-tangent *at $C_k$ is the DSS $C_{B(k),k}$. The* median half-tangent (HT) *at $C_k$ is the arithmetical line median to the two half-tangents.*

*Any DSS defines an angle between its carrying standard line and the $x$-axis (in $[0; 2\pi[$ since a DSS is oriented). This angle will be called later on the* direction *of the DSS and denoted by the symbol $\theta$.*

The preceding discrete tangent definitions, except for the FTT, are independent of the orientation chosen for the curve (Fig. 1d-e). ET can be seen as an unambiguous version of FTT. Both FTT and ET are local longest DSS, to the expense of a loss of localization around the point. FTT and ET tend to polygonalize the digital curve even for underlying smooth shapes.

On the other hand, ST and HT have a very good localization around the point (perfectly centered for ST). However they both may have a bad behavior on even very regular shapes (e.g. at the points where a circle with integer

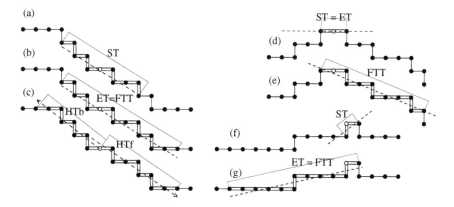

**Fig. 1.** Illustration of discrete tangents. (a) ST. (b) ET = FTT (here). (c) forward and backward HT. Subfigures d-g show specific problem raised by FTT and ST and solved by ET: (de) balanced tangent for ET and ST versus arbitrarily unbalanced tangent for FTT, (fg) false concavity detected by ST versus correct straight line for ET and FTT

radii touches the axes). They may also not locate accurately convex or concave parts of the curve (Fig. 1f-g). Note HT is also used for computing the curvature [2].

It is thus not clear which definition of discrete tangent is the best suited to a given application. A comparative evaluation of all tangent definitions is thus necessary to anticipate their behavior for given shapes and applications. This evaluation is made in Section 4. Before that, we construct a new tangent estimator which aims at mixing the qualities of the other ones: related to maximal segments as FTT and ET; computation window identical to HT; significant position of the point wrt the DSS surrounding it as ST; unambiguous definition.

## 3  Tangent Estimation Based on Maximal Segments

We define a new tangent estimator that depends on the set of maximal segments that goes through a point of the digital curve. This set is called the *pencil of maximal segments* around the point of interest. As noted by Feschet and Tougne [6], several successive points may have the same pencil. Therefore the tangent estimator takes also into account the position of the point within the pencil. More specifically, the point has a given eccentricity wrt each maximal segments. The tangent direction is estimated by a combination of the direction of each maximal segment weighted by the eccentricity. In the following subsections, we formalize the new tangent estimator, we then show it preserves convexity/concavity with minor restrictions and we explicit lastly how to compute it in optimal time.

## 3.1   Eccentricity, Maximal Segment Tangent Estimator

We index all the maximal segments of the curve by increasing indices: $M^i = C_{m_i, n_i}$ with $F(m_i) = n_i$ and $B(n_i) = m_i$. From characterizations (3) and (4) of the definition of maximal segment (Definition 3), any DSS $C_{i,j}$ and hence any point belongs to at least two maximal segments $C_{B(j), F(B(j))}$ and $C_{B(F(i)), F(i)}$. Therefore, the *pencil of maximal segments* $\mathcal{P}(k) = \{M^i, k \in M^i\}$ around any point $C_k$ is never empty. We denote by $\theta_i$ the direction of the DSS $M^i$. In the remaining of the paper, $\lambda$ is a mapping from $[0,1]$ to $\mathbb{R}^+$ with $\lambda(0) = \lambda(1) = 0$ and $\lambda > 0$ elsewhere.

The *eccentricity of* $C_k$ wrt a maximal segment $M^i$ is defined as

$$
e_i(k) = \begin{cases} \frac{\|C_k - C_{m_i}\|_1}{L_i} = \frac{k - m_i}{L_i} & \text{if } i \in \mathcal{P}(k) \\ 0 & \text{otherwise} \end{cases}, \text{ with } L_i = \|C_{n_i} - C_{m_i}\|_1. \quad (1)
$$

**Definition 5.** *The $\lambda$-maximal segment tangent direction at point $C_k$ ($\lambda$-MST) is defined as*

$$
\hat{\theta}(k) = \frac{\sum_{i \in \mathcal{P}(k)} \lambda(e_i(k))\theta_i}{\sum_{i \in \mathcal{P}(k)} \lambda(e_i(k))} = \frac{\sum_i \lambda(e_i(k))\theta_i}{\sum_i \lambda(e_i(k))}. \quad (2)
$$

Considering the properties of the eccentricity and the non-emptyness of pencils, this value is always defined and may be computed locally.

The preceding notion is extended to any real value $k$ in $[0, N[$. It is enough to consider $k$ as the curvilinear parameterization of the 4-connected contour. Any non-integer value of $k$ corresponds to a real point on the straight line linking $C_{\lfloor k \rfloor}$ and $C_{\lceil k \rceil}$. When $\lambda$ is continuous, the angle $\hat{\theta}(k)$ is continuous too and a length estimator may be derived from it, giving $\hat{l}(k) = \frac{h}{|\cos(\hat{\theta}(k))| + |\sin(\hat{\theta}(k))|}$. The length of the curve can be estimated by simple integration of $\hat{l}(k)$. Coeurjolly and Klette have reported that this method of length evaluation gives very good results [3].

## 3.2   Local Convexity or Concavity; Characterization of $\lambda$

Feschet proposes to use maximal segments for decomposing the curve into convex and concave parts [5]. The following definition shares the same idea.

**Definition 6.** *The digital curve $C$ is oriented counterclockwise wrt the discrete object it bounds. $C$ is locally convex (resp. concave) at point $C_k$ iff the angles $(\theta_i)$ of the sorted segments of $\mathcal{P}(k)$ is an nondecreasing sequence (resp. nonincreasing sequence). (Angles are brought back in $] - \pi, \pi[$ relatively to the first one.)*

We say that a tangent estimator to a digital curve satisfies the *convexity/concavity property* iff the estimated tangent direction is nondecreasing (resp. nonincreasing) on every connected subset where the curve is locally convex (resp. concave). This property holds for ET and FTT but does not hold for ST and HT (e.g. see Fig. 1). For $\lambda$-MST, it depends on the function $\lambda$ as indicated below.

**Theorem 1.** *If $\lambda$ is differentiable on $]0,1[$, then the $\lambda$-MST estimator satisfies the convexity/concavity property iff $\frac{d}{dt}(t\frac{\lambda'}{\lambda}(t)) \leq 0$ and $\frac{d}{dt}((1-t)\frac{\lambda'}{\lambda}(t)) \leq 0$ hold on this interval.*

The proof is given in appendix. It is easy to check that functions with a bell shape satisfy this constraint (e.g. functions based on binomials). This is for instance the case for the $C^2$ function $64(-x^6 + 3x^5 - 3x^4 + x^3)$ or for the $C^\infty$ function $\exp(4 - \frac{1}{x} - \frac{1}{1-x})$ extended by zeroes. One may also find functions not differentiable everywhere which satisfies the convexity/concavity. Among them, we can quote the triangle function with a peak at $\frac{1}{2}$.

## 3.3   Complexity Issues

Another interesting criterion for choosing a tangent estimation is its computational cost. Feschet and Tougne [6] showed an algorithm that computes the FTT to all points of a curve in a time linear with the number of points. We show here that all maximal segments of a curve can be computed with the same complexity. The $\lambda$-MST to all points of a curve is thus quickly computed.

Given a maximal segment $M^k = C_{m_k,n_k}$, its next maximal segment can be defined as $C_{B(n_k+1),F(B(n_k+1))}$. It is the maximal segment containing the point $n_k + 1$ and obtained from $M^k$ with a minimal number of operations (adding and removing a point). The following algorithm computes it:

> Compute_next_maximal_segment $(M^k = C_{m_k,n_k})$
>   first $\leftarrow m_k + 1$    last $\leftarrow n_k + 1$
>   while $\neg S($first, last$)$ first $\leftarrow$ first $+ 1$
>   while $S($first, last$)$ last $\leftarrow$ last $+ 1$
>   return $M^{k+1} = C_{\text{first,last}-1}$

Its principle is to remove points at the backward extremity of $M^k$ until it becomes possible to extend the resulting segment at the other end. Of course, the characteristics of the intermediate DSS must be updated at each removal or addition of a point. The time complexity of the preceding function depends on the complexity of the updates, which are proved to be $O(1)$ by:

**Theorem 2.** *Assume $S(i,j)$, and assume the characteristics $D(i,j)$ of the corresponding DSS are known. Then,*

1. *(Addition of point $C_i$ or $C_j$) deciding $S(i, j+1)$ or $S(i-1, j)$ are $O(1)$ operations and, when appropriate, computing $D(i, j+1)$ or $D(i-1, j)$ are $O(1)$ operations too (proved by Debled-Renesson and Réveillès [4]);*
2. *(Removal of point $C_i$ or $C_j$) computing $D(i+1, j)$ or $D(i, j-1)$ are $O(1)$ operations (see below).*

An immediate corollary is that all the maximal segments of a given closed digital curve are computed with a linear complexity (each point of the curve is added once to a segment and removed once). Remark that the complexity of

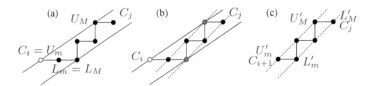

**Fig. 2.** Removal of a point from a DSS. (a) DSS $C_{i,j}$. The point $C_i$ is an upper leaning point and its removal will increase the segment slope. (b) Rotation of the leaning lines around the pivot points (in gray) during the addition/removal of $C_i$. (c) DSS $C_{i+1,j}$. Its slope and the leaning points $U'_m$ and $L'_M$ have to be recomputed

computing the pencil $\mathcal{P}(k)$ around $C_k$ depends on the local shape of the curve $(O(F(k) - B(k)))$. We now explain briefly how to update a DSS in constant time when removing a point.

Let $C_{i,j}$ be a DSS of characteristics $D(i,j) = (a, b, \mu, U_m, U_M, L_m, L_M)$. Let us recall that $U_m, U_M, L_m$ and $L_M$ are leaning points that belong to the DSS, e.g. their remainder equal $\mu$ or $\mu + |a| + |b| - 1$. $U_m$ and $U_M$ (resp. $L_m$ and $L_M$) are upper (resp. lower) leaning points, leftmost and rightmost. Without any loss in generality this digital segment belongs to the first quadrant. In the following, we denote by $(a', b', \mu', U'_m, U'_M, L'_m, L'_M)$ the characteristics $D(i+1, j)$ of the DSS $C_{i+1,j}$, which we wish to compute. Our algorithm is based on the observation that if the addition of the point $C_i$ to $C_{i+1,j}$ has changed the characteristics $D(i+1, j)$, its removal from $C_{i,j}$ should do an inverse modification to $D(i,j)$. After the examination of the incremental algorithm in [4] which explains how to update the characteristics when a point is added, this situation happens when $C_i$ is an upper or lower leaning point of the DSS $C_{i,j}$. Fig. 2 illustrates the case where $C_i$ is an upper leaning point.

We detail here the update when $C_i = U_m$ and $\overrightarrow{C_i U_M} = (b, a)$ and $L_m = L_M$. Clearly, the addition of $C_i$ to $C_{i+1,j}$ has decreased the slope of the DSS. Geometrically, it corresponds to a rotation of the upper leaning line around $U'_M$ and of the lower leaning line around $L'_m$. The two leaning points $U_M$ and $L_m$ are thus left unchanged by the removal of $C_i$. We can also easily state that the point $P = (x_{C_i} + 1, y_{C_i} - 1)$ would have extended $C_{i+1,j}$ without modifying its characteristics $D(i+1, j)$. The values $(a', b', \mu')$ are deduced from $P$. The

**Table 1.** Updates of $D(i,j)$ when removing point $C_i$

|  | $C_i = U_m \wedge \overrightarrow{C_i U_M} = (b, a) \wedge L_m = L_M$ | $C_i = L_m \wedge \overrightarrow{C_i L_M} = (b, a) \wedge U_m = U_M$ |
|---|---|---|
| $a'$ | $y_{L_m} - (y_{C_i} - 1)$ | $y_{U_m} - (y_{C_i} + 1)$ |
| $b'$ | $x_{L_m} - (x_{C_i} + 1)$ | $x_{U_m} - (x_{C_i} - 1)$ |
| $\mu'$ | $a' x_{U_M} - b' y_{U_M}$ | $a' x_{U_m} - b' y_{U_m}$ |
| $U'_m$ | $U_M - (x_{U_M} - x_{C_i} - 1)/b'(b', a')$ | $U_m$ |
| $U'_M$ | $U_M$ | $U_m + ((y_{C_j} - y_{C_i} - 1)/a' - 1)(b', a')$ |
| $L'_m$ | $L_m$ | $L_M - (y_{L_M} - y_{C_i} - 1)/a'(b', a')$ |
| $L'_M$ | $L_m + ((x_{C_j} - x_{C_i} - 1)/b' - 1)(b', a')$ | $L_M$ |

updating of $U_m$ and $L_M$ is a littl◼
the vector linking two successive ʋ
computation of the characteristics
of Table 1. Its second column corre
point and its removal decreases th◼

## 4    Experimental Evaluaı

In this section, we perform a quant
on DSS recognition. The behaviour
which monitor the estimation of t◼
ing $C^\infty$ functions requires $C^\infty$ $\lambda$ fu◼
of the underlying curve by taking
at $\frac{1}{2}$ as $\lambda$ function. This function
a circular arc when the pencil of
mal segments. Moreover it gives ◼
implementation, all tangent direct◼

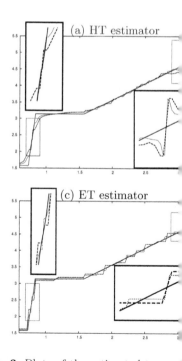

**Fig. 3.** Plots of the estimated tangen◼
shape is a circle of radius 10 with ◼
correspond to expected values, dash◼
dotted lines to estimations with a fin◼

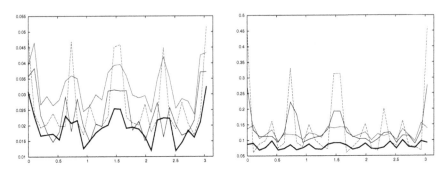

**Fig. 4.** Isotropy of tangent estimators measured with absolute error $|\hat{\theta}(t) - \theta(t)|$ (thick solid line: $\lambda$-MST, thin solid line: HT, dashed line: ST, dotted line: ET). Left: mean of absolute error. Right: Maximum of absolute error. For each estimator, 100 experiments are run on a circle of radius 50 with a center arbitrarily shifted in its pixel. The absolute error is drawn as a function of the polar angle and gathered by sectors of $\frac{5}{180}\pi$

geometric quantities are computed at curvilinear abscissa $k + \frac{1}{2}$ and all DSS includes $k$ and $k + 1$).

We first compare the behavior of tangent estimators on smooth and flat parts and on corners. The shape is a circle in three quadrants and a right angle in the fourth (see "rsquare" in Table 2). Fig. 3 displays (a subset of) the estimations of the tangent direction. Estimators that satisfies the convexity/concavity property, i.e. ET and $\lambda$-MST, create a non-decreasing sequence of directions. ST and HT

**Table 2.** Asymptotic convergence of mean and maximum absolute error for tangent estimators. Best estimator is shaded. The grid step is denoted by $h$. The maximum and minimum curvatures of the shapes are: "circle" ($\kappa_M = \kappa_m = 1$), "flower" ($\kappa_M \approx 5.8, \kappa_m \approx -26.1$), "rsquare" ($\kappa_M = 100, \kappa_m = 0$)

| Shape | T.E | mean error | | | | | maximum error | | | | |
|---|---|---|---|---|---|---|---|---|---|---|---|
| | $h$ | $\frac{1}{10}$ | $\frac{1}{20}$ | $\frac{1}{80}$ | $\frac{1}{320}$ | $\frac{1}{640}$ | $\frac{1}{10}$ | $\frac{1}{20}$ | $\frac{1}{80}$ | $\frac{1}{320}$ | $\frac{1}{640}$ |
| circle | HT | 0.0624 | 0.0411 | 0.0174 | 0.0077 | 0.0049 | 0.5432 | 0.5267 | 0.2717 | 0.1395 | 0.0935 |
| | ET | 0.0830 | 0.0565 | 0.0236 | 0.0098 | 0.0062 | 0.3887 | 0.2695 | 0.1232 | 0.0592 | 0.0422 |
| | ST | 0.0665 | 0.0443 | 0.0185 | 0.0079 | 0.0049 | 0.7700 | 0.7840 | 0.4639 | 0.2450 | 0.1651 |
| | $\lambda$-MST | 0.0541 | 0.0378 | 0.0144 | 0.0057 | 0.0035 | 0.2934 | 0.1997 | 0.0770 | 0.0383 | 0.0281 |
| flower | HT | 0.1736 | 0.1050 | 0.0364 | 0.0128 | 0.0078 | 1.2836 | 1.1753 | 0.7220 | 0.5269 | 0.5018 |
| | ET | 0.1364 | 0.0868 | 0.0369 | 0.0151 | 0.0095 | 1.4821 | 1.2028 | 0.7317 | 0.3312 | 0.2178 |
| | ST | 0.1258 | 0.0756 | 0.0311 | 0.0123 | 0.0075 | 1.2415 | 0.9831 | 0.8201 | 0.7931 | 0.7878 |
| | $\lambda$-MST | 0.1541 | 0.0881 | 0.0293 | 0.0098 | 0.0059 | 1.4821 | 1.1705 | 0.6496 | 0.2483 | 0.1479 |
| rsquare | HT | 0.0734 | 0.0501 | 0.0204 | 0.0081 | 0.0052 | 0.5228 | 0.5201 | 0.2151 | 0.1809 | 0.1359 |
| | ET | 0.0876 | 0.0572 | 0.0220 | 0.0087 | 0.0054 | 0.3858 | 0.2903 | 0.1498 | 0.0763 | 0.0543 |
| | ST | 0.0834 | 0.0560 | 0.0220 | 0.0084 | 0.0052 | 0.7496 | 0.7775 | 0.3270 | 0.3202 | 0.2452 |
| | $\lambda$-MST | 0.0529 | 0.0344 | 0.0127 | 0.0052 | 0.0032 | 0.2880 | 0.1775 | 0.0832 | 0.0440 | 0.0304 |

clearly fail, especially at points where the digital contour meets a quadrant change. Most estimators behave correctly at corners. $\lambda$-MST slightly smoothes the corner at low resolution. The tendency to polygonalize the curve of ET (and thus FTT) appears clearly on Fig. 3c.

We then evaluate the anisotropy of the estimators with the experiment described in Fig. 4. The $\lambda$-MST is more isotropic than the others, with a steady and low mean and maximal error.

We finally examine the asymptotic behavior of the absolute error for different shapes on Table 2. Both $\lambda$-MST and ET have an asymptotic convergence in mean and in maximum. It is however unclear whether the maximum error of ST and HT converges toward 0 or not for arbitrary shapes. Although the $\lambda$-MST is not always the best in mean at coarse resolution, it has the fastest asymptotic convergence in mean and in maximum whatever the shape is.

## 5   Conclusions

In this paper, we have compared several tangent estimators based on DSS recognition. After a first qualitative analysis, we have proposed a new estimator which takes the best out of the existing ones. We have checked that it satisfies the convexity/concavity property and we have shown how to compute it efficiently. After experimental evaluation, the $\lambda$-MST appears to be the most robust tangent estimator and very often the most accurate. The results are summed up in Table 3. Future work will focus on curvature estimators based on maximal segments and their properties.

**Table 3.** Comparison of discrete tangent estimators. The $\lambda$-MST estimator has an average behaviour on corners and seems to be the best elsewhere

| tangent estimator | straight parts | smooth parts | corners | convexity /concavity | isotropy | mean error | maximal error |
|---|---|---|---|---|---|---|---|
| $\lambda$-MST | + | + | = | Yes* | + | ++ | ++ |
| HT | = | +/− | + | No | − | + | − |
| ET | + | = | + | Yes | = | + | + |
| ST | = | +/− | = | No | − | + | − |

(*) For $\lambda$ functions satisfying conditions of Theorem 1

## References

1. I.M. Anderson and J.C. Bezdek. Curvature and tangential deflection of discrete arcs: a theory based on the commutator of scatter matrix pairs and its application to vertex detection in planar shape data. *IEEE Transactions on Pattern Analysis and Machine Intelligence*, 6:27–40, 1984.
2. D. Coeurjolly. *Algorithmique et géométrie discrète pour la caractérisation des courbes et des surfaces.* PhD thesis, Université Lumière Lyon 2, France, 2002.

3. D. Coeurjolly and R. Klette. A comparative evaluation of length estimators of digital curves. *IEEE Transactions on Pattern Analysis and Machine Intelligence*, 26(2):252–258, 2004.
4. I. Debled-Renesson and J.-P. Réveillès. A linear algorithm for segmentation of discrete curves. *International Journal of Pattern Recognition and Artificial Intelligence*, 9:635–662, 1995.
5. F. Feschet. Canonical representations of discrete curves. *Pattern Analysis and Application*. Submitted.
6. F. Feschet and L. Tougne. Optimal time computation of the tangent of a discrete curve: Application to the curvature. In *Proc. DGCI'99*, volume 1568 of *LNCS*, pages 31–40. Springer, 1999.
7. J.-O. Lachaud and A. Vialard. Discrete deformable boundaries for the segmentation of multidimensional images. In *Proc. 4th Int. Workshop on Visual Form*, volume 2059 of *LNCS*, pages 542–551. Springer, 2001.
8. J.-O. Lachaud and A. Vialard. Geometric measures on arbitrary dimensional digital surfaces. In *Proc. DGCI'03*, volume 2886 of *LNCS*, pages 434–443. Springer, 2003.
9. A. Lenoir, R. Malgouyres, and M. Revenu. Fast computation of the normal vector field of the surface of a 3D discrete object. In *Proc. DGCI'96*, volume 1176 of *LNCS*, pages 101–112. Springer, 1996.
10. J. Matas, Z. Shao, and J. Kittler. Estimation of curvature and tangent direction by median filtered differencing. In *Proc. of 8th International Conference on Image Analysis and Processing*, pages 83–88, 1995.
11. A. Vialard. Geometrical parameters extraction from discrete paths. In *Proc. DGCI'96*, volume 1176 of *LNCS*, pages 24–35. Springer, 1996.
12. M. Worring and A. W. M. Smeulders. Digital curvature estimation. *CVGIP: Image Understanding*, 58:366–382, 1993.

# A    Proof of Theorem 1

We show here a necessary and sufficient condition for the $\lambda$ function to define a $\lambda$-MST tangent estimator satisfying the convexity/concavity property.

**Theorem 1.** *If $\lambda$ is differentiable on $]0, 1[$, then the $\lambda$-MST estimator satisfies the convexity/concavity property iff $\frac{d}{dt}(t\frac{\lambda'}{\lambda}(t)) \leq 0$ and $\frac{d}{dt}((1-t)\frac{\lambda'}{\lambda}(t)) \leq 0$ hold on this interval.*

These two conditions once put together entail $\lambda$ is necessarily log-concave (i.e. $\ln \lambda$ is a concave function or $\frac{d^2}{dt^2}(\ln \lambda(t)) \leq 0$). Furthermore, it is enough to check $\frac{d}{dt}(t\frac{\lambda'}{\lambda}(t)) \leq 0$ for functions symmetric around $\frac{1}{2}$.

*Proof.* We first rewrite $\hat{\theta}'(k)$ as

$$\frac{\sum_{i<j}(\theta_i - \theta_j)\left(\frac{\lambda(e_j(k))\lambda'(e_i(k))}{L_i} - \frac{\lambda(e_i(k))\lambda'(e_j(k))}{L_j}\right)}{(\sum_j \lambda(e_j(k)))^2}. \tag{3}$$

We assume for instance that the angles $(\theta_i)$ of the segment in the pencil around $k$ are nondecreasing. We must thus prove $\hat{\theta}'(k)$ is nonnegative, whatever is the curve under examination. Since some curves have points with exactly two maximal segments going through, Eq. (3) may be reduced to one pair. It is thus necessary to show that each term of this sum is nonnegative. It is also a sufficient condition. Otherwise said, we have to prove for any $i < j$,

$$\forall k, m_j < k < n_i, \frac{\lambda(e_j(k))\lambda'(e_i(k))}{L_i} - \frac{\lambda(e_i(k))\lambda'(e_j(k))}{L_j} \leq 0. \tag{4}$$

Let $R_{ij} = n_i - m_j$ be the size of the common part of both segments. Setting $t = \frac{k-m_j}{R_{ij}}$, we define two analogs of the eccentricities $e_i(k)$ and $e_j(k)$ as $\epsilon_i(t) = e_i(k) = 1 - \frac{R_{ij}}{L_i}(1-t)$ and $\epsilon_j(t) = e_j(k) = \frac{R_{ij}}{L_j}t$. Eq. (4) is then equivalent to

$$\forall t \in ]0,1[, \quad \lambda(\epsilon_j(t))\frac{\lambda'(\epsilon_i(t))}{L_i} \leq \lambda(\epsilon_i(t))\frac{\lambda'(\epsilon_j(t))}{L_j} \tag{5}$$

$$\Leftrightarrow \frac{R_{ij}}{L_i}\frac{\lambda'}{\lambda}(\epsilon_i(t)) \leq \frac{R_{ij}}{L_j}\frac{\lambda'}{\lambda}(\epsilon_j(t)) \Leftrightarrow \frac{d}{dt}(\ln\lambda(\epsilon_i(t))) \leq \frac{d}{dt}(\ln\lambda(\epsilon_j(t))) \tag{6}$$

It is easy to see that $\epsilon_i(t) > t > \epsilon_j(t)$ which gives the idea to break Eq. (6) in two parts as follows, for all $t \in ]0,1[$:

$$\frac{d}{dt}(\ln\lambda(\epsilon_i(t))) \leq \frac{d}{dt}(\ln\lambda(t)) \text{ and } \frac{d}{dt}(\ln\lambda(t)) \leq \frac{d}{dt}(\ln\lambda(\epsilon_j(t))) \tag{7}$$

Eq. (7) clearly implies Eq. (6), but the converse is also true by letting $L_i$ or $L_j$ tend toward $R_{ij}$.

We focus on the right part of Eq. (7). Letting $\delta = \frac{R_{ij}}{L_j}$ and $f = \ln\lambda$, we get

$$\forall \delta, 0 < \delta < 1, \frac{d}{dt}(f(t)) \leq \frac{d}{dt}(f(\delta t)), \text{ otherwise said } f'(t) \leq \delta f'(\delta t). \tag{8}$$

We now show that Eq. (8) is equivalent to

$$\frac{d}{dt}(tf'(t)) \leq 0. \tag{9}$$

Indeed, integrating both terms of the last inequality between $\delta t$ and $t$ shows sufficiency. It is also necessary since Eq. (8) can be rewritten with $h = (1-\delta)t$ as:

$$f'(t) \leq (1 - \frac{h}{t})f'(t-h) \tag{10}$$

$$\frac{f'(t) - f'(t-h)}{h} + \frac{f'(t-h)}{t} \leq 0 \tag{11}$$

Getting the limit when $h$ tends toward 0 and multiplying both sides by $t$ give $tf''(t) + f'(t) \leq 0$, which is exactly Eq. (9). Same reasoning applied to left part of Eq. (7) bring

$$\frac{d}{dt}((1-t)f'(t)) \leq 0, \tag{12}$$

which concludes the proof.                                                    □

# Surface Volume Estimation of Digitized Hyperplanes Using Weighted Local Configurations

Joakim Lindblad

Centre for Image Analysis, Uppsala University, Uppsala, Sweden
joakim@cb.uu.se

**Abstract.** We present a method for estimating the surface volume of four-dimensional objects in discrete binary images. A surface volume weight is assigned to each $2 \times 2 \times 2 \times 2$ configuration of image elements. The total surface volume of a digital 4D object is given by a summation of the local volume contributions. Optimal volume weights are derived in order to provide an unbiased estimate with minimal variance for randomly oriented digitized planar hypersurfaces. Only 14 out of 64 possible boundary configurations appear on planar hypersurfaces. We use a marching hypercubes tetrahedrization to assign surface volume weights to the non-planar cases. The correctness of the method is verified on four-dimensional balls and cubes digitized in different sizes. The algorithm is appealingly simple; the use of only a local neighbourhood enables efficient implementations in hardware and/or in parallel architectures.

**Keywords:** surface volume estimation, marching cubes, digital hyperplanes, 4D, cell tiling.

## 1 Introduction

In many applications of digital image analysis, quantitative geometrical measures, such as length and area of objects, are of foremost interest. When working with three-dimensional (3D) digital images, an often desired measure is the surface area of a digitized object. With modern imaging techniques and powerful computers, it has become interesting to look at higher dimensional data volumes. The four-dimensional (4D) counterpart to surface area is surface volume. In this paper we present a method to perform accurate surface volume estimations of 4D objects in binary digital images using a technique based on local cell tiling.

In [11, 12] we presented a surface area estimator for 3D images that utilises only local computations and a small local neighbourhood to obtain an estimate that is very fast to calculate and still exhibits good performance in terms of accuracy, precision and robustness. In this paper we extend this methodology to four dimensions, and derive optimal surface volume weights for the hyxel (hyper volume picture element) configurations that appear on planar hypersurfaces.

E. Andres et al. (Eds.): DGCI 2005, LNCS 3429, pp. 252–262, 2005.
© Springer-Verlag Berlin Heidelberg 2005

## 2    Previous and Related Work

Visualization methods for high dimensional data are relatively well developed. There exist tools for 4D plane-tracing (a 4D extension of ray-tracing) and many techniques based on splatting. The popular Marching Cubes algorithm [13], for generating a triangulated iso-surface from voxel data, has also been extended to higher dimensions [1]. Far less is available in the field of image analysis and the task of extracting quantitative data from high-dimensional images. Concerning the geometry of digital 4D objects, work on, e.g., distance transforms [2] and skeletonization [6] in 4D has been presented. To the best of our knowledge, no surface volume estimation technique for discrete 4D data has previously been presented in literature. However, many similarities with surface area estimates of 3D objects and perimeter estimates of 2D objects do exist.

The perimeter of a digitized 2D object can be estimated as the cumulative distance from pixel centre to pixel centre along the border of the object, where an isothetic step is given weight 1 and a diagonal one is given weight $\sqrt{2}$. This is straightforward to accomplish using the Freeman chain code [5], but results in rather big over-estimates. Starting from an assumption that the boundary of an object is locally linear, optimal weights for the local steps have been derived, leading to an unbiased estimator with a minimal mean squared error (MSE) [10, 14]. A similar approach can be taken in order to estimate the surface area of digitized 3D objects. By counting the local configurations of voxels that appear on the boundary of a digital object a fast and accurate area estimate is achieved. In [11, 12] optimal surface area weights were derived, providing an unbiased estimator with minimal MSE. The method described in this paper is a direct extension of this technique to the 4D case.

In addition to the local type of estimators mentioned above, different multigrid convergent perimeter and surface area estimators exist, see, e.g., [4] for an overview of perimeter estimators, and [7, 9, 3] for examples of multigrid convergent surface area estimators. This class of estimators ensure convergence toward the true value as the grid resolution increases [8]. Many multigrid convergent estimators are based on finding straight line/plane segments. However, in order to do so we can no longer use local algorithms. Coeurjolly et al. [3] have presented efficient algorithms based on discrete normal vector field integration, where the problem of perimeter/surface area estimation is transformed into a problem of normal vector estimation. It seems that this approach may be extended to higher dimensions. To our knowledge no one has so far attempted to do so.

## 3    Surface Area Estimation

To introduce the methodology, we present, in this section, a brief derivation of the 3D version of the method; for measuring the surface area of a digitized 3D object. For a more detailed description see [12].

The estimation is based on counting local configurations of $2 \times 2 \times 2$ voxels. In a binary image, the number of possible configurations of the eight voxels is

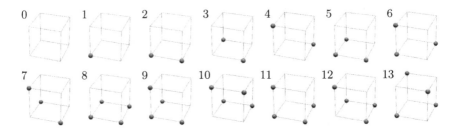

**Fig. 1.** The 14 major 3-cubes of $2 \times 2 \times 2$ voxels. Marked voxel centres are inside the object. The complementary cases are classified to be the same as the original cases. Only cases $c_1$, $c_2$, $c_5$, $c_8$, and $c_9$, appear for planar surfaces

$2^{2^3} = 256$. Using rotation, mirror, and complement symmetry, the 256 configurations can be grouped into 14 major cases $c_i$, see Fig. 1.

A surface area contribution $A_i$ is assigned to each case ($c_0$ does not represent a boundary situation, and therefore has zero area contribution). The number of occurrences $N_i$ of each of the 13 surface configurations is computed for the digitized object, and the surface area estimate $\hat{A}$ of the object is calculated as

$$\hat{A} = \sum_{i=1}^{13} A_i N_i. \tag{1}$$

The area contributions $A_i$ are optimally selected so as to provide an unbiased estimate with a minimal MSE when the method is applied to infinite planes (the surfaces of half-spaces) digitized over an isotropic distribution of normal directions. This optimization can be justified by the fact that the surface of an object with limited curvature becomes locally planar as the sampling density increases. Only five of the 13 possible surface configurations appear for planar surfaces. We call these five cases the planar cases. They dominate the boundary of most objects digitized at a resolution high enough to capture the details of the surface structure.

When performing the optimization, we can, due to the symmetry of the sampling grid, without loss of generality, restrict the study to planes that can be expressed as a function $z(x, y) = z'_x x + z'_y y + k$, $0 \le z'_y \le z'_x < 1$. Voxels with a centre on, or below, the plane are included in the object. We vary the offset term $k$ and observe the configurations that appear when a plane of a given normal direction cuts a column of cubes (an infinite stack of cubes in the z-direction). Depending on if $z'_x + z'_y$ is less or greater than 1, two different sets are observed. This is shown in Figs. 2 and 3. We keep track of the intersections between the surface and all cubes in the column. For example, in Fig. 2(b) the lower cube is of type $c_5$ and the upper one is of type $c_1$.

For a plane in general position the offset term is uniformly distributed. Given a specific normal direction $n$, the expected number of occurrences of each case $E_i = E[N_i|n]$ per column of intersected cubes, can be directly calculated from Figs. 2 and 3.

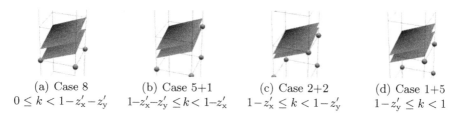

(a) Case 8
$0 \leq k < 1 - z'_x - z'_y$

(b) Case 5+1
$1 - z'_x - z'_y \leq k < 1 - z'_x$

(c) Case 2+2
$1 - z'_x \leq k < 1 - z'_y$

(d) Case 1+5
$1 - z'_y \leq k < 1$

**Fig. 2.** The different cases appearing for $z'_x + z'_y \leq 1$ as $k$ is varied

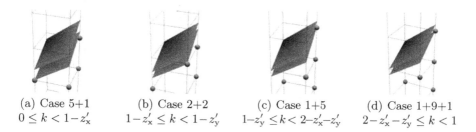

(a) Case 5+1
$0 \leq k < 1 - z'_x$

(b) Case 2+2
$1 - z'_x \leq k < 1 - z'_y$

(c) Case 1+5
$1 - z'_y \leq k < 2 - z'_x - z'_y$

(d) Case 1+9+1
$2 - z'_x - z'_y \leq k < 1$

**Fig. 3.** The different cases appearing for $z'_x + z'_y > 1$ as $k$ is varied

For $z'_x + z'_y \leq 1$,
$$E_1 = 2z'_y, \quad E_2 = 2(z'_x - z'_y), \quad E_5 = 2z'_y, \quad E_8 = 1 - z'_x - z'_y,$$
and for $z'_x + z'_y > 1$,
$$E_1 = 2z'_y, \quad E_2 = 2(z'_x - z'_y), \quad E_5 = 2(1 - z'_x), \quad E_9 = (z'_x + z'_y - 1).$$
The number of intersected columns of cubes, for a planar surface segment of area $A$ and normal direction $n$, is $N_{\text{col}}(n) = \frac{1}{\sqrt{1 + z'^2_x + z'^2_y}} A$. The estimated surface area is given by

$$\hat{A}(n) = \sum_{i=1}^{13} A_i E_i(n) N_{\text{col}}(n). \qquad (2)$$

The MSE of (2) is minimized over all normal directions, while keeping zero bias, in order to find optimal values for $A_i$. On a planar surface cases 1, 5, and 9 always appear together, which leads to a non-unique solution. The reason for this is that the planar surface continues into the neighbouring cubes and divides them in a way that creates the co-appearing cases. Leaving $A_1$ as a variable we obtain the following solution:

$$A_2 = 0.669, \quad A_5 = 1.190 - A_1, \quad A_8 = 0.927, \quad A_9 = 1.694 - 2A_1. \qquad (3)$$

These weights provide, independent on how we choose $A_1$, an unbiased area estimate with a coefficient of variation (CV=$\mu/\sigma$) of 1.40% for planar surfaces.

In order to estimate the surface area of general object boundaries, area contributions have to be assigned to all the 13 surface cases. Since the non-planar

**Table 1.** Table of elementary areas assigned to the different 3-cube cases. $A_1$ is left undefined

| | | | | |
|---|---|---|---|---|
| $A_1 =$ undefined | $A_2 = 0.6690$ | $A_3 = 2A_1$ | $A_4 = 2A_1$ | |
| $A_5 = 1.1897 - A_1$ | $A_6 = A_1 + A_2$ | $A_7 = 3A_1$ | $A_8 = 0.9270$ | |
| $A_9 = 1.6942 - 2A_1$ | $A_{10} = 2A_2$ | $A_{11} = 1.5731$ | $A_{12} = A_1 + A_5$ | $A_{13} = 4A_1$ |

cases are, in general, scarcely appearing in real object volumes, the area contribution assigned to these cases will have a limited impact on the overall surface area estimate. Dividing the cubes into (locally) face-connected components, all but one of the additional cases can be decomposed into the five planar ones. This way we introduce a minimal number of new values. The only truly new configuration is $c_{11}$, which we here assign an area of 1.5731 (derived from a Marching Cubes triangulation of that case). The surface area weights of all 13 cases are summarized in Table 1, where $A_1$ is left undefined. Note, however, that this subdivision of cases is not uniquely determined. For example, $c_7$ can either be split into $3c_1$ or $c_9 + c_1$ depending on if we look at the original or the complementary case.

For curved surfaces, the relation between cases $c_1$, $c_5$, and $c_9$ no longer holds, and the specific choice of the free parameter $A_1$ will affect the estimation result. In [12] the freedom to choose the value of $A_1$ is used to minimize the estimation error of the method when applied to a distribution of digitized balls of increasing radii.

## 4   Surface Volume Estimation

For 4D surface volume estimation we use configurations of $2 \times 2 \times 2 \times 2$ hyxels, which in a binary image gives $2^{2^4} = 65\,536$ different configurations. Using rotation, mirror, and complement symmetry, they can be grouped into 222 major cases. 14 of these, shown in Fig. 4, appear on the surfaces of planar volumes. Just as in the 3D case, we restrict the study to hyperplanes that can be expressed as a function $w(x, y, z) = w'_x x + w'_y y + w'_z y + k,\ 0 \leq w'_z \leq w'_y \leq w'_x < 1$. We vary

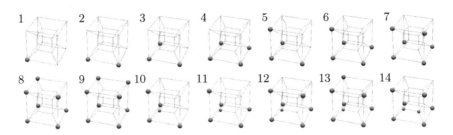

**Fig. 4.** The 14 planar 4-cubes, appearing on the surface of planar volumes. Marked hyxel centres are inside the object

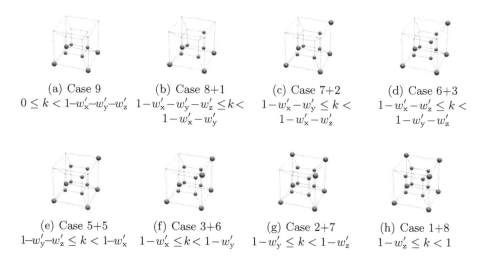

(a) Case 9
$0 \leq k < 1{-}w'_x{-}w'_y{-}w'_z$

(b) Case 8+1
$1{-}w'_x{-}w'_y{-}w'_z \leq k <$
$1{-}w'_x{-}w'_y$

(c) Case 7+2
$1{-}w'_x{-}w'_y \leq k <$
$1{-}w'_x{-}w'_z$

(d) Case 6+3
$1{-}w'_x{-}w'_z \leq k <$
$1{-}w'_y{-}w'_z$

(e) Case 5+5
$1{-}w'_y{-}w'_z \leq k < 1{-}w'_x$

(f) Case 3+6
$1{-}w'_x \leq k < 1{-}w'_y$

(g) Case 2+7
$1{-}w'_y \leq k < 1{-}w'_z$

(h) Case 1+8
$1{-}w'_z \leq k < 1$

**Fig. 5.** The different cases appearing for $w'_x + w'_y + w'_z \leq 1, w'_x \leq w'_y + w'_z$ as $k$ is varied

the offset term $k$ and study the configurations that appear when a hyperplane of a given normal direction cuts a column (in the $w$-direction) of 4-cubes.

Depending on the slope of the hyperplane, one out of 14 different sets of configurations appear. The set, $w'_x + w'_y + w'_z \leq 1, w'_x \leq w'_y + w'_z$, is shown in Fig. 5. Just as in the 3D case, it is straightforward, albeit a bit tedious, to calculate the expected number of cells of each type that appear when a column of 4-cubes is intersected by a hyperplane of a given normal direction. The configurations are listed in the Appendix. The number of intersected columns of 4-cubes, for a planar surface segment of volume $V$ and normal direction $\boldsymbol{n}$, is $N_{col}(\boldsymbol{n}) = \frac{1}{\sqrt{1+w'^2_x+w'^2_y+w'^2_z}} V$. The estimated surface volume is given by

$$\hat{V}(\boldsymbol{n}) = \sum_{i=1}^{14} V_i E_i(\boldsymbol{n}) N_{col}(\boldsymbol{n}). \qquad (4)$$

We minimize the MSE of (4) over all normal directions, while keeping zero bias, in order to find optimal values for $V_i$. Just as in the 3D case, we do not get enough information from using only planar objects to find a unique solution. Leaving $V_1, V_2, V_3, V_5$ as variables the optimization leads to the weights presented in Table 2. These weights provide an unbiased volume estimate with a CV of

**Table 2.** Table of elementary volumes assigned to the planar 4-cubes

| | | | |
|---|---|---|---|
| $V_1 =$ undefined | $V_2 =$ undefined | $V_3 =$ undefined | $V_4 = 0.668$ |
| $V_5 =$ undefined | $V_6 = 0.609 - V_3 + V_5$ | $V_7 = 1.194 - V_2$ | $V_8 = 1.707 - V_1 - V_5$ |
| $V_9 = 0.920$ | $V_{10} = 0.972 - V_1$ | $V_{11} = 1.558 - V_1 - V_3$ | |
| $V_{12} = 2.113 - V_1 - V_2 - V_5$ | $V_{13} = 2.630 - 2V_1 - 2V_5$ | $V_{14} = 1.680 - 2V_2$ | |

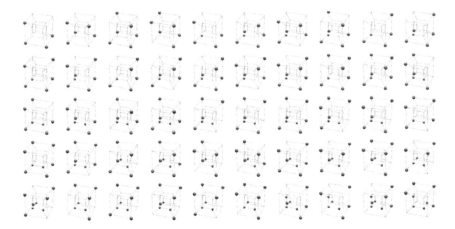

**Fig. 6.** The 50 non-planar 4D sub-cases

0.77% for planar hypersurfaces, independent on how we choose $V_1, V_2, V_3, V_5$. The maximum absolute error, AD=7.97% is reached for hyperplanes aligned with the digitization grid.

### 4.1    Curved Volumes

In order to estimate the surface volume of curved 4D objects, we need to assign volume contributions to all 221 surface cases. Similar as for the 3D case, we can reduce this number by dividing the 4-cubes into (locally) volume connected components. This way the 222 major cases can be reduced to 65 sub-cases. Note that this is less than the number given by Roberts [15]. The reason for this is that we consider it valid to look at the complement also when splitting the cells into components (since complement is used anyway to reach the 222 major cases). This is similar to the optional splitting of $c_7$ into $c_9 + c_1$ in the 3D case. It is an open question how to assign an optimal surface volume weight to the 50 non-planar sub-cases (shown in Fig. 6).

## 5    Simulations

To verify and evaluate the performance of the estimator, we test the method on synthetic objects of known surface volumes. The used test objects are 4-balls of radii 1–80 hyxels and 4-cubes of side lengths 2–160 hyxels. We generate 30 000 instances of each object in the continuous space and digitize them using Gauss digitization, with a random orientation and position in the digitization grid. We have assigned surface volume weights derived from a marching hypercubes tetrahedrization [1] to the non-planar cases that appear on the edges of the 4-cubes, and to the unassigned planar cases, $c_1$, $c_2$, $c_3$, and $c_5$.

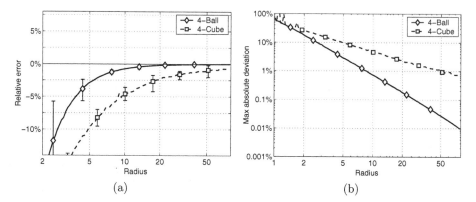

**Fig. 7.** Relative error (a) and absolute deviation (b) of the surface volume estimates for digitized objects of increasing size. The error bars indicate minimum and maximum values of the estimate

Average relative error and maximum absolute deviation for the surface volume estimation method are shown in Fig. 7. The surface of a large 4-ball is a good sampling of hyperplanes of all normal directions and the described surface volume estimator is therefore expected to exhibit very low variance on such objects. This is verified by the simulations, where superlinear convergence $O(r^{-\alpha})$, $\alpha \approx 2$, is observed. Improved low resolution performance can be achieved by adjusting the weights $V_1$, $V_2$, $V_3$, and $V_5$. This is beyond the scope of this paper.

## 6  Summary

We have presented a method for estimating the surface volume of binary 4D objects using local computations. The surface volume is computed as a sum of local volume contributions. Optimal volume weights for the $2 \times 2 \times 2 \times 2$ configurations of hyxels that appear on digital planar hypersurfaces are derived. The method gives an unbiased estimate with minimum variance for randomly oriented planar hypersurfaces. Theoretic worst case CV for the suggested surface volume estimator is 0.77%, and the maximum absolute error is 7.97%. The maximum error is reached for planar hypersurfaces aligned with the digitization grid. The solution for planar volumes is not unique and freedom in the choice of parameters may be used to improve the performance at lower resolutions in a manner similar to what was done for the 3D case in [12]. For curved volumes additional cases appear. It is an open question how to assign optimal volume weights to these cases.

**Acknowledgements.** We thank Nataša Sladoje for valuable scientific support, Dr. Rephael Wenger for making his programs for generating iso-surface lookup tables freely available on the Internet, and Prof. Nahun Kiryati for (gently) pushing us out into the fourth dimension.

# References

[1] P. Bhaniramka, R. Wenger, and R. Crawfis. Isosurface construction in any dimension using convex hulls. *IEEE Trans. on Vision and Computer Graphics*, 10(2):130–141, 2004.

[2] G. Borgefors. Weighted digital distance transforms in four dimensions. *Discrete Applied Mathematics*, 125(1):161–176, 2003.

[3] D. Coeurjolly, F. Flin, O. Teytaud, and L. Tougne. Multigrid convergence and surface area estimation. In *Theoretical Foundations of Computer Vision*, volume 2616 of *LNCS*, pages 101–119. Springer-Verlag, 2003.

[4] D. Coeurjolly and R. Klette. A comparative evaluation of length estimators. In *Proceedings of the 16th International Conference on Pattern Recognition (ICPR)*, pages IV: 330–334, Quebec, 2002. IEEE Computer Science.

[5] H. Freeman. Boundary encoding and processing. In B. S. Lipkin and A. Rosenfeld, editors, *Picture Processing and Psychopictorics*, pages 241–266, New York, 1970. Academic Press.

[6] P. P. Jonker. Skeletons in N dimensions using shape primitives. *Pattern Recognition Lett*, 23(4):677–686, 2002.

[7] Y. Kenmochi and R. Klette. Surface area estimation for digitized regular solids. In L. J. Latecki, R. A. Melter, D. M. Mount, and A. Y. Wu, editors, *Vision Geometry IX*, pages 100–111. Proc. SPIE 4117, 2000.

[8] R. Klette. Multigrid convergence of geometric features. In G. Bertrand, A. Imiya, and R. Klette, editors, *Digital and Image Geometry*, volume 2243 of *LNCS*, pages 314–333. Springer-Verlag, 2001.

[9] R. Klette and H. J. Sun. Digital planar segment based polyhedrization for surface area estimation. In C. Arcelli, L. P. Cordella, and G. Sanniti di Baja, editors, *Visual Form 2001*, volume 2059 of *LNCS*, pages 356–366, Capri, Italy, 2001. Springer-Verlag.

[10] Z. Kulpa. Area and perimeter measurement of blobs in discrete binary pictures. *Computer Graphics and Image Processing*, 6:434–454, 1977.

[11] J. Lindblad. Surface area estimation of digitized planes using weighted local configurations. In I. Nyström, G. Sanniti di Baja, and S. Svensson, editors, *DGCI*, volume 2886 of *LNCS*, pages 348–357, Naples, Italy, 2003. Springer-Verlag.

[12] J. Lindblad. Surface area estimation of digitized 3D objects using weighted local configurations. *Image and Vision Computing*, 23(2):111–122, 2005. Special issue on Discrete Geometry for Computer Imagery.

[13] W. E. Lorensen and H. E. Cline. Marching Cubes: A high resolution 3D surface construction algorithm. In *Proceedings of the 14th ACM SIGGRAPH on Computer Graphics*, volume 21, pages 163–169, 1987.

[14] D. Proffit and D. Rosen. Metrication errors and coding efficiency of chain-encoding schemes for the representation of lines and edges. *Computer Graphics and Image Processing*, 10:318–332, 1979.

[15] J. C. Roberts and S. Hill. Piecewise linear hypersurfaces using the marching cubes algorithm. In R. F. Erbacher and A. Pang, editors, *Visual Data Exploration and Analysis VI, Proceedings of SPIE*, volume 3643, pages 170–181, 1999.

# Appendix

The expected number of occurrences of the 14 planar 4D-configurations when a column of 4-cubes is intersected by hyperplanes of different slopes. Cases which are not mentioned for a specific slope do not appear. Cases $c_1$ and $c_2$ appear with the same frequency for all slopes; $E_1 = 2w'_z$, $E_2 = 2(w'_y - w'_z)$.

---

1a: $w'_x + w'_y + w'_z \leq 1$, $\quad w'_x \leq w'_y + w'_z$.

| | | | |
|---|---|---|---|
| $E_3 = 2(w'_x - w'_y)$, | $E_5 = 2(w'_y + w'_z - w'_x)$, | $E_6 = 2(w'_x - w'_y)$, | |
| $E_7 = 2(w'_y - w'_z)$, | $E_8 = 2w'_z$, | $E_9 = 1 - w'_x - w'_y - w'_z$. | |

---

1b: $w'_x + w'_y + w'_z \leq 1$, $\quad w'_x > w'_y + w'_z$.

| | | | |
|---|---|---|---|
| $E_3 = 2w'_z$, | $E_4 = 2(w'_x - w'_y - w'_z)$, | $E_6 = 2w'_z$, | |
| $E_7 = 2(w'_y - w'_z)$, | $E_8 = 2w'_z$, | $E_9 = 1 - w'_x - w'_y - w'_z$. | |

---

2a: $w'_x + w'_y + w'_z > 1$, $\quad w'_x + w'_y \leq 1$, $\quad w'_x \leq w'_y + w'_z$.

| | | | |
|---|---|---|---|
| $E_3 = 2(w'_x - w'_y)$, | $E_5 = 2(w'_y + w'_z - w'_x)$, | $E_6 = 2(w'_x - w'_y)$, | |
| $E_7 = 2(w'_y - w'_z)$, | $E_8 = 2(1 - w'_x - w'_y)$, | $E_{13} = w'_x + w'_y + w'_z - 1$. | |

---

2b: $w'_x + w'_y + w'_z > 1$, $\quad w'_x + w'_y \leq 1$, $\quad w'_x > w'_y + w'_z$.

| | | | |
|---|---|---|---|
| $E_3 = 2w'_z$, | $E_4 = 2(w'_x - w'_y - w'_z)$, | $E_6 = 2w'_z$, | |
| $E_7 = 2(w'_y - w'_z)$, | $E_8 = 2(1 - w'_x - w'_y)$, | $E_{13} = w'_x + w'_y + w'_z - 1$. | |

---

3a: $w'_x + w'_y > 1$, $\quad w'_x + w'_z \leq 1$, $\quad w'_x \leq w'_y + w'_z$, $\quad w'_x + w'_y - 1 \leq w'_z$.

| | | | |
|---|---|---|---|
| $E_3 = 2(w'_x - w'_y)$, | $E_5 = 2(w'_y + w'_z - w'_x)$, | $E_6 = 2(w'_x - w'_y)$, | |
| $E_7 = 2(1 - w'_x - w'_z)$, | $E_{12} = 2(w'_x + w'_y - 1)$, | $E_{13} = 1 + w'_z - w'_x - w'_y$. | |

---

3b: $w'_x + w'_y > 1$, $\quad w'_x + w'_z \leq 1$, $\quad w'_x > w'_y + w'_z$, $\quad w'_x + w'_y - 1 \leq w'_z$.

| | | | |
|---|---|---|---|
| $E_3 = 2w'_z$, | $E_4 = 2(w'_x - w'_y - w'_z)$, | $E_6 = 2w'_z$, | |
| $E_7 = 2(1 - w'_x - w'_z)$, | $E_{12} = 2(w'_x + w'_y - 1)$, | $E_{13} = 1 + w'_z - w'_x - w'_y$. | |

---

3c: $w'_x + w'_y > 1$, $\quad w'_x + w'_z \leq 1$, $\quad w'_x \leq w'_y + w'_z$, $\quad w'_x + w'_y - 1 > w'_z$.

| | | | |
|---|---|---|---|
| $E_3 = 2(w'_x - w'_y)$, | $E_5 = 2(w'_y + w'_z - w'_x)$, | $E_6 = 2(w'_x - w'_y)$, | |
| $E_7 = 2(1 - w'_x - w'_z)$, | $E_{12} = 2w'_z$, | $E_{14} = w'_x + w'_y - 1 - w'_z$. | |

---

3d: $w'_x + w'_y > 1$, $\quad w'_x + w'_z \leq 1$, $\quad w'_x > w'_y + w'_z$, $\quad w'_x + w'_y - 1 > w'_z$.

| | | | |
|---|---|---|---|
| $E_3 = 2w'_z$, | $E_4 = 2(w'_x - w'_y - w'_z)$, | $E_6 = 2w'_z$, | |
| $E_7 = 2(1 - w'_x - w'_z)$, | $E_{12} = 2w'_z$, | $E_{14} = w'_x + w'_y - 1 - w'_z$. | |

---

4a: $w'_x + w'_z > 1$, $\quad w'_y + w'_z \leq 1$, $\quad w'_x \leq w'_y + w'_z$, $\quad w'_x + w'_y - 1 \leq w'_z$.

| | | | |
|---|---|---|---|
| $E_3 = 2(w'_x - w'_y)$, | $E_5 = 2(w'_y + w'_z - w'_x)$, | $E_6 = 2(1 - w'_y - w'_z)$, | |
| $E_{11} = 2(w'_x + w'_z - 1)$, | $E_{12} = 2(w'_y - w'_z)$, | $E_{13} = 1 + w'_z - w'_x - w'_y$. | |

---

4b: $w'_x + w'_z > 1$, $\quad w'_y + w'_z \leq 1$, $\quad w'_x > w'_y + w'_z$, $\quad w'_x + w'_y - 1 \leq w'_z$.

| | | | |
|---|---|---|---|
| $E_3 = 2w'_z$, | $E_4 = 2(w'_x - w'_y - w'_z)$, | $E_6 = 2(1 - w'_x)$, | |
| $E_{11} = 2(w'_x + w'_z - 1)$, | $E_{12} = 2(w'_y - w'_z)$, | $E_{13} = 1 + w'_z - w'_x - w'_y$. | |

4c: $w'_x + w'_z > 1, \quad w'_y + w'_z \leq 1, \quad w'_x \leq w'_y + w'_z, \quad w'_x + w'_y - 1 > w'_z.$

$E_3 = 2(w'_x - w'_y), \qquad E_5 = 2(w'_y + w'_z - w'_x), \qquad E_6 = 2(1 - w'_y - w'_z),$
$E_{11} = 2(w'_x + w'_z - 1), \quad E_{12} = 2(1 - w'_x), \qquad E_{14} = w'_x + w'_y - 1 - w'_z.$

4d: $w'_x + w'_z > 1, \quad w'_y + w'_z \leq 1, \quad w'_x > w'_y + w'_z, \quad w'_x + w'_y - 1 > w'_z.$

$E_3 = 2w'_z, \qquad\qquad E_4 = 2(w'_x - w'_y - w'_z), \quad E_6 = 2(1 - w'_x),$
$E_{11} = 2(w'_x + w'_z - 1), \quad E_{12} = 2(1 - w'_x), \qquad E_{14} = w'_x + w'_y - 1 - w'_z.$

5a: $w'_y + w'_z > 1, \quad w'_x + w'_y - 1 \leq w'_z.$

$E_3 = 2(w'_x - w'_y), \qquad E_5 = 2(1 - w'_x), \qquad E_{10} = 2(w'_y + w'_z - 1),$
$E_{11} = 2(w'_x - w'_y), \qquad E_{12} = 2(w'_y - w'_z), \qquad E_{13} = 1 + w'_z - w'_x - w'_y.$

5c: $w'_y + w'_z > 1, \quad w'_x + w'_y - 1 > w'_z.$

$E_3 = 2(w'_x - w'_y), \qquad E_5 = 2(1 - w'_x), \qquad E_{10} = 2(w'_y + w'_z - 1),$
$E_{11} = 2(w'_x - w'_y), \qquad E_{12} = 2(1 - w'_x), \qquad E_{14} = w'_x + w'_y - 1 - w'_z.$

# Rectification of the Chordal Axis Transform and a New Criterion for Shape Decomposition

Lakshman Prasad

Space and Remote Sensing Sciences Group (ISR-2),
International, Space, and Response Division,
Los Alamos National Laboratory,
Los Alamos, NM 87545, USA.
prasad@lanl.gov

**Abstract.** In an earlier work we proposed the chordal axis transform (CAT) as a more useful alternative to the medial axis transform (MAT) for obtaining skeletons of discrete shapes. Since then, the CAT has benefited various applications in 2D and 3D shape analysis. In this paper, we revisit the CAT to address its deficiencies that are artifacts of the underlying constrained Delaunay triangulation (CDT). We introduce a valuation on the internal edges of a discrete shape's CDT based on a concept of approximate co-circularity. This valuation provides a basis for suppression of the role of certain edges in the construction of the CAT skeleton. The result is a rectified CAT skeleton that has smoother branches as well as branch points of varying degrees, unlike the original CAT skeleton whose branches exhibit oscillations in tapered sections of shapes and allows only degree-3 branch points. Additionally, the valuation leads to a new criterion for parsing shapes into visually salient parts that closely resemble the empirical decompositions of shapes by human subjects as recorded in experiments by M. Singh, G. Seyranian, and D. Hoffman.

**Keywords:** Shape, Delaunay triangulation, chordal axis transform, medial axis, skeleton, shape decomposition, morphology, co-circularity, shape graph, grouping, chord strength.

## 1 Introduction

The skeleton of a shape is an important descriptor that provides structural information about the shape. Skeletons are used to compare shapes, identify shape parts, and, in case of thin objects such as textual characters, even represent the shapes themselves. Blum [1] defined the skeleton of a two dimensional shape with a continuous closed contour as the locus of centers of maximal discs (i.e., discs touching the shape contour at two or more points) interior to the shape, with each center attributed the radius of the corresponding maximal disc. This definition of a shape's skeleton is known as the medial axis transform (MAT) of the shape. While the MAT is an elegant characterization of the skeleton of a shape with a continuous boundary, it has proved to be difficult to use as a practical tool to analyze shapes. Indeed, for example, minor oscillations in shape contours due to insignificant features or noise result in skeletal

E. Andres et al. (Eds.): DGCI 2005, LNCS, pp. 263–275, 2005.

branches that are not easy to isolate, a skeletal feature may be spatially far-removed from the contour feature it represents, and a skeleton part may greatly exaggerate or diminish the importance of the contour feature that gave rise to it (Figs. 1 & 2). The medial axis transform is not defined for shapes specified by discretely sampled contours, as typically encountered in digital imagery. Several extensions of the MAT to discrete shapes have been formulated using pixel morphology [3] and geometry [4, 5]. These methods, however, require uniform or well-sampled representations of the shape to yield satisfactory skeletons (Fig. 8d).

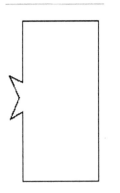 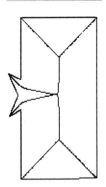

**Fig. 1.** Rectangle with boundary feature          **Fig. 2.** MAT skeleton exaggerating feature

## 2 Background

In earlier works [6, 7, 8] we proposed the chordal axis transform (CAT) as a more useable and stable definition of the skeleton of a shape that is robust to sparse and uneven sampling of shape boundary. Since then it has gained currency among researchers in the area of 2D and 3D shape analysis and modeling [12-15]. In this section, we will review the CAT, its strengths, and drawbacks to set the context for this paper.

**Definition 1:** A *maximal chord of tangency* (Fig. 3) connects two points of tangency of a maximal disc inscribed in a shape such that at least one of the two arcs of the maximal disc's bounding circle subtended by the chord is free of points of tangency with the shape's boundary.

**Definition 2:** The *Chordal Axis Transform* (CAT) of a planar shape is the set of all ordered pairs $(p,\delta)$, where $p$ and $\delta$ are either the midpoint and half the length, respectively, of a maximal chord of tangency, or the center and radius, respectively, of a maximal disc with three or more maximal chords of tangency.

Although the definition of the CAT appears to be a variation of that of the MAT, there are important differences between the two transforms. First, the CAT, as defined, yields a piecewise smooth disconnected protoskeleton (Fig. 5). By joining the midpoints of the maximal chords of a maximal disc with three or more chords to the center of the maximal disc if the center lies within the polygon determined by the chords, or to the center of the longest chord otherwise, we obtain a connected skeleton of the shape (Fig. 6).

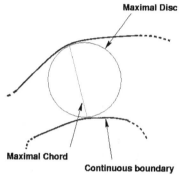

**Fig. 3.** A maximal disc and the associated maximal chord of tangency of a shape with continuous boundary

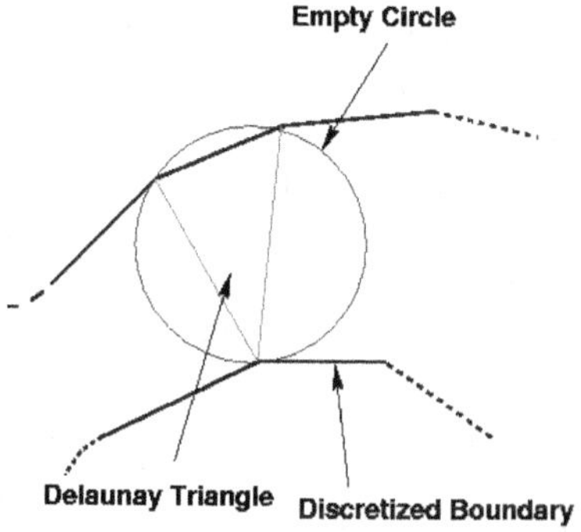

**Fig. 4.** An empty circle and the associated Delaunay triangle of a shape with discrete boundary

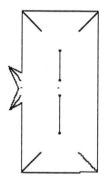

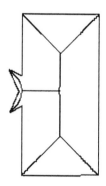

**Fig. 5.** CAT proto-skeleton of shape in Fig. 1

**Fig. 6.** Connected CAT Skeleton

Second, the CAT can be stably defined for a shape whose boundary is discrete (i.e., specified as sequences of points separated in space) (Fig. 8). This is done by replacing maximal discs by empty circles that pass through three or more points of the shape's discrete boundary such that no circle contains a boundary point in its interior that is visible to two boundary points lying on the circle (Two vertices $u$ and $v$ of a simple polygon are *visible* to each other, if the line segment joining $u$ and $v$ does not intersect the exterior of the polygon). Each such empty circle identifies a triangle whose edges lying in the shape's interior replace maximal chords of tangency in the discrete version of the CAT (Fig. 4). The triangles so formed are indeed the Delaunay triangles of a constrained Delaunay triangulation (CDT) [2] of the shape's interior. It is worth noting here that this extension of the CAT to discrete shapes is natural from the point of view of constructing skeletons. This is because the constrained Delaunay triangulation is the geometric dual of the generalized Voronoi axis [2] of the contour point set. Indeed, the MAT is essentially the Voronoi skeleton of a shape. In using the dual of the Voronoi axis, we can define a more robust and manipulable skeleton than the MAT that applies to discrete shapes whose boundaries are sparsely and unevenly sampled. We can also ensure strong invertibility of the skeleton to recover the shape [8]. The CDT of a shape's interior gives rise to three kinds of triangles, namely *Junction* triangles (J) that have all their edges inside the shape and signify shape bifurcations, *Sleeve* triangles (S) that have one edge in common with the shape boundary and signify shape prolongations, and *Terminal* triangles (T) that have two edges in common with the shape boundary and signify shape terminations. The connected CAT skeleton for a discretized shape is obtained from its CDT by i) joining the midpoints of the internal edges of each S-triangle by a line segment, ii) joining the midpoints of the internal edges of each J-triangle to its circumcenter if the triangle is acute, or to the midpoint of its longest side if it is obtuse (Figs. 7, 8). These localized constructions ensure that the CAT skeleton does not cross the boundary of the shape irrespective of the sparsity of sampling. Henceforth we will restrict ourselves to the structure of the CAT skeleton in the rest of the paper and direct the interested reader to [8, 9] for other details and implications of the CAT.

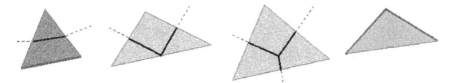

**Fig. 7.** Construction of skeletal segments in the various types of triangles in the CDT of a shape: from left to right, skeleton segments in a sleeve triangle, an obtuse junction triangle, an acute junction triangle, and a terminal triangle

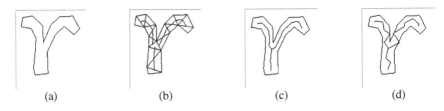

(a)          (b)          (c)          (d)

**Fig. 8.** Construction of the CAT skeleton of a sparsely and unevenly sampled shape and comparison with its discrete MAT skeleton: (a) A discrete shape, (b) CDT and construction of sleeve skeletal segments, (c) Connected skeleton after construction of skeletal segment in the junction triangle, (d) Discrete MAT skeleton obtained by connecting adjacent Voronoi vertices (circumcenters of triangles) of the contour points. The comparison shows the stability of the CAT skeleton over that of the MAT of a shape with sparsely sampled contour

The CAT skeleton of a discrete shape is robust in the face of sparse and irregular presentations of the shape boundary. It also allows easy excision of insignificant and noisy features via a simple pruning criterion [6, 7, 8]. Finally, it enables parts-based decomposition of shapes into structurally meaningful components [8, 9].

## 3  Drawbacks of the Discrete CAT Skeleton

However, the CAT skeleton, as defined above, has certain structural deficiencies (which are also present in discrete realizations of the MAT skeleton) that warrant rectification. The CAT skeleton exhibits oscillations through shape regions that are tapered (Fig. 9). The CAT skeleton allows only branches of degree three to represent shape ramifications even when higher degree branches are more natural to represent them (Fig. 10). These deficiencies are artifacts of giving equally important roles to all chords (i.e., internal triangle edges of the shape's CDT) in constructing the CAT skeleton. Indeed, in a previous work [8], we considered special cases where more than three points on a shape boundary are co-circular with respect to an empty circle. We noted that the triangle edges that form the chords of the polygon determined by the co-circular points are not uniquely defined (i.e., any triangulation of the interior of the co-circular polygon will be consistent with the Delaunay criterion of triangulation.) We proposed that the restriction of the shape's skeleton to such a polygon be constructed by joining all the midpoints of the polygon's edges that are internal to the shape to the circumcenter of the polygon or, if the latter falls outside the polygon, to the midpoint of longest edge of the polygon that does not lie on the shape boundary. In effect, we discarded internal shape edges that are common to two co-circular triangles in the CDT of a shape. We will generalize this notion of co-circularity to define a valuation on the chords of a shape that will help filter chords which are common edges of nearly co-circular triangles. The motivation for this is to prevent common edges of nearly co-circular triangles from participating in the construction of the skeleton. This will greatly reduce skeleton oscillations in tapered regions of shapes. Indeed, consider two adjacent, nearly co-circular, sleeve triangles. Unless the external edges of the triangles are parallel, the midpoints of the internal edges of the two triangles will not lie on a straight line, thus producing an oscillation in the skeleton. If the common internal edge of this triangle pair is discounted, then the skeleton of the polygon determined by the triangle pair is given by the line segment joining the midpoints of the remaining two internal edges, thus locally rectifying the CAT skeleton of the shape (Fig. 12).

## 4  A Measure of Chord Strength

We introduce a valuation on the chords of a discrete shape's CAT. The chords of the CAT are edges of the CDT of the shape that are shared by two triangles. Let the angles opposite a chord $c$ in its two flanking triangles be $\theta$ and $\varphi$. We then define the strength of $c$ by

$$S(c) = 1-( \theta + \varphi)/\pi. \tag{1}$$

Thus $S$ is a valuation on the set of all chords of a shape, with values in the half-open interval $[0, 1)$. This is because the empty circle condition of the CDT ensures that the sum of the angles across from a chord does not exceed $\pi$ radians. We will refer to this valuation as the chord strength. $S$ takes the value 0 on chords that are flanked by co-circular triangles. This observation is based upon an elementary fact of Euclidean geometry that the opposite angles of a cyclic quadrilateral add up to $\pi$ radians. Hence, the smaller the strength of a chord, the closer its flanking triangles are to being co-circular, and vice versa. We are now ready to suppress the chords of low strength in the construction of the CAT skeleton. In what follows, we will address shapes without holes to keep the discussion simple. The techniques described can easily be extended to shapes with holes as well.

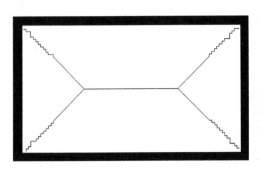

**Fig. 9.** CAT skeleton showing oscillations in tapered regions of a rectangular shape

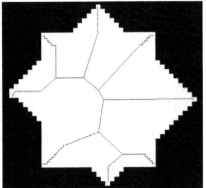

**Fig. 10.** CAT skeleton showing degree three branching for a star-like shape

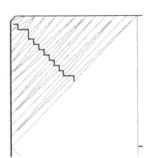

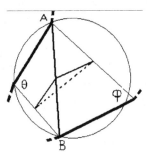

**Fig. 11.** Corner detail of Fig 9. showing CDT and skeleton oscillations

**Fig. 12.** Rectification of skeleton (dotted line) by suppressing weak chord AB

**Fig. 13.** Skeleton in Fig. 11 rectified by suppressing weak chords

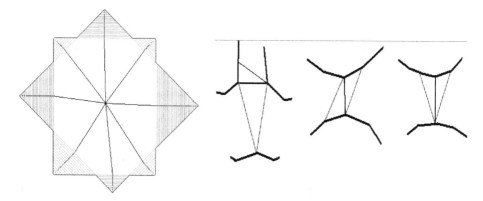

**Fig. 14.** Rectified CAT skeleton of shape in Fig. 10, with a degree 8 branch point

**Fig. 15.** Schematic of CDT in the neighborhood of chords at junction of part with shape (left) and shape necks (center and right)

## 5 Chord Strength Profile Extrema and Shape Decomposition

The directed chords of a shape can be ordered by traversing the boundary of the shape and recording in sequence the chords encountered. For instance, at boundary point **p,** a directed chord **pq** joining **p** to another boundary point **q** will be recorded, as will the chord **qp** on arrival at point **q**. Thus each chord will be visited twice in traversing the shape contour. We call the plot of chord strengths versus chord numbers the chord strength profile (CSP). The CSP is a circular function in that the first and last chords in this enumeration are neighbors. Using the CSP, we select chords whose strengths are strictly greater than that of at least one neighbor and greater than or equal to the strengths of both neighbors. More precisely, the chord $c_i$ is selected if and only if

$$[ (S(c_i) > S(c_{i-1})) \ \& \ (S(c_i) \geq S(c_{i+1})) ] \ | \ [ (S(c_i) \geq S(c_{i-1})) \ \& \ (S(c_i) > S(c_{i+1})) ] \tag{2}$$

i.e., the selected chords' strengths are at least *one-sided* local maxima of the CSP. Only these selected chords will be allowed to play a role in determining the skeleton of the shape.

### 5.1 Polygonal Decomposition of Shapes

Next, we construct a triangle grouping graph whose vertices are the triangles of the CDT of the shape, and with an edge between two vertices corresponding to adjacent triangles if and only if their common edge is a suppressed chord. A connected component analysis via a depth-first-search traversal of this grouping graph yields a polygonal decomposition of the shape, with each polygon comprised of pairwise adjacent, approximately co-circular triangles belonging to the same connected component. We refer to such polygons decomposing a shape as Delaunay polygons. As in the case of the CDT of a shape, these polygons can be classified into terminal, sleeve, and junction Delaunay polygons depending on whether they have one, two, or more chords, respectively, among their bounding edges. Again, as in the case of triangles, skeletal segments are constructed in each Delaunay polygon to obtain a skeleton of the shape. The midpoints of chords of a sleeve Delaunay polygon are joined together to yield a sleeve skeletal segment. In the case of a junction Delaunay polygon, we define its *barycenter* as the weighted average of the midpoints of its chords, where the weight of each chord's midpoint is the (normalized) length of the chord. The midpoints of the chords are then joined to this barycenter to yield a skeletal segment of the junction Delaunay polygon. The collection of all the skeletal segments, with their adjacencies inherited from the adjacencies of the parent Delaunay polygons, form a connected rectified CAT skeleton of the shape. The suppression of weak chords remedies not only the skeletal oscillations (Figs. 9, 11, 12, 13), but also the purely degree-3 branch points (Figs. 10, 14) forced by the CDT in the original CAT skeleton prior to rectification.

## 5.2  Skeleton Pruning

A pruning criterion for excising skeleton segments corresponding to insignificant shape features is easily specified. For each chord of a junction Delaunay polygon, the length of the shape boundary arc subtended by it (and not including the polygon,) as a fraction of total shape boundary length is computed. In our experiments we have found that this fraction provides a satisfactory measure of the saliency of the part bounded by the chord and the boundary arc that is also efficient to compute. If this fraction falls below a predetermined threshold, the chord is an external boundary segment of a pruned shape. Accordingly, a new barycenter of the junction polygon is computed with the remaining chords taken into consideration if the remaining chords number greater than two. Otherwise, the junction Delaunay polygon is demoted to a sleeve or terminal Delaunay polygon and appropriate skeleton segments are constructed anew. Other pruning criteria such as excising parts based on the ratio of distance of furthest point of part from subtending chord to chord length, or alternately, to the distance of chord from barycenter, may also be implemented at additional computational cost.

## 5.3  Visually Salient Shape Decomposition

The selection criterion for a shape's chords, specified in condition (**2**), may be applied repeatedly to the CSPs of successive generations of selected chords, yielding chords whose strengths are higher order maxima in the original CSP (Fig. 18). These strong chords correspond to cuts of the shape into visually salient parts. They typically occur at the intersection of limbs with the shapes and necks of the shape where there is a narrowing of the shape girth. The reason for this is intuitively captured in Fig. 15 where the structure of the shape around the chord at part junctions and necks forces the sum of the angles opposite the chord in the flanking triangles of the shape's CDT to be smaller, and hence the chord to have greater strength than in other places of the shape. The strength of a chord weakens with increase in its length for the same boundary geometry in the vicinity of its endpoints. Interestingly, this behavior is consistent with the *shortcut rule* of parsing shapes using shortest cuts, proposed in

[11] based on experiments in human vision. However, in order to ensure the saliency of the parts resulting from cuts induced by the maxima of the CSP, we bias the CSP by multiplying the strength of each chord with the ratio of the (smaller) arclength of the shape subtended at the chord to the length of the chord. This has the effect of enhancing the strength of chords that subtend visually salient parts of the shape, while diminishing the strengths of chords that subtend minor or noisy protuberances of the shape. It is important to note that we do not process the boundaries of shapes by removing noise or smoothing.

Thus, the CSP provides not only a means of rectifying the CAT skeleton (Fig. 16), but also a means of decomposing shapes into visually meaningful parts (Fig. 17). A well known work in this area is that of K. Siddiqi et al [10]. Their approach to shape decomposition is also motivated by considerations of visual saliency and yields good results. In contrast, our approach addresses obtaining good shape skeletons as well as good shape decompositions in a unified manner by proposing a single criterion for solving both problems. The CSP maps the two-dimensional problem of shape analysis to the analysis of a one-dimensional function's extrema. This opens up a host of well known techniques for analyzing 1-D signals such as wavelet transforms to obtain hierarchical decompositions of shapes. We note that the shape decompositions obtained by our method closely resembles the outcomes of experiments in shape decomposition by M. Singh, G. Seyranian, and D. Hoffman [11] using human subjects. We believe that the property of strong chords of CDTs to yield visually meaningful decomposition of shapes has the potential to be developed into a useful and elegant tool in investigating and understanding shapes.

**Fig. 16.** Shapes and their rectified CAT skeletons based on 1$^{st}$ order maxima of their CSPs

**Fig. 17.** Shapes and their decomposition based on higher order (order 4) maxima of their CSP. Adjacent parts are shown in alternating shades

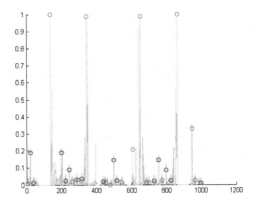

**Fig. 18.** The normalized chord strength profile vs. chord visit number of the top left shape in Fig. 17. indicating 4th order maxima (circled). The four tallest peaks correspond to the two horizontal cuts (two peaks for each cut, as each chord is visited twice) decomposing the shape into three salient parts. The shorter circled peaks correspond to cuts across small noisy protuberances

# 6    Shapes with Holes

In the case of shapes with $g$ holes, one can define $g+1$ CSPs, one for each hole contour, and one for the outer shape contour. The CSPs may then be analyzed individually for $n^{th}$ order maxima just as in the case of a single CSP of a shape without holes. In the case of a CSP of a shape with holes, a chord will be encountered only once in the CSP if it joins two different contours and twice if it joins two points on the same contour. A chord is retained as a strong chord of the shape if and only if it is a strong chord ($n^{th}$ order maximum) of at least one CSP. Based on the strong chords obtained thus, one can construct rectified CAT skeletons and obtain visually salient decompositions of shapes with holes as well.

Alternatively, one can formulate the problem of finding strong chords of a discrete shape on the dual graph of its CDT, where the vertices of the graph represent triangles, and with an edge between each pair of vertices corresponding to adjacent triangles in the CDT. Each graph edge can be weighted with the strength of the chord that separates the triangles whose representative vertices the edge connects. Edge weight maxima can be computed along minimal cycles corresponding to hole contours and the cycle corresponding to the outer contour. One can then group vertices that are path connected by edges whose weights correspond to non-maximal chord strengths to obtain a new graph whose vertices correspond to Delaunay polygons described before. Iterations of this process yield higher order maxima and hierarchical shape decompositions.

# 7    Computational Issues

The constrained Delaunay triangulation of a planar shape specified by $n$ points runs in o($n\log n$) time. The construction of the CSP and detection of maxima is linear in the number of edges which are $2n-3+3g$ in number, where $g<n/3$ is the number of holes. The connected component analysis on the grouping graph is linear in the number of vertices (representing the triangles of the shape's CDT which are $n-2+2g$ in number). Shape pruning is, again, linear in the number of edges. Finally, skeletonization is also linear in the number of edges. Thus the overall efficiency of our shape decomposition and skeletonization scheme is high and amenable to real-time applications.

# 8    Conclusion

In this paper we have demonstrated a property of the chords of constrained Delaunay triangulations of 2-D shapes that induces a hierarchy of visually salient decompositions by defining a valuation on the chords. This valuation, which we call chord strength, along with the ordering induced by the shape boundary on the chords, maps the two-dimensional problem of shape decomposition into a one-dimensional problem of analyzing a function's extrema. We have briefly demonstrated how one can obtain rectified shape skeletons as well as visually meaningful shape decompositions using successive selection of strong chords. The search for chords that best decompose a shape using the extrema of the chord strength profile function introduced here can be improved upon significantly and is part of our ongoing work in shape analysis. The unified approach provided by our method to both skeletonization and decomposition of shapes is the key contribution of this paper.

In three dimensions, one can define a valuation characterizing approximate co-sphericity of two face-adjacent Delaunay tetrahedra by defining the strength of their interface as the ratio of the distance between their circumcenters to the mean of their circumsphere diameters. However, the 3-D case is more complex. For instance, two adjacent co-spherical tetrahedra need not form a convex polytope, and therefore their interface may be an important structural partition. Thus, additional conditions will be required in order to extend our ideas presented here to three-dimensional shapes.

## Acknowledgement

This work has been fully supported by the U. S. DOE under contract No. W-7405-ENG-36 through an LDRD ER (#20030162) research grant. We would like to thank the reviewers for their helpful comments.

## Reference

1. H. Blum.: A Transformation for Extracting New Descriptors of Shape. Symp. Models for Speech and Visual Form Weiant Whaten-Dunn (Ed) MIT Press (1967)
2. J. E. Goodman, J. O'Rourke. (Eds.): Handbook of Discrete and Computational Geometry, CRC Press (1997)
3. J. R. Parker: Algorithms for Image Processing and Computer Vision, John Wiley & Sons (1997)
4. R.L. Ogniewicz: Skeleton-space: A multiscale shape description combining region and boundary information," in Proc. IEEE CVPR, Seattle, WA, June 1994
5. D. Attali, A. Montanvert.: Computing and Simplifying 2D and 3D Continuous Skeletons, Computer Vision and Image Understanding, Vol. 67, Sept. 1997
6. L. Prasad: Morphological Analysis of Shapes, CNLS Newsletter, No. 139, July 1997, LALP-97-010-139, Center for Nonlinear Studies, Los Alamos National Laboratory
7. L. Prasad, R. L. Rao & G. Zweig: Skeletonization of shapes using Delaunay triangulations, Fifth SIAM Conference on Geometric Design, Nashville, TN, Nov. 1997
8. L. Prasad, R. L. Rao.: A Geometric Transform for Shape Feature Extraction, Proc. SPIE, vol. 4117, Vision Geometry IX (2000)
9. L. Prasad, A. Skourikhine, B. Schlei.: Feature-based Syntactic and Metric Shape Recognition, Proc. SPIE, vol. 4117, Vision Geometry IX (2000)
10. K. Siddiqi, B. B. Kimia: Parts of Visual Form: Computational Aspects, IEEE Trans. on PAMI, vol. 17, No. 3, March 1995
11. M. Singh, G. Seyranian, D. Hoffman: Parsing Silhouettes: The Short-cut Rule, Perception and Psychophysics, 61, 636-660 (1999)
12. T. Igarashi, S. Matsuoka, H. Tanaka.: Teddy: A Sketching Interface for 3D Freeform Design, ACM SIGGRAPH'99, 409-416, Los Angeles, CA (1999)

13. P. Felzenszwalb.: Representation and detection of deformable shapes. Proc. CVPR, vol.1, pp. 102—108 (2003)
14. J. Arvo, K. Novins : Smart Text: A Synthesis of Recognition and Morphing *AAAI Spring Symposium on Smart Graphics*, Stanford, California, pp 140-147, March 2000
15. S. Yamakawa, K. Shimada.: Quad-Layer: Layered Quadrilateral Meshing of Narrow Two-Dimensional Domains by Bubble Packing and Chordal Axis Transform, ASME/DETC/DAC (2001)

# Generalized Functionality for Arithmetic Discrete Planes

Valerie Berthé, Christophe Fiorio, and Damien Jamet

LIRMM, Université Montpellier II,
161 rue Ada, 34392 Montpellier Cedex 5 - France
{berthe, jamet, fiorio}@lirmm.fr

**Abstract.** The discrete plane $\mathfrak{P}(a, b, c, \mu, \omega)$ is the set of points $(x, y, z) \in \mathbb{Z}^3$ satisfying $0 \leq ax+by+cz+\mu < \omega$. In the case $\omega = \max(|a|, |b|, |c|)$, the discrete plane is said naive and is well-known to be functional on a coordinate plane. The aim of our paper is to extend the notion of functionality to a larger family of arithmetic discrete planes by introducing a suitable orthogonal projection direction $(\alpha, \beta, \gamma)$ satisfying $\alpha a + \beta b + \gamma c = \omega$. We then apply this functionality property to the enumeration of some local configurations, that is, the $(m, n)$-cubes such as introduced in [VC99].

**Keywords:** digital planes; arithmetic planes; local configurations; functionality of discrete planes.

The discrete plane $\mathfrak{P}(a, b, c, \mu, \omega)$ is the set of integer points $(x, y, z) \in \mathbb{Z}^3$ satisfying $0 \leq ax + by + cz + \mu < \omega$. In the case $\omega = \max(|a|, |b|, |c|)$, the discrete plane is said naive and is well-known to be functional on one of the coordinate planes, that is, for any point of $P$ of this coordinate plane, there exists a unique point in the discrete plane obtained by adding to $P$ a third coordinate. Naive planes have been widely studied, see for instance [Rev91, DRR94, DR95, AAS97, VC97, Col02, BB02].

The present paper extends the notion of functionality for naive discrete planes to a larger family of arithmetic discrete planes. For that purpose, instead of projecting on a coordinate space, we introduce a suitable orthogonal projection on a plane along a direction $(\alpha, \beta, \gamma)$, in some sense dual to the normal vector of the discrete plane $\mathfrak{P}(a, b, c, \mu, \omega)$, that is, $\alpha a + \beta b + \gamma c = \omega$, so that the projection of $\mathbb{Z}^3$ and the points of the discrete plane are in one-to-one correspondence. One interest of the notion of functionality is that it reduces a three-dimensional problem to a two-dimensional one, allowing a better understanding of the combinatorial and geometric properties of discrete planes. We thus apply this functionality property to the enumeration of some local configurations, the $(m, n)$-cubes, for a large family of arithmetic discrete planes, following the approach of [Vui99, BV01].

For clarity issues, we have chosen to work here in a three-dimensional space but all the results and methods presented extend in a natural way to $\mathbb{R}^n$, with $n \geq 2$, as well as to arithmetic discrete lines.

E. Andres et al. (Eds.): DGCI 2005, LNCS 3429, pp. 276–286, 2005.
© Springer-Verlag Berlin Heidelberg 2005

# 1    Basic Notions and Arithmetic Discrete Planes

Let $(a, b, c) \in \mathbb{R}^3$, $\mu \in \mathbb{R}$ and $\omega \in \mathbb{R}_+^\star$; the *arithmetic discrete plane* $\mathfrak{P}(a, b, c, \mu, \omega)$ is defined as follows:

$$\mathfrak{P}(a, b, c, \mu, \omega) = \{(x, y, z) \in \mathbb{Z}^3 \mid 0 \leq ax + by + cz + \mu < \omega\}.$$

Moreover, if $\omega = \max\{|a|, |b|, |c|\}$ (resp. $\omega = |a| + |b| + |c|$) then $\mathfrak{P}(a, b, c, \mu, \omega)$ is said to be *naive* (resp. *standard*).

In the present paper, in order to simplify the notation and to facilitate the generalization of our results to higher dimensions, we use a vector-based representation. Let $\{\overrightarrow{e_1}, \overrightarrow{e_2}, \overrightarrow{e_3}\}$ be the canonical basis of the $\mathbb{R}$-vector space $\mathbb{R}^3$. Let $\overrightarrow{v}$ and $\overrightarrow{v'}$ be two vectors of $\mathbb{R}^3$. The notation $(\overrightarrow{v}, \overrightarrow{v'})$ stands for the usual scalar product in $\mathbb{R}^3$. Let $i \in \{1, 2, 3\}$, we denote by $v_i = (\overrightarrow{v}, \overrightarrow{e_i})$ the *i-th coordinate* of $\overrightarrow{v}$ related to the basis $\{\overrightarrow{e_1}, \overrightarrow{e_2}, \overrightarrow{e_3}\}$.

Hence, for any arithmetic discrete plane $\mathfrak{P}$, there exist a vector $\overrightarrow{v} \in \mathbb{R}^3$ and two real numbers $\mu \in \mathbb{R}$ and $\omega \in \mathbb{R}_+^\star$ such that

$$\mathfrak{P} = \{\overrightarrow{x} \in \mathbb{Z}^3 \mid 0 \leq (\overrightarrow{x}, \overrightarrow{v}) + \mu < \omega\}.$$

In the sequel of this paper, we denote such a plane by $\mathfrak{P}(\overrightarrow{v}, \mu, \omega)$. For a given $\overrightarrow{\alpha} \in \mathbb{Z}^3$, let $\Pi_{\overrightarrow{\alpha}} : \mathbb{R}^3 \to \{\overrightarrow{x} \in \mathbb{R}^3 \mid (\overrightarrow{\alpha}, \overrightarrow{x}) = 0\}$ stand for the orthogonal projection map onto the plane $(\overrightarrow{\alpha}, \overrightarrow{x}) = 0$. We furthermore use the notation $\pi_{\overrightarrow{\alpha}}$ when we consider the restriction of the projection $\Pi_{\overrightarrow{\alpha}}$ to a subset of $\mathbb{R}^3$, as for instance $\pi_{\overrightarrow{\alpha}} : \mathfrak{P} \to \{\overrightarrow{x} \in \mathbb{R}^3 \mid (\overrightarrow{\alpha}, \overrightarrow{x}) = 0\}$, for a discrete plane $\mathfrak{P}$.

Let us recall a classical property of naive discrete planes having a positive normal vector:

**Theorem 1.** *[DRR94] Let $\mathfrak{P} = \mathfrak{P}(\overrightarrow{v}, \mu, \omega)$ be a naive discrete plane. If $v_i = \omega$, for $i = 1, 2$ or $3$, then $\mathfrak{P}$ is in bijection with the integer points of the plane $(\overrightarrow{e_i}, \overrightarrow{x}) = 0$ by the projection map $\Pi_{\overrightarrow{e_i}}$, that is, the restriction map $\pi_{\overrightarrow{e_i}} : \mathfrak{P} \longrightarrow \Pi_{\overrightarrow{e_i}}(\mathbb{Z}^3)$ is a bijection. The plane $(\overrightarrow{e_i}, \overrightarrow{x}) = 0$ is called a* functional plane *of $\mathfrak{P}$.*

An analogous result holds for standard discrete planes:

**Theorem 2.** *[BV00] Let $\mathfrak{P} = \mathfrak{P}(\overrightarrow{v}, \mu, \omega)$ be a standard discrete plane. Let $\overrightarrow{\alpha} = \overrightarrow{e_1} + \overrightarrow{e_2} + \overrightarrow{e_3}$. Then, the restriction map $\pi_{\overrightarrow{\alpha}} : \mathfrak{P} \longrightarrow \Pi_{\overrightarrow{\alpha}}(\mathbb{Z}^3)$ is a bijection.*

# 2    Generalized Functionality

First, let us notice that in each of the two cases investigated in Theorem 1 and 2, the following property holds: let $\mathfrak{P}$ be a naive or a standard discrete plane with normal vector $\overrightarrow{v}$ and with thickness $\omega$; then there exists a vector $\overrightarrow{\alpha}$ in $\mathbb{Z}^3$ such that the restriction map $\pi_{\overrightarrow{\alpha}} : \mathfrak{P} \longrightarrow \Pi_{\overrightarrow{\alpha}}(\mathbb{Z}^3)$ is a bijection, and $(\overrightarrow{\alpha}, \overrightarrow{v}) = \omega$.

In this section, we extend this property to any discrete plane $\mathfrak{P}(\overrightarrow{v}, \mu, \omega)$ whatever its thickness $\omega$ by introducing a dual vector $\overrightarrow{\alpha} \in \mathbb{Z}^3$

such that $(\overrightarrow{v}, \overrightarrow{\alpha}) = \omega$. Furthermore, we improve this result by showing that the projections $\pi_{\overrightarrow{\alpha}}$ are the only ones which provide a one-to-one correspondence between the discrete plane $\mathfrak{P}(\overrightarrow{v}, \mu, \omega)$ and the projection of $\mathbb{Z}^3$; this will then yield a one-to-one correspondence between a discrete plane and a two-dimensional lattice.

## 2.1   A Bijective Projection for Arithmetic Discrete Planes

**Theorem 3.** *Let $\mathfrak{P} = \mathfrak{P}(\overrightarrow{v}, \mu, \omega)$ be a discrete plane where $\overrightarrow{v} \in \mathbb{R}^3$ is a non-zero vector, $\mu \in \mathbb{R}$ and $\omega \in \mathbb{R}_+^\star$. Let $\overrightarrow{\alpha} \in \mathbb{Z}^3$ such that $\gcd(\alpha_1, \alpha_2, \alpha_3) = 1$ and $(\overrightarrow{\alpha}, \overrightarrow{v}) \neq 0$. Then, $\pi_{\overrightarrow{\alpha}} : \mathfrak{P} \longrightarrow \Pi_{\overrightarrow{\alpha}}(\mathbb{Z}^3)$ is a bijection if and only if $|(\overrightarrow{\alpha}, \overrightarrow{v})| = \omega$.*

The proof of Theorem 3 first requires a technical lemma:

**Lemma 1.** *Let $\mathfrak{P} = \mathfrak{P}(\overrightarrow{v}, \mu, \omega)$ be a discrete plane with $(\overrightarrow{v}, \mu, \omega) \in \mathbb{R}^3 \times \mathbb{R} \times \mathbb{R}_+^\star$.*

1. *If $\dim_{\mathbb{Q}}(v_1, v_2, v_3) = 1$, then there exists $(\overrightarrow{v'}, \mu', \omega') \in \mathbb{Z}^3 \times \mathbb{Z} \times \mathbb{N}$ such that $\mathfrak{P} = \mathfrak{P}(\overrightarrow{v'}, \mu', \omega')$ and $\gcd(v_1', v_2', v_3') = 1$.*
2. *If $\dim_{\mathbb{Q}}(v_1, v_2, v_3) > 1$, then the family $((\overrightarrow{x}, \overrightarrow{v}) + \mu)_{\overrightarrow{x} \in \mathfrak{P}}$ is dense in $[0, \omega[$.*

*Proof.* 1. Let us suppose that $\dim_{\mathbb{Q}}(v_1, v_2, v_3) = 1$. Then, there exists $\zeta \in \mathbb{R}_+^\star$ such that $(\zeta v_1, \zeta v_2, \zeta v_3) \in \mathbb{Z}^3$. Let $\overrightarrow{v'} = \zeta \overrightarrow{v}$, $\mu' = \lceil -\zeta \mu \rceil$ and $\omega' = \lceil \zeta \omega - \zeta \mu \rceil - \lceil -\zeta \mu \rceil$. An easy computation gives $\mathfrak{P}(\overrightarrow{v}, \mu, \omega) = \mathfrak{P}(\overrightarrow{v'}, \mu', \omega')$. Finally, according to [AAS97], $\overrightarrow{v}$ can be chosen with $\gcd(v_1, v_2, v_3) = 1$.
2. If $\dim_{\mathbb{Q}}(v_1, v_2, v_3) > 1$, then we conclude by the classical following result: the set $\{m + n\alpha \mid (m, n) \in \mathbb{Z}^2\}$ is dense in $\mathbb{R}$ if $\alpha \notin \mathbb{Q}$. ∎

With the hypothesis of Lemma 1, let us observe that $\mathfrak{P}(\overrightarrow{v'}, \mu', \omega')$ is a naive (resp. standard) discrete plane, if so is $\mathfrak{P}(\overrightarrow{v}, \mu, \omega)$.

*Proof of Theorem 3.* We assume w.l.o.g that $(\overrightarrow{\alpha}, \overrightarrow{v}) > 0$. Let $\overrightarrow{x} = (x_1, x_2, x_3)$, $\overrightarrow{x'} = (x_1', x_2', x_3') \in \mathbb{Z}^3$; $\pi_{\overrightarrow{\alpha}}(\overrightarrow{x}) = \pi_{\overrightarrow{\alpha}}(\overrightarrow{x'})$ if and only if there exists $(k, k') \in \mathbb{Z}^2$ such that $k'(\overrightarrow{x'} - \overrightarrow{x}) = k\overrightarrow{\alpha}$. With no loss of generality we can suppose that $\gcd(k, k') = 1$; then, $k'$ divides $\gcd(\alpha_1, \alpha_2, \alpha_3)$ and $|k'| = 1$. In other words, $\pi_{\overrightarrow{\alpha}}(\overrightarrow{x}) = \pi_{\overrightarrow{\alpha}}(\overrightarrow{x'})$ if and only if there exists $k \in \mathbb{Z}$ such that $\overrightarrow{x'} = \overrightarrow{x} + k\overrightarrow{\alpha}$. Moreover, $\overrightarrow{x} + k\overrightarrow{\alpha} \in \mathfrak{P}$ if and only if

$$\frac{-((\overrightarrow{x}, \overrightarrow{v}) + \mu)}{(\overrightarrow{\alpha}, \overrightarrow{v})} \leq k < \frac{\omega - ((\overrightarrow{x}, \overrightarrow{v}) + \mu)}{(\overrightarrow{\alpha}, \overrightarrow{v})}.$$

1) Let us first assume that $(\overrightarrow{\alpha}, \overrightarrow{v}) = \omega$. Then,

$$\# \left[\!\left[ \frac{-((\overrightarrow{x}, \overrightarrow{v}) + \mu)}{(\overrightarrow{\alpha}, \overrightarrow{v})}, \frac{\omega - ((\overrightarrow{x}, \overrightarrow{v}) + \mu)}{(\overrightarrow{\alpha}, \overrightarrow{v})} \right[\!\right[ = 1,$$

and we have proved that $\pi_{\overrightarrow{\alpha}} : \mathfrak{P} \longrightarrow \Pi_{\overrightarrow{\alpha}}(\mathbb{Z}^3)$ is a bijection.

2) Conversely, let us assume that $\pi_{\overrightarrow{\alpha}} : \mathfrak{P} \longrightarrow \Pi_{\overrightarrow{\alpha}}(\mathbb{Z}^3)$ is a bijection.

   i. If $\dim_{\mathbb{Q}}(v_1, v_2, v_3) = 1$, then, thanks to Lemma 1, we can suppose that $\overrightarrow{v} \in \mathbb{Z}^3$, with $\gcd(v_1, v_2, v_2) = 1$, and $(\mu, \omega) \in \mathbb{Z} \times \mathbb{N}^\star$. Let $\overrightarrow{x} \in \mathbb{Z}^3$ such that $(\overrightarrow{x}, \overrightarrow{v}) + \mu = 0$. Then $\overrightarrow{x} \in \mathfrak{P}$ and $(\overrightarrow{x} + \overrightarrow{\alpha}, \overrightarrow{v}) + \mu = (\overrightarrow{x}, \overrightarrow{v}) + (\overrightarrow{\alpha}, \overrightarrow{v}) + \mu = (\overrightarrow{\alpha}, \overrightarrow{v}) > 0$. Moreover, $\pi_{\overrightarrow{\alpha}}(\overrightarrow{x} + \overrightarrow{\alpha}) = \pi_{\overrightarrow{\alpha}}(\overrightarrow{x})$. Since $\pi_{\overrightarrow{\alpha}}$ is injective then $\overrightarrow{x} + \overrightarrow{\alpha} \notin \mathfrak{P}$, and hence $(\overrightarrow{\alpha}, \overrightarrow{v}) \geq \omega$. On the other hand, let $\overrightarrow{x'} \in \mathbb{Z}^3$ such that $(\overrightarrow{x'}, \overrightarrow{v}) + \mu = -1$. Then, $(\overrightarrow{x'} + \overrightarrow{\alpha}, \overrightarrow{v}) + \mu = (\overrightarrow{x'}, \overrightarrow{v}) + (\overrightarrow{\alpha}, \overrightarrow{v}) + \mu = (\overrightarrow{\alpha}, \overrightarrow{v}) - 1 \geq 0$. Since $\pi_{\overrightarrow{\alpha}}$ is surjective and $(\overrightarrow{\alpha}, \overrightarrow{v}) > 0$, then $\overrightarrow{x} + \overrightarrow{\alpha} \in \mathfrak{P}$, that is, $(\overrightarrow{\alpha}, \overrightarrow{v}) - 1 < \omega$, or equivalently, $(\overrightarrow{\alpha}, \overrightarrow{v}) \leq \omega$.

   ii. Let us suppose that $\dim_{\mathbb{Q}}(v_1, v_2, v_3) \geq 2$. Then, each interval $\left[ \frac{-((\overrightarrow{x}, \overrightarrow{v}) + \mu)}{(\overrightarrow{\alpha}, \overrightarrow{v})}, \frac{\omega - ((\overrightarrow{x}, \overrightarrow{v}) + \mu)}{(\overrightarrow{\alpha}, \overrightarrow{v})} \right[$, with $\overrightarrow{x} \in \mathfrak{P}$, contains one and exactly one integer if and only if $(\overrightarrow{\alpha}, \overrightarrow{v}) = \omega$ by Lemma 1.

                                                                           ■

Projecting according to $\overrightarrow{\alpha}$ corresponds to looking at the plane along a direction parallel to $\overrightarrow{\alpha}$. Moreover, Theorem 3 states that, looking at the discrete plane $\mathfrak{P}(\overrightarrow{v}, \mu, \omega)$ along this direction, one can see all points of $\mathfrak{P}(\overrightarrow{v}, \mu, \omega)$ as if they were on the plane $(\overrightarrow{\alpha}, \overrightarrow{x}) = 0$. In Section 2.3, we show that a natural regular lattice structure emerges from this point of view.

As a generalization of functional planes for naive discrete planes, we define:

**Definition 1.** *Let $\mathfrak{P} = \mathfrak{P}(\overrightarrow{v}, \mu, \omega)$ be a discrete plane with $\overrightarrow{v} \in \mathbb{R}^3$ a non-zero vector, $\mu \in \mathbb{R}$ and $\omega \in \mathbb{R}_+^\star$. Let $\overrightarrow{\alpha} \in \mathbb{Z}^3$ such that $\pi_{\overrightarrow{\alpha}} : \mathfrak{P} \longrightarrow \Pi_{\overrightarrow{\alpha}}(\mathbb{Z}^3)$ is a bijection. The plane $(\overrightarrow{\alpha}, \overrightarrow{v}) = 0$ is called a (generalized) functional plane of $\mathfrak{P}$.*

## 2.2 Existence of a Dual Vector

In the case of an arithmetic discrete plane with normal vector $\overrightarrow{v} \in \mathbb{R}^3$ and thickness $\omega \in \mathbb{R}_+^\star$, there is no reason for a vector $\overrightarrow{\alpha} \in \mathbb{Z}^3$ to exist satisfying $(\overrightarrow{\alpha}, \overrightarrow{v}) = \omega$ (consider the case $(v_1, v_2, v_3, \omega)$ is $\mathbb{Q}$-free). However, if $\mathfrak{P}(\overrightarrow{v}, \mu, \omega)$ is an arithmetic discrete plane with normal vector $\overrightarrow{v} \in \mathbb{Z}^3$, then, according to Lemma 1, we can suppose that $\omega \in \mathbb{Z}$ and $\gcd(v_1, v_2, v_3) = 1$. We then deduce from Bezout's Lemma that there exists a vector $\overrightarrow{\alpha} \in \mathbb{Z}^3$ such that $(\overrightarrow{\alpha}, \overrightarrow{v}) = \omega$. Let us prove now that $\overrightarrow{\alpha} \in \mathbb{Z}^3$ can be chosen such that $\gcd(\alpha_1, \alpha_2, \alpha_3) = 1$.

**Theorem 4.** *Let $\mathfrak{P}(\overrightarrow{v}, \mu, \omega)$ be an arithmetic discrete plane with $(\overrightarrow{v}, \mu, \omega) \in \mathbb{Z}^3 \times \mathbb{Z} \times \mathbb{Z}_+^\star$ and $\gcd(v_1, v_2, v_3) = 1$. There exists $\overrightarrow{\alpha} \in \mathbb{Z}^3$ such that $(\overrightarrow{\alpha}, \overrightarrow{v}) = \omega$ and $\gcd(\alpha_1, \alpha_2, \alpha_3) = 1$. In other words, there exists $\overrightarrow{\alpha} \in \mathbb{Z}^3$ such that $\pi_{\overrightarrow{\alpha}} : \mathfrak{P}(\overrightarrow{v}, \mu, \omega) \longrightarrow \Pi_{\overrightarrow{\alpha}}(\mathbb{Z}^3)$ is a bijection.*

*Proof.* Let $\overrightarrow{\beta} \in \mathbb{Z}^3$ such that $(\overrightarrow{\beta}, \overrightarrow{v}) = 1$. Then, $(\omega \overrightarrow{\beta}, \overrightarrow{v}) = \omega$. Let $\overrightarrow{u} \in \{\overrightarrow{x} \in \mathbb{Z}^3 \mid (\overrightarrow{x}, \overrightarrow{v}) = 0\}$, let $d = \gcd(u_1, u_2, u_3)$ and let $\overrightarrow{\alpha} = \omega \overrightarrow{\beta} + d^{-1} \overrightarrow{u}$. Then, an easy computation gives $(\overrightarrow{\alpha}, \overrightarrow{v}) = \omega$ and $\gcd(\alpha_1, \alpha_2, \alpha_3) = 1$. We end the proof by applying Theorem 3. ■

We have illustrated Theorem 4 in Figure 1 in the case of a discrete line for a better visualisation of the situation.

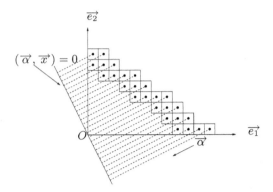

**Fig. 1.** Generalized functionality: the orthogonal projection of the discrete line $0 \leq 7x_1 + 10x_2 + \mu < 24$ onto the line $2x + y = 0$

## 2.3    Functional Regular Lattice Associated to an Arithmetic Discrete Plane

Let us see now how any arithmetic discrete plane $\mathfrak{P}$ can be recoded in a functional way on a regular two-dimensional lattice, despite its three-dimensional structure.

Let $\mathfrak{P} = \mathfrak{P}(\overrightarrow{v}, \mu, \omega)$ be an arithmetic discrete plane. Let $\overrightarrow{\alpha} \in \mathbb{Z}^3$ such that $\gcd(\alpha_1, \alpha_2, \alpha_3) = 1$ and $(\overrightarrow{\alpha}, \overrightarrow{v}) = \omega$ (in case $(\overrightarrow{v}, \mu, \omega) \in \mathbb{Z}^3 \times \mathbb{Z} \times \mathbb{N}^*$, and $\gcd(v_1, v_2, v_3) = 1$, the existence of such a vector $\overrightarrow{\alpha}$ comes from Theorem 4). One of the coefficients $\alpha_i$, for $i \in \{1, 2, 3\}$ being non-zero, we assume in this section that $\alpha_3 \neq 0$ with no loss of generality.

First, let us notice that since $\Pi_{\overrightarrow{\alpha}}(\overrightarrow{\alpha}) = \overrightarrow{0}$, then, for all $\overrightarrow{x} \in \mathbb{Z}^3$,

$$\Pi_{\overrightarrow{\alpha}}(\overrightarrow{e}_3) = -\frac{\alpha_1}{\alpha_3} \Pi_{\overrightarrow{\alpha}}(\overrightarrow{e_1}) - \frac{\alpha_2}{\alpha_3} \Pi_{\overrightarrow{\alpha}}(\overrightarrow{e_2}).$$

Then, for all $\overrightarrow{x} \in \mathbb{Z}^3$,

$$\Pi_{\overrightarrow{\alpha}}(\overrightarrow{x}) = x_1 \Pi_{\overrightarrow{\alpha}}(\overrightarrow{e_1}) + x_2 \Pi_{\overrightarrow{\alpha}}(\overrightarrow{e_2}) + x_3 \Pi_{\overrightarrow{\alpha}}(\overrightarrow{e_3})$$
$$= \left( \frac{\alpha_3 x_1 - \alpha_1 x_3}{\alpha_3} \right) \Pi_{\overrightarrow{\alpha}}(\overrightarrow{e_1}) + \left( \frac{\alpha_3 x_2 - \alpha_2 x_3}{\alpha_3} \right) \Pi_{\overrightarrow{\alpha}}(\overrightarrow{e_2})$$

and

$$\Pi_{\overrightarrow{\alpha}} : \mathbb{R}^3 \longrightarrow \{ \overrightarrow{x} \in \mathbb{R}^3 \mid (\overrightarrow{\alpha}, \overrightarrow{x}) = 0 \}$$
$$\overrightarrow{x} \longmapsto \frac{\alpha_3 x_1 - \alpha_1 x_3}{\gcd(\alpha_1, \alpha_3)} \overrightarrow{f_1} + \frac{\alpha_3 x_2 - \alpha_2 x_3}{\gcd(\alpha_2, \alpha_3)} \overrightarrow{f_2}, \tag{1}$$

with

$$\overrightarrow{f_1} = \frac{\gcd(\alpha_1, \alpha_3)}{\alpha_3} \Pi_{\overrightarrow{\alpha}}(\overrightarrow{e_1}) \text{ and } \overrightarrow{f_2} = \frac{\gcd(\alpha_2, \alpha_3)}{\alpha_3} \Pi_{\overrightarrow{\alpha}}(\overrightarrow{e_2}).$$

We thus deduce that $\Gamma_{\overrightarrow{\alpha}} = \Pi_{\overrightarrow{\alpha}}(\mathbb{Z}^3) = \Pi_{\overrightarrow{\alpha}}(\mathfrak{P})$ is a sub-lattice of the two-dimensional lattice $\mathbb{Z}\overrightarrow{f_1} + \mathbb{Z}\overrightarrow{f_2}$. The lattice $\Gamma_{\overrightarrow{\alpha}}$ is called a *functional lattice* of $\mathfrak{P}$. This generalizes the concept of functionality defined for naive discrete planes as a projection onto the integer points of one of the coordinate planes.

# 3    From a Functional Lattice to the Associated Discrete Plane

Let $\mathfrak{P} = \mathfrak{P}(\overrightarrow{v}, \mu, \omega)$ be an arithmetic discrete plane and $\Gamma_{\overrightarrow{\alpha}}$ be a functional lattice of $\mathfrak{P}$ (see Section 2.3). A natural question is: "given an element $\overrightarrow{y} \in \Gamma_{\overrightarrow{\alpha}}$, how can we recover the unique vector $\overrightarrow{x} \in \mathfrak{P}$ such that $\pi_{\overrightarrow{\alpha}}(\overrightarrow{x}) = \overrightarrow{y}$? " In the following, we investigate this question for the classical classes of arithmetic discrete planes, namely the naive, the standard [Rev91, DRR94, DR95] and the graceful ones [BB99, BB02].

## 3.1    Generalized Functionality for a Particular Class of Discrete Planes

Let $\mathfrak{P}(\overrightarrow{v}, \mu, \omega)$ be an arithmetic discrete plane and let $\overrightarrow{\alpha} \in \mathbb{Z}^3$ such that $(\overrightarrow{\alpha}, \overrightarrow{v}) = \omega$. In this section, we assume that there exists $i \in \{1, 2, 3\}$ such that $\alpha_i = 1$. This condition includes the set of naive, standard and graceful arithmetic planes (see Section 3.2). Let us thus suppose that $\alpha_3 = 1$. In this case, let us notice that $\Gamma_{\overrightarrow{\alpha}} = \mathbb{Z}\overrightarrow{f_1} + \mathbb{Z}\overrightarrow{f_2}$.

Let $\overrightarrow{y} \in \Gamma_{\overrightarrow{\alpha}}$. From now on, if no confusion is possible with the representation of $\overrightarrow{y}$ related to the basis $\{\overrightarrow{e_1}, \overrightarrow{e_2}, \overrightarrow{e_3}\}$, we will denote $(y_1, y_2)$ the unique pair of integers such that $\overrightarrow{y} = y_1 \overrightarrow{f_1} + y_2 \overrightarrow{f_2}$.

Let $\overrightarrow{x} \in \mathfrak{P}$ and let $\overrightarrow{y} = \pi_{\overrightarrow{\alpha}}(\overrightarrow{x}) \in \Gamma_{\overrightarrow{\alpha}}$. According to (1), one has $x_1 = y_1 + \alpha_1 x_3$ and $x_2 = y_2 + \alpha_2 x_3$. Hence, $(\overrightarrow{x}, \overrightarrow{v}) + \mu = y_1 v_1 + y_2 v_2 + x_3(\alpha_1 v_1 + \alpha_2 v_2 + v_3) + \mu$ and

$$0 \leq (\overrightarrow{x}, \overrightarrow{v}) + \mu = v_1 y_1 + v_2 y_2 + x_3 \omega + \mu < \omega. \tag{2}$$

Thus, given any $\overrightarrow{y} \in \mathbb{Z}^2$, we can easily recover the unique vector $\overrightarrow{x} \in \mathfrak{P}$ such that $\pi_{\overrightarrow{\alpha}}(\overrightarrow{x}) = \overrightarrow{y}$. Indeed, let us first note that (2) yields an explicit formula for the *height* $x_3$ of the points of $\mathfrak{P}$, that is, $x_3 = -\left\lfloor \frac{v_1 y_1 + v_2 y_2 + \mu}{\omega} \right\rfloor$. Let us call $H_{\mathfrak{P}, \overrightarrow{\alpha}} : \Gamma_{\overrightarrow{\alpha}} \longrightarrow \mathbb{Z}$ the function which to any point $y_1 \overrightarrow{f_1} + y_2 \overrightarrow{f_2} \in \Gamma_{\overrightarrow{\alpha}}$ associates the height $x_3$ of the corresponding point $\overrightarrow{x} \in \mathfrak{P}$, that is, the unique point $\overrightarrow{x} \in \mathfrak{P}$ such that $\pi_{\overrightarrow{\alpha}}(\overrightarrow{x}) = \overrightarrow{y}$:

$$H_{\mathfrak{P}, \overrightarrow{\alpha}} : \overrightarrow{y} \mapsto -\left\lfloor \frac{v_1 y_1 + v_2 y_2 + \mu}{\omega} \right\rfloor.$$

One thus obtains:

**Proposition 1.** *If $\alpha_3 = 1$, then the function $\pi_{\overrightarrow{\alpha}}^{-1} : \Gamma_{\overrightarrow{\alpha}} \longrightarrow \mathfrak{P}$ is defined by, for all $\overrightarrow{y} \in \Gamma_{\overrightarrow{\alpha}}$ :*

$$\pi_{\overrightarrow{\alpha}}^{-1}(\overrightarrow{y}) = {}^t \begin{pmatrix} y_1 \\ y_2 \\ 0 \end{pmatrix} + H_{\mathfrak{P}, \overrightarrow{\alpha}}(y_1, y_2) \; {}^t \begin{pmatrix} \alpha_1 \\ \alpha_2 \\ 1 \end{pmatrix}. \tag{3}$$

## 3.2    Classical Examples

Let us suppose that $\vec{v} \in \mathbb{N}^3$, and $v_3 = \max\{v_1, v_2, v_3\}$. If $\mathfrak{P}$ is a naive or a standard discrete plane, then we can suppose $\alpha_3 = 1$, since $v_i \geq 0$ for $i \in \{1, 2, 3\}$. In the special case of naive discrete planes, we recover the already known formula:

**Corollary 1.** *If $\mathfrak{P}$ is a naive discrete plane, then $\vec{\alpha} = \vec{e_3}$, for all $\vec{x} \in \mathfrak{P}$, $\pi_{\vec{\alpha}}(\vec{x}) = x_1 \vec{e_1} + x_2 \vec{e_2}$ and for all $\vec{y} \in \Gamma_{\vec{\alpha}}$,*

$$\pi_{\vec{e_3}}^{-1}(\vec{y}) = y_1 \vec{e_1} + y_2 \vec{e_2} - \left\lfloor \frac{v_1 y_1 + v_2 y_2 + \mu}{v_3} \right\rfloor \vec{e_3}.$$

Concerning the case of the standard discrete planes, we obtain, as a direct consequence of Proposition 1:

**Corollary 2.** *If $\mathfrak{P}$ is a standard discrete plane, then $\vec{\alpha} = \vec{e_1} + \vec{e_2} + \vec{e_3}$, for all $\vec{x} \in \mathfrak{P}$, $\pi_{\vec{\alpha}}(\vec{x}) = (x_1 - x_3)\vec{e_1} + (x_2 - x_3)\vec{e_2}$, and for all $\vec{y} \in \Gamma_{\vec{\alpha}}$,*

$$\pi_{\vec{e_3}}^{-1}(\vec{y}) = {}^t\begin{pmatrix} y_1 \\ y_2 \\ 0 \end{pmatrix} - \left\lfloor \frac{v_1 y_1 + v_2 y_2 + \mu}{v_1 + v_2 + v_3} \right\rfloor {}^t\begin{pmatrix} 1 \\ 1 \\ 1 \end{pmatrix}.$$

Let us suppose now that $\mathfrak{P} = \mathfrak{P}(\vec{v}, \mu, \omega)$ is a graceful plane, that is, $0 \leq v_1, \leq v_2 \leq v_3$ and $\omega = \max(v_1 + v_2, v_3)$. If $v_1 + v_2 \leq v_3$, then $\mathfrak{P}$ is a naive discrete plane and this case has already been studied. Let us then assume that $\omega = v_1 + v_2$. Let $\vec{\alpha} = \vec{e_1} + \vec{e_2}$. Then, for all $\vec{x} \in \mathfrak{P}$, $\pi_{\vec{\alpha}}(\vec{x}) = (x_1 - x_2)\vec{e_1} + x_3 \vec{e_3}$.

Up to a permutation on the set $\{\alpha_1, \alpha_2, \alpha_3\}$, we recover the following from Proposition 1:

**Proposition 2.** *If $\mathfrak{P}$ is a graceful plane. Then $\vec{\alpha} = \vec{e_1} + \vec{e_2}$ and the function $\pi_{\vec{\alpha}}^{-1} : \Gamma_{\vec{\alpha}} \longrightarrow \mathfrak{P}$ is defined by, for all $\vec{y} \in \Gamma_{\vec{\alpha}}$,*

$$\pi_{\vec{\alpha}}^{-1}(\vec{y}) = {}^t\begin{pmatrix} 0 \\ y_1 \\ y_2 \end{pmatrix} + \left\lceil \frac{v_2 y_1 - v_3 y_2 + \mu}{v_1 + v_2} \right\rceil {}^t\begin{pmatrix} 1 \\ 1 \\ 0 \end{pmatrix}.$$

## 4    Plane Partitions and Local Configurations

The aim of this section is to apply the previous results to the study of $(m, n)$-cubes and local configurations, generalizing the study performed for naive planes in [VC97, Sch97, Gér99, VC99, Col02]. For the sake of consistency, we call them here $\vec{m}$-cubes rather than $(m, n)$-cubes.

Let $\mathfrak{P} = \mathfrak{P}(\vec{v}, \mu, \omega)$ be an arithmetic discrete plane and let $\vec{\alpha} \in \mathbb{Z}^3$ such that $\gcd(\alpha_1, \alpha_2, \alpha_3) = 1$ and $(\vec{\alpha}, \vec{v}) = \omega$ (recall that if $\vec{v} \in \mathbb{Z}^3$ and $\gcd(v_1, v_2, v_3) = 1$, then the existence of $\vec{\alpha}$ is ensured by Theorem 4). Let us assume furthermore that $\alpha_3 = 1$.

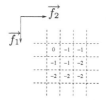

**Fig. 2.** From left to right: a local configuration of the discrete plane $\mathfrak{P}(4\overrightarrow{e_1} + 2\overrightarrow{e_2} + 5\overrightarrow{e_3}, 0, 7)$ and its corresponding preimage by $\pi_{\overrightarrow{e_2}+\overrightarrow{e_3}}^{-1}$

Let $\overrightarrow{m} \in (\mathbb{N}^\star)^2$ be given. By $\overrightarrow{m}$-cube, we mean a local configuration in the discrete plane that can be observed thanks to $\pi_{\overrightarrow{\alpha}}$ through an $\overrightarrow{m}$-window in the projection lattice $\Gamma_{\overrightarrow{\alpha}}$. More precisely,

**Definition 2.** *Let $\overrightarrow{m} \in (\mathbb{N}^\star)^2$. The $\overrightarrow{m}$-cube $\mathcal{C}(\overrightarrow{y}, \overrightarrow{m})$, with $\overrightarrow{y} \in \Gamma_{\overrightarrow{\alpha}}$, is defined as the following subset of $\mathfrak{P}$:*

$$\mathcal{C}(\overrightarrow{y}, \overrightarrow{m}) = \left\{\pi_{\overrightarrow{\alpha}}^{-1}(\overrightarrow{y} + \overrightarrow{i}),\ \overrightarrow{i} \in [\![0, m_1[\![\times[\![0, m_2[\![ \right\}.$$

In order to enumerate the different types of $\overrightarrow{m}$-cubes that occur in $\mathfrak{P}$, we represent them as local configurations as follows.

**Definition 3.** *The $\overrightarrow{m}$-local configuration $LC(\overrightarrow{y}, \overrightarrow{m})$, with $\overrightarrow{y} \in \mathbb{Z}^2$ and $m \in (\mathbb{N}^\star)^2$, is defined as follows:*

$$LC(\overrightarrow{y}, m) = [H_{\mathfrak{P},\overrightarrow{\alpha}}(\overrightarrow{z}) - H_{\mathfrak{P},\overrightarrow{\alpha}}(\overrightarrow{y})]_{\overrightarrow{z} \in [\![0, m_1-1]\!]\overrightarrow{f_1} + [\![0, m_2-1]\!]\overrightarrow{f_2}}.$$

*We say that $\overrightarrow{y}$ is an* index *of occurrence of the local configuration $LC(\overrightarrow{y}, \overrightarrow{m})$.*

Let us note that a local configuration is a plane partition.

*Example 1.* For instance, let us consider the arithmetic discrete plane $\mathfrak{P} = \mathfrak{P}(\overrightarrow{v}, \mu, \omega)$ with $\overrightarrow{v} = 4\overrightarrow{e_1} + 2\overrightarrow{e_2} + 5\overrightarrow{e_3}$, $\mu = 0$ and $\omega = 9$. Let $\overrightarrow{\alpha} = \overrightarrow{e_1} + \overrightarrow{e_3}$. We illustrate the local configuration $LC(\overrightarrow{f_1} + \overrightarrow{f_2}, 3(\overrightarrow{e_1} + \overrightarrow{e_2}))$ of $\mathfrak{P}$ and its preimage by $\pi_{\overrightarrow{\alpha}}^{-1}$ in Fig. 2.

We follow here the approach developed in [Vui99]. For a naive discrete plane $\mathfrak{P}$, it is well known that, given two points $\overrightarrow{x}$ and $\overrightarrow{x'}$ of $\mathfrak{P}$ such that their projections by $\pi_{\overrightarrow{\alpha}}$ are 4-connected in the functional plane, then $|x_3 - x_3'| \leq 1$. In other words, the difference between the height of $\overrightarrow{x}$ and $\overrightarrow{x'}$ is at most 1. A quite unexpected fact is that this property holds for any arithmetic discrete plane with $\alpha_3 = 1$. More precisely, it is easy to see that, for all $\overrightarrow{y} \in \Gamma$, $H_{\mathfrak{P},\overrightarrow{\alpha}}\left(\overrightarrow{y} + \overrightarrow{f_1}\right) - H_{\mathfrak{P},\overrightarrow{\alpha}}(\overrightarrow{y}))$ takes only two values, namely $-\lfloor\frac{v_1}{\omega}\rfloor$ and $-\lfloor\frac{v_1}{\omega}\rfloor - 1$. In the same way, $H_{\mathfrak{P},\overrightarrow{\alpha}}\left(\overrightarrow{y} + \overrightarrow{f_2}\right) - H_{\mathfrak{P},\overrightarrow{\alpha}}(\overrightarrow{y}))$ takes only the values $-\lfloor\frac{v_2}{\omega}\rfloor$ and $-\lfloor\frac{v_2}{\omega}\rfloor - 1$.

In each case, one of these values is odd, whereas the other one is even; we define $E_h$ and $O_h$ to be respectively the even and the odd value taken by $-\lfloor \frac{v_1}{\omega} \rfloor$ and $-\lfloor \frac{v_1}{\omega} \rfloor - 1$; we similarly define $E_v$ and $O_v$. It is now natural to introduce the following two-dimensional sequence:

$$U = (U_{\overrightarrow{y}})_{\overrightarrow{y} \in \Gamma_{\overrightarrow{\alpha}}} = (H_{\mathfrak{P}, \overrightarrow{\alpha}}(\overrightarrow{y}) \mod 2)_{\overrightarrow{y} \in \Gamma_{\overrightarrow{\alpha}}} \in \{0, 1\}^{\mathbb{Z}^2}.$$

By definition, it is easily seen that the sequence $U$ satisfies:

$$\forall \overrightarrow{y} \in \Gamma_{\overrightarrow{\alpha}}, \ U_{\overrightarrow{y}} = 0 \text{ if and only if } -\frac{y_1 v_1 + y_2 v_2 + \mu}{\omega} \mod 2 \in [0, 1[.$$

Let $w = [w_y]_{y \in [0, m_1 - 1] \times [0, m_2 - 1[}$ be a word of size $m_1 \times m_2$ over $\{0, 1\}$. We define the *complement* $\overline{w}$ of $w$ as follows: $\overline{w} = [\overline{w_y}]_{y \in [0, m_1 - 1[ \times [0, m_2 - 1[}$, where $\overline{1} = 0$ and $\overline{0} = 1$. Let us recall [Vui99, BV00] that the set of factors of the sequence $U$ is stable under complementation. We thus introduce the following equivalence relation:

$$v \sim w \text{ if and only if } v \in \{w, \overline{w}\}.$$

We have the following theorem, inspired by [Vui99]:

**Theorem 5.** *There is a natural bijection between the equivalence classes of the relation $\sim$ of the factors of the sequence $U$ and the $\overrightarrow{m}$-local configurations of $\mathfrak{P}$.*

*Proof.* Consider the local configuration $L = LC(\overrightarrow{y}, \overrightarrow{m})$; we can associate to it the $m_1 \times m_2$ word

$$[L(\overrightarrow{z}) \mod 2]_{\overrightarrow{z} \in [0, m_1 - 1[ \overrightarrow{f_1} + [0, m_2 - 1[ \overrightarrow{f_2}},$$

that we denote for short $L \mod 2$. If $H_{\mathfrak{P}, \overrightarrow{\alpha}}(\overrightarrow{y})$ is even, then $L \mod 2$ is a factor of the two-dimensional sequence $U$; otherwise, $H_{\mathfrak{P}, \overrightarrow{\alpha}}(\overrightarrow{y})$ is odd and $\overline{L} \mod 2$ is a factor of $U$ and so is $L \mod 2$, by stability of the set of factors of $U$ by complementation.

Conversely, let us show how we can canonically reconstruct a $\overrightarrow{m}$-local configuration, with $\overrightarrow{m} \in (\mathbb{N}^\star)^2$, from a given $m_1 \times m_2$-factor $w$ of the two-dimensional sequence $U$. Let us first assume that $w_{\overrightarrow{0}} = 0$. We define a plane partition $H = [H(\overrightarrow{z})]_{\overrightarrow{z} \in [0, m_1 - 1[ \overrightarrow{f_1} + [0, m_2 - 1[ \overrightarrow{f_2}}$ by induction as follows: we set $H(\overrightarrow{0}) = 0$; let $\overrightarrow{z} \in [0, m_1 - 1[ \overrightarrow{f_1} + [0, m_2 - 1[ \overrightarrow{f_2}$ be a non-zero vector. If $w_{\overrightarrow{z} + \overrightarrow{f_1}} = w_{\overrightarrow{z}}$, then we set $H(\overrightarrow{z} + \overrightarrow{f_1}) = H(\overrightarrow{z}) + E_h$. Otherwise, we set $H(\overrightarrow{z} + \overrightarrow{f_1}) = H(\overrightarrow{z}) + O_h$. Similarly, if $w_{\overrightarrow{z} + \overrightarrow{f_2}} = w_{\overrightarrow{z}}$, then we set $H(\overrightarrow{z} + \overrightarrow{f_2}) = H(\overrightarrow{z}) + E_v$. Otherwise, we set $H(\overrightarrow{z} + \overrightarrow{f_2}) = H(\overrightarrow{z}) + O_v$.

The plane partition $H$ is a local configuration of $\mathfrak{P}$; indeed, if $w$ occurs at index $\overrightarrow{y}$ in $U$, then $H = LC(\overrightarrow{y}, \overrightarrow{m})$ and $w = (H \mod 2)$ since $H(\overrightarrow{y})$ is even (we have $w_{\overrightarrow{0}} = 0$). Now, if $w_{\overrightarrow{0}} = 1$, we apply the same reconstruction process to $\overline{w}$. We recover again a local configuration $LC(\overrightarrow{y}, \overrightarrow{m})$ such that $\overline{w} = (LC(\overrightarrow{y}, \overrightarrow{m}) \mod 2)$. ∎

One deduces, in particular, from Theorem 5 that any local configuration of the discrete plane $\mathfrak{P}$ occurs at least twice: once at an index $\overrightarrow{y}$ with $H(\overrightarrow{y})$ even and second, at an index $\overrightarrow{y}$ such that $H(\overrightarrow{y}')$ is even.

Let us now investigate the enumeration of $\overrightarrow{m}$-cubes occuring in a given arithmetic plane. The number of $(3,3)$-cubes included in a given naive arithmetic discrete plane has been proved to be at most 9 in [VC97]. More generally, in [Rev95, Gér99], the authors proved that, given a naive arithmetic discrete plane $\mathfrak{P}$, $\mathfrak{P}$ contains at most $m_1 m_2$ $\overrightarrow{m}$-cubes. In the following theorem, we show that this property also holds for $\overrightarrow{m}$-local configurations in an arithmetic discrete plane $\mathfrak{P}(\overrightarrow{v}, \mu, \omega)$, which is non-necessarily naive.

**Theorem 6.** *Let $\mathfrak{P} = \mathfrak{P}(\overrightarrow{v}, \mu, \omega)$ be a discrete plane, $\overrightarrow{\alpha} \in \mathbb{Z}^3$ such that $(\overrightarrow{\alpha}, \overrightarrow{v}) = \omega$ and $\alpha_3 = 1$, and let $\overrightarrow{m} \in (\mathbb{N}^\star)^2$. Then, $\mathfrak{P}$ contains at most $m_1 m_2$ $\overrightarrow{m}$-local configurations.*

*Proof.* According to [Vui99, BV00], the factors of size $m_1 \times m_2$ of the sequence $U$ are in one-to-one correspondence with the intervals of $\mathbb{R}/2\mathbb{Z}$ of extremal points $-\frac{i_1 v_1 + i_2 v_2}{\omega}$ and $-\frac{i_1 v_1 + i_2 v_2}{\omega} + 1$ with $(i_1, i_2) \in [0, m_1 - 1] \times [0, m_2 - 1]$. There are at most $2 m_1 m_2$ such points and the result follows from Theorem 5. ∎

## 5    Conclusion and Perspectives

The aim of the present work was to introduce suitable tools generalizing the classical ones used in the study of arithmetic discrete planes. We have exhibited a *generalized functionality* for arithmetic discrete planes $\mathfrak{P}(\overrightarrow{v}, \mu, \omega)$ and proved that, as soon as $|(\overrightarrow{\alpha}, \overrightarrow{v})| = w$ and $\gcd(\alpha_1, \alpha_2, \alpha_3) = 1$, there is a one-to-one correspondence between $\mathfrak{P}$ and a two-dimensional lattice $\Gamma_\alpha$. Thanks to these results, we have shown for various classes of arithmetic discrete planes, how to recover $\overrightarrow{x} \in \mathfrak{P}$ in correspondence with any $\overrightarrow{y} \in \Gamma_\alpha$. We also have investigated plane partitions and local configurations and extended the well-known result on the number of $(m, n)$-configurations in a naive plane, that is, there are at most $mn$ such configurations.

This approach offers new perspectives to investigate further general properties of arithmetic discrete planes of any thickness. In particular, we plan to use it to generate arbitrarily large parts of discrete planes via symbolic substitutions following [ABS04], to recover the corresponding Farey tessellation as well as the symmetry properties of $\overrightarrow{m}$-local configurations of a discrete plane [VC99], and finally as a new approach to the recognition problem of discrete planes [FST96, FP99, VC00].

## Acknowledgements

We would like to thank Fabrice Philippe for a careful reading of this paper.

# References

[AAS97]   Éric Andres, Raj Acharya, and Claudio Sibata. The Discrete Analytical Hyperplanes. *Graph. Models Image Process.*, 59(5):302–309, 1997.

[ABS04]   Pierre Arnoux, Valérie Berthé, and Anne Siegel. Two-dimensional Iterated Morphisms and Discrete Planes. *Theoret. Comput. Sci.*, 319:145–176, 2004.

[BB99]    Valentin E. Brimkov and Reneta P. Barneva. Graceful Planes and Thin Tunnel-Free Meshes. In *DGCI, 8th International Conference*, volume 1568 of *LNCS*, pages 53–64, 1999.

[BB02]    Valentin E. Brimkov and Reneta P. Barneva. Graceful Planes and Lines. *Theoret. Comput. Sci.*, 283:151–170, 2002.

[BV00]    Valérie Berthé and Laurent Vuillon. Tilings and Rotations on the Torus: A Two-Dimensional Generalization of Sturmian Sequences. *Discrete Math.*, 223:27–53, 2000.

[BV01]    Valérie Berthé and Laurent Vuillon. Palindromes and Two-Dimensional Sturmian Sequences. *J. Autom. Lang. Comb.*, 6(2):121–138, 2001.

[Col02]   Marie Andrée Jacob-Da Col. About Local Configurations in Arithmetic Planes. *Theoret. Comput. Sci.*, 283:183–201, 2002.

[DR95]    Isabelle Debled-Renesson. *Reconnaissance des Droites et Plans Discrets*. Thèse de doctorat, Université Louis Pasteur, Strasbourg, France, 1995.

[DRR94]   Isabelle Debled-Renesson and Jean-Pierre Reveillès. A New Approach to Digital Planes. In *Vision geometry III, Proc. SPIE*, volume 2356, Boston, USA, 1994.

[FP99]    Jean Françon and Laurent Papier. Polyhedrization of the boundary of a voxel object. In *DGCI, 8th International Conference*, volume 1568 of *LNCS*, pages 425–434, 1999.

[FST96]   Jean Françon, Jean-Maurice Schramm, and Mohamed Tajine. Recognizing Arithmetic Straight Lines and Planes. In *DGCI, 6th International Workshop*, LNCS, pages 141–150, 1996.

[Gér99]   Yan Gérard. Local Configurations of Digital Hyperplanes. In *DGCI, 8th International Conference*, volume 1568, pages 65–75, 1999.

[Rev91]   Jean-Pierre Reveillès. *Calcul en Nombres Entiers et Algorithmique*. Thèse d'état, Université Louis Pasteur, Strasbourg, France, 1991.

[Rev95]   Jean-Pierre Reveillès. Combinatorial Pieces in Digital Lines and Planes. In *Vision geometry IV, Proc. SPIE, 2573*, volume 2573, pages 23–24, San Diego, CA, 1995.

[Sch97]   Jean-Maurice Schramm. Coplanar Tricubes. In *DGCI, 7th International Workshop*, volume 1347 of *LNCS*, pages 87–98, 1997.

[VC97]    Joëlle Vittone and Jean-Marc Chassery. Coexistence of Tricubes in Digital Naive Plane. In *DGCI, 7th International Workshop*, volume 1347 of *LNCS*, pages 99–110, 1997.

[VC99]    Joëlle Vittone and Jean-Marc Chassery. (n,m)-cubes and Farey Nets for Naive Planes Understanding. In *DGCI, 8th International Conference*, volume 1568 of *LNCS*, pages 76–87, 1999.

[VC00]    Joëlle Vittone and Jean-Marc Chassery. Recognition of Digital Naive Planes and Polyhedrization. In *DGCI, 9th International Conference*, volume 1953 of *LNCS*, pages 296–307. IAPR, 2000.

[Vui99]   Laurent Vuillon. Local Configurations in a Discrete Plane. *Bull. Belgian Math. Soc.*, 6:625–636, 1999.

# Complexity Analysis for Digital Hyperplane Recognition in Arbitrary Fixed Dimension

Valentin E. Brimkov[1] and Stefan S. Dantchev[2]

[1] Fairmont State University, 1201 Locust Avenue, Fairmont,
West Virginia 26554-2470, USA
vbrimkov@fairmontstate.edu
[2] University of Durham, Science Labs,
South Road, Durham DH1 3LE, England
s.s.dantchev@durham.ac.uk.

**Abstract.** We consider the following problem. Given a set of points $M = \{p^1, p^2, \ldots, p^m\} \subseteq \mathbb{R}^n$, decide whether $M$ is a portion of a digital hyperplane and, if so, determine its analytical representation. In our setting $p^1, p^2, \ldots, p^m$ may be *arbitrary* points (possibly, with rational and/or irrational coefficients) and the dimension $n$ may be any arbitrary *fixed* integer. We provide an algorithm that solves this digital hyperplane recognition problem by reducing it to an integer linear programming problem of fixed dimension within an algebraic model of computation. The algorithm performs $O(m \log D)$ arithmetic operations, where $D$ is a bound on the norm of the domain elements.

**Keywords:** *Digital hyperplane, digital plane recognition, integer programming.*

## 1 Introduction

Digital plane segment (DPS) recognition is a basic problem in image analysis, attracting a lot of interest in recent years. Several algorithms for this problem have been proposed. (See the recent survey [5] by Brimkov, Coeurjolly, and Klette). [24] suggests an algorithm based on convex hull separability. Algorithm involving plane characterization by evenness in grid adjacency models is discussed in [26]. [9] proposes an approach based on tests for existence of lower and upper supporting ("oblique") planes for the given set of points. [14] suggests recognition by least-square optimization. See also [27] for further contributions. A number of algorithms exploit the idea to reduce the problem to a relevant linear program and solve it by employing existing methods from linear programming. [10] suggests a method by converting DPS to a system of $m^2$ linear inequalities, where $m$ is the cardinality of the given set of points. The system is solved by the Fourier elimination algorithm. One can also apply Fukuda's CDD algorithm for solving systems of linear inequalities by successive intersection of half-spaces defined by the inequalities. An efficient incremental algorithm based on a similar approach is proposed in [15]. In [8] Buzer presents an incremental linear time algorithm based on

E. Andres et al. (Eds.): DGCI 2005, LNCS 3429, pp. 287–298, 2005.
© Springer-Verlag Berlin Heidelberg 2005

solving a linear program by appropriate modification of Megiddo's algorithm [18]. Most of the above-mentioned algorithms perform well in practice. However, with a few exceptions (e.g., [8]), rigorous time complexity analysis is not available.

In the present theoretical work we consider somewhat more general version of the DPS recognition problem: Given a set of points $M = \{p^1, p^2, \ldots, p^m\} \subseteq \mathbb{R}^n$, decide whether $M$ is a portion of a digital hyperplane and, if so, determine that analytical digital hyperplane. Here $p^1, p^2, \ldots, p^m$ may be *arbitrary* points, possibly with integer and/or irrational coefficients. Such kind of data may result, e.g., from certain computational processes. The considerations take place in an *arbitrary dimension* $n$, provided that $n$ is *fixed* (i.e., bounded by an arbitrary constant). We provide an algorithm that solves the above problem by reducing it to an integer linear programming problem of a fixed dimension within an algebraic model of computation. This last problem is solved by a (theoretically) efficient algorithm based on a number of well-known results from theory of algorithms and complexity (some of them earlier authors' contributions). The algorithm works on input data that are *arbitrary* real numbers. In particular, it applies to problems with integer or rational data. Our algorithm solves the problem with $O(m \log D)$ arithmetic operations, where $D$ is a bound on the norm of the domain elements. The obtained theoretical results are somewhat in the spirit of Buzer's results [8] (first reported at DGCI'02).

To our knowledge of the available literature, this is the first integer programming based algorithm for a DPS recognition problem. The reason for absence of other similar methods is that ILP was believed to be inapplicable to DPS recognition due to its NP-hardness (see, e.g., related discussion in [8]). The present paper illustrates that from a theoretical point of view, for fixed dimensions, an integer linear program is almost as easy to solve as a linear program. Moreover, in some cases the proposed integer programming approach may have certain advantages over a linear programming approach, especially in avoiding very large integers that may result from a LP formulation. It also seems to us that our algorithm is the first one for DPS in *higher* dimensions, whose description is accompanied with rigorous complexity analysis. Another purpose of this work is to demonstrate the wealth of applying knowledge and results from other branches of theoretical computer science (such as theory of algorithms and complexity) to problems of digital geometry.

The paper is organized as follows. In Section 2 we recall some basic definitions from the theory of arithmetic planes and obtain the integer linear program corresponding to the considered problem. In Section 3 we present an integer programming algorithm that solves any integer program of the considered type. We conclude with some remarks in Section 4.

## 2    Feasible Digital Plane Recognition

In order to make our further considerations clearer, we first consider the 2D version of the DPS recognition problem, that is, a digital line segment recognition.

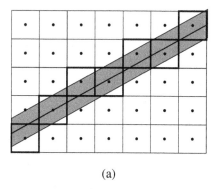 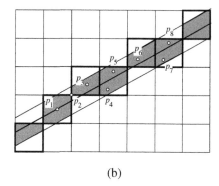

(a)                                    (b)

**Fig. 1.** Illustrations to the notions of feasibility. a) Feasible region related to a digital line. b) Feasible parts of pixels forming the feasible set of a digital line

Here we are given a set $M = \{p^1, p^2, \ldots, p^m\}$ of *integer* points in the plane, and we look for a digital line that contains these points.

Several equivalent definitions of a digital line are known (see the survey by Rosenfeld and Klette [22].) Here we conform to the analytical definition proposed by Reveillès [21].

A (naive) *digital line*[1] is a set of pixels $L(a_1, a_2, b, \max(|a_1|, |a_2|)) = \{(x_1, x_2) \in \mathbb{Z}^2 | 0 \leq a_1 x_1 + a_2 x_2 + b + \lfloor \max(|a_1|, |a_2|)/2 \rfloor < \max(|a_1|, |a_2|)\}$, where $a_1, a_2, \mu \in \mathbb{Z}$. $L(a_1, a_2, b, \max(|a_1|, |a_2|))$ can be considered as a discretization of a straight line with equation $ax_1 + ax_2 + b = 0$. It involves all pixels (unit squares centered at integer points of the plane) whose centers fall in between two parallel boundary straight lines $a_1 x_1 + a_2 x_2 + b + \lfloor \max(|a_1|, |a_2|)/2 \rfloor = 0$ and $a_1 x_1 + a_2 x_2 + b + \lfloor \max(|a_1|, |a_2|)/2 \rfloor = \max(|a_1|, |a_2|)$.[2] We will call the strip $F(a_1, a_2, b) = \{(x_1, x_2) \in \mathbb{R}^2 | 0 \leq a_1 x_1 + a_2 x_2 + b + \lfloor \max(|a_1|, |a_2|)/2 \rfloor < \max(|a_1|, |a_2|)\}$ a *feasible region* of $\mathbb{R}^2$ relative to $L(a_1, a_2, b, \max(|a_1|, |a_2|))$. See Fig. 1a.

Now consider a pixel $p \in L(a_1, a_2, b, \max(|a_1|, |a_2|))$. As Fig. 1b suggests, a part of $p$ is inside the feasible region $F(a_1, a_2, b)$, while the rest of it is outside $F(a_1, a_2, b)$. The former will be called the *feasible part* of $p$ relative to the line $L(a_1, a_2, b, \max(|a_1|, |a_2|))$ and denoted $F_{a_1, a_2, b}(p)$. The points of $F_{a_1, a_2, b}(p)$ will be referred to as *feasible points* of $p$. Finally, the union of all feasible parts of all pixels in a segment of a digital line $L$ will be called the *feasible set* of the digital line segment and denoted $F_L(a_1, a_2, b)$ (see Fig. 1b).

All above definitions and notions trivially extend to arbitrary dimension $n$. Thus a (naive) *digital hyperplane* is a set of $n$-cells[3]

---

[1] also called "arithmetic line."

[2] Because of the strict right inequality in the definition, pixels' centers cannot lie on the second line.

[3] $n$-dimensional counterparts of pixels.

$$H(a_1, a_2, \ldots, a_n, b, |a|_{\max})$$

$$= \left\{ (x_1, x_2, \ldots, x_n) \in \mathbb{Z}^n \Big| 0 \leq a_1 x_1 + a_2 x_2 + \ldots + a_n x_n + b + \left\lfloor \frac{|a|_{\max}}{2} \right\rfloor < |a|_{\max} \right\},$$

where $|a|_{\max} = \max(|a_1|, |a_2|, \ldots, |a_n|)$. (See [1, 2] for basic definitions and facts and [4] for further studies.) Its feasible region is

$$F(a_1, a_2, \ldots, a_n, b)$$

$$= \left\{ (x_1, x_2, \ldots, x_n) \in \mathbb{R}^n \Big| 0 \leq a_1 x_1 + a_2 x_2 + \ldots + a_n x_n + b + \left\lfloor \frac{|a|_{\max}}{2} \right\rfloor < |a|_{\max} \right\}.$$

A feasible part $F_{a_1,a_2,\ldots,a_n,b}(p)$ of an $n$-cell $p$ and a feasible set $F_H(a_1, a_2, \ldots, a_n, b)$ of a digital hyperplane $H$ are defined analogously to the 2D case.

With this preparation, we are able to state the following generalization of a digital hyperplane segment recognition problem, which we call the *feasible digital hyperplane segment* recognition problem and abbreviate FeasDHS.

## FeasDHS Recognition:

Given a set of points $M = \{p^1, p^2, \ldots, p^m\} \subseteq \mathbb{R}^n$, decide whether $M$ is included in the feasible part $F_H(a_1, a_2, \ldots, a_n, b)$ of some digital hyperplane $H(a_1, a_2, \ldots, a_n, b, |a|_{\max})$, and, if so, determine its coefficients $a_1, a_2, \ldots, a_n, b$.

Note that in this setting more than one point $p^i$ may belong to the same pixel of the discrete space. Moreover, a point $p^i$ may have irrational coordinates, such as the point $p^2$ in Fig. 1b.

We now obtain formulation of FeasDHS in terms of an integer programming program.

It is not hard to realize that an element $p^i$ of $M$ is a feasible point of some $n$-cell $v$ (i.e., $p^i \in F_{a_1,a_2,\ldots,a_n,b}(v)$) if and only if there exist integers $a_1, a_2, \ldots, a_n$, and $b$, such that the following conditions are met:

1. $0 \leq a_1 p_1^i + a_2 p_2^i + \ldots + a_n p_n^i + b + \left\lfloor \frac{|a|_{\max}}{2} \right\rfloor < |a|_{\max}$, and

2. $0 \leq a_1 \lceil p_1^i \rfloor + a_2 \lceil p_2^i \rfloor + \ldots + a_n \lceil p_n^i \rfloor + b + \left\lfloor \frac{|a|_{\max}}{2} \right\rfloor < |a|_{\max}$.

($\lceil . \rfloor$ denotes the operator "the closest integer" to a given real number. If $x$ is a "half-integer", we set $\lceil x \rfloor = \lceil x \rceil$, e.g., $\lceil 3.5 \rfloor = 4$.)

The first condition causes $p^i$ to belong to the feasible region relative to a digital hyperplane with coefficients $a_1, a_2, \ldots, a_n$ and $b$, while the second one ensures that $p^i$ belongs to an $n$-cell from the same digital hyperplane. Note that both conditions are essential: If Condition 1 is missing, then $p^i$ may be outside the feasible region. If Condition 2 does not hold, then $p^i$ may not belong to all $n$-cells from the digital hyperplane with coefficients $a_1, a_2, \ldots, a_n, b$.

When $i$ runs from 1 to $m$, we get an integer linear problem with $n + 1$ unknowns $a_1, a_2, \ldots, a_n, b$ and $4m$ linear constraints.

As already mentioned, we will deal with the case when the dimension $n$ is an arbitrary fixed integer. We will also suppose that the coefficients $a_1, a_2, \ldots, a_n, b$

we look for are bounded in size, i.e. $|a_1| \leq d_1, |a_2| \leq d_2, \ldots, |a_n| \leq d_n, |b| \leq d_{n+1}$, as the bounds $d_1, d_2, \ldots, d_{n+1}$ are a part of the problem input. In the next section we will see that this condition is dictated by the very nature of the problem, especially by the fact that some of the coefficients may be irrational numbers. From a practical point of view, this condition does not restrict the generality, as we can always suppose that the absolute value of the largest coefficient is bounded by, e.g., the largest positive integer that we may use in practice. Moreover, by assuming bounds on the plane coefficients one can avoid occurrence of very large numbers in the problem solution.

# 3 Algorithm for Integer Programming of Fixed Dimension

In this section we describe an efficient algorithm for integer linear programming programs as those corresponding to FeasDHS. Consider the following integer linear program:[4]

> (ILP)  Given a matrix $A \in \mathbb{R}^{m \times n}$ and vectors $b \in \mathbb{R}^m, d \in \mathbb{R}^n$,
> find $x \in \mathbb{Z}^n$ such that $Ax \leq b$, where $\mathbf{0} \leq x \leq d$.

To simplify our further considerations, we have assumed that the coordinates of a domain element $x = (x_1, \ldots, x_n)$ satisfy the conditions $0 \leq x_i \leq d_i$ rather than $|x_i| \leq d_i$, $1 \leq i \leq n$. Clearly, a problem with constraints of the first type is equivalent to one with constraints of the second type up to a change of the variables.

The input entries are arbitrary real numbers and the adopted model of computation is an algebraic computation model. This kind of model has been traditionally used in scientific computing, algebraic complexity, computational geometry, and (although not explicitly) numerical analysis (see, e.g., [19, 20, 25]). In that model, the assumption is that all the real numbers in the input have unit size, and the basic algebraic operations $+, -, *, /$ and the relation $\leq$ are executable at unit cost. Thus the algebraic complexity of a computation on a problem instance is the number of operations and branchings performed to solve the instance.

At this point it is important to mention that the requirement in the ILP formulation for bounded domain (i.e., $\mathbf{0} \leq x \leq d$) is essential and predetermined by the intrinsic nature of the problem, namely by the fact that the coefficients may be *irrational* numbers. In such a case, a problem with unbounded domain may be, in general, *undecidable*, as shown in [6].

In the rest of this paper we present an algorithm for ILP when the value of $n$ is fixed. The algorithm consists of two stages: a reduction of the given real

---

[4] In the feasDHS formulation we have certain rounding operations. It is well-known [3] that rounding of a real number $x$ can be performed in $\log |x|$ basic arithmetic operations. Thus the coefficients of the second inequality in the feasDHS definition can be computed in $O(m \log |x|_{\max})$ time overall, where $|x|_{\max} = \max(|x_1|, |x_2|, \ldots, |x_n|)$.

input to an integer input determining the same admissible set, followed by an application of Lenstra's algorithm [16]. The first stage involves simultaneous Diophantine approximation techniques, while the second employs two well-known algorithms: the Lovász' basis reduction algorithm [17] and the Hermite normal form algorithm (see, e.g., [13]).

### 3.1     Subroutines to the Main Algorithm

**Lovász Lattice Basis Reduction Algorithm.** The input to Lovász algorithm consists of linearly independent vectors $b_1, b_2, \ldots b_n \in \mathbb{Q}^n$, considered as a basis for a lattice $L$. The algorithm transforms them iteratively. At the end, they form a basis for $L$ which is reduced in the Lovász sense. First we recall some definitions, then describe the Lovász lattice basis reduction algorithm itself, following [11].

With a basis $b_1, b_2, \ldots b_n$, we associate the orthogonal system $b_1^*, b_2^*, \ldots b_n^*$, where $b_i^*$ is the component of $b_i$ which is orthogonal to $b_1, b_2, \ldots b_{i-1}$. The vectors $b_1^*, b_2^*, \ldots b_n^*$ can be computed by Gram-Schmidt orthogonalization:
$$b_1^* = b_1, \quad b_i^* = b_i - \sum_{j=1}^{i-1} \mu_{i,j} b_j^*, \quad 2 \leq i \leq n, \quad \mu_{i,j} = \langle b_i, b_j^* \rangle \big/ \|b_j^*\|^2.$$

The basis $b_1, b_2, \ldots b_n$ is *size-reduced* if all $|\mu_{i,j}| \leq \frac{1}{2}$. Given an arbitrary basis $b_1, b_2, \ldots b_n$, we can transform it into a size-reduced basis with the same Gram-Schmidt orthogonal system, as follows:

For every $i$ from 2 to $n$; for every $j$ from $i-1$ to 1;
Set $b_i := b_i - \lceil \mu_{i,j} \rfloor b_j$ and update $\mu_{i,k}$ for $1 \leq k \leq i-1$, by setting $\mu_{i,k} = \mu_{i,k} - \lceil \mu_{i,j} \rfloor \mu_{j,k}$.

We outline a variant of the Lovász lattice basis reduction algorithm next.

1. *Initiation.* Compute the Gram-Schmidt quantities $\mu_{i,j}$ and $b_i^*$ for $1 \leq j < i \leq n$. Size-reduce the basis.
2. *Termination condition.* If $\|b_i^*\|^2 \leq 2 \|b_{i+1}^*\|^2$ for $1 \leq i \leq n-1$, then stop.
3. *Exchange step.* Choose the smallest $i$ such that $\|b_i^*\|^2 > 2 \|b_{i+1}^*\|^2$. Exchange $b_i$ and $b_{i+1}$. Update the Gram-Schmidt quantities. Size-reduce the basis. Go to 2.

Gram-Schmidt quantities in Step 3 are updated as follows:
$$\|b_i^*\|^2_{new} = \|b_{i+1}^*\|^2 + \mu_{i+1,i}^2 \|b_i^*\|^2, \quad \|b_{i+1}^*\|^2_{new} = \|b_i^*\|^2 \|b_{i+1}^*\|^2 \big/ \|b_i^*\|^2_{new}$$
$$\mu_{i+1,i}^{new} = \mu_{i+1,i} \|b_i^*\|^2 \big/ \|b_i^*\|^2_{new}$$
$$\begin{pmatrix} \mu_{i,j}^{new} \\ \mu_{i+1,j}^{new} \end{pmatrix} = \begin{pmatrix} \mu_{i+1,j} \\ \mu_{i,j} \end{pmatrix} \text{ for } 1 \leq j \leq i-1$$
$$\begin{pmatrix} \mu_{j,i}^{new} \\ \mu_{j,i+1}^{new} \end{pmatrix} = \begin{pmatrix} 1 & \mu_{i+1,i}^{new} \\ 0 & 1 \end{pmatrix} \begin{pmatrix} 0 & 1 \\ 1 & -\mu_{i+1,i} \end{pmatrix} \begin{pmatrix} \mu_{j,i} \\ \mu_{j,i+1} \end{pmatrix} \text{ for } i+2 \leq j \leq n.$$

The other $\|b_i^*\|^2$'s and $\mu_{i,j}$'s do not change.

After termination of the above algorithm, we have a size-reduced basis for which $\|b_i^*\|^2 \leq 2 \|b_{i+1}^*\|^2$, $1 \leq i \leq n-1$. We call such a basis *reduced in the Lovász sense*. The following lemma was proved in [7].

**Lemma 1.** *The algebraic complexity of Lovász' basis reduction algorithm applied to an $n \times n$ rational matrix with entries of size $O(S)$, is $O(Sn^5 \log n)$, and the bit-size of the entries in the reduced basis is $O(Sn^3)$.*

**Hermite Normal Form Algorithm.** In the algorithm's description we follow [23]. The input for the algorithm is an $m \times n$ $(m \le n)$ integer matrix $A$ of full rank. The algorithm uses a matrix of the form

$$A' = \left( A \ \middle| \ \begin{matrix} M & & \\ & \ddots & \\ & & M \end{matrix} \right),$$

where $M$ is the absolute value of some nonsingular $m \times m$ minor of $A$. $A'$ has the same Hermite normal form as $A$. The algorithm consists of the following five steps:

1. Cause all the entries of the matrix $A$ to fall into the interval $[0, M)$, by adding to the first $n$ columns of $A'$ proper integer multiples of the last $n$ columns;
2. For $k$ from 1 to $m$ do 3-4;
3. If there are $i \neq j$, $k \le i, j \le n + k$, such that $a'_{k,i} \ge a'_{k,j} > 0$, then subtract from the $i$th column the $j$th one multiplied by $\left\lfloor \frac{a'_{k,i}}{a'_{k,j}} \right\rfloor$. Then reduce the $i$th column modulo $M$. Go to 3;
4. Exchange the $k$th column and the only column with $a'_{k,i} > 0$;
5. For every $i$ from 2 to $n$; for every $j$ from 1 to $i - 1$, add an integer multiple of the $i$th column to the $j$th one, to get $a'_{i,i} > a'_{i,j} \ge 0$.

We have the following lemma [7].

**Lemma 2.** *Let $A$ be an $m \times n$ $(m \le n)$ integer matrix of full rank with entries of size $O(S)$. Then the algebraic complexity of the Hermite normal form algorithm that reduces $A$ into its Hermite normal form, is $O(m^2 n(\log m + S))$, and the bit-size of all resulting integers is $O(Smn)$.*

Since the above lemma admits a short proof, we sketch it next in order to provide the reader with an idea how statements of this kind can be demonstrated.

We introduce the function

$$F\left(a'_{k,k}, a'_{k,k+1}, \ldots a'_{k,n+k}\right) := \prod_{k \le i \le n+k} a'_{k,i}$$

for $a'_{k,i} > 0$. After one iteration of Step 3, we have

$$F_{new}/F = (a'_{k,i} - \lfloor a'_{k,i}/a'_{k,j} \rfloor \, a'_{k,j})/a'_{k,i},$$

which implies both $F_{new}/F < 1/2$ and $F_{new}/F < a'_{k,j}/a'_{k,i}$. It is not hard to see that one iteration of Step 3 can be performed in time $O\left(m \log(a'_{k,i}/a'_{k,j})\right) =$

$O\left(m \log(F/F_{new})\right)$. So, Step 3 takes $O\left(m \log \frac{F_{start}}{F_{end}}\right)$ time. We have that $F_{start} < M^{n+1}$, $F_{end} \geq 1$. Moreover, we have the following simple fact: if $a$ is a non-zero rational number of bit-size at most $S$, then $1/2^S \leq |a| \leq 2^S$. This last fact implies the following property of matrices: given a non-singular $n \times n$ rational matrix $B$ whose entries are of bit-size at most $S$, then $1/2^{n^2 S} \leq |\det(B)| \leq n!2^{nS}$. From here we obtain $M = O\left(m!2^{mS}\right)$. Hence, the overall running time of Step 3 is $O\left(nm\left(\log m + S\right)\right)$. Then, the complexity of the Hermite normal form algorithm is $O\left(nm^2\left(\log m + S\right)\right)$. Since all the resulting integers are smaller than $M$, their bit-size is $O\left(Smn\right)$.

## 3.2   Simultaneous Diophantine Approximation

Our algorithm employs in one of its steps the well-known algorithm for finding a simultaneous Diophantine approximation to a given rational vector. Specifically, we will use the following lemma.

**Lemma 3.** *(see, e.g., [23–Corollary 6.4c]) There exists a polynomial algorithm which, given a vector $a \in \mathbb{Q}^n$ and a rational number $\varepsilon$, $0 < \varepsilon < 1$, finds an integral vector $p$ and an integer $q$ such that $\|a - \frac{1}{q}p\| < \varepsilon/q$, and $1 \leq q \leq 2^{n(n+1)/4}\varepsilon^{-n}$.*

We will also need an algorithm that reduces the constraints with real coefficients to constraints with integer coefficients, determining the same admissible set. The first phase of this reduction is a substitution of a given *real* vector with an appropriate *rational* vector, justified by the following lemma.

**Lemma 4.** *Given a vector $\alpha \in \mathbb{R}^n$ with $|\alpha_j| \leq 1, j = 1, 2, \ldots, n$, and $D \in \mathbb{Z}_+$, there exists an $O(n^4 \log n(n + \log D))$ algorithm that finds $p \in \mathbb{Z}^n$ and $q \in \mathbb{Z}_+$ such that $|\alpha_j - p_j/q| < 1/(qD)$, $j = 1, 2, \ldots, n$, and $1 \leq q \leq \lceil 2^{n(n+5)/4}D^n \rceil$.*

The required $p \in \mathbb{Z}^n$ and $q \in \mathbb{Z}_+$ can be found as follows.

### Diophantine Approximation to a Real Vector

1. For each $\alpha_j$, $1 \leq j \leq n$, find the closest rational fraction $a_j$ with denominator $G = \lceil 2^{n(n+5)/4}D^{n+1} \rceil$.
2. Apply the algorithm of Lemma 3 with input $a = (a_1, \ldots, a_n) \in \mathbb{Q}^n$ and $\varepsilon = 1/(2D)$. □

By Lemma 3, the output is a vector $p \in \mathbb{Z}^n$ and an integer $q \in \mathbb{Z}_+$ with

$$\|a - (1/q)p\| < 1/(2qD) \text{ and } 1 \leq q \leq \lceil 2^{n(n+5)/4}D^n \rceil.$$

Clearly, $|\alpha_j - a_j| \leq 1/(2G)$. Then we have

$$\left|\alpha_j - \frac{p_j}{q}\right| \leq |\alpha_j - a_j| + \left|a_j - \frac{1}{q}p_j\right|$$

$$\leq |\alpha_j - a_j| + \left\|a - \frac{1}{q}p\right\| < \frac{1}{2G} + \frac{1}{2qD}$$

$$\leq \frac{1}{2.\lceil 2^{n(n+5)/4}D^n\rceil.D} + \frac{1}{2qD} \leq \frac{1}{qD},$$

i.e., the obtained vector $p$ and integer $q$ are as desired.

Consider first Step 1. For a given real number $\alpha_j$, the closest rational fraction with denominator $G = \lceil 2^{n(n+5)/4}D^{n+1}\rceil$ can be found in time $O(\log G) = O(n^2 + n\log D)$. Thus the overall time complexity of Step 1 is $O(n^3 + n^2\log D)$.

Step 2 involves the simultaneous Diophantine approximation algorithm applied to the particular class of inputs $a \in \mathbb{Q}^n$, $\varepsilon = 1/(2D)$ obtained in Step 1. As a matter of fact, this is a specialization of the Lovász basis reduction algorithm, applied to a certain matrix. It has been proved in [6–Lemma 4.4] that the *number of iterations* performed in this step is $O(n^4\log n(n + \log D))$. Then the overall time complexity of the algorithm of Lemma 4 is $O(n^4\log n(n + \log D))$, as well.

The algorithm of Lemma 4 can be used to substitute any *real* constraint $ax \leq b$ with an *integer* one, preserving the same admissible integer points $x$ with $0 \leq x \leq d$, $d \in \mathbb{R}^n$. More precisely, we have the following lemma.

**Lemma 5.** *Let* $T = \{x \in \mathbb{Z}^n : ax \leq b; 0 \leq x \leq d\}$, *where* $a \in \mathbb{R}^n$, $b \in \mathbb{R}$, $d \in \mathbb{Z}_+^n$. *Then there exists an algorithm which finds a vector* $r \in \mathbb{Z}^n$ *and a number* $r_0 \in \mathbb{Z}$ *such that* $T = \{x \in \mathbb{Z}^n : rx \leq r_0; 0 \leq x \leq d\}$. *The algorithm involves at most $n$ applications of the algorithm from Lemma 4, with* $D = ||d||$.

Proof of the above fact is available in [6–Lemma 5.1]. Now we are able to complete the algebraic complexity analysis of integer programming of fixed dimension, which we do in the next section.

## 3.3    Algorithm for ILP

In this section we use the results from the previous section to obtain an $O(m\log D)$ algorithm for ILP, where $D = ||d||$, as defined in Lemma 5.

As already mentioned, the algorithm consists of two stages. In the *first stage*, it reduces the constraints with real coefficients to constraints with integer coefficients which determine the same admissible set of integer points. In the *second stage*, the Lenstra's algorithm [16] is applied to the integer data problem obtained as an output of the first stage.

From Lemmas 4 and 5, we obtain that the overall time complexity of the reduction stage is $O(mn^5\log n(n + \log D))$. Furthermore, the bit-size of the generated integers is $O(n^2(n + \log D))$. Therefore, the overall bit-size of the reduced problem is $O(mn^3(n + \log D))$.

We now complete the complexity analysis of the second stage of the algorithm. That stage is an application of the Lenstra's [16] algorithm to the integer linear problem obtained as output of the first stage. A recursive step of Lenstra's algorithm reduces an $n$-dimensional problem to a set of subproblems of dimension $n - 1$, whose number is exponential but depending only on $n$. The basic algorithms used in this reduction are the Lovász basis reduction algorithm and the Hermite normal form algorithm. In addition, in order to compute a homothetic approximation to the underlying polyhedron with constant homothety ratio, a number of linear programming problems of dimension $(m + 2n) \times n$ have to be solved.

The Lovász basis reduction algorithm and the Hermite normal form algorithm are both applied to matrices of dimension depending only on $n$. Moreover, all entries of these matrices are of bit-size $O(\log D)$, as the value of $n$ is fixed. Then, by Lemmas 1 and 2, the complexity of the two algorithms as well as the bit-size of the integers they generate, are bounded by $O(\log D)$.

During the execution of the Lenstra's algorithm, there are $O(\log D)$ linear programming problems to be solved. Each of them can be solved in time $O(m+n)$ (i.e., *linear* in $m$) using the well-known Megiddo's algorithm [18]. Hence, if $n$ is fixed, the overall complexity of this stage is $O(m \log D)$. This completes the proof of the following theorem.

**Theorem 1.** *There is an $O(m \log D)$ algorithm for ILP with a fixed number of variables, where $D = ||d||$.*

### 3.4     Theoretical Versus Practical Efficiency

The proposed algorithm solves the considered problem ILP within an algebraic computation model by performing $O(m \log D) = O(m \log ||d||)$ arithmetic operations for any fixed dimension $n$. Usually, algorithms of such kind of complexity are considered as theoretically efficient. However, in order to make a reasonable foresight about the practical efficiency of the computation, one also has to evaluate the implicit constant hidden in the big-$O$ notation.

Specifically, in order to solve an ILP (and thus the original hyperplane recognition problem) the algorithm uses as subroutines a number of well-known algorithms for some basic combinatorial problems. Keeping in mind the algorithm description, it is not hard to realize that the overall number of these problems is exponential in $n$. In practice, however, it might not be a problem for two reasons. First, $n$ is a constant (usually a small one) and, second, the average-case time complexity of ILP is believed to be much lower than the worst-case time complexity. Moreover, the Lovász lattice basis reduction algorithm and the Hermite normal form algorithm are polynomial in $n$ and therefore very efficient even for relatively large dimensions. The only exception is the Megiddo's linear programming algorithm, whose time complexity involves an implicit constant factor of the order $\Omega(2^{n^2})$. For relatively small dimensions Megiddo's algorithm is known to perform well in practice. For moderately large dimensions one can use instead of Megiddo's algorithm some recent more practical algorithms that have better theoretical running time,[5] are easier to implement, and perform well in practice.

For large dimensions $n$, however, the algorithm is clearly inefficient, like all other algorithms involving the Megiddo's method (in particular, Buzer's digital plane recognition algorithm mentioned above). Nevertheless, results of this kind

---

[5] For example, [12] provides a randomized linear programming algorithm whose running time involves an implicit constant factor that is subexponential in $n$.

provide useful insight on certain limitations that an efficient computation may feature.

The ultimate test for our algorithm is, of course, an efficient implementation that would allow us to run it on real data and compare it with the existing algorithms. We see this as an important direction for future research.

# 4     Concluding Remarks

We have presented an $O(m \log D)$ algorithm for solving the digital hyperplane segment recognition problem in arbitrary fixed dimension, where $D = ||d||$ is a bound on the norm of the domain elements (possible hyperplane coefficients). The input may be a set of points with arbitrary real coordinates. The algorithm also applies to classical digital plane recognition where the given points have integer coefficients.

The algorithm works on an integer linear program formulation and solves it theoretically efficiently. We believe that this result, together with some other theoretical results, will contribute to the better understanding structural, algorithmic, and complexity issues of digital plane recognition.

# Acknowledgements

The authors thank the two anonymous referees for their useful remarks and suggestions.

# References

1. Andres, E., *Modélisation Analytique Discrète d'Objets Géométriques*, Thèse de habilitation à diriger des recherches, Universit"e de Poitiers, Poitiers, France, 2001
2. Andres, E., R. Acharya, C. Sibata, Discrete analytical hyperplanes, *Graphical Models Image Processing* **59**, 302–309 (1997)
3. Blum, L., M. Shub, S. Smale, On a Theory of Computation and Complexity over the Real Numbers: NP-Completeness, Recursive Functions and Universal Machines, *Bull. Amer. Math. Soc.* (NS) **21**, 1–46 (1989)
4. Brimkov, V.E., E. Andres, R.P. Barneva, Object Discretizations in Higher Dimensions, *Pattern Recognition Letters*, **23**, 623–636 (2002)
5. Brimkov, V.E., D. Coeurjolly, R. Klette, Digital Planarity - A Review, CITR-TR 142, 2004
6. Brimkov, V.E., S.S. Danchev, Real Data – Integer Solution Problems within the Blum-Shub-Smale Computational Model, *J. of Complexity* **13**, 279–300 (1997)
7. Brimkov, V.E., S.S. Dantchev, On the Complexity of Integer Programming in the Blum-Shub-Smale Computational Model, In: Theoretical Computer Science. Exploring New Frontiers of Theoretical Informatics, van Leeuwen, J., O. Watanabe, M. Hagiya, P.D. Mosses, T. Ito (Eds.), LNCS-1872, 286-300 (2000)
8. Buzer, L., A Linear Incremental Algorithm for Naive and Standard Digital Dines and Planes Recognition, *Graphical Models* **65** 61–76 (2003)

9. Debled-Rennesson, I., J.-P. Reveillès, A New Approach to Digital Planes, *Vision Geometry III*, SPIE-2356, 12–21 (1994)
10. Françon, J., J.M. Schramm, M. Tajine, Recognizing Arithmetic Straight Lines and Planes, 6th Int. Conf. *Discrete Geometry for Computer Imagery*, Springer, LNCS-1176, 141–150 (1996)
11. Hastad, J., B. Just, J.C. Lagarias, C.P. Schnoor, Polynomial Time Algorithms for Finding Integer Relations among Real Numbers, *SIAM J. Comput.* **18**, 859–881 (1989)
12. Kalai, G., A Subexponential Randomized Simplex Algorithm, *24th Annual ACM Symposium on the Theory of Computation*, ACM Press, 475-482 (1992)
13. Kannan, R., A. Bachem, Polynomial Algorithms for Computing the Smith and Hermite Normal Forms of an Integer Matrix, *SIAM J. Comput.* **8**, 499–507 (1979)
14. Klette, R., I. Stojmenović, J. Žunić, A Parametrization of Digital Planes by Least Square Fits and Generalizations, *Graphical Models Image Processing* **58**, 295–300 (1996)
15. Klette, R., H.-J. Sun, Digital Planar Segment Based Polyhedrization for Surface Area Estimation, In: Arcelli, C., L.P. Cordella, and G. Sanniti di Baja, editors, *Visual Form 2001*, Springer, Berlin, pages 356–366 (2001)
16. Lenstra, H.W., Jr., Integer Programming with a Fixed Number of Variables, *Math. Oper. Res.* **8**, 538–548 (1983)
17. Lenstra, A.K., H.W. Lenstra, Jr., L. Lovász, Factoring Polynomials with Rational Coefficients, *Math. Ann.* **261**, 515–534 (1982)
18. Megiddo, N., Linear Programming in Linear Time when the Dimension is Fixed, *J. of ACM* **31** (1), 114–127 (1984)
19. Novak, E., The Real Number Model in Numerical Analysis, *J. of Complexity* **11**, 57–73 (1994)
20. Preparata, F.P., M.I. Shamos, *Computational Geometry*, Springer-Verlag, Berlin Heidelberg New York, 1985
21. Reveillès, J.-P., Géométrie Discrète, Calcul en Nombres Entiers et Algorithmique, Thèse d'état, Univ. Louis Pasteur, Strasbourg, 1991
22. Rosenfeld, A., R. Klette, Digital Straightness, In: *Electronic Notes in Theoretical Computer Science* **46** (2001)
23. Schrijver, A., *Theory of Linear and Integer Programming*, Wiley, Chichester New York Brisbane Toronto Singapore, 1986
24. Stojmenović, I., R. Tosić, Digitization Schemes and the Recognition of Digital Straight Lines, Hyperplanes and Flats in Arbitrary Dimensions, *Vision Geometry, Contemporary Mathematics Series*, **119** 197–212 (1991)
25. Strassen, V., Algebraic Complexity Theory, In: van Leeuwen, J. (Ed.), *Handbook of Theoretical Computer Science*, Vol. A, Elsevier, Amsterdam, 633–672 (1990)
26. Veelaert, P., Digital Planarity of Rectangular Surface Segments, *IEEE Pattern Analysis and Machine Int*, **16**, 647–652 (1994)
27. Vittone, J., J.-M. Chassery, Recognition of Digital Naive Planes and Polyhedrization, 9th Int. Conf. *Discrete Geometry for Computer Imagery*, Springer, LNCS-1953, 296–307 (2000)

# An Elementary Algorithm for Digital Line Recognition in the General Case

Lilian Buzer

A2SI Laboratory, ESIEE, 2 bd Blaise Pascal,
Cité Descartes, - BP 99, 93162 Noisy-Le-Grand Cedex, France
`Buzerl@esiee.fr`
Institut Gaspard Monge, Unité Mixte CNRS-ESIEE, UMR 8049

**Abstract.** This paper is concerned with the naive and, more generally, $\alpha$-thick digital line recognition problem. Previous incremental algorithms deal with the 8-connected case [DR95] or with sophisticated machinery coming from Linear Programming [Buz03]. We present the first elementary method [Buz02] that works with any set of points (not necessarily 8-connected) and we propose a linear time algorithm under some restrictions (which were implicitly assumed in [DR95]). This paper deals with implementation details giving pseudo-code of our method. We insist on linking the recognition problem to the intrinsic properties of convex hulls.

**Keywords:** Digital line, incremental, recognition, convex hull, thickness, implementation.

## 1 Introduction

When one processes digital images, one also sometimes wants to know how to recognize basic geometric entities. In this way appears the recognition problem of digital lines of variable thickness (see Fig. 1). In this paper, we set up the definition of $\alpha$-thick digital line which allows to represent a wide variety of digital line segments. We show that the knowledge of some basic information on the convex hull of a set of points $S$ (thickness or critical supporting lines) is sufficient to determine if $S$ is a subset of an $\alpha$-thick digital line segment. Then, we adapt this remark in order to create an elementary algorithm for the recognition problem. We exhibit its optimal complexity relatively to many different configurations (static, incremental, dynamic). Thus, we propose a short implementation of the linear incremental version which works when points are inserted in a given direction. This new method and the previous one based on Megiddo algorithm [Buz03] are able to incrementaly process $\alpha$-thick digital line with non-connected pixels. In [Buz03] we only deal with naive digital line (where $\alpha = 1$), but this previous method is able to recognize $\alpha$-thick digital line by replacing the thickness of 1 by $\alpha$. Conversely, the new approach can operate naive digital line when we set $\alpha$ equal to 1. If the two algorithms recognize the same objects using the same thickness notion, the way they work is completely

E. Andres et al. (Eds.): DGCI 2005, LNCS 3429, pp. 299–310, 2005.
© Springer-Verlag Berlin Heidelberg 2005

different. The Megiddo approach does not explicitly compute the new thickness value when a point is inserted. We can only know if the current thickness is below a given threshold. With the new method, we are able to determine the current thickness value at any moment and to add other criteria that can be estimated from the thickness : density of pixels or curvature. This entails a higher complexity $O(n\log n)$ but in most cases we will work in configurations that allow to use a linear and elementary version of this new approach. Throughout this paper, we will denote by $O_x$ and $O_y$ the axes of the Cartesian coordinate system, by $e_x$ and $e_y$ the associated units vectors and by $(x, y)$ the Cartesian components of points.

## 2    $\alpha$-Thick Digital Line

We hereafter present $\alpha$-thick digital line. Its seminal definition was given by Reveillès in [Rev91]. Any digital line $D$ in $\mathbb{Z}^2$ is described by a set of parameters: the *normal vector* $N = (a, b)$ in $\mathbb{Z}^2 \backslash \{0\}$ with $gcd(a, b)$ equal to 1, the *inferior bound* $\gamma$ in $\mathbb{Z}$ and the *arithmetic thickness* $w$ in $\mathbb{Z}$. A point $(x, y)$ of a digital line with parameters $(a, b), \gamma, w$ verifies the following diophantine inequality:

$$\gamma \le a.x + b.y < \gamma + w \tag{1}$$

If we choose the arithmetic thickness to be equal to the infinity norm of the normal vector: $||N||_\infty = \sup\{|a|, |b|\}$, we obtain an 8-connected object called *a naive digital line* (see Fig. 1). We extend this definition to the $\alpha$-*thick digital line* where $\alpha$ corresponds to a thickness ratio whose reference ($\alpha = 1$) is the naive digital line. A point $(x, y)$ belongs to such a line if it verifies:

$$\gamma \le a.x + b.y < \gamma + \alpha. \sup\{|a|, |b|\} \tag{2}$$

We can consider three different subclasses of digital lines depending on the components of the normal vector $N = (a, b)$. When $|a| = |b| = 1$, the digital line has a slope of $\pm 45°$. When $|a| < |b|$ (resp. $|b| < |a|$), the resulting slope relative to the $O_x$ axis (resp. $O_y$ axis) is always comprised between $-45°$ and $+45°$ (see Fig. 2).

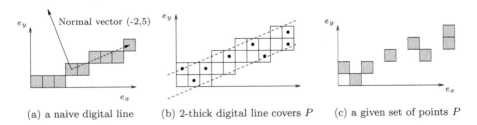

(a) a naive digital line        (b) 2-thick digital line covers $P$        (c) a given set of points $P$

**Fig. 1.** The recognition problem using different thickness for the digital lines

# 3   Thickness Criterion

## 3.1   Introducing the Notion of Thickness

The definition of a digital line is intrinsically algebraic. We use an equivalent characterization which is more linked to the field of Euclidian geometry:

**Lemma 1.** *A set of points is a piece of an $\alpha$-thick digital line if and only if these points can be covered by a strip of rational slope and of horizontal or vertical thickness strictly inferior to $\alpha$.*

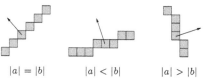

$|a| = |b|$      $|a| < |b|$      $|a| > |b|$

**Fig. 2.** Different types of digital line orientations

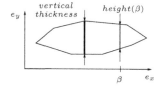

**Fig. 3.** Height and vertical thickness of a convex hull

## 3.2   Thickness and Convex Hull

In this subsection, we show that the notion of thickness (more precisely the thickness of the convex hull of the input points) plays an important role in the recognition problem.

**Definition 1.** *The height at abscissa $\beta$ of a convex set $C$ is defined to be the length of the segment resulting from the intersection of $C$ with the vertical line $x = \beta$. We call the vertical thickness of $C$ the maximum reached by $height(x)$ (see Fig. 3). The width and the horizontal thickness are symmetrically defined.*

**Lemma 2.** *A convex polygon $N$ has a vertical thickness strictly less than $\alpha$ iff there exists a strip of vertical thickness strictly less than $\alpha$ that covers $N$.*

▶ Proof: let $x$ denote the abscissa which corresponds to the maximum height of $N$. The upper and lower border of $N$ at abscissa $x$ can be linked to either a vertex or an edge. We can consider three different configurations:

1. edge-edge: this case only appears if both edges are parallel (see Fig. 4.b). If this were not the case, it would exist a greater value for the vertical thickness of $N$ (see Fig. 4.a). As $N$ is convex, it is included in the strip defined by these two edges. So $N$ can be covered by a strip of correct thickness.
2. edge-vertex: as the maximum is achieved at this abscissa, the line passing through this vertex and parallel to this edge is tangent to $N$ (see Fig. 4.c). We use this line and the previous edge to build a valid strip.

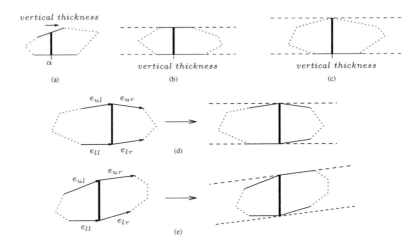

**Fig. 4.** The different configurations of the vertical thickness location

3. vertex-vertex: this case is a bit more complicated. Let $e_{ul}, e_{ur}, e_{ll}$ and $e_{lr}$ denote the emerging edges from these two vertices. As the polygon is convex, we have $slope_{e_{ur}} \leq slope_{e_{ul}}$ and $slope_{e_{lr}} \geq slope_{e_{ll}}$. The maximum height reached at this abscissa implies $slope_{e_{ul}} \geq slope_{e_{ll}}$ and $slope_{e_{ur}} \leq slope_{e_{lr}}$. If we choose $e_{ll}$ as a border of our strip, $e_{ul}$ and $e_{lr}$ lies inside. Assume that $slope_{e_{ur}} \leq slope_{e_{ll}}$. In this case (see Fig. 4.d) $N$ is included in a valid strip. In the opposite case where $slope_{e_{ur}} > slope_{e_{ll}}$ (see Fig. 4.e), this edge can not be chosen as a border for our strip. Hopefully, $e_{ur}$ is a correct choice. Indeed, we have $slope_{e_{ur}} \leq slope_{e_{ul}}$ and $slope_{e_{ur}} \leq slope_{e_{lr}}$ by assumptions; as $slope_{e_{ur}} > slope_{e_{ll}}$ this finally implies that $N$ is included in a strip of vertical thickness strictly less than $\alpha$. ◀

Using lemmas 1 and 2, we finally obtain the next property which links the thickness of a convex hull to the $\alpha$-thick digital line recognition problem:

*Property 1.* A set of points is a piece of an $\alpha$-thick digital line if and only if its convex hull has an horizontal or vertical thickness strictly inferior to $\alpha$.

## 3.3    The Importance of Convex Hull

A convex set $C$ has a vertical thickness strictly less than $\alpha$ if and only if $C$ and its translation by $\alpha.e_y$ have an empty intersection. Consider the case where a point $P$ is inserted on the right of $C$ (see Fig. 5.a). Let $u$ (resp. $l$) denote the point lying on the upper (resp. lower) border through which a *supporting line* (a line passing through $P$ and tangent to $C$) passes. If the triangle $ulP$ intersects $C + \alpha.e_y$, then the vertical thickness of the new convex set is greater or equal to $\alpha$. Conversely, if the intersection is empty, the vertical thickness of $ulP$ is strictly less than $\alpha$, so the intersection between $ulP$ and $ulP + \alpha.e_y$ is empty. It follows:

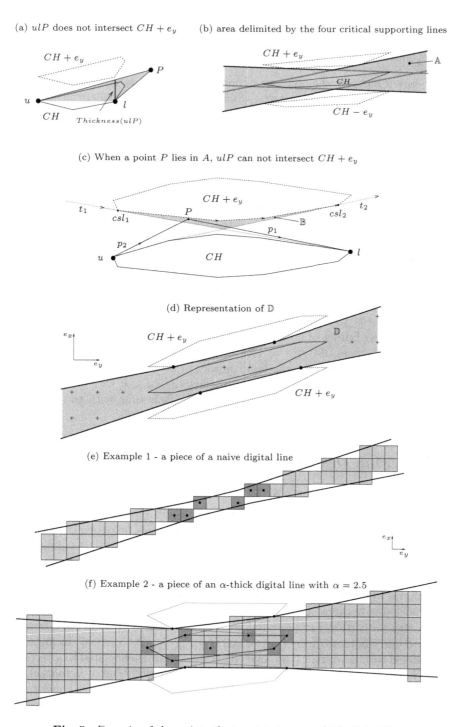

(a) $ulP$ does not intersect $CH + e_y$

(b) area delimited by the four critical supporting lines

(c) When a point $P$ lies in $A$, $ulP$ can not intersect $CH + e_y$

(d) Representation of $\mathbb{D}$

(e) Example 1 - a piece of a naive digital line

(f) Example 2 - a piece of an $\alpha$-thick digital line with $\alpha = 2.5$

**Fig. 5.** Domain of the points that maintain an $\alpha$-thick digital line

*Property 2.* We can insert a point to a convex set $C$ while maintaining its vertical thickness strictly below $\alpha$ if and only if the two supporting lines do not intersect $C + \alpha.e_y$ and $C - \alpha.e_y$.

For the convex hull $CH$ of a given set of points, we want to determine the domain $\mathbb{D}$ of the points that can be inserted while preserving the vertical thickness of $CH$ strictly below $\alpha$. We call *critical supporting lines (CSL)* between two convex objects $C_1$ and $C_2$ the two lines tangent to both convex objects such that $C_1$ and $C_2$ lie on opposite sides of each line. We show that the area $\mathbb{A}$ (see 5.b) delimited by the four CSL between $CH + \alpha.e_y$ and $CH$ and between $CH$ and $CH - \alpha.e_y$ plays an important role. Indeed, according to the previous property, when a point lying in $\mathbb{A}$ is inserted the new vertical thickness is always valid. We now examine the area $\mathbb{B}$ above the two CSL $t_1$ and $t_2$ and under $CH + e_y$ (see Fig. 5.c). Let $csl_1$ and $csl_2$ denote the vertices of $CH + e_y$ that support $t_1$ and $t_2$. We show that when a point $P$ lying in $\mathbb{B}$ is inserted, the associated supporting segments $p_1$ and $p_2$ between $P$ and $CH$ do not intersect $CH + e_y$. By assumptions, $P$ is above $t_1$ and $t_2$. As $p_1$ (resp. $p_2$) is tangent to $CH$, it must cross $t_1$ (resp. $t_2$). So $slope_{p_1} \leq slope_{t_1}$ and $slope_{p_2} \geq slope_{t_2}$. By convexity properties, the slopes of all the edges of $CH$ between $csl_1$ and $csl_2$ are ranging from the slope of $t_1$ to the slope of $t_2$. Thus $p_1$ and $p_2$ can not cross $CH + e_y$. Finally, the upper border of $\mathbb{D}$ is delimited by the CSL lines emerging from $csl_1$ and $csl_2$ and the edges of $CH + e_y$ located between these two vertices (see Fig. 5.d). We symmetrically define the lower border of $\mathbb{D}$. We present this result for a naive digital line (see Fig. 5.e) and an $\alpha$-thick digital line with $\alpha = 2.5$ (see Fig. 5.f). The CSL can be computed in optimal linear time with the rotating calipers approach [Tou83].

## 4    Algorithm Design

### 4.1    Computing the Vertical Thickness of a Convex Hull

We recall that our definition of thickness is different from the one usually used in Computational Geometry. In this field, the thickness is linked to the strip of minimal width (computed with the $L_2$-norm) that encloses all the points (see Fig. 6). The rotating caliper algorithm [HoT88] can be used to compute this value in optimal linear time in the number of vertices. In the same way, we can determine the vertical thickness in linear time by using a double-traversal of the upper and lower borders of the convex hull. We set up a new approach based on convexity properties. Let $f(x)$ and $g(x)$ denote the functions defined by the upper and lower borders of the convex hull (see Fig. 7). The vertical thickness is always linked to a hull vertex (the edge-edge case is always associated to a vertex-edge or vertex-vertex case). So we only have to check the value of $height(x)$ at the abscissas of the hull vertices. Notice that $f$ is concave and $g$ is convex, so the difference $height(x) = f(x) - g(x)$ is a concave function and we can apply a binary search to find its optimum. For example, suppose we have $k$ vertices on $U$ the upper border of the hull. We compute the value $height(U_{k/2})$ and

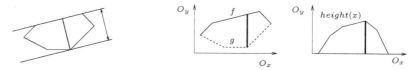

**Fig. 6.** The common thickness definition

**Fig. 7.** The vertical thickness

$height(U_{k/2+1})$. By another binary search on the lower border, we determine which vertices are faced to $U_{k/2}$ and $U_{k/2+1}$ in $O(\log n)$ time. Suppose that $height(U_{k/2}) < height(U_{k/2+1})$. Then by convexity property all the vertices of indices less than $k/2$ can not define the optimum. So we are able to reject one half of the upper vertices. We perform in the same way for the vertices of the lower hull. So we have removed at least one half of all the vertices in $O(\log n)$ time. By continuing the binary search on $height(x)$, we finally obtain the vertical thickness in $O(log^2 n)$ time.

## 4.2   The Incremental Approach

This is an on-line version where no information about insertions are known in advance. At each iteration, thickness and covering strip are computed. When a new point is inserted, we compute the resulting convex hull using classical incremental convex hull algorithm [PS85,BKS95] in $O(\log n)$ time. When the point lies inside the current strip, the thickness is unchanged. When the point lies outside, we have to compute the new thickness. Using the previous method, we would obtain an $O(n \log^2 n)$ algorithm, but we improve the global complexity by simplifying the binary search. We only consider the case where the point is inserted on the right of the hull (see Fig. 8.a), others cases are similar (see Fig. 8.b). The light grey area corresponds to a triangle; its thickness can be computed in constant time. The other area is delimited by a piece of the upper border and a tangent. Thus, all the vertices of this area are linked to the same segment and we can perform a simple binary search to find in $O(\log n)$ time the maximum height of this area. Finally, we can update the vertical thickness in $O(\log n)$ time per insertion.

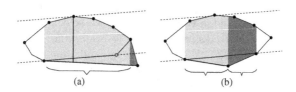

**Fig. 8.** Computing the new thickness in the incremental method

### 4.3    A Special Case

We often know in real applications that points are inserted in the same direction. Previous algorithms [DR95] take advantage of this in order to obtain a linear time complexity. We present an elementary implementation designed for this special case. First, we recall that the incremental convex hull under this assumption can be processed in $O(n)$ time using classical methods [Mlk87]. We choose to insert points by increasing abscissa (see Fig. 9). Notice that insertions on the right of the convex hull shifts the vertical thickness to the right. Let $\beta$ denote the abscissa where the previous vertical thickness was located. The dimmed area is delimited by the line $x = \beta$, a border of the previous strip and a tangent. It covers the left part of the new convex hull. In this triangle, all values of $height(x)$ are less or equal to $height(\beta)$. So the vertical thickness must be located on the right of this area. We do not use a binary search anymore. We just traverse each vertex on the right up to the maximal height. Each trial has a $O(1)$ cost, and the total number of trials is bounded by the total amount of inserted points. So the incremental recognition of a digital line of thickness $d$ can be computed in $O(n)$ time if we insert points in a given order.

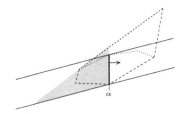

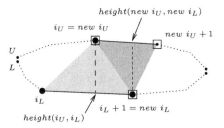

**Fig. 9.** Insertion on the right shifts vertical thickness to the right

**Fig. 10.** Notations of the algorithm

### 4.4    Pseudo-Code

We present (see Fig. 13) the main function used to compute the vertical thickness of a convex hull: *UPDATE_VERTICAL_THICKNESS*. The hull will consist of two lists $U$ and $L$ corresponding to its upper and lower borders. Let $n_U$ and $n_L$ denote the size of $U$ and $L$ respectively. We call $i_U$ and $i_L$ the indices of the couple of vertices lying on $U$ and $L$ respectively that define the vertical thickness. When the case edge-vertex appears, we select the left vertex of the edge. The other vertex is necessarily the one of $U_{i_U+1}$ and $L_{i_L+1}$ with minimum abscissa (function *HEIGHT*). We call *new* $i_U$ and *new* $i_L$ (see Fig. 10) the indices of vertices that are associated with the next discontinuity of $height(x)$. The implementation presented here processes points entered from left to right (the insertion order in the vertical direction is unimportant). The same function can be used in the three others directions by symmetrizing the $(x, y)$ coordinates.

## 4.5    Complexity of the Algorithms

We may use *dynamic* convex hull (incremental and decremental) that induces an $O(n \log^2 n)$ overhead [PS85]. We recall the complexities of our methods:

| case | convex hull | incremental | in one direction | dynamic |
|------|-------------|-------------|------------------|---------|
| thickness computation | $O(\log^2 n)$ | $O(n \log n)$ | $O(n)$ | $O(n \log^2 n)$ |

## 4.6    Applications of $\alpha$-Thick Lines

We are able to convert any set of points into a polygonal chain of digital line segments, even if no assumptions on connectedness can be done and no information about thickness is known a priori. We present in Fig. 11 a first test on a low quality digitization. We may adjust the thickness in order to obtain a compromise between the number of segments and the conversion quality.

**Pixel Traversal Order.** First, we have to choose the pixel order that will be used in our recognition process. Pixels are stored in a data structure $L$ that allows efficient range searching. Suppose we have a function $STARTING\_POINT$ that arbitrary selects a starting pixel. This point represents the first level and it can be removed from $L$. To fill the next level we only have to select pixels that are in the neighborhood of the pixels in the current level (using a particular distance function). When no pixels are found, we widen the current neighborhood until we find one or more points. When a maximal thickness $\delta$ is allowed for the recognition, it is judicious to limit the maximal neighborhood to a width $\delta$. Thus, the holes of at most $\delta$ pixels can be filled. The pixels of the same level are arbitrary ordered. We recall the entire process in the following algorithm and present an example in figure 12.

**The Overall Method.** We process pixels one after the other relative to the predefined order. When the thickness of the current digital segment exceeds the given threshold $\delta$, we backtrack to the previous pixel and select the corresponding digital segment and the associated thickness (that may be between 0 and $\delta$). When we want to obtain a polygonal chain, we may insert the last pixel of the previous digital segment in the next one. The method presented here is quite elementary and is based on a greedy approach. Thus the result is completely dependent of the chosen pixel traversal order and no uniqueness can be obtained.

**Open Problems.** When we have worked with digital segments, no considerations have been done about the quality of the jonction between two consecutive digital segments. In fact, pieces of them can cover themselves at the jonction. Optimizations relatively to the number of digital segments and to the visual quality remain to solve. Thus many questions are left untreated and their management would require an improved method that goes beyond the scope of this paper.

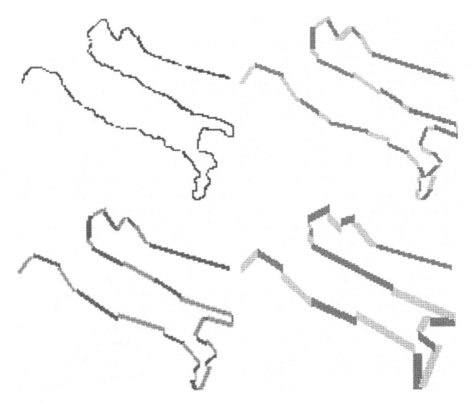

**Fig. 11.** A low quality digitization of the Italian coast converted using $\alpha$-thick digital segments ($\alpha = 2.5, 3$ and $4.5$). We successively obtain 37, 30 and 20 digital segments

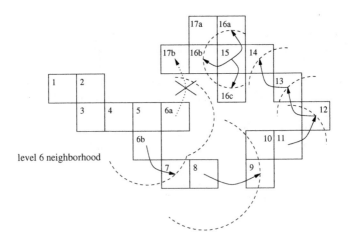

**Fig. 12.** Computing the pixel order traversal

**Function ORDER_PIXELS** (L,Level,δ)

$i_{CL} = 1$     − *current level index*
**while**   ( L.non_empty )
          $Level[i_{CL}] = STARTING\_POINT(L)$
          **while**   ( $Level[i_{CL}].non\_empty$ )
                    $d = 1$
                    **repeat**     $Level[i_{CL} + 1] = \text{Neighborhood}(L, Level[i_{CL}], d)$
                                $d = d + 1$
                    **until**     ( $Level[i_{CL} + 1].non\_empty$   OR   $d > \delta$ )
                    $L.remove(Level[i_{CL}])$
                    $i_{CL} = i_{CL} + 1$

**Function UPDATE_VERTICAL_THICKNESS** (U,$n_U$,L,$n_L$,$i_U$,$i_L$,h,P)
   **if** $(n_U = 0)$   $\{ n_L = n_U = 2; U_1 = U_2 = L_1 = L_2 = P; h = 0; \textbf{return}; \}$

               − *backtracking   to   update   the   convex   hull* −
   **while**   $(n_U > 1)$ and $(U_{n_U-1} - U_{n_U}) \wedge (P - U_{n_U}) \leq 0$   $\{ n_U = n_U - 1\}$
   **while**   $(n_L > 1)$ and $(L_{n_L-1} - L_{n_L}) \wedge (P - L_{n_L}) \geq 0$   $\{ n_L = n_L - 1\}$

      − *when a vertex that defines the previous thickness is removed* −
   **if** $(i_U > n_U)$   $\{ i_U = n_U \}$
   **if** $(i_L > n_L)$   $\{ i_L = n_L \}$

                     − *insertion of the new point P* −
   $n_L = n_L + 1$;   $L_{n_L} = P$;
   $n_U = n_U + 1$;   $U_{n_U} = P$;

                        − *update Height(x)* −
   $h = HEIGHT(U, L, i_U, i_L)$;

                     − *traverse the hull while the height increases* −
   **while**   $(i_L \neq n_L - 1)$ or $(i_U \neq n_U - 1)$
      **if**   $(U_{i_U+1}.x < L_{i_L+1}.x)$         $\{$ new $i_U = i_U + 1$; new $i_L = i_L; \}$
      **else**                       $\{$ new $i_U = i_U$; new $i_L = i_L + 1; \}$
      $h_{next} = HEIGHT(U, L, \text{new } i_U, \text{new } i_L)$;
      **if**   $(h_{next} > h)$   $\{ h = h_{next}; i_U = \text{new } i_U; i_L = \text{new } i_L; \}$
      **else**                 **return**;

**Function HEIGHT** (U,L,$i_U$,$i_L$)
   **if**   $(U_{i_u}.x < L_{i_l}.x)$     **return**   $VERT\_DIST(U_{i_u}, U_{i_u+1}, L_{i_l})$;
   **else**                **return**   $VERT\_DIST(L_{i_l}, L_{i_l+1}, U_{i_u})$;

**Fig. 13.** The core function of the recognition method based on thickness computation

# 5   Conclusion and Future Works

We introduce a new digital line recognition method based on the vertical and horizontal thickness. Our method is well adapted to recognize any set of points where no assumptions on connectedness can be done. The internal machinery uses classical convex hull algorithms.

The knowledge of the convex hull of the current set of points allows to increase the thickness computation efficiency, to make easier any implementation in the two-dimensional case, to determine the points that can be inserted while preserving a valid digital line, to obtain an incremental or dynamic method and to extend our approach to points with real coordinates. We do not deal with the strategy used to order points during the recognition. At each step, the thickness is computed and we may suppose that the associated normal vector is close to the local curvature. Thus, we may take advantage of this information in order to enhance the overall conversion quality. Other criteria may be inserted like density. Indeed, we may consider the ratio between the number of points inside the hull and the hull surface. This would be an automatic criterion that could replace the usual predefined thickness threshold. Such strategies combined with the dynamic approach could lead to a new generation of adaptive algorithms...

The source code in C++ of our method as well as a basic geometry kernel, some examples and some tests can be downloaded from the web page of the author: www.esiee.fr/~buzerl/DG/.

# References

[BKS95]  M. de Berg, M. van Kreveld, and S. Schirra. *A new approach to subdivision simplification*. In Proc. of Auto-Carto 12, pages 79-88, 1995.

[Buz02]  L. Buzer. *Reconnaissance des plans discrets - Simplification polygonale*. Thèse, Université d'Auvergne, Clermont-Ferrand, 2002.

[Buz03]  L. Buzer. *A linear incremental algorithm for naive and standard digital lines and planes recognition*, Graphical Models, 65(1-3), pp. 61-76, 2003.

[DR95]   I. Debled-Rennesson, J.-P. Reveillès. *A linear algorithm for segmentation of digital curves*. International Journal of Pattern Recognition and Artificial Intelligence, Volume **9**, N. 6, dec. 1995.

[Hot88]  M. E. Houle and G. T. Toussaint. *Computing the width of a set*. IEEE Trans. Pattern Anal. Mach. Intell., PAMI-10, 761-765, 1988.

[Mlk87]  A.A. Melkman. *On-line construction of the convex hull of a simple polyline*. Information Processing Letters, **25**:11-12, 1987.

[PS85]   F.P. Preparata, M.I. Shamos. *Computational Geometry*. Springer-Verlag, New-York, 1985.

[Rev91]  J.P. Reveillès. *Géometrie discrète, calculs en nombre entiers et algorithmique*. Thèse d'état, Université Louis Pasteur, Strasbourg, 1991.

[Tou83]  G.T. Toussaint. *Solving geometric problems with the rotating calipers*. Proceedings of IEEE MELECON'83, Athens, Greece, May 1983.

# Supercover Model and Digital Straight Line Recognition on Irregular Isothetic Grids

David Coeurjolly

Laboratoire LIRIS - CNRS FRE 2672,
Université Claude Bernard Lyon1,
43 Bd du 11 novembre 1918,
Villeurbanne, France
dcoeurjo@liris.cnrs.fr

**Abstract.** On the classical discrete grid, the analysis of digital straight lines (DSL for short) has been intensively studied for nearly half a century. In this article, we are interested in a discrete geometry on irregular grids. More precisely, our goal is to define geometrical properties on irregular isothetic grids that are tilings of the Euclidean plane with different sized axis parallel rectangles.

## 1  Introduction

When a straight line is digitized on a square grid, we obtain a sequence of grid points defining a digital straight-line segment. This computer representation of such a simple Euclidean object has drawn considerable attention in many applications (drawing [3], shape characterization [13, 14, 17, 7], ...). The structure of DSL is now well known and links have been illustrated between DSL and objects from number theory or theory of words (see Rosenfeld and Klette [26] for a survey on digital straightness). Beyond this characterization, an important task in computer vision consists in the recognition of DSL segments. More precisely, given a set of pixels, we have to decide if there exists a DSL segment that contains the given pixels. Many efficient algorithms exist to implement such a recognition process [15, 18, 11, 4]. Based on a digital straight line recognition algorithm, we can also define a segmentation process that decomposes a discrete curve into maximal DSL segments.

In this article, we are interested in defining a geometry on irregular isothetic grids. More precisely, we consider grids defined by a tiling of the plane using axis parallel rectangles. Such a grid model includes, for example, the classical discrete grid, the elongated grids [27] and the quadtree based grids [16]. In the following sections, we focus a general supercover digitization model on the irregular isothetic grids which is consistent with the classical one if the discrete space is considered. A previous work can be found in [8] in which irregular grids with squares are considered. In this model, the chosen digitization scheme is the naive model and it suffers from some inconsistencies. For example, the digitization of an Euclidean straight line may be a disconnected set of pixels. In the following,

E. Andres et al. (Eds.): DGCI 2005, LNCS 3429, pp. 311–322, 2005.

we generalize the model to irregular tilings of rectangles with an appropriate digitization model.

Many applications may benefit from these developments. For example, we can cite the analysis of quadtree compressed shape, or the use of geometrical properties in objects represented by interval or affine arithmetics [21, 22, 10, 5]. In this last example, we talk about *data driven grids*.

Section 2 presents more formal definitions of the irregular grids which allow to define the supercover model in Section 3. Then, we present the definition and the recognition algorithm of digital straight lines in these grids (Section 4). Experiments and results are shown in Section 5. Finally, we briefly illustrate the application of the irregular model in interval arithmetic (Section 6).

## 2   Definitions

First of all, we define an *irregular isothetic grid*, denoted $\mathbb{I}$, as a tiling of the plane with isothetic rectangles. In this framework, the rectangles have not necessarily the same size but we can notice that the classical digital space is a particular irregular isothetic grid. In that case, all squares are centered in $\mathbb{Z}^2$ points and have a border size equal to 1. Figure 1 illustrates some examples of irregular isothetic grids.

In the following, a rectangle of an isothetic grid is called a *pixel*. Each pixel $P$ is defined by its center $(x_P, y_P) \in \mathbb{R}^2$ and a size $(l_P^x, l_P^y) \in \mathbb{R}^2$. Before we introduce objects and straight lines in such grids, we need adjacency relations between pixels.

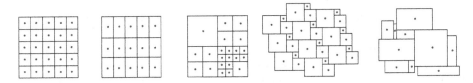

**Fig. 1.** Examples of irregular isothetic grids: *(from left to right)* the classical discrete grid $((x_P, y_P) \in \mathbb{Z}^2$ and $l_P^x = l_P^y = 1)$, an elongated grid $(l_P^x = \lambda, l_P^y = \mu$ and $(x_P, y_P) = (\lambda i, \mu j)$ with $(i, j) \in \mathbb{Z}^2)$, a quadtree decomposition (for a cell of level $k$, $(x_P, y_P) = (\frac{m}{2^k}, \frac{n}{2^k})$ and $l_P^x = l_P^y = \frac{1}{2^{k-1}}$ for some $m, n \in \mathbb{Z}$); a unilateral and equitransitive tiling by squares: the size of the biggest square is equal to the sum of the two other square sizes; finally a general irregular isothetic grid

**Definition 1 (ve−adjacency, e−adjacency).** *Let $P$ and $Q$ be two pixels. $P$ and $Q$ are ve-adjacent if:*

$$|x_P - x_Q| = \frac{l_P^x + l_Q^x}{2} \text{ and } |y_P - y_Q| \leq \frac{l_P^y + l_Q^y}{2},$$

*or*

$$|y_P - y_Q| = \frac{l_P^y + l_Q^y}{2} \text{ and } |x_P - x_Q| \leq \frac{l_P^x + l_Q^x}{2}.$$

*P and Q are* e-adjacent *if we consider an exclusive "or" and strict inequalities in the above* ve-adjacent *definition.*

In the following definitions, we use the notation *k-adjacency* in order to express either the *ve-adjacency* or the *e-adjacency*. Using these adjacency definitions, several basic objects can be defined:

**Definition 2 (*k*−path).** *Let us consider a set of pixels $\mathcal{E} = \{P_i, i \in \{1, \ldots, n\}\}$ and a relation of $k$−adjacency. $\mathcal{E}$ is a $k - path$ if and only if for each element $P_i$ of $\mathcal{E}$, $P_i$ is $k$−adjacent to $P_{i-1}$.*

**Definition 3 (*k*−object).** *Let $\mathcal{E}$ be a set of pixels, $\mathcal{E}$ is a $k$−object if and only if for each couple of pixels $(P, Q)$ belonging to $\mathcal{E} \times \mathcal{E}$, there exists a $k$−path between $P$ and $Q$ in $\mathcal{E}$.*

**Definition 4 (k-arc).** *Let $\mathcal{E}$ be a set of pixels, $\mathcal{E}$ is a $k$−arc if and only if for each the element of $\mathcal{E} = \{P_i, i \in \{1, \ldots, n\}\}$, $P_i$ has exactly two $k$−adjacent pixels, except $P_1$ and $P_n$ which are called the extremities of the $k$−arc.*

**Definition 5 (k-curve).** *Let $\mathcal{E}$ be a set of pixels, $\mathcal{E}$ is a $k$-curve if and only if $\mathcal{E}$ is a k-arc and $P_1 = P_n$.*

If we consider pixels such that $l_P^x = l_P^y = 1$ and $(x_P, y_P) \in \mathbb{Z}^2$ (*i.e.* a 2D digital space), all these definitions coincide with the classical ones [24, 25]. More precisely, the *ve*−adjacency (resp. *e*−adjacency) is exactly the 8-adjacency (resp. the 4-adjacency).

In the following, we only consider geometrical properties of such objects. A complete topological analysis of $k$−curves and $k$−objects is not addressed here.

# 3    Supercover Model on the Irregular Isothetic Grids

Before defining the digital straight lines on the irregular isothetic grids, we have to consider a digitization model. In the following, we choose to extend the supercover model. This model was first introduced by Cohen-Or and Kaufman in [9] on the classical discrete grid and then widely used since it provides an analytical characterization of basic supercover objects (*e.g.* lines, planes, 3D polygons, ...) [2, 1].

**Definition 6 (Supercover on irregular isothetic grids).** *Let $F$ be an Euclidean object in $\mathbb{R}^2$. The supercover $\mathbb{S}(F)$ is defined on an irregular isothetic grid $\mathbb{I}$ by:*

$$\mathbb{S}(F) = \{P \in \mathbb{I} \mid \mathbb{B}^{\infty}(P) \cap F \neq \emptyset\} \tag{1}$$

$$= \{P \in \mathbb{I} \mid \exists (x, y) \in F, |x_P - x| \leq \frac{l_P^x}{2} \text{ and } |y_P - y| \leq \frac{l_P^y}{2}\} \tag{2}$$

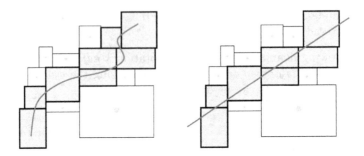

**Fig. 2.** Illustration of the supercover digitization of a curve *(left)* and of a straight line *(right)*

*where* $\mathbb{B}^\infty(P)$ *is the rectangle centered in* $(x_P, y_P)$ *of size* $(l_P^x, l_P^y)$ *(if* $l_P^x = l_P^y$, $\mathbb{B}^\infty(P)$ *is the ball centered in* $(x_P, y_P)$ *of size* $l_P^x$ *for the* $L_\infty$ *norm).*

This model has got several properties:

**Proposition 1.** *Let* $F$, $G$ *be Euclidean objects in* $\mathbb{R}^2$, $\alpha \in \mathbb{R}^2$ *and an* $\mathbb{I}$-*grid, we have:*

$$\mathbb{S}(F \cup G) = \mathbb{S}(F) \cup \mathbb{S}(G), \tag{3}$$

$$\mathbb{S}(F) = \bigcup_{\alpha \in F} \mathbb{S}(\alpha), \tag{4}$$

$$\mathbb{S}(F \cap G) \subseteq \mathbb{S}(F) \cap \mathbb{S}(G), \tag{5}$$

$$\text{if } F \subseteq G \text{ then } \mathbb{S}(F) \subseteq \mathbb{S}(G). \tag{6}$$

*Proof.* All these statements can be easily proved by definition of the supercover model. For example, we prove the first one as follows:

$$\begin{aligned}
\mathbb{S}(F \cup G) &= \{p \in \mathbb{I} \mid \mathbb{B}^\infty(P) \cap (F \cup G) \neq \emptyset\} \\
&= \{p \in \mathbb{I} \mid (\mathbb{B}^\infty(P) \cap F) \cup (\mathbb{B}^\infty(P) \cap G) \neq \emptyset\} \\
&= \mathbb{S}(F) \cup \mathbb{S}(G)
\end{aligned}$$

Figure 2 illustrates some examples of the supercover digitization of Euclidean objects.

If $\mathbb{I}$ is the classical digital space (*i.e.* $(x_P, y_P) \in \mathbb{Z}^2$ and $l_P^x = l_P^y = 1$), many links exist between the supercover of an Euclidean straight line and classical digital straight line definitions [1, 26]. Since we have not any assumption on the irregular grid, no strong topological property can be stated on the supercover of an Euclidean straight line.

**Proposition 2.** *Let* $l$ *be an Euclidean straight line and a* $\mathbb{I}$-*grid, the* $\mathbb{S}(l)$ *is a single ve−object.*

*Proof.* The proof is direct since $\mathbb{I}$ is a tiling of the plane with closed pattern. The digitization of $l$ is necessarily a connected set of pixels.

# 4 Digital Straight Line Definition and Recognition

## 4.1 Definitions

**Definition 7 (Irregular Isothetic Digital Straight Line).** *Let $S$ be a set of pixels in $\mathbb{I}$, $S$ is called a piece of irregular digital straight line (IDSL for short) iff there exists an Euclidean straight line $l$ such that:*

$$S \subseteq \mathbb{S}(l). \tag{7}$$

*In other words, $S$ is a piece of IDSL iff there exists $l$ such that:*

$$\forall P \in S, \quad \mathbb{B}^{\infty}(P) \cap l \neq \emptyset. \tag{8}$$

To detect if $\mathbb{B}^{\infty}(P) \cap l$ is empty or not, we use the notations presented in Figure 3. Hence, $\mathbb{B}^{\infty}(P) \cap l$ is not empty iff $l$ crosses either (or both) the diagonals $d_1$ or $d_2$ of $P$.

Without loss of generality, we suppose that $l$ is given by $y = \alpha x + \beta$ with $(\alpha, \beta) \in \mathbb{R}^2$ (an appropriate treatment can be design to handle the straight lines $x = k$ with $k \in \mathbb{R}$). To solve the recognition problem, we use the following statement:

$$\mathbb{B}^{\infty}(P) \cap l \neq \emptyset \quad \Leftrightarrow \quad l \cap d_1 \neq \emptyset \text{ and } \alpha \geq 0 \tag{9}$$
$$\text{or} \quad l \cap d_2 \neq \emptyset \text{ and } \alpha < 0 \tag{10}$$

During a recognition process, it is convenient to consider the set of Euclidean straight lines whose digitization contains the set of pixels $S$: if such a set is empty, we can conclude that $S$ is not a discrete straight line segment. In the literature, the set of Euclidean straight lines whose digitization contains $S$ is called the *preimage* of $S$. Many works have been done concerning the preimage analysis in the classical discrete grid [12, 18, 19].

We first consider Equation (9): given the pixel $P$, the straight line containing $d_1$ is

$$y = -\frac{l_P^y}{l_P^x} \cdot x + y_P + \frac{l_P^y}{l_P^x} \cdot x_P. \tag{11}$$

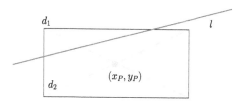

**Fig. 3.** Notations used to detect if the pixel of center $(x_P, y_P)$ belongs to the supercover of a straight line $l$ ($d_1$ and $d_2$ are the diagonals of the rectangle $P$)

If we compute the intersection point between this straight line and $l$, the abscissa $x$ is given by:

$$x \cdot \left( \alpha + \frac{l_P^y}{l_P^x} \right) = y_P + \frac{l_P^y}{l_P^x} \cdot x_P - \beta. \tag{12}$$

Hence, to ensure that $l$ intersects $d_1$, $x$ must be such that:

$$x_P - \frac{l_P^x}{2} \leq x \leq x_P + \frac{l_P^x}{2}. \tag{13}$$

Since $\alpha \geq 0$, we have:

$$\left( x_P - \frac{l_P^x}{2} \right) \cdot \left( \alpha + \frac{l_P^y}{l_P^x} \right) \leq y_P + \frac{l_P^y}{l_P^x} \cdot x_P - \beta \leq \left( x_P + \frac{l_P^x}{2} \right) \cdot \left( \alpha + \frac{l_P^y}{l_P^x} \right) \tag{14}$$

Finally, the condition given in Equation (9) can be represented by the two following inequalities in the $(\alpha, \beta)$-parameter space:

$$\mathcal{E}^+(P) = \begin{cases} \alpha \left( x_P - \frac{l_P^x}{2} \right) + \beta - y_P - \frac{l_P^y}{2} \leq 0 \\ \alpha \left( x_P + \frac{l_P^x}{2} \right) + \beta - y_P + \frac{l_P^y}{2} \geq 0 \end{cases}. \tag{15}$$

If we consider Equation (10) and using similar arguments, we obtain the following inequalities:

$$\mathcal{E}^-(P) = \begin{cases} \alpha \left( x_P - \frac{l_P^x}{2} \right) + \beta - y_P + \frac{l_P^y}{2} \geq 0 \\ \alpha \left( x_P + \frac{l_P^x}{2} \right) + \beta - y_P - \frac{l_P^y}{2} \leq 0 \end{cases}. \tag{16}$$

$\mathcal{E}^+(P)$ is defined for $\alpha \geq 0$ and $\mathcal{E}^-(P)$ for $\alpha < 0$. We can now define the preimages of a piece of IDSL:

**Definition 8 (Preimages of an IDSL).** *Let $S$ be a piece of IDSL, the two preimages $\mathcal{P}^+$ and $\mathcal{P}^-$ of $S$ are given by:*

$$\mathcal{P}^+(S) = \bigcap_{P \in S} \mathcal{E}^+(P), \tag{17}$$

$$\mathcal{P}^-(S) = \bigcap_{P \in S} \mathcal{E}^-(P). \tag{18}$$

Hence, the recognition process can be described as follows:

**Proposition 3.** *Let $S$ be a set of pixels in a $\mathbb{I}$-grid. $S$ is a piece of IDSL iff $\mathcal{P}^+(S) \neq \emptyset$ or $\mathcal{P}^-(S) \neq \emptyset$.*

## 4.2    Recognition

Using Proposition 3, the recognition on an IDSL leads to a linear programming problem: we have to decide whether a linear inequality system has a solution or not. More precisely, two different classes of algorithms exist: the IDSL *identification* algorithms which decide if $S$ is an IDSL or not, and the IDSL *recognition* algorithms which return the complete preimages (maybe empty) of the recognized IDSL.

Given a linear inequality system, several algorithms can be found to test if a solution exists, *i.e.* to *identify* an IDSL. For example, the Meggido's algorithm [20] can decide if a solution exists in $O(n)$ time if $n$ is the number of linear constraints. An incremental version of this algorithm with the same computational cost can also be found [6].

If we need a complete description of the feasible region, *i.e.* a *recognition* of the IDSL, Preparata and Shamos [23] proposed an optimal $O(n \log n)$ time online algorithm. This algorithm can easily be implemented since basic computational geometry tools are used (convex hull and dual-space transform).

In Algorithm 1, we present the simple IDSL segmentation algorithm of a $ve-$arc. The preimage update in lines 6 and 7 can be performed using either an identification or a recognition algorithm.

---

**Algorithm 1** IDSL segmentation algorithm

---
1: Let $S = \{P_i, i = 1, ..., n\}$ be the $ve-$arc
2: $\mathcal{P}^+ := \{\alpha \geq 0\}$
3: $\mathcal{P}^- := \{\alpha < 0\}$
4: Mark $P_1$ as the starting point of an IDSL
5: **for** $i$ from 1 to $n$ **do**
6:     Update the preimage $\mathcal{P}^+$ with $\mathcal{E}^+(P_i)$
7:     Update the preimage $\mathcal{P}^-$ with $\mathcal{E}^-(P_i)$
8:     **if** $\mathcal{P}^+ = \emptyset$ and $\mathcal{P}^- = \emptyset$ **then**
9:        {*the IDSL recognition fails*}
10:        Mark $P_i$ as a starting point of a new segment
11:        $\mathcal{P}^+ := \{\alpha \geq 0\} \cap \mathcal{E}^+(P_i)$
12:        $\mathcal{P}^- := \{\alpha < 0\} \cap \mathcal{E}^-(P_i)$
13:     **end if**
14: **end for**

---

If we consider the classical discrete model (*i.e.* $(x_P, y_P) \in \mathbb{Z}^2$ and $l_P^x = l_P^y = 1$), Algorithm 1 implements a recognition of classical supercover digital straight lines.

## 5    Experiments and Results

In our experiments we have implemented Algorithm 1 in MAPLE using the built-in linear programming procedures. First, Figure 4 illustrates the recognition of

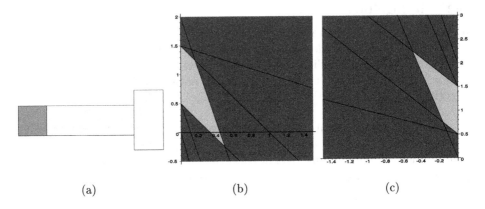

**Fig. 4.** Illustration of the preimages associated to a piece of IDSL: $(a)$ a sequence of $(x_P, y_P, l_P^x, l_P^y)$ in $\{(1,1,1,1),(3,1,3,1),(5,1,1,2)\}$, $(b)$ its preimage $P^+$ and $(c)$, its preimage $P^-$ in the $(\alpha, \beta)$-parameter space

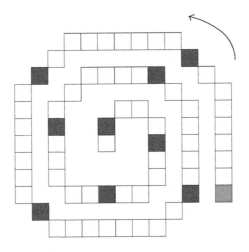

**Fig. 5.** Segmentation of a classical 8−connected curve into IDSL

a simple piece of IDSL with its associated preimages. Each point $(\alpha_0, \beta_0)$ in the $P^+$ or $P^-$ (gray areas in Figure 4-$(b-c)$), is an Euclidean straight $(a)$.

In Figure 5, we illustrate an IDSL segmentation of a classical discrete curve.

$x = k$ (with $k \in \mathbb{R}$) are excluded from the preimages,

$(a)$).

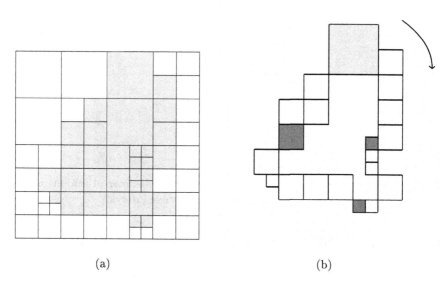

(a)                                                    (b)

**Fig. 6.** (*a*) Quadtree decomposition of a binary object and (*b*), the segmentation of its boundary pixels into IDSL

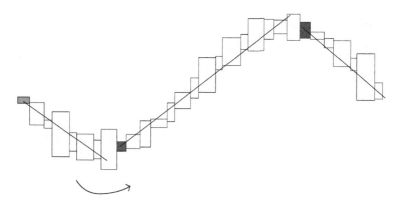

**Fig. 7.** Illustration of the segmentation algorithm on a general irregular curve. The Euclidean straight lines are manually extracted from the preimages associated to each IDSL segment

Finally, Figure 7 shows the segmentation result on a general $ve$−curve. In this illustration, we have superposed to the curve an Euclidean straight line per IDSL segment manually extracted from the preimages.

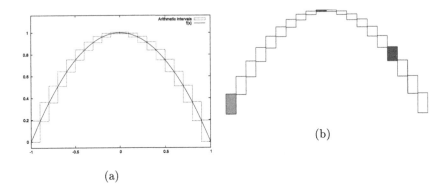

(b)

(a)

**Fig. 8.** The approximation of the function $f(x)$ using the interval arithmetic $(a)$ and the result of the IDSL segmentation algorithm on these intervals

## 6    An Application to Interval Arithmetic Analysis

In this section, we briefly present an application for which the irregular isothetic model has been developed: the analysis of interval arithmetic objects. The interval arithmetic is a widely used range based model for numerical computation where each quantity $x$ is represented by an interval $\bar{x}$ of floating point numbers [21, 22, 10, 5]. On those intervals, arithmetical operations are defined in such way that each resulting interval $\bar{x}$ is guaranteed to contain the unknown value corresponding to the real $x$ quantity.

Briefly, an interval $\bar{x}$ is represented by $[\bar{x}.lo, \bar{x}.hi]$ and the basic operations are given by:

$$\bar{x} + \bar{y} = [\bar{x}.lo + \bar{y}.lo, \bar{x}.hi + \bar{y}.hi] \tag{19}$$

$$\bar{x} - \bar{y} = [\bar{x}.lo - \bar{y}.hi, \bar{x}.hi - \bar{y}.lo] \tag{20}$$

$$\bar{x} \cdot \bar{y} = [\min(\bar{x}.lo \cdot \bar{y}.lo, \bar{x}.lo \cdot \bar{y}.hi, \bar{x}.hi \cdot \bar{y}.lo, \bar{x}.hi \cdot \bar{y}.hi), \tag{21}$$

$$\max(\bar{x}.lo \cdot \bar{y}.lo, \bar{x}.lo \cdot \bar{y}.hi, \bar{x}.hi \cdot \bar{y}.lo, \bar{x}.hi \cdot \bar{y}.hi)] \tag{22}$$

Many other operations can be designed in an interval arithmetic form. In Figure 8-$(a)$, we have the approximation using interval arithmetic representation of the function $f(\bar{x}) = -(\bar{x} - 1) \cdot (\bar{x} + 1)$ where the intervals $\bar{x}$ are given by a uniform subdivision of $[-1, 1]$ into 20 subintervals. Each rectangle in Figure 8-$(a)$ is given by $[\bar{x}.lo, \bar{x}.hi] \times [f(\bar{x}).lo, f(\bar{x}).hi]$. By construction of the intervals and since each operation guarantees that the real function belongs to each interval, we can prove that such an interval arithmetic object is a $ve-$curve. Hence, the IDSL segmentation algorithm can be applied to obtain a polygonal approximation of the intervals (see Figure 8-$(b)$). In that case, the entire grid $\mathbb{I}$ is unknown, only a subset of it is given. Furthermore, since the set of isothetic irregular pixels are defined by $f(\bar{x})$, we talk about a *data driven grid model*.

The use of IDSL on objects given by intervals permits to have a first geometrical analysis of the unknown underlying Euclidean curve.

# 7   Conclusion

In this article, we have presented a global digitization framework on irregular isothetic grids: the supercover model. Based on this digitization scheme, we have defined the digital straight lines in such grids and algorithmic solutions to solve the recognition and segmentation problem. All these developments allow us to characterize and analyze border of quadtree objects for example. Note that all the proposed definitions and algorithms match perfectly with the classical ones in the particular case of the discrete model. We have also illustrated the interest of such a model to analyze object boundaries of quadtree encoded shapes and in the interval arithmetic field.

Future works can be decomposed into two main problems : first the topological and arithmetical analysis of IDSL when a small class of irregular grids is considered (*e.g.* the quadtree based model). Then, the use of these discrete geometry tools in the interval analysis field.

# References

1. E. Andrès. Modélisation analytique discrète d'objets géométriques. Master's thesis, Laboratoire IRCOM-SIC, Université de Poitiers, 2000.
2. E. Andrès, P. Nehlig, and J. Françon. Tunnel-free supercover 3D polygons and polyhedra. *Computer Graphics Forum*, 16(3):C3–C13, September 1997.
3. J. Bresenham. An incremental algorithm for digital plotting. In *Proc. ACM Natl. Conf.*, 1963.
4. V. E. Brimkov and S. S. Dantchev. Digital hyperplane recognition in arbitrary fixed dimension. Technical report, CITR-TR-154 Center for Image Technology and Robotics, University of Auckland, New Zealand, 2004.
5. K. Bühler. Linear interval estimations for parametric objects: Theory and application. pages 522–531.
6. L. Buzer. An incremental linear algorithm for digital line and plane recognition using a linear incremental feasibility problem. In *10th International Conference on Discrete Geometry for Computer Imagery*, number 2301 in LNCS, pages 372–381. Springer, 2002.
7. D. Coeurjolly and R. Klette. A comparative evaluation of length estimators of digital curves. *IEEE Transactions on Pattern Analysis and Machine Intelligence*, 26(2):252–258, feb 2004.
8. D. Coeurjolly and L. Tougne. Digital straight line recognition on heterogeneous grids. In *SPIE Vision Geometry XII*, volume 5300, pages 108–116, san Jose, USA, 2004.
9. D. Cohen-Or and A. Kaufman. Fundamentals of surface voxelization. *Graphical models and image processing: GMIP*, 57(6):453–461, November 1995.
10. L.H. de Figueiredo and J. Stolfi. Self-validated numerical methods and applications. 1997. http://www.dcc.unicamp.br/~stolfi/EXPORT/projects/affine-arith/.
11. I. Debled and J.P. Reveillès. A linear algorithm for segmentation of digital curves. In *Third International Workshop on parallel Image Analysis*, June 1994.
12. L. Dorst and A. W. M. Smeulders. Discrete representation of straight lines. *IEEE Transactions on Pattern Analysis and Machine Intelligence*, 6:450–463, 1984.

13. L. Dorst and A. W. M. Smeulders. Length estimators for digitized contours. *Computer Vision, Graphics, and Image Processing*, 40(3):311–333, December 1987.
14. L. Dorst and A. W. M. Smeulders. Discrete straight line segments: Parameters, primitives and properties. In R. Melter, P. Bhattacharya, and A. Rosenfeld, editors, *Vision Geometry, series Contemporary Mathematics*, volume 119, pages 45–62. American Mathematical Society, 1991.
15. A. Hübler E. Creutzburg and V. Wedler. Decomposition of digital arcs and contours into a minimal number of digital straight line segments. In *Proc. 6th Intl. Conf. on Pattern Recognition*, page 1218, 1982.
16. R. A. Finkel and J. L. Bentley. Quad trees: a dsta structure for retrieval on composite key. *Acta Informatica*, 4(1):1–9, 1974.
17. R. Klette and B. Yip. The length of digital curves. *Machine Graphics & Vision*, 9:673–703, 2000. extended version of: R. Klette, V. V. Kovalevsky, B. Yip. Length estimation of digital curves. In *Proc. Vision Geometry VIII*, SPIE 3811, pages 117–129.
18. M. Lindenbaum and A. M. Bruckstein. On recursive, o(n) partitioning of a digitized curve into digital straigth segments. *IEEE Transactions on Pattern Analysis and Machine Intelligence*, 15(9):949–953, september 1993.
19. M. D. McIlroy. A note on discrete representation of lines. *AT&T Technical Journal*, 64(2):481–490, February 1985.
20. N. Meggido. Linear programming in linear time when the dimension is fixed. *Journal of the ACM*, 31(1):114–127, 1984.
21. Ramon E. Moore. *Interval Analysis*. Prentice-Hall, Englewood Cliffs, N.J., 1966.
22. Ramon E. Moore. *Methods and Applications of Interval Arithmetic*. Studies in Applied Mathematics. SIAM, Philadelphia, 1979.
23. F. P. Preparata and M. I. Shamos. *Computational Geometry : An Introduction*. Springer-Verlag, 1985.
24. A. Rosenfeld. Connectivity in digital pictures. *Journal of the ACM*, 17(1):146–160, January 1970.
25. A. Rosenfeld. Digital straight lines segments. *IEEE Transactions on Computers*, pages 1264–1369, 1974.
26. A. Rosenfeld and R. Klette. Digital straightness. In S. Fourey, G.T. Herman, and T.Y. Kong, editors, *International Workshop on Combinatorial Image Analysis*, volume 46 of *Electronic Notes in Theoretical Computer Science*, Temple University, Philadelphia, U.S.A., August 2001. Elsevier Science Publishers.
27. I.-M. Sintorn and G. Borgefors. Weighted distance transforms for images using elongated voxel grids. In *10th International Conference on Discrete Geometry for Computer Imagery*, number 2301 in LNCS, pages 244–254. Springer, 2002.

# Local Point Configurations of Discrete Combinatorial Surfaces

Yukiko Kenmochi[1] and Yusuke Nomura[2]

[1] UMR 8049 - IGM, CNRS/University of Marne-la-Vallée/ESIEE, France
[2] Department of Information Technology, Okayama University, Japan
y.kenmochi@esiee.fr

**Abstract.** Representing discrete objects by polyhedral complexes, we study topological properties of boundary points and surface points. We then obtain the local point configurations of discrete surfaces which are also considered to be boundaries of discrete objects.

## 1 Introduction

Surface structures are often used in the three-dimensional image analysis, for example, an active balloon which is a deformable surface for the image segmentation [2]. In many cases, a continuous surface such as a spline surface is applied to a deformable surface and after its deformation it is rediscretized to be a discrete object in a digital image. Obviously, this is not efficient if the input and output are both discrete. To construct completely discrete methods similarly to reference [4], we study discrete surfaces and their local point configurations such as those in the $3 \times 3 \times 3$ region in this paper as our first step.

In our previous work [10], we presented a boundary tracking algorithm which provides a triangulation of a set of boundary points given by

$$Br_m(\mathbf{V}) = \{\mathbf{x} \in \mathbf{V} : \mathbf{N}_m(\mathbf{x}) \cap \overline{\mathbf{V}} \neq \emptyset\}$$

where $\mathbf{V}$ is the input, i.e., a discrete object in a three-dimensional image and $\overline{\mathbf{V}}$ is the complement. $\mathbf{N}_m(\mathbf{x})$ is the $m$-neighborhood of a point $\mathbf{x} = (i, j, k)$ in a three-dimensional discrete space $\mathcal{Z}^3$, consisting of lattice points whose coordinates are all integers in $\mathcal{R}^3$, defined by

$$\mathbf{N}_m(\mathbf{x}) = \{(p, q, r) \in \mathcal{Z}^3 : (i - p)^2 + (j - q)^2 + (k - r)^2 \leq t\}$$

where $t = 1, 3$ for $m = 6, 26$ respectively. Because we apply discrete polyhedral complexes [10] based on combinatorial topology [1, 12, 13] to object representation, we can obtain topologies for boundary points. With a help of the topologies, we found that our boundary points include not only surface points, i.e., points on combinatorial 2-manifold, but also non-surface points, i.e., singular points, as shown in Fig. 1. In this paper, we use local topological notions similarly to our work [8] to discriminate boundary points and also surface points. Such notions enable us to present an algorithm to count the local point configurations of discrete surfaces for the 6- and 26-neighborhood systems.

E. Andres et al. (Eds.): DGCI 2005, LNCS 3429, pp. 335–346, 2005.

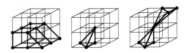

**Fig. 1.** Examples of point configurations in the $3 \times 3 \times 3$ region so that the central point is considered to be a boundary point [10]; a surface point (left), a surface point but not a simplicity surface point [3] (center), and a non-surface point, i.e., a singular point (right)

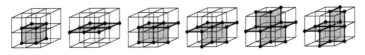

**Fig. 2.** The 6 local configurations of $3 \times 3 \times 3$ points for discrete combinatorial surfaces in the 6-neighborhood system

For the 6-neighborhood system, the definition of discrete combinatorial are given by Françon in [5] and he showed that there are 6 local configurations of discrete surfaces for the 6-neighborhood system as illustrated in Fig. 2. Note that the similar results are obtained by using different approaches, for example, in [7]. The discrete deformation model based on such discrete surface configurations for 6-neighborhood system is also presented in [4].

In [5], however, the 26-neighborhood is not practically treated so that we do not see how to generate discrete combinatorial surfaces for the 26-neighborhood system, Morgenthaler *et al.* defined discrete surfaces by using the point connectivity based on the Jordan surface theorem; any Jordan surface divides the space into two [11]. In [3], Couprie *et al.* pointed out that, for the 26-neighborhood system, Morgenthaler's discrete surfaces have only 13 local point configurations while their discrete surfaces, called simplicity surfaces, have 736 configurations. However, we see that even simplicity surfaces do not give enough configurations if we would like to treat our boundary points. For example, we obtain a boundary point as illustrated in Fig. 1 (center) which is not considered to be a simplicity surface.

In this paper, we show that the surface configurations which appear on our boundaries for the 6-neighborhood system are the same 6 configurations as illustrated in Fig. 2 and derive new results of surface configurations for the 26-neighborhood system. We also discuss the utilities of such study on local point configurations of discrete surfaces for three-dimensional shape analysis.

## 2    Discrete Polyhedral Complexes

### 2.1    Definitions

In combinatorial topology, any object in a 3-dimensional Euclidean space $\mathcal{R}^3$ is represented by a set of simplexes [1, 12] or more generally convex polyhedra [1, 13]. We construct convex polyhedra whose vertices are lattice points in $\mathcal{Z}^3$. Such

**Table 1.** All possible discrete convex polyhedra for the 6- and 26-neighborhood systems

convex polyhedra are called discrete convex polyhedra, and a discrete polyhedral complex is constructed as a set of discrete convex polyhedra combined together without contradiction [10].

Let us consider a set of eight lattice points which are the vertices of a unit cube. Setting a value of each point to be 1 or 0 and calling such a point 1- or 0-point, we obtain $2^8$ different 1-point configurations in a unit cube. Up to symmetries and rotations, the number of different configurations becomes 23. For each configuration which has at least one 1-point in a unit cube, we make a convex hull of 1-points and call it a discrete convex polyhedron.

**Definition 1.** *If a convex hull of 1-points in a unit cube has n dimensions where $n = 0, 1, 2, 3$ and its adjacent vertices are m-neighboring for $m = 6, 26$, such a convex hull is called an n-dimensional discrete convex polyhedron for the m-neighborhood system.*

Table 1 shows $n$-dimensional discrete convex polyhedra, $n = 0, 1, 2, 3$, for the 6- and 26-neighborhood systems. Hereafter, we abbreviate $n$-dimensional discrete convex polyhedra to $n$-polyhedra. The face of an $r$-polyhedron $\sigma$ is defined to be the set of all $s$-polyhedra which are included in the boundary of $\sigma$ where $s < r$, denoted by $face(\sigma)$. For example, we see in Table 1 that a 3-polyhedron for the 6-neighborhood system includes eight 0-polyhedra, twelve 1-polyhedra, and six 2-polyhedra. By using faces, we define discrete polyhedral complexes.

**Definition 2.** *A discrete polyhedral complex **K** is the set of discrete convex polyhedra satisfying the following conditions:*

1. $\emptyset \in \mathbf{K}$;
2. *if $\sigma \in \mathbf{K}$, $face(\sigma) \subset \mathbf{K}$;*
3. *if $\sigma, \tau \in \mathbf{K}$, $\sigma \cap \tau = \bigcup_{a \in face(\sigma) \cap face(\tau)} a$.*

**Fig. 3.** Non-pure (left) and pure (right) 3-complexes

The dimension of $\mathbf{K}$ is defined as the maximum dimension of discrete convex polyhedra belonging to $\mathbf{K}$. Hereafter, we abbreviate an $n$-dimensional discrete polyhedral complex to an $n$-dimensional discrete complex or simply an $n$-complex. We will present several properties of discrete complexes.

**Definition 3.** *Let $\mathbf{K}$ be an $n$-complex. If we have at least one $n$-polyhedron $\sigma \in \mathbf{K}$ for every $s$-polyhedron $\tau \in \mathbf{K}$ where $s < n$ satisfying $\tau \in face(\sigma)$, then $\mathbf{K}$ is said to be pure.*

Figure 3 shows examples of pure and non-pure discrete complexes. The 3-complex in Fig. 3 (left) is not pure because it includes 0-, 1- and 2-polyhedra which do not belong to any 3-polyhedra. If we remove these 0-, 1- and 2-polyhedra from Fig. 3 (left), we obtain a pure 3-complex in Fig. 3 (right).

**Definition 4.** *Let $\mathbf{K}$ be a discrete complex. The combinatorial closure of a subset $\mathbf{K}_0 \subset \mathbf{K}$ is defined as*

$$Cl(\mathbf{K}_0) = \mathbf{K}_0 \cup (\underset{a \in \mathbf{K}_0}{\cup} face(a)).$$

**Definition 5.** *Let $\mathbf{K}$ be a discrete complex, and $\sigma, \tau$ be arbitrary elements in $\mathbf{K}$. We say that $\mathbf{K}$ is connected, if we have a path $\sigma = a_1, a_2, \ldots, \tau = a_n$ which satisfies the following conditions:*

*1. $a_i \in \mathbf{K}$ for every $i = 1, 2, \ldots, n$;*
*2. $Cl(\{a_i\}) \cap Cl(\{a_{i+1}\}) \neq \emptyset$ for every $i = 1, 2, \ldots, n - 1$.*

### 2.2    Discrete Complex Construction from a Lattice Point Set

The goal of this paper is to count the number of local configurations of boundary points which form discrete surfaces, in a $3 \times 3 \times 3$ region $\mathbf{N}_{26}(\boldsymbol{x})$ of $\mathcal{Z}^3$ by using topological properties of discrete complexes presented in the next section. Before obtaining topological properties, thus, we need to construct a discrete complex $\mathbf{K}_m$ for $m = 6, 26$ from a subset $\mathbf{V} \subseteq \mathbf{N}_{26}(\boldsymbol{x})$ where all points in $\mathbf{V}$ (resp. the complement $\overline{\mathbf{V}}$) are 1-points (resp. 0-points). Given a $\mathbf{V} \subseteq \mathbf{N}_{26}(\boldsymbol{x})$, we briefly explain how to construct a discrete complex $\mathbf{K}_m$ for each $m = 6, 26$.

Let us first consider the case of $m = 26$. From Definition 1, dividing a $3 \times 3 \times 3$ region $\mathbf{N}_{26}(\boldsymbol{x})$ into eight $2 \times 2 \times 2$ unit cubic regions, we obtain a discrete convex polyhedron by constructing a convex hull of 1-points in each unit cubic region. From Definition 2, collecting all discrete convex polyhedra for eight unit cubic regions, we finally obtain a discrete complex $\mathbf{K}_{26}$ such as a set of all discrete convex polyhedra and their faces.

For the case of $m = 6$, we only consider convex hulls whose adjacent vertices are all 6-neighboring, and obtain a discrete complex $\mathbf{K}_6$ similarly to $\mathbf{K}_{26}$. The details and the precise algorithm are presented in reference [10].

# 3    Point Classification by Combinatorial Topology

In $\mathcal{Z}^3$, discrete complexes whose dimensions are from zero to three can exist. In this section, we first present topological properties of discrete complexes for each dimension from one to three by using the notions of star and link [13] similarly to the previous work [8]. By the topological properties, we classify all points in the skeleton of $\mathbf{K}$, denoted by $Sk(\mathbf{K})$, which is the set of vertices of all discrete convex polyhedra in $\mathbf{K}$, and find out the topological type of boundary points which are considered to be on discrete surfaces.

## 3.1    Star and Link

The star and the link are defined for each vertex in a discrete complex as follows.

**Definition 6.** *For a discrete complex* $\mathbf{K}$*, the star of a point* $x \in Sk(\mathbf{K})$ *is defined so that*

$$star(x) = \{\sigma \in \mathbf{K} : x \in \sigma\}.$$

**Definition 7.** *For a discrete complex* $\mathbf{K}$*, the link of a point* $x \in Sk(\mathbf{K})$ *is defined so that*

$$link(x) = Cl(star(x)) \setminus star(x).$$

The star and link with respect to $\mathbf{K}$ are denoted by $star(x : \mathbf{K})$ and $link(x : \mathbf{K})$, respectively, when we emphasize $\mathbf{K}$. Similarly to discrete complexes, we define the dimension of $star(x : \mathbf{K})$ as the maximum dimension of discrete convex polyhedra belonging to $star(x : \mathbf{K})$ and denoted by $dim(star(x : \mathbf{K}))$. Note that $star(x : \mathbf{K})$ is not always a discrete complex because it may not satisfy the second condition in Definition 2. On the other hand, $link(x : \mathbf{K})$ always becomes a discrete complex and even pure if $\mathbf{K}$ is pure.

## 3.2    Topological Properties in One Dimension

If a 1-complex $\mathbf{K}$ is pure, stars of points in $Sk(\mathbf{K})$ are classified into the following three types: linear stars, semi-linear stars, and neither of them. Let $|\mathbf{A}|$ be the number of elements in a set $\mathbf{A}$.

**Definition 8.** *Let* $\mathbf{K}$ *be a pure 1-complex. If* $|link(x)| = 2$*,* $star(x)$ *is called linear.*

**Definition 9.** *Let* $\mathbf{K}$ *be a pure 1-complex. If* $|link(x)| = 1$*,* $star(x)$ *is called semi-linear.*

Figure 4 illustrates points whose stars are linear, semi-linear and neither of them. We see that a point is an endpoint of a curve if its star is semi-linear, and an intermediate point of a curve if its star is linear. If the star of a point is neither linear nor semi-linear, it is an intersection of a curve. By using the above definitions, we define discrete curves in $\mathcal{Z}^3$.

**Fig. 4.** Examples of points whose stars are linear, semi-linear and neither of them, illustrated as white, grey and black points, respectively

**Definition 10.** *Let* **K** *be a connected and pure 1-complex. If the star of every point in* $Sk(\mathbf{K})$ *is either linear or semi-linear and at least one point whose star is semi-linear exists in* $Sk(\mathbf{K})$, **K** *is called a discrete curve with endpoints.*

**Definition 11.** *Let* **K** *be a connected and pure 1-complex. If the star of every point in* $Sk(\mathbf{K})$ *is linear,* **K** *is called a discrete closed curve.*

### 3.3    Topological Properties in Two Dimensions

If a 2-complex **K** is pure, stars of points in $Sk(\mathbf{K})$ are classified into the following three types: cyclic stars, semi-cyclic stars, and neither of them.

**Definition 12.** *Let* **K** *be a pure 2-complex. If* $link(\mathbf{x})$ *is a discrete closed curve,* $star(\mathbf{x})$ *is cyclic.*

**Definition 13.** *Let* **K** *be a pure 2-complex. If* $link(\mathbf{x})$ *is a discrete curve with endpoints,* $star(\mathbf{x})$ *is semi-cyclic.*

Figure 5 illustrates points of stars which are cyclic, semi-cyclic and neither of them. We see that a point is an edge point of a surface if its star is semi-cyclic, and an interior point of a surface if its star is cyclic. If the star of a point is neither cyclic nor semi-cyclic, the point is at an intersection of surfaces. What we call local surface structures or configurations in this paper are, thus, cyclic stars. By using these properties, we define discrete surfaces in $\mathcal{Z}^3$.

**Definition 14.** *Let* **K** *be a connected and pure 2-complex. If the star of every point in* $Sk(\mathbf{K})$ *is either cyclic or semi-cyclic, and at least one point whose star is semi-cyclic exists in* $Sk(\mathbf{K})$, **K** *is called a discrete surface with edges.*

**Definition 15.** *Let* **K** *be a connected and pure 2-complex. If the star of every point in* $Sk(\mathbf{K})$ *is cyclic,* **K** *is called a discrete closed surface.*

We see that discrete closed surfaces are considered to be 2-dimensional discrete combinatorial manifolds as presented in reference [5].

**Fig. 5.** Examples of points whose stars are cyclic, semi-cyclic and neither of them, illustrated as white, grey and black points, respectively

**Fig. 6.** Examples of points whose stars are spherical, semi-spherical and neither of them, illustrated as white, grey and black points, respectively

### 3.4    Topological Properties in Three Dimensions

If a 3-complex **K** is pure, stars of points in $Sk(\mathbf{K})$ are classified into the following three types: spherical stars, semi-spherical stars, and neither of them.

**Definition 16.** *Let* **K** *be a pure 3-complex. If* $link(\boldsymbol{x})$ *is a discrete closed surface,* $star(\boldsymbol{x})$ *is spherical.*

We define semi-spherical stars by using the notion of combinatorial boundary given by Definition 4.

**Definition 17.** *Let* **K** *be a pure n-complex and* **H** *be the set of all* $(n-1)$-*polyhedra in* **K** *each of which is a face of exactly one n-polyhedron in* **K**. *The combinatorial boundary of* **K** *is then defined as a pure* $(n-1)$-*complex such that*

$$\partial\mathbf{K} = Cl(\mathbf{H}).$$

We now see that the endpoints of a discrete curve and the edges of a discrete surface in Definitions 10 and 14 are their combinatorial boundaries.

**Definition 18.** *Let* **K** *be a connected and pure 3-complex. If* $link(\boldsymbol{x})$ *is a discrete surface with edges, and the edges, i.e., the combinatorial boundary* $\partial(link(\boldsymbol{x}))$ *is a discrete closed curve, then* $star(\boldsymbol{x})$ *is semi-spherical.*

Figure 6 illustrates points whose stars are spherical, semi-spherical and neither of them. It shows that points whose stars are spherical and non-spherical considered to be interior and boundary points of a 3-complex, respectively. If the stars of boundary points are not semi-spherical, such points are considered to be singular points, i.e., intersection points of the boundaries.

In the previous work [8], $star(\boldsymbol{x})$ is simply defined to be semi-spherical if $link(\boldsymbol{x})$ is a discrete surface with edges. However, we found a counter-example such as a white central point in Fig. 7 (left); its link is a discrete surface with two combinatorial boundaries, and its star should not be regarded as a semi-spherical star because it is not topologically equivalent to a semi-sphere. We therefore modify our definition of semi-spherical stars.

Such modification enables us to prove the following important proposition.

**Proposition 1.** *Let* **K** *be a pure 3-complex. If* $star(\boldsymbol{x} : \mathbf{K})$ *is semi-spherical, then* $star(\boldsymbol{x} : \partial\mathbf{K})$ *is cyclic.*

**Fig. 7.** An example of a point whose star is considered to be semi-spherical with our previous definition in [8], but not semi-spherical with that in this paper, illustrated as the white point in the left figure. Its link and its combinatorial boundary are illustrated as a half-tone region and two black bold closed curves in the right figure

Before proving this proposition, we first derive the next lemma.

**Lemma 1.** *Let* **K** *be a pure 3-complex. If* $star(\boldsymbol{x} : \mathbf{K})$ *is semi-spherical, then*

$$\partial(link(\boldsymbol{x} : \mathbf{K})) = link(\boldsymbol{x} : \partial\mathbf{K}).$$

*(Proof)* From Definitions 7 and 17 (or 18), we see that both $\partial(link(\boldsymbol{x} : \mathbf{K}))$ and $link(\boldsymbol{x} : \partial\mathbf{K})$ are pure 1-complexes. Thus, we need to prove simply for 1-polyhedra $\sigma$ that $\sigma \in link(\boldsymbol{x} : \partial\mathbf{K})$ if and only if $\sigma \in \partial(link(\boldsymbol{x} : \mathbf{K}))$.

Let $\sigma$ be each 1-polyhedron in $\partial(link(\boldsymbol{x} : \mathbf{K}))$. Then $\sigma$ is a face of some 2-polyhedron $\tau$ in $star(\boldsymbol{x} : \mathbf{K})$. In $star(\boldsymbol{x} : \mathbf{K})$, we have two different types of 2-polyhedra $\tau$ such that $\tau \in \partial\mathbf{K}$ or $\tau \notin \partial\mathbf{K}$. Now, we show that $\tau \in \partial\mathbf{K}$ if $\sigma \in \partial(link(\boldsymbol{x} : \mathbf{K}))$ by proving its contraposition. If $\tau \notin \partial\mathbf{K}$, $\tau$ becomes a shared face of two 3-polyhedra in $star(\boldsymbol{x} : \mathbf{K})$, therefore $\sigma$ becomes a shared face of two 2-polyhedra in $link(\boldsymbol{x} : \mathbf{K})$. This means that $\sigma \notin \partial(link(\boldsymbol{x} : \mathbf{K}))$. Because $\tau \in star(\boldsymbol{x} : \mathbf{K})$ and $\tau \in \partial\mathbf{K}$, $\tau \in star(\boldsymbol{x} : \partial\mathbf{K})$. Therefore, if $\sigma \in \partial(link(\boldsymbol{x} : \mathbf{K}))$, $\sigma \in link(\boldsymbol{x} : \partial\mathbf{K})$.

Next, we show that $\sigma \in \partial(link(\boldsymbol{x} : \mathbf{K}))$ if $\sigma \in link(\boldsymbol{x} : \partial\mathbf{K})$ contrarily. For any 1-polyhedron $\sigma \in link(\boldsymbol{x} : \partial\mathbf{K})$, we have a 2-polyhedron $\tau \in \partial\mathbf{K}$ such that $\sigma \in face(\tau)$. Such $\tau$ is in $star(\boldsymbol{x} : \mathbf{K})$, and it is not a shared face of two 3-polyhedra in $star(\boldsymbol{x} : \mathbf{K})$ but there is exactly one 3-polyhedron $\upsilon$ such that $\tau \in face(\upsilon)$. Therefore, $\sigma$ is not a shared face of two 2-polyhedra in $link(\boldsymbol{x} : \mathbf{K})$, and from Definition 17, $\sigma \in \partial(link(\boldsymbol{x} : \mathbf{K}))$.    *(Q.E.D.)*

*(Proof of Proposition 1)* From Definition 18, if $star(\boldsymbol{x} : \mathbf{K})$ is semi-spherical, $\partial(link(\boldsymbol{x} : \mathbf{K}))$ is a discrete closed curve. We then obtain from Lemma 1 that $link(\boldsymbol{x} : \partial\mathbf{K})$ is also a discrete closed curve. Thus, from Definition 12, $star(\boldsymbol{x} : \partial\mathbf{K})$ is cyclic.    *(Q.E.D.)*

### 3.5    Topological Point Classification

By using topological properties defined above, we classify points $\boldsymbol{x} \in Sk(\mathbf{K})$ into the following twelve types. Note that each point in $Sk(\mathbf{K})$ is classified into one of them.

**Type 0:** $dim(star(\boldsymbol{x})) = 0$;
**Type 1a:** $dim(star(\boldsymbol{x})) = 1$ and $star(\boldsymbol{x})$ is linear;
**Type 1b:** $dim(star(\boldsymbol{x})) = 1$ and $star(\boldsymbol{x})$ is semi-linear;

**Type 1c:** $dim(star(x)) = 1$ and $star(x)$ is neither linear nor semi-linear;

**Type 2a:** $dim(star(x)) = 2$, $Cl(star(x))$ is pure and $star(x)$ is cyclic;

**Type 2b:** $dim(star(x)) = 2$, $Cl(star(x))$ is pure and $star(x)$ is semi-cyclic;

**Type 2c:** $dim(star(x)) = 2$, $Cl(star(x))$ is pure and $star(x)$ is neither cyclic nor semi-cyclic;

**Type 2d:** $dim(star(x)) = 2$ and $Cl(star(x))$ is not pure;

**Type 3a:** $dim(star(x)) = 3$, $Cl(star(x))$ is pure and $star(x)$ is spherical;

**Type 3b:** $dim(star(x)) = 3$, $Cl(star(x))$ is pure and $star(x)$ is semi-spherical;

**Type 3c:** $dim(star(x)) = 3$, $Cl(star(x))$ is pure and $star(x)$ is neither spherical nor semi-spherical;

**Type 3d:** $dim(star(x)) = 3$ and $Cl(star(x))$ is not pure.

Our discrete surfaces are considered to appear at the 2-dimensional combinatorial boundaries of 3-complexes, that is, $\partial \mathbf{K}$ where $dim(\mathbf{K}) = 3$. From Proposition 1, we see that our points of interest whose local point configurations form discrete surfaces have type 3b.

# 4    Local Point Configurations of Discrete Surfaces

Let us consider a 1-point $x \in \mathcal{Z}^3$ with a set $\mathbf{V}$ of its neighboring 1-points in $\mathbf{N}_{26}(x)$ and a discrete complex $\mathbf{K}_m$, for $m = 6, 26$, constructed from $\mathbf{V} \subseteq \mathbf{N}_{26}(x)$ as explained in the subsection 2.2. In this section, we present an algorithm for examining whether $x$ is a boundary point considered to be on a discrete surface, i.e., $star(x : \partial \mathbf{K})$ is cyclic or $star(x : \mathbf{K})$ is semi-spherical, for a given set $\mathbf{V}$. It means that we check if the type of $x$ is 3b. We apply it for all possible configurations of $\mathbf{V}$ to count the number of local point configurations which form discrete surfaces.

## 4.1    Algorithm

Setting the value of $x$ to be 1, for any configuration of a 1-point set $\mathbf{V} \subseteq \mathbf{N}_{26}(x)$, we apply the following algorithm which returns 1 if $x$ is type 3b.

**Algorithm 1**

**Input:**   *A 1-point set* $\mathbf{V}$ *in a* $3 \times 3 \times 3$ *region of* $\mathcal{Z}^3$.
**Output:**   *If the central point* $x$ *has type 3b, return 1, otherwise, return 0.*
**begin**

    1. *construct* $\mathbf{K}_m$ *from* $\mathbf{V}$ *choosing* $m = 6$ *or* $26$;
    2. *obtain* $star(x : \mathbf{K}_m)$;
    3. **if** $dim(star(x : \mathbf{K}_m)) \neq 3$ **then return** 0;
    4. **if** $Cl(star(x : \mathbf{K}_m))$ *is not pure* **then return** 0;
    5. **if** $star(x : \mathbf{K}_m)$ *is semi-spherical* **then return** 1;
    **else return** 0;

**end**

**Fig. 8.** Two different discrete complexes $\mathbf{K}_{26}$ (left) and $\mathbf{K}'_{26}$ (right) around the central points $\boldsymbol{x}$ so that $star(\boldsymbol{x} : \partial\mathbf{K}_{26}) = star(\boldsymbol{x} : \partial\mathbf{K}'_{26})$

### 4.2    Experiments

By using Algorithm 1, we count the 1-point configurations of $\mathbf{V}$, namely $Sk(\mathbf{K}_m)$, where the central point $\boldsymbol{x}$ is type 3b. They are called surface complicial configurations. Up to rotations and symmetries, the number of all 1-point configurations of $\mathbf{V}$ is reduced to $1,426,144$ from $2^{26} = 67,108,864$. Among them, we obtain $14,031$ and $290,979$ surface complicial configurations for the 6- and 26-neighborhood systems, respectively. We also consider the point configurations of $Sk(Cl(star(\boldsymbol{x} : \partial\mathbf{K}_m))$ where $\boldsymbol{x}$ is type 3b to see only the surface structures around $\boldsymbol{x}$, that is $star(\boldsymbol{x} : \partial\mathbf{K}_m)$. They are called surface star configurations. We then see that there are different discrete complexes $\mathbf{K}_m$ and $\mathbf{K}'_m$ such that $star(\boldsymbol{x} : \partial\mathbf{K}_m) = star(\boldsymbol{x} : \partial\mathbf{K}'_m)$, as illustrated in Fig. 8, and they have the same surface star configurations. We count surface star configurations and finally obtain 6 and $6,028$ for $m = 6, 26$, respectively.

## 5    Conclusions

Given a subset $\mathbf{V} \subseteq \mathbf{N}_{26}(\boldsymbol{x})$, we presented a method for classifying the central point $\boldsymbol{x}$ into one of the twelve types by the topological property of its star after obtaining a complex $\mathbf{K}_m$. With our conclusion such that the type of points which are boundary points having local point configurations of discrete surfaces is 3b, we counted surface complicial configurations $Sk(\mathbf{K}_m)$ and obtained 14,031 and 290,979 configurations for $m = 6, 26$ up to symmetries and rotations. We also obtained 6 and 6,028 surface star configurations, namely, $Sk(Cl(star(\boldsymbol{x} : \partial\mathbf{K}_m)))$, for $m = 6, 26$. The same result for $m = 6$ is already presented in reference [5] and they are illustrated in Fig. 2. We see that a boundary point illustrated as the central point in Fig. 1 (center) has a surface star configuration. This explains why our discrete surfaces have more configurations than that of simplicity surfaces [3].

References [6, 9] show that there are 5 and 32 different configurations of stars which appear in discrete combinatorial planes for $m = 6, 26$, respectively. Such planar stars for $m = 6$ are shown as the five left configurations in Fig. 2. We also illustrate the 32 configurations of planar stars for $m = 26$ in Fig. 9. Note that oriented surfaces are considered in [6, 9] and the 8 and 34 configurations are obtained for $m = 6, 26$. While there is only one non-planar star for $m = 6$, we see that, for $m = 26$, most of the $6,028$, namely $5,994$ surface stars are non-planar and they do not appear on discrete planes but appear on discrete non-planar surfaces. Figure 10 shows that, for example, every boundary point

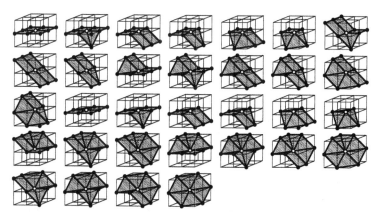

**Fig. 9.** The 32 local star configurations which appear in discrete combinatorial planes for the 26-neighborhood system

**Fig. 10.** Boundary points of three-dimensional digitized objects, such as a cube and a sphere are classified into two types in the 26-neighborhood system: they are illustrated as white and black points if the stars are planar and non-planar, respectively

appearing at the faces of a digitized cube has one of the 32 planar stars illustrated in Fig. 9. On the other hand, around the vertices and edges of a digitized cube, boundary points have non-planar stars. Figure 10 also shows that many boundary points on non-planar surfaces such as a sphere have planar stars rather than non-planar stars. From such experiments, we consider that the study on local configurations of boundary points in the 26-neighborhood system might be useful for shape analysis of three-dimensional images. We remark that the similar shape analysis may not be worth doing for the 6-neighborhood system because most of all boundary points of three-dimensional digitized objects have planar stars.

**Acknowledgements.** A part of this work was supported by Grant-in-Aid for Scientific Research of the Ministry of Education, Culture, Sports, Science and Technology of Japan under the contract of 15700152 and 16650040.

# References

1. P. S. Alexandrov, *Combinatorial Topology*, Vol. 1, Graylock Press, Rochester, New York, 1956.
2. L. D. Cohen, "On active contour models and balloons," Computer Vision, Graphics and Image Processing: Image Understanding, Vol. 53, No. 2, pp. 211-218, 1991.
3. M. Couprie, G. Bertrand, "Simplicity surface: a new definition of surfaces in $\mathbf{Z}^3$," In Vision Geometry VII, Proceedings of SPIE, Vol. 3454, pp. 40-51, 1998.
4. A. Esnard, J-O. Lachaud, A. Vialard, "Discrete deformable boundaries for 3D image segmentation," Technical Report No. 1270-02, LaBRI, University of Bordeaux 1, 2002.
5. J. Françon, "Discrete combinatorial surfaces," Graphical Models and Image Processing, Vol. 57, No. 1, pp. 20-26, 1995.
6. J. Françon, "Sur la topologie d'un plan arithmétique," Theoretical Computer Science, Vol. 156, pp. 159-176, 1996.
7. P. P. Jonker, S. Svensson, "The generation of N dimensional shape primitives," in LNCS 2886 : Discrete Geometry for Computer Imagery, Proceedings of 11th International Conference, DGCI2003, pp.420-433, Springer-Verlag, Heidelberg, 2003.
8. Y. Kenmochi, A. Imiya, A. Ichikawa, "Discrete combinatorial geometry," Pattern Recognition, Vol. 30, No. 10, pp. 1719-1728, 1997.
9. Y. Kenmochi, A. Imiya, "Combinatorial topologies for discrete planes", in LNCS 2886 : Discrete Geometry for Computer Imagery, Proceedings of 11th International Conference, DGCI2003, pp.144-153, Springer-Verlag, Heidelberg, 2003.
10. Y. Kenmochi, A. Imiya, "Combinatorial boundary tracking of a 3D lattice point set," submitted to Journal of Visual Communication and Image Representation.
11. D. G. Morgenthaler, A. Rosenfeld, "Surfaces in three-dimensional digital images," Information and Control, Vol. 51, pp. 227-247, 1981.
12. J. Stillwell, *Classical Topology and Combinatorial Group Theory*, Springer, New York, 1993.
13. G. M. Ziegler, *Lectures on Polytopes*, Springer, New York, 1995.

# Reversible Polygonalization of a 3D Planar Discrete Curve: Application on Discrete Surfaces

Isabelle Sivignon[1], Florent Dupont[2], and Jean-Marc Chassery[1]

[1] Laboratoire LIS,
Domaine universitaire Grenoble - BP46,
38402 St Martin d'Hères Cedex, France
{sivignon, chassery}@lis.inpg.fr
[2] Laboratoire LIRIS - Université Claude Bernard Lyon 1,
Bâtiment Nautibus - 8, boulevard Niels Bohr,
69622 Villeurbanne cedex, France
florent.dupont@liris.cnrs.fr

**Abstract.** Reversible polyhedral modelling of discrete objects is an important issue to handle those objects. We propose a new algorithm to compute a polygonal face from a discrete planar face (a set of voxels belonging to a discrete plane). This transformation is reversible, *i.e.* the digitization of this polygon is exactly the discrete face. We show how a set of polygons modelling exactly a discrete surface can be computed thanks to this algorithm.

## 1   Introduction

Since a few years now, many methods have been proposed in order to compute a polyhedral representation of a discrete object, or more precisely of its surface. Indeed, such a transformation has many useful properties such as:

- a compression of the discrete data: a polyhedral representation suppresses the redundancy of the discrete representation;
- modelling of the discrete object
- better visualization, . . .

Two kinds of reconstruction exist: either we only need an approximation of the discrete surface, or the reversibility of the transformation is desired.

A first naive method consists in computing the convex hull of the discrete points composing the discrete object. This approach simply needs an efficient convex hull algorithm, which is a very well known problem (see [1] for instance). Nevertheless, the polyhedral surface reconstructed is close to the discrete object only in case of convex, or nearly convex objects.

The Marching-Cubes algorithms [2, 3] are the most popular methods to get a polyhedral (triangulated in this case) surface from a discrete object. This transformation is reversible along a classical digitization scheme (OBQ). Nevertheless, this construction is based on local configurations: this induces that the number

E. Andres et al. (Eds.): DGCI 2005, LNCS 3429, pp. 347–358, 2005.
© Springer-Verlag Berlin Heidelberg 2005

of faces of the polyhedral surface depends directly on the number of surface voxels of the discrete object. In order to compute a surface composed of a number of faces not related to the number of discrete points, a global study of the discrete object geometry is needed.

Then, another class of methods is based on the following outline: first, the discrete object surface is decomposed into pieces of discrete planes, and second, a polyhedral representation of the discrete object based on those discrete faces is computed. Such a framework has been used for instance by Borianne and Françon in [4] where a pair digitization/reconstruction is proposed. The authors conjecture that this pair defines a reversible transformation, but this has not been proven yet. In [5], Françon and Papier also proposed an algorithm based on this scheme. Nevertheless, in this case, they directly transform the discrete faces into polygonal non coplanar faces, which is not a satisfactory modelling of the object.

The last two methods we recall in this short state of the art compute an approximation of the discrete surface. The first one was proposed by Burguet and Malgouyres in [6] and uses a "topological Voronoi Diagram". This diagram is used in order to decompose the discrete surface into regions, which are triangulated to get the polyhedral representation. Finally, Yu and Klette [7] use the minimum length polygon algorithm on each slice of the discrete object and sue those polygons together to obtain an approximation of the discrete object surface.

The algorithm presented in this paper is based on a segmentation of the discrete object surface into pieces of discrete planes. We present an algorithm that computes, for each discrete face, a planar polygon containing the voxels of the discrete face in its digitization. Such a transformation is achieved *via* an analytical modelling of the discrete face, which defines a compact description of the discrete object itself.

This paper is composed of three sections. In Section 2, we present the general framework of our algorithm, defining the notions of discrete plane, surface, connectivity and segmentation we use together with the dual spaces. The third section deals with the description of our algorithm and finally, application results of this algorithm over each discrete face of discrete surfaces are proposed in Section 4.

## 2    Framework and Tools

### 2.1    Preliminaries

First of all, we define the framework used in this paper. Our algorithm takes in input a discrete surface that has already been segmented into pieces of discrete planes.

The definitions of discrete plane and discrete surface used for the segmentation process are induced by the reconstruction we propose. Indeed, our transformation defines a discrete polygon (analytical description) from a discrete face (defined by a set of discrete points). The notion of discrete polygon was introduced by Andrès in [8] using the standard digitization model. This digitization

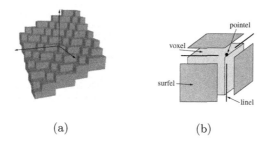

**Fig. 1.** (a) Example of a discrete standard plane. (b) Decomposition of a voxel into lower dimensional elements

scheme is based on the supercover digitization which states that any pixel (or voxel) crossed by the object belongs to the digitized object. Supercovers may contain "bubbles", *i.e.* many digitization pixels for one point (points with half-integer coordinates for instance). To cope with that problem, an orientation convention is defined and leads to the standard digitization scheme.

Consequently, standard planes must be used for the segmentation process. A standard plane is the thinnest 6-connected discrete plane without tunnels: any path (6-connected, 18-connected or 26-connected) joining the two background sides of a plane contains at least one voxel of the plane. An illustration of a standard plane is given in Figure 1(a).

Concerning the definition of discrete surfaces, two main approaches exist: the surface elements are either object voxels or object voxels' faces. In this work, we will define the object surface as the set of voxels' faces (called *surfels*, see Figure 1(b)) belonging to an object and a background voxel. In other words, the surface is composed of the faces visible when the object is displayed. This definition of surface is well-adapted for standard planes segmentation: in this case, discrete (lattice) points are not the object voxels but the vertices of those voxels (called *pointels*, see Figure 1(b)). It is easy to see that those vertices are linked along the 6-connectivity, which is consistent with the use of standard planes.

Using standard planes also induces the connectivities we consider for the object. Standard planes have a combinatorial structure of 2-dimensional manifolds [9, 10]. Thus, the discrete surface we work on should have the properties of a $2D$ combinatorial manifold as well, which implies that 6-connectivity has to be considered for the discrete object.

At this point, we have defined all the elements needed for a segmentation process: discrete plane and surface, connectivity. Now let us describe the properties the segmentation must fulfill:

1. pointels adjacent to a common surfel belong to a common discrete face
2. the projection of each discrete face along its main direction (direction given by the maximum parameter of its normal vector) is a set of 4-connected pixels
3. each discrete face is homeomorphic to a topological disk.

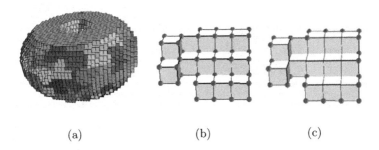

**Fig. 2.** Example of a segmentation result required for the reconstruction: (a) on a torus, (b) detail of a discrete face and (c) of its boundary

Those three conditions imply that any discrete face is a combinatorial 2-manifold with boundary, and that this boundary can be described as a 6-connected $3D$ discrete curve.

Figure 2(a) gives an example of the result we get with a segmentation algorithm fulfilling those three conditions (see [11]). Note that the top of the torus is decomposed into two discrete faces instead of one, so that each discrete face is homeomorphic to a disk. On Figure 2(b), a discrete face is depicted, and each pointel belonging to this face is marked by a small sphere. The boundary of this discrete face can be described as a 6-connected curve, as illustrated in Figure 2(c).

## 2.2   Principle

Since each discrete face is composed of discrete points belonging to a standard plane, there exist a Euclidean plane crossing all the discrete face points. The overall algorithm consists in computing for each discrete face $f$ crossed by a plane $p$, a polygonal line embedded in $p$ and crossing all the boundary points of $f$. This polygonal line defines a polygon containing exactly the discrete points of $f$ in its standard digitization.

The result of such an algorithm is a set of polygons, one for each discrete face, such that the standard digitization of each polygon is exactly a standard face, and finally, the standard digitization of the set of polygons is exactly the initial discrete object surface.

To compute a polygonal line from a discrete face boundary, we propose the following outline, that we present in [12] in the case of 2D discrete curves and non coplanar 3D discrete curves. Consider a 6-connected discrete curve $S$ described as an ordered set of discrete points $\{v_1, v_2, \ldots v_n\}$. A Euclidean point $r_1$ is chosen inside the first point $v_1$, and the following voxels are added one by one (they define a discrete segment $s_1$) while there exist a Euclidean line going through $r_1$, through all the voxels of $s_1$ and embedded in the carrier plane $p$. In other words, $s_1$ is incrementally extended while:

- $s_1$ is a 3D discrete segment
- among the lines which contain $s_1$ in their digitization, there exist at least one line that is embedded in $p$ and that goes through the fixed point $r_1$.

When one of those two conditions is no more fulfilled, the first real segment endpoint $r_2$ is computed as a common point of the computed line and $s_1$'s last pixel. The fixed extremity of the next real segment is set to $r_2$ and this process starts over.

## 2.3    Dual Spaces and Preimages

*Dual Spaces.* In order to polygonalize discrete faces boundaries, the question " does this set of voxels belong to a discrete segment ?" will need an answer, and one method to solve this problem is to rewrite it in a dual space. The main idea is that a line in the Euclidean space is represented by a point in the dual space, and conversely, a point in the Euclidean space corresponds to a line in the dual space.

This tranformation is very similar to the Hough Transform [13, 14] that is classically used in image analysis for shape detection problems. The main difference between the Hough transform and the transform we use here is that the uncertainty related to the discrete nature of the data is not handled during a quantification step but during the tranform itself. Many works in discrete geometry use this transform for 2D discrete line or plane recognition [15, 16, 17] but we will not provide a full state of the art on this point here.

An illustration of the mapping we use is given in Figure 3. Note that in this figure, the dual representation of the line defined by the equation $ax - by + r = 0$ is the point $(\frac{a}{b}, \frac{r}{b})$, and thus, that this representation is based on a normalization along one direction (direction $y$ in this case). Consequently, two dual spaces can

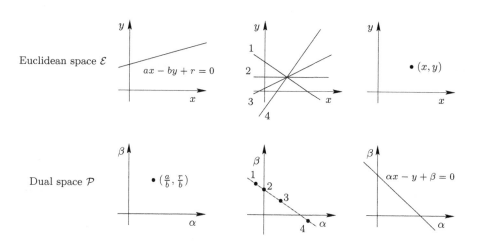

**Fig. 3.** Representation of the links between the Euclidean (top) and the dual spaces (bottom) for elementary geometric objects

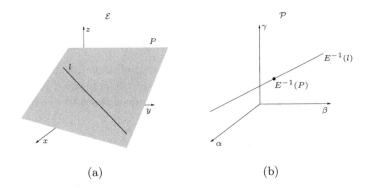

(a)                                    (b)

**Fig. 4.** Representation of a 3D line embedded in a plane in the Euclidean (a) and the dual (b) spaces

be defined in the 2D space, one for each direction. Since line parameters are represented in dual spaces, those spaces are also called parameter spaces.

Similarly, three dual spaces can be defined in 3D. One plane in the Euclidean space is represented by one point in the dual space and conversely. One 3D line in the Euclidean space $\mathcal{E}$ is represented by another 3D line in the parameter space $\mathcal{P}$. In the following, we will denote by $E$ the operator which transforms one element of the parameter space into its corresponding element in the Euclidean space.

One important point for this work is how to represent in the parameter space the embedding of a 3D line into a given plane. It is actually easy to see that since a 3D line $l$ maps to another 3D line $E^{-1}(l)$ (each point of $E^{-1}(l)$ corresponds to a plane containing $l$), and since a plane $P$ maps to the point $E^{-1}(P)$, then $l$ is embedded in $P$ if and only if $E^{-1}(l)$ goes through $E^{-1}(P)$ (see Figure 4).

*Preimages.* Consider a set of pixels $\epsilon$ and a digitization scheme $D$. We call preimage of $\epsilon$ the set of Euclidean lines containing $\epsilon$ in their digitization. This set is represented in the dual space as a set of points.

Let us consider for instance the line $l$ defined by $ax - by + r = 0$ where $a > 0$ and $b > 0$. Then, the standard digitization of $l$ is the set of discrete points $(x, y)$ fulfilling the inequalities $-\frac{a+b}{2} \leq ax - by + r < \frac{a+b}{2}$. The lines containing the point $(x_0, y_0)$ in their digitization are those of parameters $(\alpha, \beta)$ (defined by the equation $\alpha x - y + \beta = 0$) fulfilling the inequalities $-\frac{\alpha+1}{2} \leq \alpha x_0 - y_0 + \beta < \frac{\alpha+1}{2}$. Thus, each discrete point defines two half-spaces in the parameter space, and the intersection of those half-spaces is the set of parameters of the lines containing this discrete point in their digitization. Given a set of discrete points, we call preimage of this set the convex polygon of the parameter space defined by the intersection of the constraints related to the discrete points (see Figure 5).

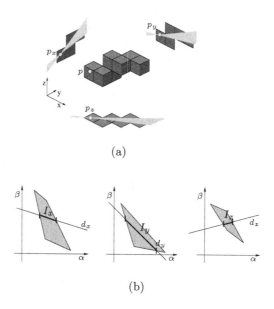

(a)

(b)

**Fig. 5.** Recognition of the three projections of a given set of voxels with a fixed point $p$

# 3    Reversible Polygonalization of a Planar 3D Discrete Curve

In this section, we present a new algorithm to compute a polygonal planar curve from a 3D discrete planar curve in a reversible way. This algorithm is based on three steps, described in the following three paragraphs.

## 3.1    Recognition of a 3D Standard Segment

The first step is to define an algorithm to recognize a 3D standard segment, *i.e.* the standard digitization of a 3D line segment. A standard discrete line can be defined in an analytical way:

**Definition 1.** *Let us consider a 3D straight line of directional vector $(a, b, c)$ going through the point $(x_0, y_0, z_0)$. Then the standard digitization of this line is the set of integer points fulfilling the conditions given by the following double inequalities:*

$$-\frac{|a|+|b|}{2} \le bx - ay + ay_0 - bx_0 < \frac{|a|+|b|}{2}$$
$$-\frac{|a|+|c|}{2} \le cx - az + az_0 - cx_0 < \frac{|a|+|c|}{2}$$
$$-\frac{|b|+|c|}{2} \le cy - bz + bz_0 - cy_0 < \frac{|b|+|c|}{2}$$

*where $(b > 0$ or $(b = 0$ and $a > 0))$ and $(c > 0$ or $(c = 0$ and $a > 0))$ and $(c > 0$ or $(c = 0$ and $b > 0))$ (otherwise, the strict and large inequalities of those equations are exchanged, see [8]).*

From this definition, we derive that if a set of voxels is a 3D standard segment, then the three projections of this set of voxels are 2D standard segments.

Consequently, in order to ensure that the three projections of the set of voxels are 2D standard segments, we compute the three preimages of those sets of pixels. If the three preimages are not empty, then this condition is fulfilled, otherwise, the set of voxels is not a 3D standard segment. Moreover, we said in Paragraph 2.2 that before the recognition step, a point is fixed inside one voxel of the set considered. Thus, the only lines which are interesting for us are the ones which go through this fixed point. As illustrated in Figure 5(a), the projection of this fixed point $p$ onto the three coordinate planes defines three points $p_x$, $p_y$ and $p_z$ that are represented by three lines in the parameter spaces. Thus, the preimages we work one are no more polygons but simply segments denoted by $I_x$, $I_y$ and $I_z$ (see Figure 5(b)).

Nevertheless, this condition over the three projections is not sufficient to define a 3D standard segment, and a compatibility condition between the parameters of the three projections needs to be added (see [12], where non planar curves are studied, for details). In the case of planar curves, this compatibility condition does not need to be handled since we will see that it is ensured while considering the embedding of the curve into a plane.

## 3.2    Ensure Coplanarity

In the following, we consider the general case where the carrier plane $P$ is defined by the equation $ax + by + cz + \mu = 0$, with $a$, $b$ and $c$ not equal to zero.

It is easy to see that any of the three preimage segments $I_x$, $I_y$ and $I_z$ can be represented in two out of the three dual spaces $\mathcal{P}_x$, $\mathcal{P}_y$ and $\mathcal{P}_z$. Indeed, those preimage segments are embedded in the planes $\alpha = 0$ or $\beta = 0$ in those dual spaces. For instance, consider the dual space $\mathcal{P}_x$ where the two segments $I_z$ and $I_y$ can be represented. In this space, the carrier plane $P$ is represented by a point $E^{-1}(P)$. Thus, the 3D lines $l$ embedded in $P$ and containing the set of voxels considered in their digitization are those such that $E^{-1}(l)$ crosses $E^{-1}(P)$, $I_y$ and $I_z$.

A reduction process of the segments $I_z$ and $I_y$ according to $E^{-1}(P)$ is done, as illustrated in Figure 6. The grey cone drawn on this figure represents all the lines going through one point of $I_z$ and the point $E^{-1}(P)$. Thus, the points of $I_y$ which do not belong to this cone must be deleted since there does not exist a line going through $I_z$, $E^{-1}(P)$ and those points (depicted between brackets on Figure 6). After the reduction of $I_y$, the reduction of $I_z$ is computed.

This pair of reductions is computed in each dual space $\mathcal{P}_x$, $\mathcal{P}_y$ and $\mathcal{P}_z$, such that each preimage $I_x$, $I_y$ and $I_z$ is reduced twice. Finally, we have the following result:

**Proposition 1.** *After the six reductions presented above, the preimages of the three projections of the set of voxels $S$ represent exactly the set of 3D lines that are solution for $S$ and embedded in the carrier plane.*

The proof of this proposition is directly derived from the fact that one point in the preimage of one of $S$' projections defines a unique 3D line in the Euclidean

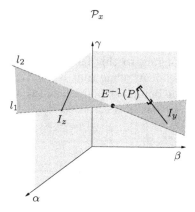

**Fig. 6.** Example of dual segments reductions according to the carrier plane $P$

space. Indeed, such a point represents a 2D line which, together with the carrier plane, defines a 3D line.

Moreover, those reductions ensure that the preimages of the projections are compatible, *i.e.* that there exist a 3D line such that its projections contain the projected set of pixels in their standard digitization (see previous paragraph).

### 3.3    Choice of Fixed Extremities

In Section 2.2, we saw that the first step of our algorithm is to choose an initial point in the first voxel of the curve and in the carrier plane $P$. This point belongs to the intersection between a voxel and a Euclidean plane. From the definition of standard plane, we know that the voxels cut by a given plane are those belonging to the standard digitization of this plane. The geometry of this intersection has been studied by Reveillès [18] and Andrès et al. [19] who show that the only five geometric shapes possible are a triangle, a trapezoid, a pentagon, a parallelogram or an hexagon (Figure 7). They moreover characterize completely the shape of the intersection between a plane and a voxel according to the position of the voxel in the corresponding standard plane. In [18], Reveillès gives the arithmeticalexpression of intersection vertices coordinates. Thus, the initial point chosen in our algorithm is simply the barycentre of the vertices computed thanks to Reveillès [18] and Andrès et al. [19] results.

**Fig. 7.** The five possible intersections between a voxel and a plane

## 4    Application on Discrete Faces: Results

The result of this algorithm over a single discrete face is represented in Figure 8. In (a), the pointels belonging to this face are labelled. In (b), a polygonal curve embedded in the Euclidean plane that is solution for the discrete face is computed: on this figure, the boundary pointels of (a) are replaced by unit cubes centered on the pointels, so that it is easier to check that the computed line is

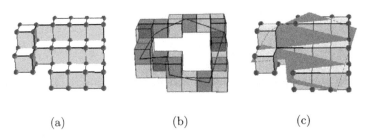

(a)                    (b)                    (c)

**Fig. 8.** Different steps of the polygonalization of a discrete face: (a) the discrete face pointels, (b) computation of the polygonal line, (c) final polygon

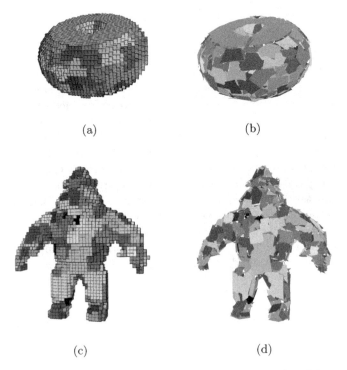

(a)                                (b)

(c)                                (d)

**Fig. 9.** Results of the polygonalization algorithm over complex objects surfaces

entirely included in the curve. Finally, the polygon computed is represented in (c), together with the discrete face it corresponds to: the standard digitization of the polygon is exactly the set of pointels of the discrete face.

On the complexity point of view, each discrete point of the face boundary is seen only once, computing the reduction of the preimages intervals according to a new discrete point is done in logarithmic time (using Grabiner algorithm, see [17]) and the reductions of the intervals presented in Proposition 1 is done in constant time. Finally, the overall complexity of this algorithm is in $\mathcal{O}(n \log n)$ where $n$ denotes the number of boundary points for a given curve.

If we consider complex objects, this algorithm is applied over each discrete face computed by the segmentation algorithm. A set of polygons (one for each face) is computed such that the standard digitization of this set of polygons is exactly the surface pointels of the initial discrete object. Figure 9 presents two results over a torus and the image named "Al": the initial discrete object is depicted on the left, and the set of polygons computed is represented on the right.

## 5   Conclusion

In this paper we described a new algorithm to transform a discrete face into a discrete polygon and a Euclidean polygon. The discrete polygon is defined analytically thanks to the decomposition of the discrete face boundary into discrete analytical segments. The transformation into a Euclidean polygon is done incrementally at the same time, fixing the first extremity of each segment as the last extremity of the previous segment. Next we applied this algorithm onto each discrete face defined by a discrete surface segmentation algorithm. We get a set of polygons modelling the discrete surface in a reversible way: the standard digitization of each polygon is exactly a discrete face of the segmentation.

An interesting future work would be to improve this modelling by sewing the different polygons in order to get a surface while preserving the reversibility property. This problem may be related to the polygonal reconstruction of several adjacent discrete regions in 2D.

## References

1. (Qhull) http://www.qhull.org.
2. Lorensen, W.E., Cline, H.E.: Marching cubes: a high resolution 3D surface construction algorithm. In Stone, M.C., ed.: SIGGRAPH '87 Conference Proceedings, Computer Graphics, Volume 21, Number 4 (1987) 163–170
3. Kenmochi, Y., Imiya, A., Ezquerra, N.F.: Polyhedra generation from lattice points. In Miguet, S., Montanvert, A., Ubéda, S., eds.: Discrete Geometry for Computer Imagery. Volume 1176 of LNCS., Springer-Verlag (1996) 127–138
4. Borianne, P., Françon, J.: Reversible polyhedrization of discrete volumes. In: Discrete Geometry for Computer Imagery, Grenoble, France (1994) 157–167

5. Françon, J., Papier, L.: Polyhedrization of the boundary of a voxel object. In Bertrand, G., Couprie, M., Perroton, L., eds.: Discrete Geometry for Computer Imagery. Volume 1568 of LNCS., Springer-Verlag (1999) 425–434

6. Burguet, J., Malgouyres, R.: Strong thinning and polyhedrization of the surface of a voxel object. In Borgefors, G., Nyström, I., Sanniti di Baja, G., eds.: Discrete Geometry for Computer Imagery. Volume 1953 of LNCS., Uppsala, Sude, Springer-Verlag (2000) 222–234

7. Yu, L., Klette, R.: An approximative calculation of relative convex hulls for surface area estimation of 3D digital objects. In Kasturi, R., Laurendeau, D., Suen, C., eds.: IAPR International Conference on Pattern Recognition. Volume 1., Québec, Canada, IEEE Computer Society (2002) 131–134

8. Andrès, E.: Defining discrete objects for polygonalization : the standard model. In Braquelaire, A., Lachaud, J.O., Vialard, A., eds.: Discrete Geometry for Computer Imagery. Volume 2301 of LNCS., Springer-Verlag (2002) 313–325

9. Françon, J.: Discrete combinatorial surfaces. Graphical Models and Image Processing 51 (1995) 20–26

10. Françon, J.: Sur la topologie d'un plan arithmétique. Theoretical Computer Science 156 (1996) 159–176

11. Sivignon, I., Dupont, F., Chassery, J.M.: Discrete surface segmentation into discrete planes. In Klette, R., Zunic, J., eds.: International Workshop on Combinatorial Image Analysis. Volume 3322 of LNCS., Springer-Verlag (2004) 458–473

12. Sivignon, I., Breton, R., Dupont, F., Andrès, E.: Discrete analytical curve reconstruction without patches. Image and Vision Computing 23 (2005) 191–202

13. Hough, P.: Method and means for recognizing complex patterns. United States Patent, $n°3$, 069, 654 (1962)

14. Maître, H.: Un panorama de la transformation de Hough. Traitement du Signal 2 (1985) 305–317

15. McIlroy, M.D.: A note on discrete representation of lines. AT&T Technical Journal 64 (1985) 481–490

16. Lindenbaum, M., Bruckstein, A.: On recursive, $\mathcal{O}(n)$ partitioning of a digitized curve into digital straight segments. IEEE Trans. on Pattern Anal. and Mach. Intell. 15 (1993) 949–953

17. Vittone, J., Chassery, J.M.: Recognition of digital naive planes and polyhedrization. In Borgefors, G., Nyström, I., Sanniti di Baja, G., eds.: Discrete Geometry for Computer Imagery. Volume 1953 of LNCS., Springer-Verlag (2000) 296–307

18. Reveillès, J.P.: The geometry of the intersection of voxel spaces. In Fourey, S., Herman, G.T., Kong, T.Y., eds.: International Workshop on Combinatorial Image Analysis. Volume 46 of Electronic Notes in Theoretical Computer Science., Philadeplhie, Elsevier (2001)

19. Andrès, E., Sibata, C., Acharya, R., Shin, K.: New methods in oblique slice generation. In: SPIE Medical Imaging. Volume 2707 of Proceedings of SPIE. (1996) 580–589

# Uncertain Geometry in Computer Vision

Peter Veelaert

Hogeschool Gent, Dept. INWE, Belgium
peter.veelaert@hogent.be

**Abstract.** We give an overview of the main ideas and tools that have been employed in uncertain geometry. We show how several recognition problems in computer vision can be translated into combinatorial optimization problems that involve intersection hypergraphs, and how we can obtain approximate solutions for these problems when we replace the hypergraphs by intersection graphs. The statistical properties of these graphs are important when we design algorithms for the extraction of geometric primitives from images. We illustrate the use of uncertain geometry with examples involving the detection of circles and the computation of transformations between images.

## 1 Introduction

The computation of the uncertainty of geometric primitives and transformations is an important problem in computer vision. In the classical approach, the uncertainty of geometric parameters is estimated by statistical analysis. This approach has the disadvantage, however, that the computation soon becomes too complicated unless one introduces simplifying, but often unrealistic assumptions with regard to the statistical model, e.g. that uncertainty can be represented by uncertainty ellipsoids.

Since a purely statistical approach has many shortcomings, other methods have been proposed. The way in which geometric uncertainty has been modeled and handled, however, depends very much on the field of application. Several methods are aimed at obtaining numerical robustness in geometric modeling: robust geometric computation [32], interval geometry [16,17], rounded geometry [19]. One of the prototype problems in this field is the robust intersection of line segments [32]. Also in robotics and mechanical design there is a need to deal with impreciseness. The models proposed there include the use of finite precision arithmetic [12], the use of probability density functions [5,6], and the use of tolerance zones for mechanical parts [7].

When the application is the extraction of geometric primitives and relationships, uncertainty involves more than numerical errors. Some ad-hoc approaches have been quite successful. Lowe as well as Nacken introduce significance measures for geometric relations, e.g. a function that measures how far two line segments are from being collinear [11,13]. A recurring problem, however, is that such a function does not necessarily satisfy the basic rules of geometry, such as the transitivity of collinearity [13]. In this paper we advocate a more uniform approach to geometric uncertainty, consisting of one common methodology applied to many geometric problems, which also takes care of transitivity and other properties [29].

E. Andres et al. (Eds.): DGCI 2005, LNCS 3429, pp. 359–370, 2005.
© Springer-Verlag Berlin Heidelberg 2005

This paper gives an overview of the main ideas and tools that have been employed in uncertain geometry. Positional and geometric uncertainty is modeled by polytopes instead of ellipsoids. Such an approach has several advantages. First, in applications regarding positional or geometric uncertainty, polytopes can give a much better idea about the shape of the uncertainty region. Second, when using polytopes to capture uncertainty, many recognition problems in computer vision can be reformulated as either convex or combinatorial optimization problems for which standard algorithms are known. The paper also mentions more recent work on geometric transformations, where the positions of the points are not precisely known, a topic inspired by registration problems in image processing.

Section 2 describes how geometric uncertainty is modeled in this paper. The advantages of the approach become clear in Section 3 in which we translate the extraction of geometric primitives into a combinatorial optimization problem involving hypergraphs. Section 4 describes how intersection hypergraphs can be replaced by intersection graphs. The statistical properties of intersection graphs are the subject of Section 5.

## 2     Modeling Geometric Uncertainty

What are the geometric problems that must be solved in computer vision, and for which of these problems uncertainty is an important issue? Uncertainty is almost unavoidable in one of the most common tasks in computer vision, i.e. when we extract geometric primitives from real digital images, look for relations between these primitives, or look for geometric transformations between points or primitives. The positional uncertainty of the extracted features has several causes: distortion by lenses, which, when measured in terms of pixels actually gets larger when the resolution of a digital camera gets better, the positional error made by a feature detector, or the geometric imperfectness of the real world, i.e. straight edges are not perfectly straight, round objects do not form perfect circles, hand-drawn figures are loosely drawn.

Another kind of uncertainty is due to the misclassification of features. A feature detector may report features that are not really there (false positives), may fail to report features (false negatives), or in the best case return a likelihood measure for the presence of a feature. Feature classification errors are mainly caused by noise and illumination conditions, combined with the often ill-posed nature of feature detection, in particular when we want to detect features characterized by high intensity variations. Therefore, while extracting geometric primitives from an image we must cope with different kinds of uncertainty at different stages of the process.

The examples used most often to illustrate the concepts of uncertain geometry are digitized straight lines and planes [24, 27, 25, 26, 29, 28]. Two less-known applications are circles and transformations.

*Example 1. Circles.* Suppose we don't know the exact location of the center $(a, b)$ and radius $r$ of a circle in an image, but we are given a set $S_i$ of image points $(x_j, y_j)$ that lie inside the circle, and a set $S_o$ of image points $(x_k, y_k)$ that lie outside a circle, as shown in Figure 1. What is known about the position of the center $(a, b)$ of the circle? The parameters $a, b$ and $r$ must satisfy

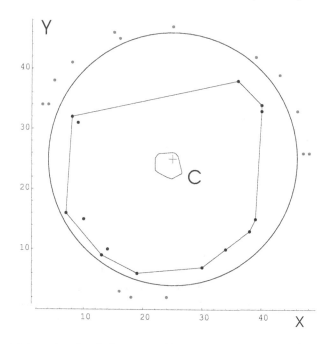

**Fig. 1.** Uncertainty region for circle centers. The gray dots represent the points that must lie outside each circle; the black dots must lie inside

$$(x_j - a)^2 + (y_j - b)^2 < r^2, \tag{1}$$
$$r^2 < (x_k - a)^2 + (y_k - b)^2 \tag{2}$$

for all $(x_j, y_j) \in S_i$ and $(x_k, y_k) \in S_o$. Such a solution exists if and only if

$$(x_j - a)^2 + (y_j - b)^2 < (x_k - a)^2 + (y_k - b)^2 \tag{3}$$

for all possible point pairs $(x_j, y_j) \in S_i$ and $(x_k, y_k) \in S_o$. Or equivalently, we must have

$$2a(x_k - x_j) + 2b(y_k - y_j) < x_k{}^2 - x_j{}^2 + y_k{}^2 - y_j{}^2, \tag{4}$$

which determines a half-plane bounded by the perpendicular bisector of the points $(x_j, y_j)$ and $(x_k, y_k)$. The intersection of all half-planes determines the uncertainty region $C$ of the center, which is also shown in Figure 1, together with one of the circles that separates $S_i$ and $S_o$ and has its center (+) in $C$. Likewise we can compute an uncertainty interval for the radius of the circles separating $S_i$ and $S_o$. The upper bound for this interval is the radius of the largest circle with center in $C$ and not containing any point of $S_o$. The lower bound is the radius of the smallest circle with center in $C$ and containing all points of $S_i$.

*Example 2. Uncertainty of geometric transformations.* Here we want to model geometric transformations for which the position of the image points is not known precisely.

More specifically, to model the uncertainty of the image $(x', y')$ of a point $p = (x, y)$ in $\mathbb{R}^2$, we define the uncertainty region $R$ as a convex polygon bounded by $n$ halfplanes,

$$r_1 x' + s_1 y' \geq 1 \tag{5}$$
$$r_2 x' + s_2 y' \geq 1$$
$$\cdots$$
$$r_n x' + s_n y' \geq 1$$

in the $x'y'$-plane. We let $T(p, R)$ denote the set of all transformations $T$ that map $p$ into the uncertainty region $R$. In this example we restrict ourselves to affine transformations, defined as

$$x' = ax + by + e \tag{6}$$
$$y' = cx + dy + f$$

with $(x', y')$ as image-point, $(x, y)$ as source-point and the 6 parameters of the transformation: $a, \ldots, f$. Then the uncertainty of the transformations $T(p, R) : p \to R$ can be described as a convex polyhedron in 6 dimensions (one dimension for each transformation parameter) by substituting the equations of (6) in (5), yielding

$$r_1(ax + by + e) + s_1(cx + dy + f) \geq 1 \tag{7}$$
$$r_2(ax + by + e) + s_2(cx + dy + f) \geq 1$$
$$\cdots$$
$$r_n(ax + by + e) + s_n(cx + dy + f) \geq 1.$$

The notion of an uncertainty transformation can be extended to sets of points: let $S$ be a finite set of points $p_i \in \mathbb{R}^2$, $\mathcal{R}$ a collection of subsets $R_j$ of $\mathbb{R}^2$, and $f$ a mapping that assigns each point in $S$ to its corresponding subset in $\mathcal{R}$, then $T(S, \mathcal{R}, f)$ denotes the set of all affine transformations that map each point $p_i$ into the set $R_j = f(p_i)$. Clearly, we have $T(S, \mathcal{R}, f) = \cap_i T(p_i, f(p_i))$. Thus $T(S, \mathcal{R}, f)$ is a convex polytope in a 6 dimensional space.

Once a transformation polytope $T$ has been determined, we can use it to find the image of all other points by uncertain transformations [20]. To be precise, let $p$ be a point not in $S$ and $T$ a given polytope, then we let $R(p, T)$ denote the uncertainty region resulting from mapping $p$ by the transformations in $T$; that is, $R(p, T) = \{q \in \mathbb{R}^2 : q = T(p) \text{ for some } T \in T\}$. One can show that $R(p, T)$ is the convex hull of the points $T_k(p)$ where the transformations $T_k$ denote the vertices of the polytope $T$ [20]. Figure 2 shows how three points $p_1, p_2, p_3$ and their image regions $R_1, R_2, R_3$ restrict the set of possible transformations to the transformation uncertainty polytope $T = \cap_{i=1,\ldots,3} T(p_i, R_i)$, which consists of all transformations that map $p_i$ into $R_i$, for $i = 1, \ldots, 3$. With $T$ we can compute the uncertainty region $R(q, T)$ for any point $q$ in the plane. Some of these regions are shown in Figure 2.

Examples 1 and 2 involve inequalities that are linear. Unfortunately, this is not always true, even in the case of a straightforward generalization of Example 2.

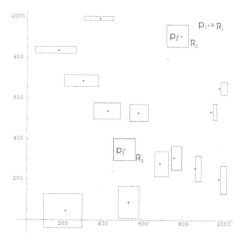

**Fig. 2.** The uncertainty regions $R(p, \mathcal{T})$ for a number of points where the transformation uncertainty polytope has been determined by $(p_i, R_i)$ for $i = 1, \ldots, 3$. The size of the uncertainty regions varies with the position of the point $q$

*Example 3. Uncertain source and image points.* Here we consider transformations where the positions of the source points as well as the positions the image points are uncertain. If we define the uncertainty region of the image point by inequalities of the form $r_i x' + s_i y' \geq 1$, the uncertainty of the source point by inequalities of the form $p_j x + q_j y \geq 1$ and the affine transformation as in (6), then the parameters of the transformation satisfy a system of inequalities of the form

$$r_i(ax + by + e) + s_i(cx + dy + f) \geq 1 \tag{8}$$
$$p_j x + q_j y \geq 1 \tag{9}$$

in which the unknowns $a, \ldots, f, x, y$ appear in a non-linear form (i.e. the product $ax$).

## 3 Turning Geometry into Combinatorial Optimization

Although, as illustrated in Example 3, we cannot always describe uncertainty in terms linear inequalities, Examples 1 and 2, and the work done on digital lines and planes, show that it is possible for many geometric problems. When positional uncertainty is formulated in terms of linear inequalities, the detection of a geometric primitive becomes a problem in convex optimization. For example, to find a minimal number of line segments covering as much points as possible we must find the smallest partitioning of a system such that each subsystem has a solution [27]. Is there a general method to solve such problems? One approach is to attack the problem directly e.g. by applying the simplex algorithm to the system, or if necessary to a large number of its subsystems. A more fruitful approach, however, is to employ Helly's Theorem which turns a linear optimization problem into a combinatorial optimization problem. Helly's Theorem can be stated as follows [9, 14, 18]:

**Theorem 1 (Helly).** *Let $\mathcal{F}$ be a family of convex subsets of $\mathbf{R}^d$ with at least $d + 1$ elements. If $\mathcal{F}$ satisfies the following two conditions:*

1. *the intersection of any $d + 1$ sets in $\mathcal{F}$ is non-empty,*
2. *$\mathcal{F}$ is finite or all elements of $\mathcal{F}$ are compact,*

*then the intersection of all the elements in $\mathcal{F}$ is non-empty.*

Helly's Theorem can be immediately applied to the examples of the previous section.

*Example 4. A theorem on circles.* Let $S_i$ and $S_o$ be defined as in Example 1. Each inequality in (4) defines a convex set in the $ab$-parameter plane. According to Helly's Theorem the finite system (4) will have a solution if and only if each 3-inequality subsystem of (4) has a solution. The results is an elegant property: the given sets $S_i$ and $S_o$ can be separated by a circle if and only if each 3-point subset of $S_i$ can be separated from each 3-point subset of $S_o$ by a circle.

The property derived in Example 4 is similar to the numerous properties that have been proven for digital straight lines and digital planes, of which the chord property is the most well-known [15]. Although establishing such properties is challenging from a mathematical viewpoint, in computer vision the interesting step is to use Helly's Theorem to reformulate detection problems as combinatorial problems.

*Example 5. Circle Detection.* We are given two arbitrary sets $S_i$ and $S_o$ of points in the plane that, as in Example 1, must lie either inside or outside a circle. In this example, however, the given sets are not necessarily separable by a circle. To determine the maximal subset of each set, such that the two subsets can be separated by a circle we first write down the system with $|S_i| \times |S_o|$ inequalities as in (4). Next, we construct a 3-uniform intersection hypergraph $H$ with $|S_i| \times |S_o|$ vertices, where each vertex corresponds to one of the inequalities. The hyperedges of $H$ are formed by the triples of vertices that correspond to those 3-inequality subsystems of (4) that have a solution. The largest complete 3-uniform subhypergraph in $H$ corresponds to the subsets $S_1 \subseteq S_i$, $S_2 \subseteq S_o$ such that $S_1$ and $S_2$ can be separated by a circle, and $|S_1| + |S_2|$ is maximal.

A further extension of the detection problem is the grouping problem, where multiple instances of a geometric primitive must be grouped according to some criterion, e.g. size.

*Example 6. Grouping Circles of Similar Size.* We are given a collection of sets $S^k$ of points, where the points of each set $S^k$ have been partitioned into a set $S_i^k$ of points that must lie inside a circle, and a set of points $S_o^k$ that lie outside it. Then, for each set $S^k$ we can determine the uncertainty interval for the radius of the circles that contain the inside points and exclude the outside points. According to Helly's Theorem, if the interval graph of the radii contains a clique of size $N$, then there are $N$ circles with a common radius corresponding to the sets $S^k$ represented by the vertices of the clique. Thus the interval graph can be used to extract the largest clique of circles that have the same radius, or likewise the graph can be used to partition the circles by a minimum clique covering algorithm into a minimal number of groups so that each group consists of circles that have a similar radius.

The reformulation of a geometric problem as a combinatorial problem has even more advantages, because it becomes easier to add other criteria, for example, the requirement that the points inside a circle should not be too far apart from each other.

*Example 7. Additional Constraints for Circles.* When we look for a circle that separates the maximum number of points, as in Example 5, we may exclude circles whose inside points are too far apart from each other. To be precise, if we have a circle $C$ then we may require that for each point $p \in S_i$ lying inside $C$ there is at least one other point $q \in S_i$ lying inside $C$ such that the distance between $p$ and $q$ is less than a given threshold. This can be modeled by a graph $G$ whose vertices represent the points in $S_i$, and in which two vertices are adjacent when the corresponding points lie close enough to each other. The additional constraint states that a circle is only accepted when the vertices that lie inside it form a connected component of the graph $G$.

In Example 4 Helly's original theorem was used to derive a property for circles in images. Helly's Theorem has been the subject of further extensions and variations. Danzer *et al* and Hadwiger *et al* give extensive overviews of what was known in 1963 [4,8]. More recent advances can be found in [1,2,3,10,23,30], a recent overview in [31]. Some examples in which Helly's extensions are used in uncertain geometry are given in [24]. The application of Helly's Theorem to geometric transformations is described in [20, 21, 22]. Interval graphs have also been used to extract groups of parallel line segments from images [25, 26, 29].

# 4   Intersection Graphs as Approximations for Intersection Hypergraphs

Only in $\mathbf{R}^1$ the application of Helly's Theorem leads to intersection graphs. In $\mathbf{R}^d$, $d > 1$, it leads to hypergraphs. Examples illustrating why intersection graphs are not sufficient for $d > 1$ are easy to find. In $\mathbf{R}^2$ the three edges of a triangle are convex sets. Each pair of edges has a non-empty intersection at a corner, the intersection of the three edges, however, is empty. Likewise, in 3-dimensional space a tetrahedron has four 2-dimensional sides. For each triple of sides there is a non-empty intersection at a corner point, but the intersection of the four sides is empty.

In detection problems, however, not only the correctness of the result but also the computational effort needed to obtain a result matters. Even though an intersection graph may give wrong information about common intersections of sets, it may still be useful as long as we can verify afterwards whether an intersection is non-empty. A maximum clique in an intersection graph may not correspond to a non-empty intersection. If the non-emptiness is easy to verify, however, and if the probability of false information is low, then the worst that can happen is that one must try the second largest clique. In fact, by using new types of intersection graphs we can improve the chance that a clique in an intersection graph corresponds to a non-empty intersection of sets, even in $\mathbf{R}^d$, $d > 1$.

The general idea is the following. Let $H$ be a 3-uniform intersection hypergraph, with $N$ vertices $v_i$ each representing a set $S_i$, and a set of hyperedges $\{v_i, v_j, v_k\}$ such that $S_i \cap S_j \cap S_k \neq \emptyset$. We construct a graph $G$ with $N(N-1)/2$ vertices denoted as $v_i v_j$. The

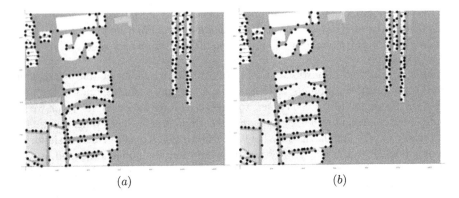

(a)                                    (b)

**Fig. 3.** Image ($a$) shows the feature points detected in the reference image, while the feature points detected for the second, transformed test image are depicted in image ($b$)

edges of $G$ consist of all vertex pairs $\{v_i v_j, v_k v_l\}$ satisfying $S_i \cap S_j \cap S_k \cap S_l \neq \emptyset$. The motivation to replace $H$ by $G$ is as follows. Suppose we find a clique in this graph with $N$ vertices $v_i v_j, v_k v_l, \dots, v_m v_n$. For each edge $\{v_i v_j, v_k v_l\}$ we have $S_i \cap S_j \cap S_k \cap S_l \neq \emptyset$ and therefore $S_i \cap S_j \cap S_k \neq \emptyset, \dots, S_j \cap S_k \cap S_l \neq \emptyset$ for the four combinations of three sets in a collection of four. Since the clique has $N(N-1)/2$ edges this implies that for the collection $C = \{S_i, S_j, \dots, S_n\}$ there are at least $4N(N-1)/2$ intersections among the $2N(2N-1)(2N-2)/6$ possible intersections of 3 arbitrary sets in $C$ that are known to have a non-empty intersection. That is, the ratio of triples that have been verified to the total amount of triples is $6(N-1)/((2N-1)(2N-2))$. This means that if one finds a clique in $G$ for which $N$ is not too large, there is a good chance that it also corresponds to a clique in the hypergraph $H$.

*Example 8. Intersection Graphs for Geometric Transformations.* We illustrate the use of intersection graphs for transformations. Consider the transformation polytopes of a transformation of the form

$$x' = ax + e \qquad (10)$$
$$y' = dy + f,$$

that is a transformation limited to a scaling and translation.

These transformations can be used to solve registration problems. Figure 3 shows an example where in two similar images points have been marked as feature points by a feature detector [22]. We must find a transformation that maps the left image upon the right image as good as possible. Maximal bounds are known for the transformation parameters, i.e. $0.9 \leq a, b \leq 1.1$, and $-20 \leq e, f \leq 20$. We also know that the feature detector is not completely accurate, that is even after adequate scaling and translation, a feature point in the second image may still be displaced a few pixels from the corresponding feature point in the first image.

To find the best transformation we proceed as follows. Let $\mathcal{T}_g$ denote the polytope defined by the inequalities $0.9 \leq a, b \leq 1.1$, and $-20 \leq e, f \leq 20$. We select a small

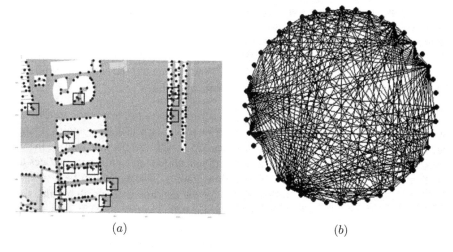

$(a)$                                                      $(b)$

**Fig. 4.** In image $(a)$, the uncertainty regions are centered on a subset of source feature points $(+)$ projected in the test image with the image feature points $(\bullet)$. All candidate feature image points are located within these regions. The intersection graph for situation $(a)$ is shown in $(b)$

number of randomly selected feature points $p_i$ in the first image (source points), marked by crosses in Figure 4(a). Around each source point we place a large rectangle $R_i$ defined by $R_i = R(p_i, \mathcal{T}_g)$. Next we construct a graph $G$ whose vertices correspond to the image points $q_{ij}$ that are located inside the regions $R_i$. Around each point $q_{ij}$ we define a small rectangle $Q_{ij}$ (only a few pixels wide) which takes into account the displacement error made by the feature detector. We compute the polytopes $\mathcal{T}_{ij} = \mathcal{T}_g \cap T(p_i, Q_{ij})$ as well as the graph $G$ shown in Figure 4(b) which is the intersection graph of the polytopes $\mathcal{T}_{ij}$. Although $G$ is an intersection graph, the existence of an edge in $G$ corresponds to a non-empty intersection of three polytopes (not two), that is, $\mathcal{T}_g \cap T(p_i, Q_{ij}) \cap T(p_k, Q_{kl}) \neq \emptyset$. Therefore, as practice shows, a clique in $G$ corresponds almost always to a non-empty transformation polytope, containing transformations that map all the feature points in the clique from the first to the second image.

Intersection graphs were also used as a replacement for 3-uniform hypergraphs to find and group collinear and concurrent line segments [28, 27].

## 5    Statistical Properties of Intersection Graphs

Real images obey statistical laws. It is almost impossible, however, to estimate the probability density function of, say, the lengths of line segments in a typical indoor scene. Nonetheless, we know for example that parallelism occurs often in real scenes and that the uncertainness of parallel relations in a digital image is mostly due to the impreciseness of the image formation process. Thus, because of transitivity, a graph representing parallel relationships in a digitized image will resemble a collection of disjoint cliques, provided the image data are sufficiently good. More generally, the graphs (and hypergraphs) that appear in uncertain geometry are far from being random graphs. This has important consequences from a computational viewpoint, since many combinatorial optimization problems are NP-complete. With standard algorithms it will take much computational effort to find a maximum clique in a large random graph (> 50 vertices). However, intersection graphs (or hypergraphs) are not random and we can use the statistical properties of the graphs to design an efficient clique finding algorithm.

*Example 9. Maximum clique for geometric transformations.* For geometric transformations, the intersection graph $G$ as constructed in Example 8 usually contains a single large clique, to which some additional vertices are loosely connected. That is, there is a subset of vertices that have large degree forming a clique, and additional vertices that have low degree. Furthermore, the intersection graph is also known to be a $r$-partite graph. This results in a considerable speed up for the maximum clique algorithm, since it must only look at subgraphs that contain at most one vertex of each subset in the partition.

This leads to the following algorithm. First, we use Turan's Theorem to compute a lower bound $b_l$ for the maximum clique size. According to Turan's Theorem a graph with $n$ vertices and without a clique of size $p$, $p > 1$ can have at most $(1 - 1/(p - 1))n^2/2$ edges. Thus we can compute a lower bound for the size of the maximum clique given $n$ and the number of edges $|E|$. Furthermore, we can improve this lower bound by eliminating one by one the vertices of minimal degree from the graph $G$ and recalculating the minimal clique size for each subgraph. Next, we derive an upper bound $b_u$ for the maximum clique size. Let $d_1, d_2, \ldots, d_m$ be the ordered, decreasing degree sequence of the graph $G$. Then an upper bound is given by the maximum value of the index $k$ such that $d_k \geq (k - 1)$, since for a clique of size $k$ we need at least $k$ vertices of degree $d_i \geq k - 1$.

We then look at all the subgraphs of order $b_l$ of $G$, and we verify whether they are cliques. Finally, if $b_u > b_l$, we try to extend each of this cliques by adding one vertex at a time. Note that the difference $b_u - b_l$ is a good indication of the computational effort that will be needed to find a maximum clique. If the difference between $b_l$ and $b_u$ is too large (typically > 2), then the graph $G$ is not what we expect and the feature points are not well chosen.

Likewise, in [27] graphs replaced hypergraphs to find collinear line segments. An algorithm based on simplicial elimination orderings was used to find clique coverings of graphs. The method works because the line segment configurations for which cliques do not correspond to non-empty intersections are very rare, and can easily be detected.

## 6    Conclusion

This paper describes a methodology, which has been applied to several recognition and detection problems in computer vision. Broadly speaking it can be summarized as follows. Linear inequalities are used to model positional and parametric uncertainty. Next, Helly's Theorem is used to reformulate a problem about inequalities as a combinatorial problem involving intersection hypergraphs, to which other combinatorial constraints can be added also in the form of graphs. Finally, to approximate the hypergraph optimization problem, we replace it by a graph optimization problem, and we use the statistics of the graph to select a good detection or grouping algorithm.

What are the limitations of this approach? First, not every form of positional or parametric uncertainty can be modeled by linear inequalities, as Example 3 shows. Nonetheless, the research done on lines, planes and transformations proves that at least some important geometric problems can be solved this way. Second, one possible limitation is the occurrence of hypergraphs as well as NP-complete optimization problems. In practice, however, this has not been an important issue yet. For every case in which this methodology was tried it has always been possible to design efficient heuristic algorithms. What are the benefits of the approach? Applications in computer vision show that it is beneficial to have a more precise model for uncertainty than the ad-hoc approaches that are often taken. Modeling uncertainty by convex sets is certainly an improvement when compared to the use of uncertainty ellipsoids, while keeping computational complexity still acceptable.

# References

1. A. B. Amenta, *Helly Theorems and Generalized Linear Programming*, PhD thesis, University of California at Berkeley, 1979.
2. M. Atallah and C. Bajaj, "Efficient algorithms for common transversals", *Inform. Process. Lett.*, vol. 25, pp. 87–91, 1987.
3. D. Avis and M. Doskas, "Algorithms for high dimensional stabbing problems", *Discrete Applied Math.*, vol. 27, pp. 39–48, 1990.
4. L. Danzer, B. Grünbaum, and V. Klee, "Helly's theorem and its relatives", *in Proceedings of the Symposium on Pure Mathematics*, vol. 7, Convexity, pp. 101–180, Providence, RI, 1963. American Mathematical Society.
5. H. F. Durrant-Whyte, "Uncertain geometry in robotics," *IEEE Trans. Robotics Automat.*, pp. 23–31, 1988.
6. H. F. Durrant-Whyte, "Uncertain geometry," in *Geometric Reasoning* (Kapur and Mundy, eds.), pp. 447–481, Cambridge: MIT Press, 1989.
7. A. Fleming, "Geometric relationships between toleranced features," in *Geometric Reasoning* (Kapur and Mundy, eds.), pp. 403–412, Cambridge: MIT Press, 1989.
8. H. Hadwiger and H. Debrunner, *Combinatorial Geometry in the Plane*, Holt, Rinehart and Winston, New York, 1964.
9. E. Helly, "Über Mengen konvexer Körper mit gemeinschaftligen Punkten", *Jahresber. D.M.V.*, vol. 32, pp. 175–176, 1923.
10. D. G. Larman, "Helly type properties of unions of convex sets", *Mathematika*, vol. 15, pp. 53–59, 1968.
11. D. Lowe, "3-d object recognition from single 2-d images," *Artificial Intelligence*, vol. 31, pp. 355–395, 1987.
12. V. J. Milenkovic, "Verifiable implementations of geometric algorithms using finite precision arithmetic," in *Geometric Reasoning* (Kapur and Mundy, eds.), pp. 377–401, Cambridge: MIT Press, 1989.
13. P. Nacken: A metric for line segments. *IEEE Trans. Pattern Anal. Machine Intell.*, **15**, 1312–1318, 1993.
14. R. T. Rockafellar, *Convex Analysis*, Princeton Univerisity Press, Princeton, 1970.
15. A. Rosenfeld, "Digital straight line segments", *IEEE Trans. Comput.*, vol. 23, pp. 1264–1269, 1974.

16. M.G.Segal and C.H.Sequin, Consistent calculations for solids modeling. *Proc. 1st Annual ACM Sympos. Comput. Geom.*, pp. 29-38, 1985.
17. M.G.Segal, Using tolerances to guarantee valic polyhedral modeling results. *Comput. Graph. (Proc. SIGGRAPH '90)*, Vol. 24, pp. 105-114, 1990.
18. J. Stoer and C. Witzgall, *Convexity and Optimization in Finite Dimensions I*, Springer, Berlin, 1970.
19. K. Sugihara, On finite-precision representations of geometric objects. *J. Comput. Syst. Sci.*, Vol. 40, pp. 2-18, 1989.
20. Teelen, K., Veelaert, P.: Uncertainty of affine transformations in digital images. Proceedings of ACIVS 2004 (Advanced Concepts for Intelligent Vision Systems), Brussels, (2004) 23–30.
21. Teelen, K., Veelaert, P.: Computing the uncertainty of geometric primitives and transformations. Prorisc, Velthoven, (2004).
22. Teelen, K., Veelaert, P.: Computing the uncertainty of transformations in digital images, accepted for SPIE's Conference on Vision Geometry XIII, San Jose (2005).
23. H. Tverberg, "Proof of Grünbaum's conjecture on common transversals for translates", *Discrete Comput. Geom.*, vol. 4, pp. 191–203, 1989.
24. P. Veelaert, "Geometric constructions in the digital plane," *J. Math. Imaging and Vision*, vol. 11, pp. 99–118, 1999.
25. P. Veelaert, "Line grouping based on uncertainty modeling of parallelism and collinearity," in *Proceedings of SPIE's Conference on Vision Geometry IX*, (San Diego), pp. 36–45, SPIE, 2000.
26. P. Veelaert, Parallel line grouping based on interval graphs, *Proc. of DGCI 2000*, vol. 1953 of Lecture Notes in Computer Science, pp. 530–541. Uppsala, Sweden: Springer, 2000.
27. P. Veelaert, "Collinearity and weak collinearity in the digital plane," *Digital and Image Geometry*, vol. 2243 of Lecture Notes in Computer Science, pp. 434–447, Springer, 2001.
28. P. Veelaert, Concurrency of line segments in uncertain geometry, *Proc. of DGCI 2002*, vol. 2301 of Lecture Notes in Computer Science, pp. 289–300. Bordeaux, France: Springer, 2002.
29. P. Veelaert, "Graph-theoretical properties of parallelism in the digital plane," Discrete Applied Mathematics, **125**, (2003), pp. 135-160.
30. R. Wenger, "A generalization of Hadwiger's transversal theorem to intersecting sets", *Discrete Comput. Geom.*, vol. 5, pp. 383–388, 1990.
31. R. Wenger, "Helly-type theorems and geometric transversals", *Handbook of Discrete and Computational Geometry*, eds. Goodman and Rourke, CRC Press, pp. 63–82, 1997.
32. C.K. Yap, "Robust geometric computation", *Handbook of Discrete and Computational Geometry*, eds. Goodman and Rourke, CRC Press, pp. 653-668, 1997.

# Optimal Blurred Segments Decomposition in Linear Time

Isabelle Debled-Rennesson[1], Fabien Feschet[2], and Jocelyne Rouyer-Degli[1]

[1] LORIA Nancy – Campus Scientifique - BP 239,
54506 Vandœuvre-lès-Nancy Cedex
{debled, rouyer}@loria.fr
[2] LLAIC - IUT Clermont-Ferrand – Campus des Cézeaux,
63172 Aubière Cedex - France
feschet@llaic3.u-clermont1.fr

**Abstract.** Blurred (previously named fuzzy) segments were introduced by Debled-Rennesson et al [1, 2] as an extension of the arithmetical approach of Reveillès [11] on discrete lines, to take into account noise in digital images. An incremental linear-time algorithm was presented to decompose a discrete curve into blurred segments with order bounded by a parameter $d$. However, that algorithm fails to segment discrete curves into a minimal number of blurred segments. We show in this paper, that this characteristic is intrinsic to the whole class of blurred segments. We thus introduce a subclass of blurred segments, based on a geometric measure of thickness. We provide a new convex hull based incremental linear time algorithm for segmenting discrete curves into a minimal number of thin blurred segments.

## 1   Introduction

Discrete (also called digital) segments are well known objects which have been thoroughly studied for more than 30 years [12]. There are many definitions of discrete segments, all equivalent for 8-connected discrete sets and 4-connected discrete sets. Discrete segments serve as building blocks for representation [16], decomposition [4] or analysis of discrete curves and more generally shapes. For instance, polygonalizations of discrete curves are widely used in shape representation [16] and can be computed in linear time [14, 3, 7]. Moreover, the use of discrete segments permits a perfect representation or reconstruction of discrete curves. However, this might result in complicated representations when discrete curves include noise or have been distorted by an acquisition process. Many polygonal approximation methods have been proposed throughout the years using different approachs [9, 13, 15, 6]. To deal with noise and as an extension of the results presented in [3], the notion of fuzzy segments was introduced in [1, 2]. From now on, we shall name these segments blurred segments rather than fuzzy segments in order to prevent any confusion with fuzzy logic and fuzzy geometry. The theorem of Debled-Rennesson and Reveillès [3] provides an incremental algorithm with linear-time complexity for the recognition of discrete segments using

E. Andres et al. (Eds.): DGCI 2005, LNCS 3429, pp. 371–382, 2005.
© Springer-Verlag Berlin Heidelberg 2005

arithmetical properties and has been extended to the case of blurred segments by Debled-Rennesson et al [1]. However, blurred segments represent supersets of the original discrete data and thus, the result obtained with the previous theorem can not be guaranteed to be optimal, in the sense that the orders of the blurred segments are not necessarily minimal. We present in this paper a study of the order of blurred segments and point out the reason why it is difficult to mimimize the order of a blurred segment in the recognition process. Hence, we present theoretical arguments to justify a restriction in the class of blurred segments in order to guarantee optimality in the recognition process. Moreover, our approach can deal with disconnected sets which was impossible with the theorem given in [1].

The paper is organized as follows. In section 2 we recall definitions and properties used in [1]. We present in section 2.3, a problem in the minimization of the order of recognized blurred segments. This problem is explained by theorem 2 and leads to the introduction of a subclass of blurred segments by adding a geometric characterization based on convex hulls. A new recognition algorithm of blurred segments is described in section 3 by the study of their equivalent characterizations in terms of convex hulls. An incremental linear-time recognition algorithm is presented which guarantees that the computed blurred segments are the thinnest possible ones. Experiments are given in section 4 to show the quality of the decomposition of the proposed algorithm. The paper ends up with some conclusions and perspectives in section 5.

## 2    Blurred Segments

### 2.1    Definitions

The notion of blurred (also called fuzzy) segments relies on the arithmetical definition of discrete lines [11] where a line, with slope $\frac{a}{b}$, lower bound $\mu$ and thickness $\omega$ (with $a$, $b$, $\mu$ and $\omega$ being integer such that $gcd(a,b) = 1$) is the set of integer points $(x, y)$ verifying $\mu \leq ax - by < \mu + \omega$. Such a line is denoted by $\mathcal{D}(a, b, \mu, \omega)$. The real lines $ax - by = \mu + \omega - 1$ and $ax - by = \mu$ are respectively named the *upper and lower leaning lines* of $\mathcal{D}(a, b, \mu, \omega)$ [3]. The integer points $(x_L, y_L)$ (resp. $(x_U, y_U)$) of the lower (resp. upper) leaning lines of $\mathcal{D}(a, b, \mu, \omega)$ are called the lower (resp. upper) leaning points of $\mathcal{D}(a, b, \mu, \omega)$. We refer to Fig. 1 for a descriptive example of these notions.

In the following, we restrict our study to points of the first octant of $\mathbb{Z}^2$, due to symmetries with respect to $Ox$, $Oy$ and the real line $x = y$. We thus always have $0 \leq y \leq x$. This hypothesis can be done without loss of generality and simplifies notations, proofs and definitions.

**Definition 1.** *[1] A set $\mathcal{S}_b$ of consecutive points ($|\mathcal{S}_b| \geq 2$) of an 8-connected curve is a **blurred segment with order d** if there is a discrete line $\mathcal{D}(a, b, \mu, \omega)$, called **bounding**, such that all points of $\mathcal{S}_b$ belong to $\mathcal{D}$ and $\frac{\omega}{max(|a|,|b|)} \leq d$.*

The notion of order of a blurred segment has been introduced to make a difference compared to the thickness of bounding lines of $\mathcal{S}$, since any sufficiently

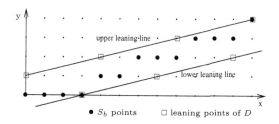

**Fig. 1.** A strictly bounding line $D$ of a blurred segment $S_b$

thick discrete line can contain $S$. Thus, two discrete lines containing $S$ can be compared with respect to their orders and this leads to a classification of the lines containing $S$. To be reasonably closed to the points of $S$, more restrictive conditions onto the discrete lines containing $S$ must be introduced as follows.

**Definition 2.** *[1] Let $S_b$ be a blurred segment with order $d$ whose abscissa interval is $[0, l-1]$ and let $\mathcal{D}(a, b, \mu, \omega)$ be a bounding line of $S_b$. $\mathcal{D}$ is named **strictly bounding for** $S_b$ if $\mathcal{D}$ possesses at least three leaning points in the interval $[0, l-1]$ and, $S_b$ contains at least one lower leaning point and one upper leaning point of $\mathcal{D}$.*

In Fig. 1, $\mathcal{D}(1, 4, -4, 8)$ is a strictly bounding line of the blurred segment $S_b$. The leaning points of $\mathcal{D}$ are the points $(4k+3, k)$ and $(4k, k+1)$, for $k \in [0, 3]$ and $S_b$ contains the lower leaning point $(3,0)$ and the upper leaning point $(12,4)$.

### 2.2 Algorithm for Segmentation into Blurred Segments with Strictly Bounding Lines

We briefly recall in this paragraph the technique used in [1] to segment a discrete curve into order $d$ blurred segments with strictly bounding lines. This segmentation relies on the following theorem, which studies the different possible cases of the growth of a blurred segment.

**Theorem 1.** *[1] Let us consider a blurred segment $S_b$ in the first octant whose abscissa interval is $[0, l-1]$ and $\mathcal{D}(a, b, \mu, \omega)$, a strictly bounding line. In this case, the order of $S_b$ is $\frac{\omega}{b}$. Let $M(x_M, y_M)$ be an integer point connected to $S_b$ whose abscissa is equal to $l$ or $l-1$. Let the **remainder** at $M$, denoted $r(M)$, be a function of $\mathcal{D}$ defined as $\mathbf{r(M)} = \mathbf{a}x_M - \mathbf{b}y_M$.*

*(i)  If $\mu \le r(M) < \mu + \omega$, then $M \in \mathcal{D}$ ;*
  *$S_b \cup M$ is a blurred segment with order $\frac{\omega}{b}$ and with $\mathcal{D}$ as strictly bounding line.*
*(ii) If $r(M) \le \mu - 1$, then $M$ is external to $\mathcal{D}$ ;*
  *$S_b \cup M$ is a blurred segment with order $\frac{\omega'}{b'}$ and the line $\mathcal{D}'(a', b', \mu', \omega')$ is strictly bounding, with*
    *$-$ $b'$ and $a'$ coordinates of the vector $\overrightarrow{P_{r(M)+1}M}$, $P_{r(M)+1}$ being the point whose remainder is $r(M)+1$ with respect to $\mathcal{D}$ and $x_{P_{r(M)+1}} \in [0, b-1]$,*

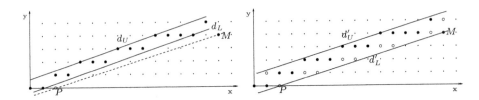

**Fig. 2.** An example of blurred segment growth relying on Theorem 1

- $\mu' = a'x_M - b'y_M$
- $\omega' = a'x_{L_L} - b'y_{L_L} - \mu' + 1$, with $L_L(x_{L_L}, y_{L_L})$ last lower leaning point of the line $\mathcal{D}$ present in $\mathcal{S}_b$.

(iii) If $r(M) \geq \mu + \omega$, then $M$ is external to $\mathcal{D}$ ;
$\mathcal{S}_b \cup \{M\}$ is a blurred segment with order $\frac{\omega'}{b'}$ and the line $\mathcal{D}'(a', b', \mu', \omega')$ is strictly bounding with
  - $b'$ and $a'$ coordinates of the vector $\overrightarrow{P_{r(M)-1}M}$, $P_{r(M)-1}$ being the point whose remainder is $r(M) - 1$ with respect to $\mathcal{D}$ and $x_{P_{r(M)-1}} \in [0, b-1]$,
  - $\mu' = a'x_{U_L} - b'y_{U_L}$ with $U_L(x_{U_L}, y_{U_L})$ last upper leaning point of the line $\mathcal{D}$ present in $\mathcal{S}_b$,
  - $\omega' = a'x_M - b'y_M - \mu' + 1$.

An example of application of this theorem is given in Fig. 2. A blurred segment $\mathcal{S}_b$ with order 2 is depicted in Fig. 2 (left). $\mathcal{D}(1, 3, -2, 4)$ is strictly bounding for $\mathcal{S}_b$, $d_U$ and $d_L$ are the leaning lines of $\mathcal{D}$. The point $M(15, 4)$ is added to $\mathcal{S}_b$. Since $r(M) = 3$, adding $M$ to $\mathcal{S}_b$ corresponds to the case (iii) of the theorem: $P$ is the point in $[0, 2]$ such that $r(P) = 2$, the slope of $\mathcal{D}'$ is computed with the vector $PM$, therefore $\mathcal{D}'(4, 13, -12, 21)$ is strictly bounding for $\mathcal{S}_b \cup \{M\}$. In Fig. 2 (right), a representation of $\mathcal{D}'$ and $\mathcal{S}_b \cup \{M\}$ (black points) is given. The points of $\mathcal{D}'$ which do not belong to $\mathcal{S}_b \cup \{M\}$ are in white, $d'_U$ and $d'_L$ are the leaning lines of $\mathcal{D}'$.

A linear time incremental algorithm of segmentation into order $d$ blurred segments was deduced from this theorem in [1, 2]. The principle was as follows: let $\mathcal{S}_b$ be the current order $d$ blurred segment, a point $M$ of $\mathcal{C}$ was added to $\mathcal{S}_b$, the characteristics of a strictly bounding line of $\mathcal{S}_b \cup M$ were computed according to Theorem 1. The current segment included the point $M$ if the value of the obtained ratio $\omega/max(|a|, |b|)$ was lower than or equal to the order $d$. Else, the current order $d$ blurred segment ended at the point located before $M$ in $\mathcal{C}$ and a new order $d$ blurred segment started at $M$.

## 2.3   Main Drawback

Theorem 1 describes an incremental method to construct a strictly bounding line. However the constructions do not garantee that the order of the built bounding line is minimal. Hence, the segmentation of a discrete curve might be done into too many blurred segments. For instance, the curve depicted in Fig. 3 (left) would be uncorrectedly segmented into several parts with $d = 1.9$, while there exists a

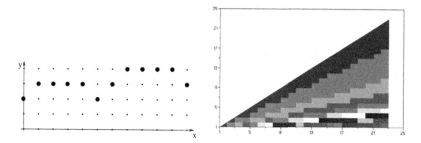

**Fig. 3.** (left) a sample discrete curve  (right) gray scale plot of $\omega/b$ with $b$ in x-axis and $a$ in y-axis

bounding line with ratio strictly lower than 1.9. The experimental investigations that we have performed, induced us to study in detail a better measure for segment thickness to garantee the order of the built bounding lines.

To understand the ratio $\omega/\max(|a|, |b|)$, we study it in the first octant using the curve in Fig. 3. For any $a$ and $b$ values, with $a \leq b$, we compute the remainder $r(M)$. The value of $\omega$ is obtained by adding 1 to the difference between the maximum and the minimum of the remainders. We denote by $\omega(a, b)$ this value. The result is depicted on Fig. 3 (right) where light grey values correspond to low values of the ratio $\omega/b$. As shown in Fig. 3 (right), the function $\omega(a, b)/b$ has a lot of local extrema. The minimalization of $\omega(a, b)/b$ seems to be a hard combinatorial problem. Moreover, $\omega(a, b)/b$ is sensitive to multiplications since the couples $(a, b)$ equal to $(1, 6)$, $(3, 18)$ and $(4, 24)$ do not produce the same value. Their values are respectively 1.83, 1.72 and 1.70. The following theorem given for the first octant without loss of generality explains why the minimization of the ratio $\omega(a, b)/b$ might be impossible to obtain due to its asymptotic behaviour.

**Theorem 2.** *For any finite set $\mathcal{S}_b$, let us denote by $\mathcal{W}$ the series $(\omega(a_k, b_k)/b_k)$ where $k$ is a positive integer, $a_k = k$ and $b_k = kb_0 + \lambda$, with $b_0$ and $\lambda$ positive integers. Then $\mathcal{W}$ is decreasing and has a limit equals to $\frac{\omega(1, b_0) - 1}{b_0}$.*

*Proof.*
We introduce the remainder $r_{(a,b)}(M) = ax_M - by_M$. It is easy to see that $r_{(a_k, b_k)}(M) = kr_{(1, b_0)}(M) - \lambda y_M$. We now introduce $\Delta_{(a,b)}(M, M')$ as follows,

$$\Delta_{(a_k, b_k)}(M, M') = r_{(a,b)}(M) - r_{(a,b)}(M') = k\Delta_{(1, b_0)}(M, M') + \lambda (y_{M'} - y_M) \tag{1}$$

Suppose that $\Delta_{(1, b_0)}(M, M') = 0$. Hence, $\Delta_{(a_k, b_k)}(M, M') = \lambda (y_{M'} - y_M)$. Since $(y_{M'} - y_M)$ is bounded on any finite set and $\lambda$ is constant, we deduce that the previous value is bounded above by a constant $\delta$.

Suppose now that $\Delta_{(1, b_0)}(M, M') \neq 0$. By using the same boundedness argument, we see that there exists a value $k_0$ such that

$$\Delta_{(a_k, b_k)}(M, M') > 0 \text{ (resp. } < 0) \iff \Delta_{(1, b_0)}(M, M') > 0 \text{ (resp. } < 0) \tag{2}$$

for any $k \geq k_0$. Hence asymptoticaly, the remainders $r_{(a_k,b_k)}(.)$ and the remainders $r_{(1,b_0)}(.)$ have the same ordering. However, the remainders $r_{(a_k,b_k)}(.)$ are diverging when $k$ tends to infinity. Thus for sufficiently large $k$, the minimum and the maximum of the remainders $r_{(a_k,b_k)}(.)$ are obtained exactly for the same points $M_{\min}$ and $M_{\max}$ as for the remainders $r_{(1,b_0)}(.)$. This permits us to deduce that for sufficiently large $k$,

$$\omega_{(a_k,b_k)} = k \left( \omega_{(1,b_0)} - 1 \right) + \lambda \left( y_{M_{\min}} - y_{M_{\max}} \right) + 1 \tag{3}$$

So,

$$\frac{\omega_{(a_k,b_k)}}{b_k} = \frac{k \left( \omega_{(1,b_0)} - 1 \right)}{k b_0 + \lambda} + \frac{y_{M_{\min}} - y_{M_{\max}} + 1}{k b_0 + \lambda} \tag{4}$$

The limits of the previous expression is given by the limit of the first term, specifically

$$\lim_{k \to +\infty} \frac{\omega_{(a_k,b_k)}}{b_k} = \frac{\left( \omega_{(1,b_0)} - 1 \right)}{b_0} \tag{5}$$

We conclude the proof by a study of a specific case obtained when the remainders $r(1, b_0)$ are all equals. In such a case, $\omega(1, b_0) = 1$ and $\omega(a_k, b_k) = 1 + \lambda \left( \max_M y_M - \min_M y_M \right)$. Thus by dividing by $b_k$ and taking the limit, we obtain 0. The result still holds and this concludes the proof.    □

To have an optimal algorithm, we slightly modify the subclass of considered blurred segments by taking the limit measure of the previous theorem as a measure for comparing blurred segments.

We start by giving a geometric description of the limit measure. The **vertical distance of a discrete line** $\mathcal{D}(a, b, \mu, \omega)$ is the vertical distance (ordinate difference) between the leaning lines of $\mathcal{D}$ and is equal to $\frac{\omega-1}{b}$. We now recall that a **supporting line** of a convex set $C$ is a line $l$ such that the intersection of $l$ with $C$ is not empty, and such that $C$ is entirely either below or above $l$ [5]. The **vertical distance of a convex set** $C$, is the minimal vertical distance of any pair of parallel supporting lines. This could also be defined as the maximal length of the intersection of $C$ with a vertical line.

**Definition 3.** *Let us consider a set of 8-connected points* $\mathcal{S}_b$. *A bounding line of* $\mathcal{S}_b$ *is said* **optimal** *if its vertical distance is minimal, i.e. if the vertical distance of* $\mathcal{S}_b$ *equals to the vertical distance of its convex hull* $\mathrm{conv}(\mathcal{S}_b)$.

This definition is illustrated in Fig. 4 and leads to the following new definition concerning blurred segments.

**Definition 4.** *A set* $\mathcal{S}_b$ *is a* **blurred segment of width** $\nu$ *if and only if its optimal bounding line has a vertical distance less or equal to* $\nu$.

The recognition of blurred segments with width $\nu$ is thus equivalent to the computation of the vertical distance of the convex set $\mathrm{conv}(\mathcal{S}_b)$.

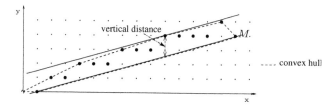

**Fig. 4.** An optimal bounding line

# 3   Convex Hulls and Blurred Segments

## 3.1   Characterization of the Vertical Distance of a Finite Convex Set

Let us denote by $L_i$ and $U_i$ respectively the lowest and highest intersection point of a finite convex set $C$ with the vertical line given by $x = i$ and by $\mathcal{V}(i)$ **the distance between $L_i$ and $U_i$**, then for any $i \neq j$, the quadrangle $(L_j U_j U_i L_i)$ is located inside $C$. It is straightforward to see that the function $\mathcal{V}(.)$ is a concave function. Hence, every local maximum of the function is also a global maximum of the function. The positions of maxima of the function $\mathcal{V}(.)$ have the following property.

**Proposition 1.** *For any finite convex set $C$, the function $\mathcal{V}(.)$ has a maximum value at a position $i$ where $L_i$ or $U_i$ can be chosen to be a vertex of the border, Bd $C$, of $C$.*

*Proof.* Let us consider a position $i$ such that neither $L_i$ nor $U_i$ are vertices of Bd $C$ and the edges of Bd $C$ containing $L_i$ and $U_i$. It is easy to see that $i$ cannot be the position of a maximum when the slopes of these edges are not equal. If the case of equality we can move on the edges until one of the two points becomes a vertex of Bd $C$.     □

From this proposition, we can deduce two facts if we are looking for the vertical distance of a convex set: first, we only have to consider points on the border of $C$ and second, extrema are obtained for some $x$ positions of the vertices of Bd $C$. This leads to three cases: (edge,vertex), (vertex,edge) or (vertex,vertex) where the first element represents the lower part of the convex set and the second element represents the upper part. It must be noticed that the previous proposition is identical to the one characterizing the width of a convex set [10].

It is clear that the edges corresponding to the position of a maximum of $\mathcal{V}(.)$ are supporting lines, taking horizontal lines for a couple (vertex,vertex). Moreover, the edge and its parallel passing through the vertex define exactly the lower and upper leaning lines of an optimal bounding line for $C$. Hence, optimal lines are deduced from the positions of maxima.

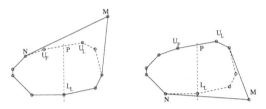

**Fig. 5.** Adding a new point either above (left) or below (right)

## 3.2    New Recognition Algorithm

Let $\mathcal{S}_b = \{(x_i, y_i),\ 0 \le i < n\}$, a blurred segment in the first octant with $\mathcal{D}(a, b, \mu, \omega)$ as optimal bounding line. We suppose that $\mathcal{S}_b$ contains two upper leaning points, $U_F$ and $U_L$, and one lower leaning point, $L_L$. $U_F$, $U_L$ and $L_L$ are vertices of the convex hull of $\mathcal{S}_b$ $conv(\mathcal{S}_b)$. Moreover, the vertical distance of $\mathcal{D}$ and of $conv(\mathcal{S}_b)$ can be calculated at the point $L_L$. To compute the convex hull of $\mathcal{S}_b$, we use Melkman's algorithm [8]. Let us recall that this algorithm incrementally computes the convex hull of $n$ points forming a simple polygonal line in $O(n)$ time based on a double ended queue (deque) list. Since points are added with increasing $x$, we are guaranteed to have a simple polygonal line.

**Growth of a Blurred Segment.** Suppose that we add a new point $M$ to $\mathcal{S}_b$, $\mathcal{S}_b' = \mathcal{S}_b \cup \{M\}$. There are three cases: $M$ is added above, inside or below $\mathcal{D}$.

If $M$ *belongs to* $\mathcal{D}$ then, after the application of Melkman's algorithm [8], the vertical distance remains the same. The new set of points, $\mathcal{S}_b'$, is a blurred segment with the same width and with the same optimal bounding line $\mathcal{D}$. In the other cases, the vertical distance must be recomputed. $P$, the intersection point between the vertical line from $L_L$ and $U_F U_L$, is strictly inside $[U_F U_L]$ (see Fig. 5).

Suppose that $M$ *is added above* $\mathcal{D}$. Let us apply Melkman's algorithm. The convex set is modified and we call $N$ the point before $M$ in the upper part of the resulting convex hull, see Fig. 5 (left). $N$ is necessarily before $U_F$ or is $U_F$. As a consequence, the vertical projection of $L_L$ is inside $[NM]$. Thus, the vertical distance of the new convex set strictly increases. The key point is to locate the new position of a maximum. It is clear that $N$ cannot represent a position of a maximum since $\mathcal{V}(N) \le \mathcal{V}(L_L)$. Moreover, $M$ does not project vertically strictly inside an edge of the lower part of $conv(\mathcal{S}_b')$. Hence, the new position of a maximum if necessarily obtained for one point at the right of $L_L$ in the lower part of $conv(\mathcal{S}_b')$. Let us recall now that local extrema are global extrema for the function $\mathcal{V}(.)$ such that we only have to test the candidate points in sequence and stop at the first local maximum, called $C$.

Suppose now that $M$ *is added below* $\mathcal{D}$. Let us apply Melkman's algorithm. The lower part of the convex set is modified and we still call $N$ the point before $M$ in the lower part of the new convex hull, see Fig. 5 (right). $N$ is necessarily to the left of $L_L$. Hence, $N$ cannot be the new position of a maximum since $\mathcal{V}(N) \le \mathcal{V}(L_L)$. We do not know precisely where $N$ is located in comparison with $[U_F U_L]$. It might be either on the left of $U_F$ or inside $[U_F U_L]$. In both cases, it is however straightforward to see that no position strictly to the left of $U_L$ can be the position of a maximum. As in the previous case, neither $N$ nor $M$ can be positions of maxima. So, the new position of a maximum is given by one point situated to the right of $U_L$ on the upper part of the new convex hull and we only have to test the candidate points in sequence and stop at the first local maximum, called $C$.

In both these cases, $S'_b$ is a blurred segment with optimal bounding line $\mathcal{D}'$ for which the points $M$, $N$, and $C$ are leaning points. Moreover, the vertical distance of $conv(S'_b)$ is equal to the vertical distance of $\mathcal{D}'$ and can be calculated at the point $C$.

We have neglected the case where $L_L$ and $P$ are vertices of $conv(S_b)$. If we choose any edge containing either $L_L$ or $P$ and keep the other point as a vertex, we obtain a couple (vertex,edge) or (edge,vertex) with the same vertical distance. This result is of course different from the one of the width of a convex set [5], but applies perfectly in our context.

**Recognition Algorithm.** From the study of the growth of a blurred segment we can deduce an incremental recognition algorithm of width $\nu$ blurred segments (see Algorithm 1) where points are taken in order of increasing $x$ values. It gives as result a boolean value equal to true if a sequence of points $S$ (input) is a width $\nu$ blurred segment. Morever the last calculated values of $a$, $b$, $\mu$ and $\omega$ are the characteristics of an optimal bounding line of $S$.

---

**Algorithm 1**: Incremental recognition of blurred segment with width $\nu$

---

**Input**        : $S$ an 8-connected sequence of integer points, $\nu$ a real value
**Output**       : $isSegment$ a boolean value, $a$, $b$, $\mu$, $\omega$ integers
**Initialization**: $isSegment = true$, $a = 0$, $b = 1$, $\omega = b$, $\mu = 0$, $M = (x_0, y_0)$.
**while** $S$ *is not entirely scanned and isSegment* **do**
  $M = $ next point of $S$;
  add $M$ to the upper and lower convex hulls of the scanned part of $S$;
  $r = ax_M - by_M$;
  **if** $r = \mu$ **then** $U_L = M$;
  **if** $r = \mu + \omega - 1$ **then** $L_L = M$;
  **if** $r \leq \mu - 1$ **then**
    | $U_L = M$;
    | Let $N$ the point before $M$ in the upper convex hull, $a_0 = y_M - y_N$,
    | $b_0 = x_M - x_N$, then $a = \frac{a_0}{gcd(a_0,b_0)}$ , $b = \frac{b_0}{gcd(a_0,b_0)}$, $\mu = ax_M - by_M$;
    | Find the first point $C$ in the lower part of the convex hull starting at
    | $L_L$, such that : slope of $[C, Cnext] > \frac{a}{b}$;
    | $L_L = C$;
  **else**
    | **if** $r \geq \mu + \omega - 1$ **then** *symmetrical case*
  **end**
  $isSegment = \frac{\omega - 1}{b} \leq \nu$;
**end**

---

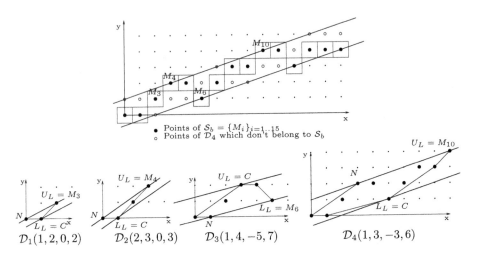

Fig. 6. An example of width 2 blurred segment incremental recognition

**Complexity.** We now study the complexity of Algorithm 1 on a set $S$ of $n$ points. The first part of the algorithm is the update of the convex hull after the insertion of the new point $M$. However, Melkman's algorithm [8] has a linear-time complexity for the whole set of insertions. Since, a point is used only for one convex hull, we deduce easily that this part still have a linear-time complexity. The second part of the algorithm concerns the update of the position of a maximum. Each time a point is added, part of the previous convex hull is examined to detect the new position of a maximum. However, the main property of the function $\mathcal{V}(.)$ is that it is a concave function. Thus as previously mentioned, we only test points for which $\mathcal{V}(.)$ is not locally maximal and stop at the first local maximum. This part of the algorithm also has a linear-time complexity. Then we can conclude that *Algorithm 1 has a linear-time complexity.*

An example of the algorithm processing is depicted in Fig. 6. Part of a discrete curve, $S_b$, is drawn at the top of the figure and the different optimal bounding lines, obtained during the incremental recognition, are given. As it can be seen, there are only four steps in the recognition process and the slopes of the supporting lines decrease or increase with respect to the added points. $S_b$ is a blurred segment of width 2 with $\mathcal{D}_4(1,3,-3,6)$ as optimal bounding line.

## 4    Experiments

The segmentation of a curve into blurred segments of width $\nu$ is done incrementally as described in [1,2], and as recalled in section 2.2. When the width of the current segment becomes strictly greater than $\nu$, a blurred segment ends at the previous point, and a new segment is initialised. In Fig. 7, we show how the curves on the left are segmented into blurred segments with width 2. The curve in Fig. 7 (top) is segmented into 18 segments. The curve in Fig. 7 (bottom) is

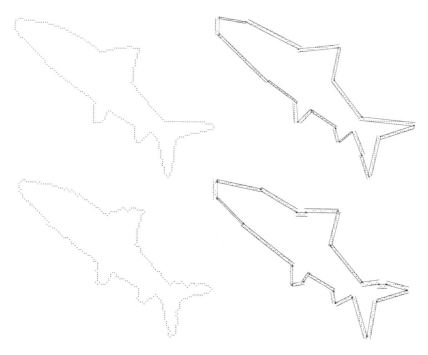

**Fig. 7.** Shark curves (top without noise, bottom with noise) decomposed into blurred segments of width 2

a noisy version of the top one; the segmentation gives more segments: 21. If we increase the width of the blurred segments, the number of segments decreases: 19 segments for the width 2.5, 18 segments for the width 3.

## 5   Conclusion

We have pursued in this paper the study of blurred segments and their applications to discrete noisy curves. Previous works of Debled-Rennesson et al [1] were not able to decompose discrete noisy curves with minimally thin blurred segments. We have identified the origin of this drawback and based on our theoretical study, we have proposed to restrict the class of blurred segments by adding a geometrical bound on their thickness. A proper measure to control the thickness of blurred segments bounded by a discrete line $\mathcal{D}(a, b, \mu, \omega)$ was demonstrated to be $(\omega - 1)/b$. Based on this modification, the recognition of blurred segments was shown to be equivalent to the computation of the vertical distance of the convex set of points of the discrete curves. We have presented an incremental linear time algorithm which solves this problem with minimality in the thickness of constructed blurred segments. Moreover, our approach also applies to disconnected sets which opens perspectives in the study of discrete curves with holes. The tools we used can be extended to 3D and this might lead to decompositions of discrete surfaces into blurred linear patches.

# References

1. I. Debled-Rennesson, J.-L. Rémy, and J. Rouyer. Segmentation of discrete curves into fuzzy segments. In *9th International Workshop on Combinatorial Image Analysis*, volume 12 of *Electronic Notes in Discrete Mathematics*, 2003.
2. I. Debled-Rennesson, J.-L. Rémy, and J. Rouyer. Segmentation of discrete curves into fuzzy segments, extended version. Technical report, INRIA Report RR-4989, http://www.inria.fr/rrrt/rr-4989.html, 2003.
3. I. Debled-Rennesson and J.P. Reveillès. A linear algorithm for segmentation of digital curves. *IJPRAI*, 9(6), December 1995.
4. F. Feschet and L. Tougne. Optimal time computation of the tangent of a discrete curve: application to the curvature. In *Discrete Geometry and Computer Imagery*, Lecture Notes in Computer Science 1568, pages 31–40. Springer Verlag, 1999.
5. M.E. Houle and G.T. Toussaint. Computing the width of a set. *IEEE Trans. on Pattern Analysis and Machine Intelligence*, 10(5):761–765, 1988.
6. A. Kolesnikov and P. Fränti. Reduced-search dynamic programming for approximation of polygonal curves. *Pattern Recognition Letter*, 24(14):2243–2254, 2003.
7. M. Lindenbaum and A. Bruckstein. On Recursive, O(N) Partitioning of a Digitized Curve into Digital Straight Segments. *IEEE Transactions on Pattern Analysis and Machine Intelligence*, 15(9):949–953, 1993.
8. A. Melkman. On-line Construction of the Convex Hull of a Simple Polygon. *Information Processing Letters*, 25:11–12, 1987.
9. J. Perez and E. Vidal. Optimum polygonal approximation of digitized curves. *Pattern Recognition Letter*, 15:743–750, 1994.
10. F.P. Preparata and M.I. Shamos. *Computational Geometry: an Introduction*. Springer-Verlag, 1985.
11. J.P. Reveillès. Géométrie discrète, calculs en nombres entiers et algorithmique. Thèse d'Etat – Université Louis Pasteur, 1991.
12. A. Rosenfeld and R. Klette. Digital Straightness – a review. *Discrete Applied Math.*, 139(1–3):197–230, 2004.
13. P. L. Rosin. Techniques for assessing polygonal approximations of curves. *IEEE Transactions on PAMI*, 19(6):659–666, 1997.
14. A.W.M. Smeulders and L. Dorst. Decomposition of discrete curves into piecewise straight segments in linear time. *Contemporary Mathematics*, 119:169–195, 1991.
15. P. Yin. A tabu search approach to polygonal approximation of digital curves. *International Journal of Pattern Recognition and Artificial Intelligence*, 14(2):243–255, 2000.
16. D. Zhang and G. Lu. Review of shape representation and description techniques. *Pattern Recognition*, 37(1):1–19, 2004.

# Shape Preserving Digitization of Binary Images After Blurring

Peer Stelldinger and Ullrich Köthe

Cognitive Systems Group, University of Hamburg,
Vogt-Köln-Str. 30, D-22527 Hamburg, Germany

**Abstract.** Topology is a fundamental property of shapes in pictures. Since the input for any image analysis algorithm is a digital image, which does not need to have the same topological characteristics as the imaged real world, it is important to know, which shapes can be digitized without topological changes. Most existing approaches do not take into account the unavoidable blurring in real image acquisition systems or use extremely simplified and thus unrealistic models of digitization with blurring. In case of the mostly used square grids we show which binary images can be digitized topologically correctly after blurring with an arbitrary non-negative radially symmetric point spread function, which is an important step forward to real digitization.

## 1   Introduction

A reliable image analysis algorithm requires a digital image having as many properties as possible in common with its continuous preimage. One intrinsically twodimensional property is the topology of shapes. There are several sampling theorems known, which describe under which circumstances the topology of some shape does not change during digitization. These theorems mostly differ in the chosen digitization model and the used sampling grid. E.g. Pavlidis showed that so-called $r$-regular shapes can be digitized with square grids without any change in topology [4]. Serra proved the same for hexagonal grids [6] and recently we extended these results to arbitrary sampling grids [1]. All of these approaches used the subset digitization where a sampling point is set if and only if it lies within the foreground region of the binary image, i.e. no blurring occurs. Unfortunately, real optical systems blur the binary image before the light reaches the optical sensors. In addition to that each sensor integrates the intensity of light over some area. Both effects can be described as blurring – a convolution of the ideal binary image with a suitable point spread function. A binary image can be recovered by considering a particular level set $L_l = \{x \in \mathbb{R}^2 | \hat{f}(x) \geq l\}$ of the blurred image $\hat{f}$, i.e. by thresholding. Of course the resulting shape heavily depends on the choice of the used point spread function. Latecki et al. [2, 3] used a point spread function which is constant in its square-shaped support and proved that $r$-regular images can be topologically correctly reconstructed after blurring and sampling with a sufficiently dense square grid. In the above mentioned previous paper [1] we proved that this is also true for point spread functions, which

E. Andres et al. (Eds.): DGCI 2005, LNCS 3429, pp. 383–391, 2005.
© Springer-Verlag Berlin Heidelberg 2005

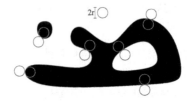

**Fig. 1.** For each boundary point of an $r$-regular set there exists an outside and an inside osculating open disc of radius $r$

are constant in their disc-shaped support. Up to now nothing has been known about blurring with point spread functions which are not constant within their support. Now we extent the results to arbitrary non-negative radially symmetric point spread functions with bounded support. We prove that given such a point spread function with a support of radius $p$ and a square grid of sampling density $r' > p$, every $r$-regular image ($r > r' + p$) will be digitized whithout any change in the topology.

## 2    Regular Sets, Sampling and Reconstruction

At first we define some basic mathematical concepts. The Complement of a set $A$ will be noted as $A^c$. The boundary $\partial A$ is the set of all common accumulation points of $A$ and $A^c$. A set $A$ is open, if it does not intersect its boundary. $\mathcal{B}_r(c) := \{x \in \mathbb{R}^2 | d(x,c) \leq r\}$ denotes the closed disc and $\mathcal{B}_r^0(c) := (\mathcal{B}_r(c))^0$ denotes the open disc of radius $r$ and center $c$. If a point $x$ has the coordinates $x_1, x_2$, we write $(x_1|x_2)$ alternatively for $x$. We denote the Euclidean distance between two points $x, y$ as $d(x,y)$ and the Hausdorff distance between two sets $A, B$ as $d_H(A, B) = \max\left(\max_{x \in A} \min_{y \in B} d(x,y), \max_{y \in B} \min_{x \in A} d(x,y)\right)$. The dilation of a set $A$ with a disc $\mathcal{B}_r$ is defined as $A \oplus \mathcal{B}_r := \{x \in \mathbb{R}^2 | d_H(A, \{x\}) \leq r\}$ and the erosion is $A \ominus \mathcal{B}_r := \{x \in \mathbb{R}^2 | d_H(A^c, \{x\}) > r\}$. $L_t(f)$ shall be the level set with threshold value $t$ of an image function $f : \mathbb{R}^2 \to \mathbb{R}$: $L_t(f) := \{x \in \mathbb{R}^2 | f(x) \geq t\}$.

Most of the existing topological sampling theorems require the binary images to be $r$-regular [1, 2, 3, 6, 7]. The concept of $r$-regular images was introduced independently by Serra [6] and Pavlidis [4]. These sets are extremely well behaved – they are smooth, round and do not have any cusps.

**Definition 1.** *A compact set $A \subset \mathbb{R}^2$ is called $r$-regular if for each boundary point of $A$ it is possible to find two osculating open discs of radius $r$, one lying entirely in $A$ and the other lying entirely in $A^c$ (see Fig. 1).*

In order to compare analog with digital images, two things are needed: First a method to compare binary images and second a formal description of the processes of sampling and reconstruction. The method for comparison we choose is *weak $r$-similarity* (see [1, 7]). If two sets are weakly $r$-similar, they are topologically equivalent (this criterion was chosen by Pavlidis [4]), have the same

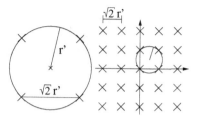

**Fig. 2.** A square grid with a distance of $\sqrt{2}r'$ between two adjacent sampling points is an $r'$-grid

homotopy tree (as used by Serra [6]) and a Hausdorff distance of at most $r$. Note that topological equivalence and identity of homotopy trees are different criteria and neither implies the other (see [1]). The usefullness of a bounded Hausdorff distance as similarity criterion is extensively discussed in the work of Ronse and Tajine (see [5] for a summary). The generality of their approach is remarkable, but it cannot directly be used for our problem, since it doesn't say anything about the topology of a digital reconstruction and it cannot be applied to images which are blurred by some point spread function.

**Definition 2.** *Two bounded sets* $A, B \subset \mathbb{R}^2$ *are called* weakly $r$-similar *if there exists a homeomorphism* $f : \mathbb{R}^2 \to \mathbb{R}^2$ *such that* $x \in A \Leftrightarrow f(x) \in B$, *and the Hausdorff distance between the set boundaries* $d_H(\partial A, \partial B) \leq r \in \mathbb{R}_+ \cup \{\infty\}$. *The used homeomorphism is called* $\mathbb{R}^2$-homeomorphism *between* $A$ *and* $B$.

In most practical cases the sampling grid is a square grid, as used in several previous sampling theorems [2,3,4]. In this paper we will restrict ourselves to this kind of sampling grid, although we used a more general approach in previous papers [1,7]. The reason is that the restriction to square grids allows us to prove a sampling theorem for a much wider class of point spread functions.

**Definition 3.** *A countable set* $S \subset \mathbb{R}^2$ *of* sampling *points with* $d_H(S, \mathbb{R}^2) \leq r'$, *i.e. the Euclidean distance from each point* $x \in \mathbb{R}^2$ *to the next sampling point is at most* $r' \in \mathbb{R}$, *is called an* $r'$-grid *if* $S \cap A$ *is finite for any bounded set* $A \in \mathbb{R}^2$. *The* pixel $\text{Pixel}_S(s)$ *of a sampling point* $s$ *is its Voronoi region, i.e. the set of all points lying at least as near to this point as to any other sampling point. If* $S = a \cdot R \cdot \mathbb{Z}^2 + b$ *for some constant* $a \in (0, \sqrt{2}r']$, *rotation matrix* $R$ *and vector* $b \in \mathbb{R}^2$, $S$ *is called* square grid *(see Fig. 2). The union of the pixels with sampling points lying in* $A$ *is the* reconstruction *of* $A$ *w.r.t.* $S$, *also called the* $S$-reconstruction *of* $A$: $\hat{A} := \bigcup_{s \in S \cap A} \text{Pixel}_S(s)$. *Two pixels are* adjacent *if they share an edge. Two pixels of* $\hat{A}$ *are* connected *if there exists a chain of adjacent pixels in* $\hat{A}$ *between them. Two sampling points are* adjacent (connected) *if their pixels are adjacent (connected). A* component *of* $\hat{A}$ *is a maximal set of connected pixels.*

This most obvious approach for sampling is to restrict the domain of the image function to the sampling grid. But this ideal digitization does not take

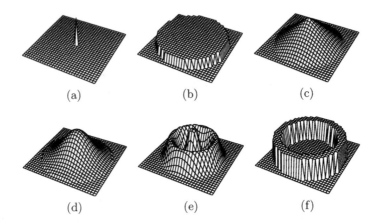

(a)          (b)          (c)

(d)          (e)          (f)

**Fig. 3.** The definition of $p$-PSFs is very broad. Examples are the dirac impulse (a), which leads to a non-blurred digitization, the disc-PSF (b) as used in [1, 7], conic PSFs (c), truncated Gaussians (d), and even non-descending PSFs (e) and (f). While (f) is an artificial example showing what kind of PSF is also allowed, (e) is of practical interest, since the camera aperture can cause such diffraction patterns

into account any blurring. This can be added by a convolution of the image with some point spread function before sampling. Digitization of a binary image has three steps: At first the image gets blurred due to the camera optic. Then the blurred grayscale preimage gets sampled and reconstructed (To reconstruct a grayscale image means to fill each pixel with the image value at the corresponding sampling point). Finally the image gets thresholded in order to get a binary result. Mathematically the last two steps commute. Thus the definition of a digitization without blurring completely determines how to digitize with some blurring. You simply have to blur the original set, apply a threshold function and digitize the result.

**Definition 4.** *A function* $k : \mathbb{R}^2 \longrightarrow \mathbb{R}$ *is called a* point spread function (PSF) *if* $\int_{\mathbb{R}^2} k(x)dx = 1$. *The PSF* $k$ *is a* blurring *PSF if it is non-negative. The PSF* $k_f$ *with* $k_f(x) := f(|x|)$ *for some function* $f : \mathbb{R}_+ \longrightarrow \mathbb{R}$ *with* $\int_0^\infty r \cdot f(r)dr = \frac{1}{2\pi}$ *is called* radially symmetric. *The function* $f$ *is called the* generator function *of* $k_f$. *If* $k_f$ *is a radially symmetric blurring PSF with* $f(r) = 0$ *for every* $r$ *greater than some* $p$, *it is called a* $p$-PSF. *Now let* $A \subset \mathbb{R}^2$ *be a binary set. Then its* characteristic function $\chi_A : \mathbb{R}^2 \to \{0, 1\}$ *is* 1 *for any* $x \in A$ *and* 0 *for any* $x \notin A$. *Given a PSF* $k$, *the* blurred image *of* $A$ *by using* $k$ *is defined as* $f_A := k \star \chi_A$ ($\star$ *denotes convolution*).

With these definitions we have everything we need to prove a sampling theorem for blurred binary images.

**Fig. 4.** After topologically correct digitization with a square grid none of the shown configurations can occur. Thus the resulting image is well-composed

## 3   Sampling-Theorem for Blurred Binary Images

In a previous paper we already proved a sampling theorem for non-blurred binary images and a theorem for binary images after blurring with a constant disc-shaped PSF [1]:

**Theorem 1.** *Let $r \in \mathbb{R}_+$ and $A$ an $r$-regular set. Then $A$ is weakly $r'$-similar to any $S$-reconstruction with some $r'$-grid $S$, $0 < r' < r$.*

In case of square grids this implies that the $S$-reconstruction of an $r$-regular set is well-composed, as defined by Latecki (see [3]). This means, the digital image does not contain any of the pixel configurations shown in Fig. 4

**Theorem 2.** *Let $r, r', p \in \mathbb{R}_+$ be positive numbers with $r' + p < r$ and let $A$ be an $r$-regular set, $k_p$ a $p$-PSF, and $f_A = k_p \star \chi_A$ the blurred image of $A$. Further let $L_l$ be the level set of $f_A$ with some level $l$ and let $S$ be an arbitrary $r'$-. Then the $S$-reconstruction $\hat{L}_l$ of $L_l$ is weakly $(r' + p)$-similar to $A$.*

In order to generalize these results to other types of point spread functions we restricted ourselves to square grids. By doing this we are able to show that any $p$-PSF with $p < r'$ can be used for digitization with an $r'$-grid, such that any $r$-regular set $(r > r' + p)$ is topologically equivalent to its digital reconstruction:

**Theorem 3.** *Let $r, r', p \in \mathbb{R}_+$ be positive numbers with $p < r'$ and $r' + p < r$ and let $A$ be an $r$-regular set, $k_p$ an arbitrary $p$-PSF, and $f_A = k_p \star \chi_A$ the blurred image of $A$. Further let $L_l$ be the level set of $f_A$ with some level $l$ and let $S$ be a square grid, which is an $r'$-grid. Then the $S$-reconstruction $\hat{L}_l$ of $L_l$ is weakly $(r' + p)$-similar to $A$.*

*Proof.* With $r' > p$ follows $2r' > r' + p$. Now let $s \in \mathbb{R}_+$ be any number in the interval $(r' + p, 2r']$. Then $A$ is $s$-regular. If we can prove the theorem for such an $s$ instead of $r$, it is also true for $r > s$. We make use of the inequality $s \le 2r'$ below.

Due to the support of the PSF, $f_A(x) = 1$ for any $x \in A \ominus \mathcal{B}_p$ and analogously $f_A(x) = 0$ for any $x \notin A \oplus \mathcal{B}_p$. Due to $s$-regularity of $A$, the sets $B := A \ominus \mathcal{B}_p$ and $C := A \oplus \mathcal{B}_p$ are both $(s - p)$-regular and weakly $p$-similar to $A$. Due to Theorem 1 their $S$-reconstructions $\hat{A}, \hat{B}$ are weakly $(r' + p)$-similar to $A$. Obviously $\hat{B} \subseteq \hat{L}_l \subseteq \hat{C}$, which implies that the Hausdorff-distance between

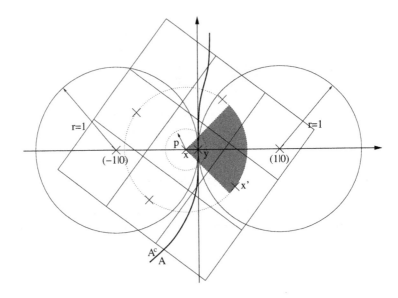

**Fig. 5.** At least one of the sampling points being adjacent to $x$ lies in the shaded sector

$\partial A$ and $\partial \hat{L}_l$ is bounded by $r' + p$. Thus we only have to show that $L_l$ is $\mathbb{R}^2$-homeomorphic to $A$. This is the case if any sampling point $x \in \hat{L}_l$ is (directly) connected in $\hat{L}_l$ with some sampling point $y \in \hat{B}$ and if any sampling point $x \notin \hat{L}_l$ is connected in $\hat{L}_l^c$ with some sampling point $y \notin \hat{C}$, because then no additional component or hole can occur. We show this by proving that for any $x \in S$ with $f_A(x) \in (0,1)$ there exists an adjacent sampling point $x_{\geq}$ with $f_A(x_{\geq}) \geq f_A(x)$ and an adjacent sampling point $x_{\leq}$ with $f_A(x_{\leq}) \leq f_A(x)$. This implies that the configurations shown in Fig. 4 cannot occur.

Let $x \in S$ be a sampling point with $f_A(x) \in (0,1)$ and let $y \in \partial A$ be the boundary point of $A$ being nearest to $x$. Due to $s$-regularity there exists a unique nearest boundary point. Without loss of generality let $x = (d|0)$, $y = (0|0)$, $s = 1$ (any other case can be derived by choosing an appropriate scale and coordinate system) and let $\mathcal{B}_1^0((1|0))$ be the inside and $\mathcal{B}_1^0((-1|0))$ be the outside osculating $s$-disc of $A$ in $y$ (see Fig. 5). Then the four sampling points being adjacent to $x$ lie on the circle with radius $\sqrt{2}r'$ and center $x$. At least one of them, which will be noted as $x'$, lies on the rightmost quarter circle which is bounded by the points $(d+r'|r')$ and $(d+r'|-r')$ (see Fig. 5). Now let $D_1 = 1+d$ be the distance between $(-1|0)$ and $x$, let $D_2 = \sqrt{(1-r'-d)^2 + r'^2}$ be the distance between $(1|0)$ and $(d + r'|r')$ and let $D_3$ be the distance between $(1|0)$ and $x'$ (see Fig. 6). Then $D3 \leq D_2$ since the center $(d|0)$ of the circle containing $x'$ is to the left of $(1|0)$.

Now let $B := \mathcal{B}_1^c$ be a binary image, which is the complement of the unit disc. By using $B$ we construct a helper function $h : [1 - p, 1 + p] \longrightarrow [0, 1]$ with $h(z) := f_B((z|0))$ (see Fig. 7). Obviously $h$ is monotonically increasing since the non-zero area $\mathcal{B}_p((z_1|0)) \cap B$ of the image $B$ covered by the PSF at postion $(z_1|0)$ is a translated superset of the same area $\mathcal{B}_p((z_2|0)) \cap B$ at position $(z_2|0)$ for

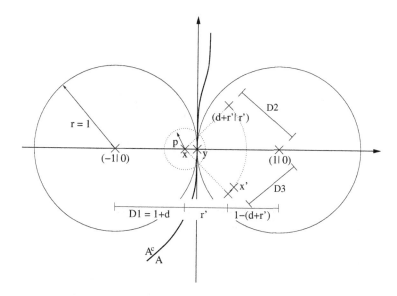

**Fig. 6.** The distance $D3$ between the center of the inside osculating disc and the adjacent sampling point $x'$ is at most equal to the distance $D_2$

any $z_1, z_2 \in [1-p, 1+p]$ with $z_1 > z_2$. The two circles $\partial \mathcal{B}_1$ and $\partial \mathcal{B}_p((z|0))$ have at most two points in common. There exists exactly one other circle of radius 1 sharing these points. This circle is centered in $(\frac{1-p^2}{z} + z|0)$. As Fig. 7 illustrates, $1 - h(z) \geq h(\frac{1-p^2}{z})$.

Since the outside osculating disc $\mathcal{B}_1((-1|0))$ is a subset of $A^c$ (see Fig. 5), $f_A(x)$ is at most equal to $h(D_1)$. Analogously since the inside osculating disc $\mathcal{B}_1((1|0))$ is a subset of $A$, $f_A(x')$ is at least equal to $1 - h(D_3)$. With $D_3 \leq D_2$ it follows that $1 - h(D_2) \leq 1 - h(D_3)$. Thus we only have to show that $h(D_1) \leq 1 - h(D_2)$ in order to prove $f_A(x) \leq f_A(x')$.

We know that $-p \leq d \leq p$ and $0 \leq 1 - r' \leq \frac{1}{2}$ (because $1 = s \leq 2r'$). Now suppose to the contrary, $h(D_1) > 1 - h(D_2)$. Then $h(D_1) > h(\frac{1-p^2}{D_2})$ and due to montony of $h$ follows $D_1 > \frac{1-p^2}{D_2} > \frac{1-(1-r')^2}{D_2}$. By substitution of $D_1$ and $D_2$ we get $1 + d > \frac{1-(1-r')^2}{\sqrt{(1-r'-d)^2 + r'^2}}$. Since both sides of the inequation are positive, we can square it, and further simplification leads to

$$\frac{(1-r') - d}{(1+d)^2}(2 + 2d + 2d^2 + d^3 - (2+d)^2(1-r') + d(1-r')^2 + (1-r')^3) < 0.$$

Since the fraction is always non-negative for the allowed $d, r'$, we only have to look at the rest of the inequation. This inequation is equivalent to both of the following inequations:

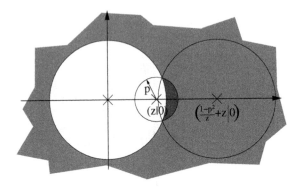

**Fig. 7.** The helper function $h$ describes the result of blurring the complement of the unit disc image with the PSF at some position with distance $z$ to the origin

$$2(1+d)(1-2(1-r')) + (1-r')((1-r')+d)^2 + \\ 2d^2(1-(1-r')) + (-d)((1-r')^2 - d^2) < 0$$

$$2(1+d)(1-2(1-r')) + d(1-r')((1-r')-d) + \\ d^3 + (1-r')^3 + 2d^2 < 0$$

In case of $d < 0$ each addend of the first inequation consists completely of nonnegative factors and in case of $d \geq 0$ each addend of the second inequation consists completely of nonnegative factors for $|d| \leq p < 1 - r' \leq \frac{1}{2}$. Thus for any $d$ one of the inequations is obviously false which implies that the assumption $h(D_1) > 1 - h(D_2)$ is not true. It follows that for any $x \in S$ with $f_A(x) \in (0,1)$ there exists an adjacent sampling point $x_\geq$ with $f_A(x_\geq) \geq f_A(x)$ and analogously there exists an adjacent sampling point $x_\leq$ with $f_A(x_\leq) \leq f_A(x)$. □

Since the class of possible point spread functions is very general, this sampling theorem can be applied to much more practical applications than previous ones. Unfortunately the restriction $p < r'$ is very strict. We conjecture that the theorem is true for any $p$ with $p + r' < r$, but we were up to now not able to prove this formally.

## 4    Conclusions

We proved a sampling theorem which can be summarized in an extremely simple statement: By using a $p$-PSF and a square grid, which is an $r'$-grid (with $r' > p$), we can digitize any $r$-regular binary image without any topological changes if only $r > r' + p$. This is true for any threshold value used for binarization.

Realistic cameras have very complicated point spread functions and often one does not know the exact PSF. Due to our result one does not have to know this, if only one can assume that it is nonnegative, radially symmetric and has a bounded support of known (or estimated) radius. Thus our result can be applied to real camera acquisition systems much better than the findings of Latecki et al. [2, 3] and some of our previous papers [1, 7], where only point spread functions were allowed which are constant in their whole support. Unfortunately our proof has a restriction to the maximal size of the PSF relatively to the sampling density. We think that this is not necessary and conjecture that our results can be generalized to any $p$-PSF with $r > p + r'$. Additionally we think that equivalent theorems can be shown for other sampling grids like hexagonal or even irregular grids, but we cannot prove this yet.

# References

1. Köthe, U., Stelldinger, P.: *Shape Preserving Digitization of Ideal and Blurred Binary Shapes.* In: I. Nyström et al. (Eds.): DGCI 2003, LNCS **2886**, pp. 82-91, Springer, 2003.
2. Latecki, L.J., Conrad, C., Gross, A.: *Preserving Topology by a Digitization Process.* Journal of Mathematical Imaging and Vision **8**, pp. 131–159, 1998.
3. Latecki, L.J.: *Discrete Representation of Spatial Objects in Computer Vision.* Kluwer, 1998.
4. Pavlidis, T.: *Algorithms for Graphics and Image Processing.* Computer Science Press, 1982.
5. Ronse, C., Tajine, M.: *Morphological Sampling of Closed Sets.* Image Analysis and Stereology **23**, pp. 89–109, 2004.
6. Serra, J.: *Image Analysis and Mathematical Morphology.* Academic Press, 1982.
7. Stelldinger, P., Köthe U.: *Shape Preservation During Digitization: Tight Bounds Based on the Morphing Distance.* In: B. Michaelis, G. Krel (Eds.): Pattern Recognition, LNCS **2781**, pp. 108-115, Springer, 2003.

# A Low Complexity Discrete Radiosity Method

Pierre Y. Chatelier and Rémy Malgouyres

LLAIC, Clermont-Ferrand
{chatelier, remy.malgouyres}@llaic3.u-clermont1.fr

**Abstract.** Radiosity in 3D scenes is usually computed using a discretization of the surfaces into patches. A discretization into voxels is possible, coupled with the use of discrete geometry. An algorithm for radiosity solving with voxels is introduced, lowering the theoretical complexity to an $O(N \log N) + O(N)$, where the $O(N)$ is largely dominant in practice, so that the apparent complexity is linear for time and space, with respect to the number of voxels in the scene. The method also fits in RAM and does not need disk storage. Instead of 3D discrete line traversal, a new algorithm is described to perform visibility computation. The voxel-based radiosity equation assumes the ideal diffuse case and uses solid angles similarly to the hemicube.

**Keywords:** Radiosity, voxels, discrete geometry, linear complexity, visibility, ideal diffuse case.

## 1 Introduction

In computer graphics, radiosity is known to globally provide smoother results than simple ray-*tracing*, since it aims at diffusing light in a better way. The main drawback is the cost involved by the new physical properties to manage, that may still be simplified to be reasonably expensive. But compared to simple ray-*casting*, the complexity remains easily controllable. Mixing radiosity and ray-tracing usually gives the best results.

Radiosity has been largely studied, but has also been limited to a discretization of the surfaces into polygonal patches. A discretization into voxels (elementary volume elements) coupled with discrete geometry can bring new possibilities of fast visibility computation, especially for complex scenes where the patches do not suit very well. The goal of this paper is to bring improvements to a previous voxel-based discrete radiosity method introduced in [5]. We present a new representation of the visibility problem, and a new data flow in computing, which lead to a quasi-linear time and space complexity in radiosity solving. The new algorithm consists in handling the visibility problem globally for each direction in space. The space is partitioned into lists of voxels, where "neighbors in the list" means "neighbors in terms of visibility". Classical ray traversal becomes useless and no time is spent in emtpy spaces.

E. Andres et al. (Eds.): DGCI 2005, LNCS 3429, pp. 392–403, 2005.

Then, the radiosity can be propagated between the voxels, using the computed visibility information. By iterating within a set of directions, the radiosity in the scene converges toward a solution of a discretized radiosity equation. A tool similar to the *hemicube* [2] is used, factorizing some computations and using the notion of solid angle.

This algorithm is designed to work on a given set of voxels. The discretization step, and the visualization with radiosity, do not belong to the algorithm, and can be derived from [5] for instance.

## 2    The Radiosity Equation

*Radiosity* is defined as the total power of light leaving a point. In practice, this notion is useful provided that some assumptions are made about the properties of the surfaces. We introduce the equation at first, then we explain how it can be interpreted. This equation comes from simplifications of a more general case, which uses the notion of *radiance* [4]. Its general continuous form is the following:

$$B(x) = E(x) + \rho_d(x) \int_{y \in scene} B(y) \frac{\cos \theta_x \cos \theta_y}{\pi \parallel x - y \parallel^2} V(x,y) dy \qquad (1)$$

An intuitive explanation of Equation (1) is: *"the total power of light leaving a point x (the B(x) term), is defined by two terms: the proper emittance of this point as a light source (the E(x) term), and some re-emission of the light it receives from its environment (the integral)"*.

- $B(x)$ and $B(y)$ makes $B(\cdot)$ present in both sides of the equality. It reflects the interdependence between a point and its environment. It also supposes that each point emits light uniformly in every direction.
- $V(x,y)$ is a visibility function, equal to 1 if $x$ and $y$ are mutually visible, and 0 otherwise. This function makes it possible to write an integral over the whole scene.
- A point re-emits only a fraction $\rho_d(x)$ of the light that it receives. Assuming that this factor depends neither on the outgoing direction, nor on the direction of the incoming light, is known as the *ideal diffuse hypothesis.*
- To quantify how much of an object is seen from a point, the term $\frac{\cos \theta_y dy}{\parallel x-y \parallel^2}$, is a direct transcription of the definition of a solid angle. We denote by $\theta_y$ the angle between the direction $(xy)$ and the normal vector at point $y$.
- Incident light is more or less important depending on whether it is received from right ahead or tangently. The more tangent a ray is, the less participation it has in the re-emitted light. This is represented by $\cos \theta_x$, where $\theta_x$ is the angle between the direction $(xy)$ and the normal vector at point $x$. Additionally, $\cos \theta_x$ is floored to 0 to prevent light coming from *below* to be re-emitted.
- The $\pi$ factor is a normalization term deriving from radiance considerations.

# 3    A Discrete Radiosity Equation

## 3.1    Discretizing the Equation

This section is a reminder of how the radiosity equation has been discretized in [5]. First, let us denote by $S_R(x)$ a virtual sphere of radius $R$ centered on $x$, by $\sigma$ a point on this sphere, by $y(x, \sigma)$ the first real surface point met from $x$ in the half-line $[x; \sigma)$. The visibility problem is, like in the hemicube [2], translated to a solid angle computation on a virtual surrounding shape: a sphere.

$$
\int_{y \in \text{scene}} B(y) \frac{\cos \theta(x, y) \cos \theta(y, x)}{\pi \parallel x - y \parallel^2} V(x, y) dy
$$

$$
= \int_{\sigma \in S_R(x)} B(y(x, \sigma)) \frac{\cos \theta(x, y(x, \sigma))}{\pi} \underbrace{\left( \cos \theta(y(x, \sigma), x) \frac{d(y(x, \sigma))}{\parallel x - y(x, \sigma) \parallel^2} \right)}_{\text{solid angle as seen from } x}
$$

$$
= \int_{\sigma \in S_R(x)} B(y(x, \sigma)) \frac{\cos \theta(x, y(x, \sigma))}{\pi} d\sigma
$$

Now, denoting by $x$ a voxel, by $\Sigma_R(x)$ a discrete sphere of radius $R$ centered on $x$, by $\sigma$ a voxel of $\Sigma_R(x)$, by $V(x, \sigma)$ the voxel of $y(x, \sigma)$, by $\hat{A}(\overrightarrow{x\sigma})$ an approximation of $d\sigma$, we can write:

$$
\int_{\sigma \in S_R(x)} B(y(x, \sigma)) \frac{\cos \theta(x, y)}{\pi} d\sigma \approx \sum_{\sigma \in \Sigma_R(x)} B(V(x, \sigma)) \frac{\cos \theta(x, V(x, \sigma))}{\pi} \hat{A}(\overrightarrow{x\sigma})
$$

We approximate Equation (1) by the following linear system:

$$
B(x) = E(x) + \rho_d(x) \sum_{\sigma \in \Sigma_R(x)} B(V(x, \sigma)) \frac{\cos \theta(x, V(x, \sigma))}{\pi} \hat{A}(\overrightarrow{x\sigma}) \tag{2}
$$

In Equation (2), the expensive information is the function $V(x, \sigma)$, (the first voxel encountered from $x$ in the direction of $\sigma$). In [5], it is precomputed, and thus is very similar to the *form factors* of the classical approach. The term $\hat{A}(\overrightarrow{x\sigma})$ can also be easily precomputed. Note that $V(x, \sigma)$ is not well defined, since several discrete rays may cross $x$ and $\sigma$ (see Fig. 1). This is not a problem since each possible ray leads to an acceptable solution.

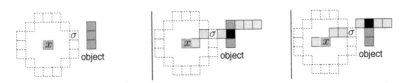

**Fig. 1.** Given $x$ and $\sigma$, several rays may be used to find a $V(x, \sigma)$

## 3.2 Solving the Equation

To solve Equation (2), a converging iterative method similar to the one used with patch-based radiosity can be used. It usually relies on Gauss-Seidel relaxation. If we consider the equation under its form $B = E + M.B$, where $B$ is a vector of elements and $M$ a matrix of factors, some properties of $M$ ensure the sequence $B_{n+1} = E + M.B_n$ to converge toward a limit, which is a solution of the discrete equation. Roughly speaking, this is a transcription of light gathering, each iteration going a step further in light re-emission. The convergence is expected since light is progressively absorbed. Technically, each iteration consists in propagating packets of radiosity between mutually visible voxels.

## 3.3 Computations Made in [5]

The goal of this paper is to introduce a new way to handle $V(x, \sigma)$. In [5], it was precomputed by a kind of discrete ray-tracing method. Since the voxelization that we use produces an octree, it was easy and efficient. This precomputation is however very expensive in terms of time and storage.

The $\hat{A}(\overline{x}\vec{\sigma})$ term has not been modified since [5]. As with the hemicube, each exterior voxel face of the surrounding sphere is a kind of screen cell. Thus, it is associated to a solid angle and each voxel is given a value summing the solid angles of its exterior faces. Moreover, the discrete sphere is considered to be global, there is not one sphere per voxel. Therefore, $\hat{A}(\overline{x}\vec{\sigma})$ can rather be written $\hat{A}(\vec{\sigma})$, and we call it a *direction factor*. A radius of 38 for the sphere produces about $15,000$ direction factors. This is a small amount of data, and requires only a few seconds to be pre-computed. It is also a good parameter for the quality of the voxel-based radiosity algorithm.

# 4 Discrete Geometry

The algorithm that we introduce uses some notions of discrete geometry. Section 4.1 is a reminder, and Section 4.2 is the proof of an interesting partitioning property.

## 4.1 Discrete Lines

A 2D discrete line [6] whose directing vector is $(a, b)$ can be represented by the set of points:

$$\{(x, y) \in \mathbb{Z}^2 / \mu \le ay - bx < \mu + \omega\}$$

where $\omega$ denotes the *arithmetical thickness* of the line, and $\mu$ sets the position of the line. The higher $\omega$, the thicker the line. If $\omega$ is too small, the set of points becomes disconnected. $\omega$ is related to the *connectivity* of the line (see Fig. 2). If $\omega = \max(|a|, |b|)$, the line is 8-connected and it is called the *naïve* case. If $\omega = |a| + |b|$, the line is 4-connected and it is called the *standard* case.

**Fig. 2.** (2D) {4-8}-neighborhood          **Fig. 3.** (3D) {6-18-26}-neighborhood

A notion of 3D discrete line has also been defined [3], where the naïve (26-connected) and the standard (6-connected) cases are also defined, as:

$$(x, y, z) \in \mathbb{Z}^3 / \begin{cases} \mu \leq cx - az < \mu + \omega \\ \mu' \leq bx - ay < \mu' + \omega' \end{cases}$$

It is noteworthy that a 3D discrete line represents the *intersection* between two *discrete planes*, each one being the orthogonal extrusion of a 2D discrete line included in one of the coordinate planes. Alternatively, one can say that a 3D discrete line projects onto two 2D discrete lines that are sufficient to retreive the 3D line. The connectivity (see Fig. 2 and Fig. 3) is related to $\omega$ and $\omega'$. If the two 2D projections are naïve (resp. standard), the 3D line is naïve (resp. standard) itself.

## 4.2   Partitioning the Space into Lines

In this section we set and prove that the space can be partitioned into parallel 3D discrete lines, along a given direction. This means that a voxel belongs to one and only one of these lines, which can be explicitly computed.

**Proposition 1.** *Let us denote by $\mathbb{Z}_*^3$ the set $\mathbb{Z}^3 \backslash \{(0,0,0)\}$. Given an integer vector $\vec{v} \in \mathbb{Z}_*^3$, a voxel space can be partitioned into a set of naïve, or a set of standard, 3D discrete lines, whose direction vector is $\vec{v}$.*

*Proof.* Let $\vec{v} = [a, b, c]$, with $(a, b, c) \in \mathbb{Z}_*^3$, $(a, b, c)$ having no common divisor other than 1. We assume without loss of generality that $a \geq b \geq c \geq 0$.

A 3D discrete line with $\vec{v}$ as directing vector is defined by two 2D projections. The connectivity of the 3D line and of its projections are related, so that we can study separately the naïve case and the standard case. Let us denote the arithmetical thicknesses by:

naïve case: $\begin{cases} \omega_{ab} = \max(|a|, |b|) \neq 0 \\ \omega_{ac} = \max(|a|, |c|) \neq 0 \\ \omega_{bc} = \max(|b|, |c|) \end{cases}$     standard case: $\begin{cases} \omega_{ab} = |a| + |b| \neq 0 \\ \omega_{ac} = |a| + |c| \neq 0 \\ \omega_{bc} = |b| + |c| \end{cases}$

Since $a \geq b \geq c \geq 0$, the relevant 2D projections are in the planes (Oxy) and (Oxz). This is why only $\omega_{bc}$ may be null. The projections are defined by:

$$\begin{cases} \{(x, y, z) \in \mathbb{Z}^3 / z = 0 \text{ and } \mu_{ab} \leq -bx + ay < \mu_{ab} + \omega_{ab}\} & \textit{projection in (Oxy)} \\ \{(x, y, z) \in \mathbb{Z}^3 / y = 0 \text{ and } \mu_{ac} \leq -cx + az < \mu_{ac} + \omega_{ac}\} & \textit{projection in (Oxz)} \end{cases}$$

Thus, the 3D discrete line is equivalent to the set of all $(x, y, z) \in \mathbb{Z}^3$ such that:

$$\begin{cases} \mu_{ab} \leq -bx + ay < \mu_{ab} + \omega_{ab} \\ \mu_{ac} \leq -cx + az < \mu_{ac} + \omega_{ac} \end{cases}$$

Since $a, b, c, \omega_{ab}, \omega_{ac}$ are fixed, only the position of the line may be chosen and we denote such a set of voxels by $L(\mu_{ab}, \mu_{ac})$. Let us introduce the set of 3D discrete lines denoted by $\{L_{i,j}\}_{(i,j)\in\mathbb{Z}^2} = \{L(i * \omega_{ab}, j * \omega_{ac})\}_{(i,j)\in\mathbb{Z}^2}$
Given $i$ and $j$, an $L_{i,j}$ 3D discrete line is defined by:

$$\begin{cases} i * \omega_{ab} \leq -bx + ay < i * \omega_{ab} + \omega_{ab} \\ j * \omega_{ac} \leq -cx + az < j * \omega_{ac} + \omega_{ac} \end{cases} \quad \text{or} \quad \begin{cases} i * \omega_{ab} \leq -bx + ay < (i+1) * \omega_{ab} \\ j * \omega_{ac} \leq -cx + az < (j+1) * \omega_{ac} \end{cases}$$

Given a voxel $(x, y, z) \in \mathbb{Z}^3$, this voxel belongs to $L_{k,l}$ with $k = \lfloor \frac{-bx+ay}{\omega_{ab}} \rfloor$ and $l = \lfloor \frac{-cx+az}{\omega_{ac}} \rfloor$ (where $\lfloor x \rfloor$ denotes the "floor" function).

Moreover, the $L_{i,j}$'s are pairwise disjoint and thus they constitute a partition. Indeed, let us consider a voxel $v = (x, y, z)$ belonging to $L_{i,j}$ and to $L_{i',j'}$, with $(i,j) \neq (i',j')$. We assume for instance $i < i'$, because if $i = i'$, the following reasoning can still be held, replacing $i$ by $j$ and say that $j < j'$.

$$v \in L_{i,j} \text{ and } v \in L_{i',j'} \Rightarrow \begin{cases} i * \omega_{ab} \leq -bx + ay < (i+1) * \omega_{ab} \\ i' * \omega_{ab} \leq -bx + ay < (i'+1) * \omega_{ab} \end{cases}$$

$$\Rightarrow \begin{cases} -bx + ay < (i+1) * \omega_{ab} \leq i' * \omega_{ab} \\ i' * \omega_{ab} \leq -bx + ay \end{cases}$$

$$\Rightarrow -bx + ay < -bx + ay \qquad \text{which is impossible}$$

As a conclusion, given a direction, and a naïve or standard connectivity, the set of corresponding $L_{i,j}$'s is a partition of the voxel space into 3D discrete lines following this direction. Then, a simple operation using the *floor* function allows to deduce, from the coordinates of a voxel, the $L_{i,j}$ line it belongs to.

## 5   Voxel-Based Radiosity

### 5.1   A New Approach of Discrete Radiosity

The main problem of the approach presented in [5] lies in the required information about visibility, stored as a precomputed set of $V(x, \vec{\sigma})$. Such information is very expensive to store for each voxel of the scene, and usually does not fit in RAM. A secondary memory is necessary. A scene with $2 \times 10^6$ voxels would basically generate about 80 GB of data. Giving full sense to discrete geometry, a new approach can be introduced, that deduces visibility on-the-fly. No pre-computations are needed, and no information has to be stored on disk. Both time and space complexity are improved by this method. This can be done by transforming the visibility problem.

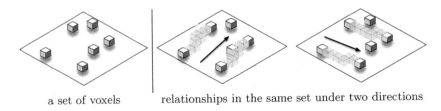

a set of voxels          relationships in the same set under two directions

**Fig. 4.** Visibility solving by finding voxels on the same 3D discrete lines

**Transforming the Visibility Problem:** Instead of iterating on each voxel $x$, and querying $V(x, \overrightarrow{\sigma})$, (*i.e.* the first voxel visible from $x$ in the direction of $\overrightarrow{\sigma}$), we can rather iterate on $\overrightarrow{\sigma}$ and compute a whole bunch of $V(x, \overrightarrow{\sigma})$ at a time for a fixed $\overrightarrow{\sigma}$. In this case, we represent a ray of light by the list of every intersected voxels, as if there was no occlusion. Thus, one list encodes several information about visibility in the given direction of the ray. If $x$ and $y$ are two consecutive voxels (not necessarily neighbors in space) in a ray (list) of direction $\overrightarrow{\sigma}$, then we can assume that $y = V(x, \overrightarrow{\sigma})$ (see Fig. 4).

Building the lists could still done by discrete ray-tracing, which has been largely studied [1] [9] [8], but we propose another approach, more adapted to the current context, where we have to compute at the same time the visibility of every couple of voxels. Our approach does not use any traversal ; in our model, the voxels of empty spaces are not represented nor traversed : they are implicitly ignored. To perform this approach, we have addressed two difficulties: how to quickly find the ray (or 3D discrete line) each voxel belongs to, and how to keep at a low cost a sorted representation of this ray to handle the analogy between consecutiveness and visibility.

## 5.2    Efficient Solving of the Visibility Problem

**Finding Which Line a Voxel Belongs to:** We have proved in Section 4.2 that the space could be partitioned into 3D discrete lines, each one characterized by a couple of integers $(i, j)$. Such a couple is extractable in constant time from the coordinates of a voxel, for a particular direction. With appropriate data structures, a linear time complexity can then be ensured to link each voxel to its line.

We represent each discrete ray by a list of the voxels it contains, not necessarily sorted at first. The set of all lists is stored in a 2D array, indexed by $i$ and $j$, which are bounded by the scene geometry. Given a direction, finding the $(i, j)$ list of a given voxel is a constant-time operation. Then, finding this list in the array of lists is also a constant-time operation. Then, adding the voxel to this list is a constant-time operation, if no sorting is done. Therefore, the overall complexity of the dispatching process, for a given direction, for one voxel, is a constant time. For $N$ voxels, the time complexity of this dispatching process is obviously $O(N)$.

$$Voxel \overset{O(1)}{\underset{extracting}{\longmapsto}} (i, j) \overset{O(1)}{\underset{finding}{\longrightarrow}} list \overset{O(1)}{\underset{inserting}{\longmapsto}} increased\ list$$

**Finding Consecutive Voxels in the Visibility Lists:** The lists holding the voxels are not sorted at first: they do not reflect the *visibility* between voxels. Hence, at first sight, the list filling step should be followed by, or mixed up with a sorting step. However, the complexity would increase up to worst case $O(N \log N)$. We show how appropriate data structures can lead to avoid this sorting step.

**Building Sorted Lists Without Extra Cost:** Sorting the lists is not required if they can be filled in a way such that the voxels are already sorted in each of them. To ensure that property, an appropriate traversing order must be found for the data structure supplying the voxels to the list-filling algorithm.

A kind of wavefront, perpendicular to the current direction, and evolving in that direction, encounters the voxels of the scene in their natural order for the given direction. In the discrete case, the wavefront can be aligned along a coordinate axis (see Fig. 5).

It is worth noting that the voxels in a 3D discrete line are ordered with respect to a lexicographic order on their coordinates $x$, $y$ and $z$, depending on the directing vector. Hence, the wavefront itself can be implemented by a lexicographic order. The relative order of $x$, $y$ and $z$ is not relevant for the discrete lines we use, since their "thickness" is at most 1, due to our choice of $\omega$. In this case, no "bubble" can appear, which would require attention (see Fig. 6).

In the 3D case, the lexicographic order is applied on $x$, $y$, and $z$ depending on the considered direction. For a direction $\overrightarrow{v} = (a, b, c)$, the lexicographic order must only fulfill the following conditions: *if a (resp. b, resp. c) $\geq 0$, then x (resp. y, resp. z) is considered in ascending order, otherwise in descending order.* For instance, with intuitive notations, the lexicographic order for $\overrightarrow{v} = (-1, 5, 2)$ can be defined as $\prec_{x\downarrow y\uparrow z\uparrow}$, $\prec_{x\downarrow z\uparrow y\uparrow}$, $\prec_{y\uparrow x\downarrow z\uparrow}$, $\prec_{y\uparrow z\uparrow x\downarrow}$, $\prec_{z\uparrow x\downarrow y\uparrow}$, or $\prec_{z\uparrow y\uparrow x\downarrow}$.

Moreover, since $\prec_{x\downarrow y\uparrow z\uparrow}$ gives the reverse order of $\prec_{x\uparrow y\downarrow z\downarrow}$ (for instance), only 4 out of 8 different lexicographic orders are to be coded to handle any direction.

So far, no conditions were required on the data structure supplying the voxels to the radiosity algorithm. But we need the ability to be given the voxels with respect to one of the four lexicographic order previously defined. In practice, without assumptions on the original data structure, it is possible to use four additional arrays, one for each lexicographic order, containing sorted pointers to the $N$ voxels. This is affordable as a precomputation, at worst in $O(N \log N)$ time, and in practice only needs a few seconds. Its practical time cost is totally negligible aside the radiosity solving step.

## 5.3    The Final Algorithm

We have shown in the previous section how to handle the visibility problem with a low complexity. To solve the voxel-based radiosity equation, a converging iterative method is used, as described in [5]. The final algorithm is described on page 401.

**Time Complexity:** The algorithm requires two parameters, which are the radius $R$ of the discrete sphere used to compute the direction factors, and a num-

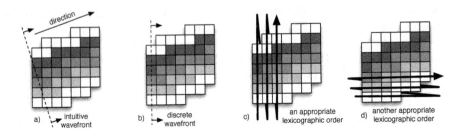

a) intuitive wavefront    b) discrete wavefront    c) an appropriate lexicographic order    d) another appropriate lexicographic order

**Fig. 5.** An appropriate traversing order of voxels respects their natural order in the discrete lines. a) A wavefront encounters the voxels in the good order. b) In the discrete case, the wavefront may be oriented along a coordinate axis. c) and d) This wavefront can be represented by a lexicographic order on the coordinates

**Fig. 6.** A 3D discrete line whose thickness is limited cannot hold such a bubble, for which the relative order of $x$, $y$ and $z$ is discriminative

ber $I$ of iterations to converge to a radiosity solution (usually less than 6). The number of directions is an $O(R^2)$ (usually a few thousands). For $N$ voxels in the scene, the time complexity is:

$$\underbrace{4 \times O(N \log N)}_{\substack{\text{preparing lexico-}\\\text{graphic orders}}} + \underbrace{I \times O(R^2) \times (O(N) + O(N))}_{\text{radiosity solving}} = O(N \log N) + O(I \times R^2 \times N)$$

with the term $O(N \log N)$ negligible in practice.

**Space Complexity:** We assume that the $N$ voxels modeling the 3D scene are already encoded in an $O(N)$ data structure. Given a direction, the set of lists we use for partitioning contains exactly one reference to each voxel, so that $O(N)$ is expected for the total space complexity. Four precomputed lexicographic arrays of voxels are needed, this is an $O(N)$. We also need an array to store the lists. Since the lists form a partition of the 3D space, the number of lists needed for a scene is related to the square of the width of the 3D scene. Only the surface of objects are discretized, so that the width of the scene is usually an $O(\sqrt{N})$. Thus, the number of lists is at most an $O(N)$, since in the worst case, where each list contains a single voxel, there are exactly $N$ lists. The lists are reset for each direction, so that the hidden constant depends solely on the geometry of the scene, not on the number of directions. Thus, the hidden constant remains small, and the space complexity needed by our algorithm is an $O(N)$ which merges with the $O(N)$ already needed by the data structure used for modeling.

Prepare the needed four lexicographic orders;

Compute a set of discrete directions;
**Pour** each direction **faire**

compute and store the associated direction factor;
**Fin Pour**

*[the number of iterations is a small constant]*
**Pour** iterations = 1 to MaxIterations **faire**

*[the number of directions is a big constant]*
**Pour** each direction **faire**

*[dispatching step: O(N)]*
Select the appropriate lexicographic order;
**Pour** each voxel (with respect to the lexicographic order) **faire**

*[The voxels are naturally sorted at insertion]*
Add it to the back of the list (3D discrete line) it belongs to;
**Fin Pour**
*[propagation (solving) step: O(N)]*
**Pour** each list **faire**

propagate radiosity between contiguous voxels;
**Fin Pour**
reset the lists;
**Fin Pour**
**Fin Pour**

**Algorithm 1:** Quasi-linear radiosity algorithm

# 6    Experimental Results

**Improvements over Previous Voxel-Based Method:** The scene presented in [5] is made of 310,000 patches, and has been discretized into about $2 \times 10^6$ voxels. It has been computed with the new algorithm in the same conditions as with the previous one: 6 iterations, about 15,000 directions, on an Athlon 900 MHz with 1.5 GB of RAM. It has shown a 60% time improvement, for identical result quality (27 hours instead of 72). A main advantage is also that no hard disk space is needed since the whole computation can be done in RAM.

**Storing Some Results on Disk to Optimize:** Each iteration uses the same set of directions and leads to the same computations to retrieve the visibility of the voxels along the rays of a given direction. This is not dramatical because there are very few iterations. However, if disk space is available, the lists computed during the *first* iteration can be dumped and recalled in the *following* iterations. Tests have shown that it results in a 20% time regression for the first iteration, and a 30% time improvement for the following ones. On one hand, more resources help optimizing, on the other hand, it does not dramatically outperform the

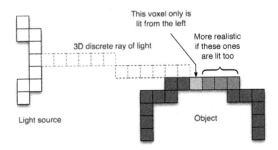

**Fig. 7.** Small correction on propagation for tangent rays

RAM-only approach, especially if we consider the huge required amount of hard disk: basically, it is about 60 GB for $2 \times 10^6$ voxels and 15,000 directions.

**Visual Artifacts:** A roughly discretized scene may contain classical artifacts, when rendering with radiosity information supplied by the present form of the algorithm: light leaks, or sharp shadows might be observed. A low number of directions in the parametrization may as well generate irregularities on surfaces that should render smooth in reality. However, an artifact specific to our algorithm has a specific workaround. Let us consider a ray almost tangent to a surface. Chances are high that many contiguous voxels of the surface belong to the same list representing the ray. With basic visibility, only the first voxel of the surface encountered from the light source would receive light, the others being occluded. With a limited number of directions, this could prevent some voxels in the middle of large flat surfaces to ever receive light directly from the light sources that are low in their horizon. To address this issue, a small correction can be done, that spread the light received by a voxel to its immediate neighbors (see Fig. 7).

## 7    Conclusion and Perspectives

A voxel-based radiosity algorithm has been presented, with a quasi-linear complexity in time and a linear complexity in space, with respect to the number of voxels encoding the scene. Many experiments remain to be done in order to improve this approach of radiosity. First, instead of limiting ourselves to the ideal diffuse case, we may add to our radiosity the management of complex Bidirectional Reflectance Distribution Functions (BRDF). We are also studying another voxelization method found in [7], a density approach showing nice results coupled with discrete ray-tracing. We are actually investigating to see if radiosity plugs well into this approach. At last, the algorithm suits well for clustering, so that our implementation let us hope for a linear improvement with respect to the number of nodes. A basic parallelization would dispatch the directions, so that each node handles a subset of directions, but we are also studying a way to dispatch the voxels on the nodes.

# References

1. John Amanatides and Andrew Woo. A fast voxel traversal algorithm for ray tracing. In *Eurographics '87*, pages 3–10. Elsevier Science Publishers, Amsterdam, North-Holland, 1987.
2. Michael F. Cohen and Donald P. Greenberg. The hemi-cube: a radiosity solution for complex environments. In *SIGGRAPH '85: Proceedings of the 12th annual conference on Computer graphics and interactive techniques*, pages 31–40. ACM Press, 1985.
3. Isabelle Debled-Rennesson. *Étude et reconnaissance des droites et plans discrets*. PhD thesis, Université Louis Pasteur, Strasbourg, 1995.
4. François X. Sillion and Claude Puech. *Radiosity & Global Illumination*. Morgan Kaufmann Publishers, Inc., 1994.
5. Rémy Malgouyres. A discrete radiosity method. In Achille Braquelaire, Jacques-Olivier Lachaud, and Anne Vialard, editors, *Discrete Geometry for Computer Imagery, 10th International Conference, DGCI 2002, Bordeaux, France*, pages 428–438. Springer, April 2002.
6. Jean-Pierre Reveillès. *Géometrie discrète, calcul en nombres entiers et algorithmique*. PhD thesis, Université Louis Pasteur, Strasbourg, 1991.
7. M. Sramek and A. Kaufman. Vxt: a c++ class library for object voxelization. *Volume Graphics*, pages 119–134, 2000.
8. Nilo Stolte and René Caubet. Discrete ray-tracing of huge voxel spaces. *Comput. Graph. Forum*, 14(3):383–394, 1995.
9. R. Yagel, D. Cohen, and A. Kaufman. Discrete ray tracing. *IEEE Computer Graphics & Applications*, 12(9):19–28, 1992.

# A Statistical Approach for Geometric Smoothing of Discrete Surfaces

Bertrand Kerautret and Achille Braquelaire

LaBRI, Laboratoire Bordelais de Recherche en Informatique,
UMR 5800, Université Bordeaux 1,
351, cours de la Libération,
33405 Talence, France
{kerautre, achille}@labri.fr

**Abstract.** In this article we propose an original method for discrete surface smoothing. This method is based on a statistical estimation of the discrete tangent plane on the voxels of the discrete surface. A geometrical constraint is used to control the recognition of the tangent plane. The resulting surface representation allows us to get both smooth normal vectors of the surface and a smooth surface mesh while preserving the geometrical properties of the surface.

**Keywords:** Digital surfaces, smoothing surface mesh, euclidean nets, discrete normals, visualization, smoothing.

## 1   Introduction

Processing data sets of three-dimensional discrete images brings up the problem of extraction and representation of geometric features, and of visualization of the surface of 3D objects. The initial volume object can be visualized as a set of 6-connected voxels (also called cuberille representation) [7]. But this representation in the discrete space is neither convenient for the analysis of geometric properties of the object nor for the visualization.

A polygonal representation of the boundary of a discrete object is usually used to represent its surface and to perform rendering. One of the first approaches to obtain such a representation was the marching cube algorithm [9]. This method has several drawbacks both from the geometrical and the topological points of view. Other algorithms exist which associate a surface mesh to a discrete surface. For example Türmer and Wütrich triangulate the surfaces by associating centers of voxels to each other [14]. Since the direct rendering of the surface obtained after such a triangulation is not smooth, normal vectors are computed in discrete space using a varying neighborhood size [12, 13]. Then the surface is rendered using Gouraud shading [6]. This rendering technique gives good results, but it smooths only the normal vector of the discrete surface and not the geometry of the surface net. Other methods use deformable models to extract a continuous surface from the original discrete surface [8, 11].

E. Andres et al. (Eds.): DGCI 2005, LNCS 3429, pp. 404–413, 2005.

An alternative consist in smoothing the object surface by moving the points of the discrete surface. In [2], Braquelaire and Pousset define Euclidean nets as a 3D extension of the model of Euclidean paths [15, 3]. In this model, each surface point may be moved inside the unit cube containing it. The smoothing is thus reversible and the original surface can be retrieved from the smoothed one. In the proposed method the points of the discrete surface were moved according to a projection onto some discrete tangent planes. This plane was estimated by searching for local geometric configurations of voxels called tricube [10, 4].

The main drawback of this method is the small neighborhood size which is used to determine the discrete tangent plane. Therefore, the precision of the final result is limited to a local analysis of the discrete surface. In recent works [5], Coeurjolly suggests to use a statistical computation of the discrete tangent plane to obtain the normal vectors of the discrete surface.

In this work we develop this approach and propose a statistical method to recognize accurate tangent plane with a varying neighborhood size. We then use this method to enhance the construction of a smoothed Euclidean net associated with a discrete surface. This method permits to obtain both smooth normal vectors and smooth surface mesh.

In the following section, we introduce the statistical estimation of the discrete tangent plane. In Section 3, we show how to apply a geometric constraint to control the recognition of the discrete tangent plane. Section 4 addresses the problem of transforming the surface from discrete space to a new surface net in continuous space. Afterward, in Section 5 experimental results on both synthetic and real data are presented. Finally, we conclude by future work and implication of this work.

# 2   Statistical Estimation of the Tangent Plane's Orientation

Let us recall some basic definitions in use in this paper. A voxel is a unit cube which the center is an integer point and a surfel is the intersection of two 6-adjacent voxels. The surface of a 6-connected object is the set of surfels adjacent to both a voxel inside and a voxel outside the object. In the same way, a linel is defined as the intersection between two 4-adjacent pixels one of wich belongs to the discrete line. From these definitions surfels and linels are be differentiated according their configurations. Fig. 1 illustrates different types of surfels and linels.

In discrete space, a discrete tangent plane can be defined as the largest discrete plane which can be reconstructed from an initial voxel. The strategy to estimate the discrete tangent plane consists in considering a random draw of surfel. Let us first consider some probalilyties on a discrete line.

**Proposition 1.** *Consider the random draw of linels of type 1 and 2 on a discrete naive line of the first quadrant (i.e. $a > 0$, $b > 0$). The probability to obtain a linel of type 1 is $\frac{b}{a+b}$, the probability to obtain a linel of type 2 is $\frac{a}{a+b}$.*

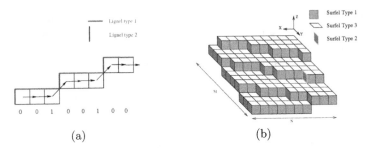

**Fig. 1.** Illustration of the different type of linel (a) and of voxels (b)

*Proof.* Let us consider the discrete naive line represented with the Freeman code. First, we suppose that the discrete line belongs to the first octant (i.e. $a \leq b$). Then the discrete line is defined by $N$ codes associated to each pixel with directions 0 and 1 (Fig.1-a). By construction, the probability to obtain the code 1 is $\frac{a}{b}$, then the number of linels of type 1 ($n_1$) is $N\frac{a}{b}$. With the same arguments, we can deduce that the number of linels of type 2 ($n_2$) is $N$. Thus the two probabilities $P_1$ and $P_2$ can be written as follows:

$$P_1 = \frac{n_1}{n_1 + n_2} = \frac{N}{N(\frac{a}{b} + 1)} = \frac{b}{a + b} \qquad (1)$$

$$P_2 = \frac{n_2}{n_1 + n_2} = \frac{N(\frac{a}{b})}{N(\frac{a}{b} + 1)} = \frac{a}{a + b} \qquad (2)$$

The other case for which the discrete line belongs to the second octant (i.e. $b > a$) can be inferred by using symmetries.

Now we consider the same process on a discrete plane:

**Proposition 2.** *Consider the process which consists of a random draw of surfels of type 1, 2 and 3 on a discrete naive plane of the first $8^{th}$ of space (i.e. $a \geq 0$, $b \geq 0$, $c \geq 0$). The probability to obtain a surfel of type 1 is $\frac{b}{a+b+c}$, the probability to obtain a surfel of type 2 is $\frac{a}{a+b+c}$ and the probability to obtain a surfel of type 3 is $\frac{c}{a+b+c}$.*

*Proof.* The proof is based on the decomposition of the plane into discrete lines. More precisely, we first suppose that $c = max(a, b, c)$. We can consider this discrete plane as composition of $N$ discrete lines $\mathcal{D}_x(a, c, \mu)$ in the direction of the x axis and $M$ discrete lines $\mathcal{D}_y(b, c, \mu)$ in the direction of the y axis (Fig. 1-b). The number of surfels of type 3 ($n_3$) is $NM$. Each discrete line $\mathcal{D}_x$ generates surfels of type 2 with probability $\frac{a}{c}$. Then the number of surfels of type 2 ($n_2$) is $NM\frac{a}{c}$. In the same way, we deduce the number of surfels of type 1 ($n_1$) equals to $NM\frac{b}{c}$. Now, we can obtain the following probabilities:

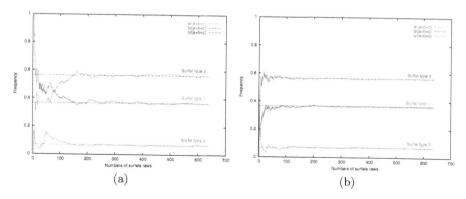

**Fig. 2.** Illustration of the convergence of a discrete plane $P(7, 17, 57, 0)$. The graphic (a) was obtained with a random raw of surfels and (b) was obtained with a cubic raw

$$P_1 = \frac{n_1}{n_1 + n_2 + n_3} = \frac{NM(\frac{b}{c})}{NM(\frac{a+b+c}{c})} = \frac{b}{a+b+c}$$
$$P_2 = \frac{n_2}{n_1 + n_2 + n_3} = \frac{NM(\frac{a}{c})}{NM(\frac{a+b+c}{c})} = \frac{a}{a+b+c} \qquad (3)$$
$$P_3 = \frac{n_3}{n_1 + n_2 + n_3} = \frac{NM}{NM(\frac{a+b+c}{c})} = \frac{c}{a+b+c}$$

The other case for which $b = max(a, b, c)$ and $a = max(a, b, c)$ can be inferred by using symmetries.

From these probability laws, the coefficients $a, b, c$ can be recovered by solving the system of equation Eq. 3. By considering the number $n_i$ of surfel of type $i$, we obtain: $c = kn_2$, $b = kn_3$, $a = kn_1$; with $k = \frac{1}{gcd(n_1, n_2, n_3)}$.

In order to estimate the parameters of the tangent plane at a point of the discrete surface we have now to traverse the neighborhood of this point in order to analyze the frequency of apparition of the different type surfels. There exists several strategies to select the surfels in the neighborhood of the considered point. We can notice that with a random selection the convergence of the frequencies is quite slow. Fig. 2 shows the evolution of the frequencies with a such draw on a discrete plane of characteristics $\mathcal{P}(7, 17, 57, 0)$. The frequency convergence can be detected after around 450 surfel draws.

A better solution consists in selecting surfels on the boundary of a cubic growing neighborhood from the initial point $P_0$. More precisely, the boundary of the cubic neighborhood of size $s$ is defined by:

$$N_s = \{P(x_0 + i, y_0 + j, z_0 + k) \mid P \in S,$$
$$P \text{ is 6-connected to } P_0, \ i, j, k \in \{-s, s\}\} \quad (4)$$

Fig. 2-b shows the frequency obtained after this selection of surfel. The frequency convergence can be detected with around 100 surfels. This selection of surfels implies a faster convergence of the estimation than the previous method. Other methods of selection were experimented, for example the surfels can be selected according to their Euclidean distances from the initial point. Therefore, since this method does not improve the convergence, we have chosen to use the cubic draw of surfels.

Now, we have to define a convergence criterion to detect when the statistical draw follows a probability law. We denote by $f_i^{(n)}$ the apparition frequency of a surfel of type $i$ with a neighborhood size equal to $n$. The criterion of recognition $K_n$ of the discrete tangent plane can be deduced from this apparition frequencies:

$$K_n = \sum_{i=1}^{6} \left| f_i^{(n)} - f_i^{(n-1)} \right| \quad \text{with: } f_i^{(n)} = \frac{s_i^{(n)}}{S_i^{(n)}},$$

where $s_i^{(n)}$ represents the number of surfels of type $i$ on the boundary of the cubic neighborhood of size $n$, and $S_i^{(n)}$ is the total number of surfels contained in $N_n$.

From this criterion, we can define the index $n_k$, for which the criterion $K$ reaches a minimum value. Then the parameters of the discrete plane are deduced from the global frequencies $F_i^{(n)}$ obtained over the whole neighborhood of size $n_k$. More precisely, we have:

$$F_i^{(n)} = \frac{\sum_{j=0}^{n} s_i^{(j)}}{S_{Tot}^{(n)}},$$

where $S_{Tot}^{(n)}$ represents the total number of all the surfels contained in cubic neighborhood of size $n$.

## 3    Position of the Tangent Plane

When estimating the discrete tangent plane, one needs to determine the parameter $\mu$ associated to the position in space of the discrete tangent plane. This parameter plays an important role to determine the position of the new points in the Euclidean space. All the discrete points $P_i$ which belong to the neighborhood of size $n$, need to verify the discrete plane equation:

$$\mu \leq ax_i + by_i + cz_i < \mu + \omega$$

From each point, an interval of possible values of $\mu_i$ can be deduced:

$$\mu_i \in [\sigma_i - \omega, \sigma_i[$$

with $\sigma_i = ax_i + by_i + cz_i$

Since all the discrete points $P_i$ do not necessary belong to the discrete plane, we compute the value of $\mu$ for which a maximum of voxels verify the discrete plane equation. If the resulting possible solutions of $\mu$ are defined by an interval $I = [\mu_{min}, \mu_{max}]$, then the solution for $\mu$ is determined as the average value of the two values $\mu_{min}$ and $\mu_{max}$.

Now since all the characteristics of the discrete tangent plane are evaluated, it is possible to compute the number of voxels which verify the discrete tangent plane equation. The percentage of pixels belonging to the discrete plane is used in association with the previous criterion to limit the neighborhood size extension when ambiguous situation appears. Let us consider for instance a point located at the junction of two planes $P_1$ and $P_2$ and suppose that each plane has only one kind of surfel. In this case the selected surfels are of two types, one from the plane $P_1$ and the other one from $P_2$. The statistical analysis does not take into account the local arrangement of surfels and will thus recognize the plane orthogonal to the bisector of $P_1$ and $P_2$. Such cases may be detected by checking the amount of points which does not belong to the recognized plane. When a tangent plane cannot be recognized the related point may be smoothed by interpolation.

## 4  Surface Reconstruction

From the binary data, a surface mesh is associated to the discrete surface by using the Türmer's algorithm [14]. The resulting surface mesh is obtained by linking together some 6-adjacent centers of voxel belonging to the discrete surface. The triangulation is determined from different topological configurations of voxel. Fig. 3-a illustrates a possible configuration of voxels with the as-

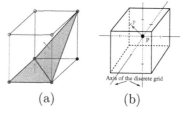

(a)                    (b)

**Fig. 3**

sociated triangulation (the voxels belonging to the discrete surface are drawn in light gray). An example of the resulting surface mesh is shown in Fig. 4-a-d. The direct rendering of the surface mesh does not looks as a continuous surface.

To obtain a smooth surface net, all the center of voxels $P$ are projected on the real tangent plane associated to the discrete tangent plane. For this purpose, we consider the upper and lower real tangent planes defined by:

$$ax + by + cz = \mu$$
$$ax + by + cz = \mu + \omega - 1 \tag{5}$$

From this two planes, we can define the real projection plane as follows:

$$ax + by + cz = \mu + \frac{\omega - 1}{2} \tag{6}$$

There are several ways to project the center of a voxel to the associated real tangent plane. A first approach is to project the discrete point according to the three axis. A better solution consists in projecting the center of voxel according the direction of the normal vector. The advantage of such a projection is the possibility to recover the normal orientation from the projected point.

Therefore, if the discrete point $P$ is the origin of the coordinate system, the new point resulting from the projection $p$ is defined by : $p = \alpha(a, b, c)$. The value of $\alpha$ is deduced from the tangent plane equation, we obtain:

$$\alpha = \frac{\mu + \frac{\omega - 1}{2}}{a^2 + b^2 + c^2}$$

This projection of discrete point $P$ is illustrated in Fig. 3-b. The new euclidean point $p$ is projected to the tangent plane in the direction of the normal vector. It is possible to show that the projection of the discrete point $P(X_p, Y_p, Z_p)$ onto the tangent plane satisfies the Euclidean net condition:

$$|x_p - X_p| < \frac{1}{2}, \ |y_p - Y_p| < \frac{1}{2} \text{ and } |z_p - Z_p| < \frac{1}{2}$$

Each point of the Euclidean net is contained in the unit voxel centered at the associated discrete point. Thus the coordinates of the discrete point can be recovered by rounding the coordinates of the projected point. As a consequence, the transformation of the surface is reversible and the surface mesh is smoothed without any loss in information.

Moreover, from the projected points we can recover the tangent plane orientation by simply computing $a = \sigma(x_p - X_p)$, $b = \sigma(y_p - Y_p)$ and $c = \sigma(z_p - Z_p)$, with $\sigma \in \{-1, 1\}$ determined according to the surface orientation of the triangulated surface. Note that a special case appears when the projected point $p$ is merged with the discrete point $P$. To avoid this situation, a solution consists in moving the projected point by an infinite small displacement in order to recover the tangent plane orientation.

Remark that the Euclidean net associated by this method with a discrete surface encode simultaneously the points of the original discrete surface, the points of the smoothed surface and the tangents estimated at each point.

## 5   Experiments

Firstly, we have experimented this method on a synthetic vase. The initial object was obtained after the rasterization of the associated function. Fig. 4-a shows the result of the initial polygonal surface obtained directly after the triangulation of the discrete surface. The surface obtained after the projection of the discrete points to the tangent plane is shown in Fig. 4-b. This surface is rendered with flat shading using only the information of the surface mesh. In Fig. 4-c the surface is rendered using the normal vectors recovered from the projection of the Euclidean points and using Gouraud shading. The final surface looks much more smooth than the initial.

To analyze the smoothing effect on the discontinuity of the surface, we have experimented the method on the surface represented in Fig. 4-d. The object was obtained by sampling a sphere of radius 30 holed by a sphere of radius 15. As shown in Fig. 4-e, the discontinuity between the two spheres is well conserved, while the small sphere is still smooth. Fig. 4-f shows the neighborhood size which is used for the statistical recognition of the discrete tangent plane. The color range from dark blue to white is associated to a size of neighborhood from 1 to 4. This repartition of the neighborhood size shows how the recognition process may detect surface discontinuity.

Then we have analyzed your method on 3D data scan from MRI images. Fig. 5-a shows the initial surface mesh of the discrete data obtained after a

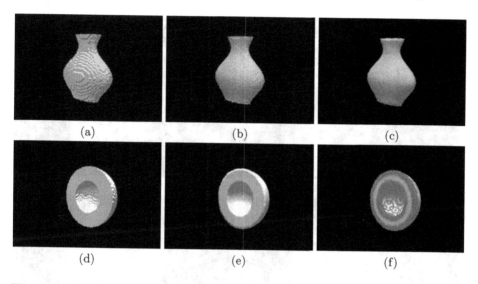

**Fig. 4.** Results obtained from synthetic objects. (a) Triangulated surface, (b) smooth surface rendered with flat shading, (c) idem with Gouraud shading, (d) triangulated surface of second object, (e) smoothed surface rendered with Gouraud shading, and (f) neighborhood size used in the statistical recognition

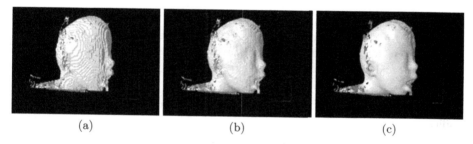

**Fig. 5.** Results obtained from the MRI scan of a child head. The binary data was obtained by a simple threshold (on an array of size $58 \times 58 \times 58$). Image (a) is the result obtained after the triangulation of the discrete surface. (b) is the resulting surface rendered with flat shading and (c) with Gouraud shading

binary segmentation. The surface was contained in an array of size $58 \times 58 \times 58$. After the application of our method the resulting surface looks very smooth compared to the initial surface and the general shape is well conserved (Fig. 5-b and c). In the same way, Fig. 6 presents results on a more complex topology as the skull. The surface was extracted from a binary array of size $49 \times 49 \times 49$.

Tab. 1 presents quantitative experiments on different discrete spheres. The area of the sphere computed from the Euclidean net was compared to the theoretical value. Another geometrical property was analyzed by the comparison of the distances of each euclidean point to the radius of each sphere.

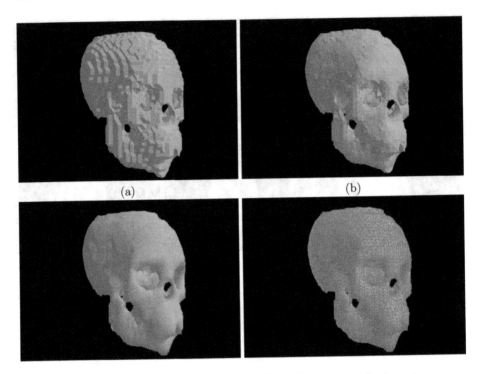

**Fig. 6.** Results obtained from the MRI scan of a skull. The binary data was obtained by a simple threshold (on an array of size $49 \times 49 \times 49$). Image (a) is the result obtained after the triangulation of the discrete surface. (b) is the resulting surface rendered with flat shading and (c) with Gouraud shading

**Table 1.** Results of geometric properties extraction on discrete sphere with different radius

| Radius | area/expected area | mean error distance |
|--------|--------------------|--------------------|
| 10 | 1.020 | 0.097 |
| 20 | 1.008 | 0.085 |
| 30 | 1.009 | 0.088 |
| 40 | 1.007 | 0.087 |
| 70 | 1.005 | 0.095 |

## 6    Conclusion and Future Work

We have introduced a new geometric and statistical method to smooth discrete surfaces. The results of experimentation on both synthetic and real objects show smooth results both on the visual and geometrical points of view. The resulting surface representation can be used for discrete surface rendering and for geometrical properties extraction.

We are currently working on other methods to define a non symmetric surfel draw. This strategy could be relevant in order to adapt the recognition of the tangent plane when discontinuities on the surface are detected. Moreover, it will be interesting to adapt this method to the inter-voxel boundaries representation. Indeed this representation gives a simple and consistent representation of the discrete surface. Finally, further investigations will concern the use of this smoothing method to improve the reconstruction of the discrete *Shape From Shading* technique introduced in [1].

# References

1. A. Braquelaire and B. Kerautret. Reconstruction of discrete surfaces from shading images by propagation of geometric features. In *Discrete Geometry for Computer Imagery*, volume 2886 of *LNCS*, pages 257–266. Springer-Verlag, 2003.
2. A. Braquelaire and A. Pousset. Automatic and reversible geometric smoothing of the boundary of a discrete 3d object. In *Discrete Geometry for Computer Imagery*, volume 1953, pages 198–209. Springer-Verlag, 2000.
3. J.P. Braquelaire and A. Vialard. Euclidean paths : a new representation of boundary of discrete regions. *Graphical Models and Images Processing*, 61:16–43, 1999.
4. J.M. Chassery and J. Vittone. Coexistence of tricubes in digital naive plane. *Lecture Notes in Computer Science*, 1347:99–110, December 1997.
5. D. Coeurjolly. *Algorithmique et géométrie discrète pour la caractérisation des courbes et des surfaces.* PhD thesis, Laboratoire ERIC, 2001.
6. H. Gouraud. Continous shading of curved surfaces. *IEEE Transaction on Computers*, 20(6):623–629, 1971.
7. G.T. Herman and H.K Liu. Three-dimensional display of human organs from computed tomograms. *Computer Graphics and Image Processing*, 9(1):1–21, 1979.
8. J.-O. Lachaud and A. Montanvert. Deformable meshes with automated topology changes for coarses-to-fine 3d surface extraction. *Medical Image Analysis*, 3(2):187–207, 1999.
9. W.E. Lorenson and H.E. Cline. Marching cubes: A height resolution 3d surface reconstruction algorithm. volume 21, pages 11–118, 1987.
10. J.M. Schramm. Coplanar tricubes. *Lecture Notes in Computer Science*, 1347:87–98, December 1997.
11. D. Terzopoulos, A. Witkin, and M. Kass. Constraints on deformable models: Recovering 3d shape and nonrigid motion. *Artificial Intelligence*, 36:91–123, 1988.
12. G. Thürmer. Smoothing normal on discrete surfaces while preserving slope discontinuities. *Computer Graphics Forum*, 20(2):103–114, 2001.
13. G. Thürmer and C. A. Wütrich. Varying neighbourhood parameters for computation of normals on surfaces in discrete space. In IEEE Computer Society Press, editor, *Computer Graphics International*, pages 616–625, 1998.
14. G. Thürmer and C.A. Wütrich. Polygon mesh generation for discrete surfaces in 3d space. In *Eighth Eurographics Workshop on Visualisation in Scientific Computing*, pages 117–126, 1997.
15. A. Vialard. *Chemins euclidiens : Un modèle de représentation des contours discrets.* Phd thesis, Université Bordeaux 1, 1996.

# Arbitrary 3D Resolution Discrete Ray Tracing of Implicit Surfaces

Nilo Stolte

École de Technologie Supérieure, 1100 rue Notre-Dame Ouest, Montréal H30 1K3, Canada
nilo.stolte@online.fr

**Abstract.** A new approach to ray tracing implicit surfaces based on recursive space subdivision is presented in this paper. Interval arithmetic, already used to calculate intersections in ray tracing and ray casting (numerically or subdividing 1D or 2D spaces), is now used here to implement a ray tracing based on reliable rays traversals into a potentially infinite octree-like subdivided space, eliminating explicit intersections. Novel, robust and efficient algorithms for ray voxelization and BSP octant ordering are used to recursively traverse rays through the space. Implicit surfaces are robustly voxelized and hierarchically stored into an octree to a certain given level. During rendering, the subdivision based voxelization of surfaces and rays continues further down until a resolution near the discrete domain of the floating point numbers is acquired. To guarantee robustness of the ray voxelization, interval arithmetic with calculations performed under appropriate rounding modes in Pentium-4 x87 and SSE2 FPUs respectively is applied. The major advantage is that the traversal algorithm is guaranteed to find reliable intersections between the rays and the scene without any explicit intersection calculation, solving a known precision problem of the ray traversal in a previous approach, used here for comparison. The precision of the traversal can be arbitrarily increased within the limitation of the floating point representation.

## 1   Introduction

Ray tracing has been relying intensively on rays-objects intersections [1] which have been persistently imputed as the cause of its low efficiency. Acceleration techniques have been proposed to reduce in one way or another the number of intersection calculations. Some are based on space subdivision in which explicit ray-object intersections are bypassed by traversing rays through the subdivided space. Nonetheless, the ray traversal itself might rely on intersections between rays and bounding volumes [2, 3] or between rays and discrete subspaces where the scene was previously voxelized [4, 5, 6, 7]. Octrees [8, 4, 6, 5, 7, 9] have been proved to have great advantages for space decomposition in these techniques, since empty regions can be efficiently skipped. However, their advantages have been considerably hindered by the fact that the intersection calculations between rays and the boundaries of the regions traversed are not only inefficient but also unreliable [6]. Sometimes, these intersection calculations were implicitly or exclusively accomplished by incremental algorithms [8, 10, 7, 9]. Even though incremental algorithms are more efficient than direct intersection calculations they lack precision and are not reliable. As the discrete traversal advances, the mismatch between the continuous

E. Andres et al. (Eds.): DGCI 2005, LNCS 3429, pp. 414–426, 2005.

ray and the discrete counterpart considerably increases due to the lack of precision, thus parts of the scene would have the tendency to disappear at the end of the traversal since what the discrete ray intersects may be different from what the floating point ray intersects. Evidently, if the discrete traversal is done in low resolution spaces, as it is the case in most acceleration algorithms, the problem is less noticeable. However, the problem is particularly worse when considerably huge discrete spaces are traversed.

Prior to the algorithm presented in section 2, robust ray traversals in discrete spaces have not yet been proposed to solve the problem, although there have been several robust techniques presented such as Lipshitz conditions or interval arithmetic. Interval arithmetic was introduced by Moore [11, 12] and by Duff and Snyder [13, 14] into computer graphics. Since the problem of rounding errors can be very serious as seen in [15], the search for reliable algorithms for rendering is of quite significant importance. Although interval arithmetic has been used in several ray casting approaches [16, 14], the algorithm in section 2 is the first complete solution of an interval arithmetic based ray tracing with 3D space subdivision using discrete ray traversal. Kalra and Barr's ray tracing in [17] adopted a guaranteed ray intersection technique, that could not be considered a reliable solution for ray traversal. Their ray tracing [17] used Lipshitz conditions to voxelize implicit surfaces. However, explicit intersections between rays and octants as well as between rays and objects were still calculated. Duff's interval arithmetic ray casting [14] is robust, but it works in the image space and applies the perspective into the surface equations, thus being not compatible with space subdivision techniques.

In this article, a new approach is shown in which the traversal is done by voxelizing implicitly represented rays using the same technique to voxelize implicit surfaces. Surfaces and rays are simultaneously voxelized to avoid all explicit intersection calculations, between rays and objects as well as between rays and octants. The correlation between the discrete ray and the continuous ray is solved here because they are exactly the same. In this sense this approach resembles a discrete ray tracing [10, 9]. The basic differences in our approach are: (1) the voxelization of surfaces and rays are robust due to the use of interval arithmetic; (2) spatial resolution is much higher, allowing reaching the discrete domain of floating point numbers as proposed in [18]; (3) the scene is voxelized to a lower resolution into the octree as in [9] but during rendering the voxelization of surfaces and rays continues on the fly until a given precision is reached. Even though voxelization plays a crucial role, methods that do not ensure robustness [19, 20, 21] cannot be used in our context. By chance, implicit surfaces can be voxelized robustly [22, 14, 13, 23, 24] and a huge variety of forms and shapes can be defined implicitly.

Robust voxelization of implicit surfaces is generally implemented using spatial recursive subdivision [22, 14, 23, 24]. The methods are all conservative, though. Thus, spurious voxels might show up, depending on the surface, the voxelization method and how it is implemented. Although Lipshitz conditions can also be used to voxelize implicit surfaces [25], interval arithmetic is preferred in our approach because it is shown to be more efficient and more reliable [23]. To avoid the spurious regions, the implicit function describing the surface is evaluated at the eight vertices of the octants to verify if there is a change in sign when the last level is reached. Since it is only done at the

```
#define NL   30
bool intersectionFound = false;
Point intersectedPoint;
int signDir = (dz<0)≪2+(dy<0)≪1+(dx<0);
bool Traversal (octant[8], level) {
    if (intersectionFound)  return true;
    if (level == NL) {
        Object obj = (surface contained in octant);
        if (!IsoValueTest(obj, octant)) {
            if (!PartialDifferentialTest(obj, octant))
                return false;
        }
        intersectionFound = true;
        intersectedPoint = mid;
        return true;
    }
    if (!(ray passes through octant)) return false;
    int izyx = ((z₀ > zₘ) ≪ 2) + ((y₀ > yₘ) ≪ 1) + (x₀ > xₘ);
    int iaux = izyx xor signDir;
    for (i=0; i<8; i++) {
        if (iaux and i) continue;  /* Octant elimination */
        int idx = izyx xor i;      /* BSP ordering */
        suboct = octant[idx];
        if (!(r passes through suboct))  continue;
        if (suboct contains a part of a surface)
            if (Traversal (suboct, level+1))  break;
    }
    return intersectionFound;
}
```

**Fig. 1.** Octree ray traversal algorithm

last level of the subdivision, the robustness throughout the process is guaranteed, but not in [17] because it is performed at each octant before testing the Lipschitz condition.

During the recursive subdivision at rendering time the order in which the octants are to be traversed is important. Our innovative BSP ordering algorithm ensures robustness in this process too. In this algorithm only the starting point of the ray and the middle point of the octant are required. The middle point is always an integer-like number that is produced by the addition of a power of two, half of the length of the octant plus the coordinate of the octant. This is guaranteed to avoid carry propagation, thus ensuring robustness as well as exactness. This calculation is always exact provided the half length of the octant is not smaller than 1 Ulp (unit in the last place) [15].

## 2    New Octree Ray Traversal Algorithm

### 2.1    Notations and Definitions

**Ray.** The notation for the ray equation starting at $(x_0, y_0, z_0)$ and with $(dx, dy, dz)$ as its the direction vector is as follows:

$$\begin{cases} x = x_0 + t \cdot dx \\ y = y_0 + t \cdot dy \\ z = z_0 + t \cdot dz \end{cases} \tag{1}$$

**Scene.** The scene is contained in an axis-aligned cube defined as the bounding box of all the surfaces. One of the cube's vertices is located at the origin $(0, 0, 0)$ and all the other ones have zero or positive coordinates. All the objects (surfaces) are defined within this cube, and previously voxelized using space subdivision.

**Octant and Splitting Planes.** The space subdivision starts splitting the scene into eight equal sized cubes, and each of these cubes is called an octant. An octant can be viewed as three intervals, each one along its respective coordinate axis:

$$[x_c, X_c] \ [y_c, Y_c] \ [z_c, Z_c] \tag{2}$$

where $(x_c, y_c, z_c)$, the vertex with lowest coordinates, is regarded as the coordinate of the octant, and $(X_c, Y_c, Z_c)$ is the vertex with highest coordinates. The subdivision of an octant is performed along the three axis-aligned splitting planes passing through the middle point of the octant, that is $(x_m, y_m, z_m) = (x_c + l/2, y_c + l/2, z_c + l/2)$, where $l$ is the size of the octant. The splitting planes equations are then $x = x_m, y = y_m$ and $z = z_m$.

The eight subdivided octants (sub-octants) are stored in memory along an order determined by their relative locations in respect to the splitting planes, with indices from 0 to 7. Actually each index consists of three bits, each corresponding to an axis. Each bit is set if the sub-octant lies on the positive side of the corresponding splitting plane, otherwise it is zero.

## 2.2   Algorithm Overview

The appealing idea of using discrete space resolutions so high to be able to reproduce the discrete domain of the floating point numbers to guarantee robustness [18] is for the first time implemented in this article. Moreover, the techniques in [24] to robustly voxelize different kinds of implicit surfaces are used as a basis in our approach to ray tracing. A conventional ray tracer is used here, while the intersection calculation is replaced by our new octree ray traversal algorithm.

During subdivision, each octant that might contain a part of a surface is further subdivided and so forth. Once a certain level is reached the octant is considered a leaf node or voxel and stored in an octree. During rendering time, the basic idea is to traverse each ray through the octree to find the first voxel containing a part of an object in the scene, which is in appearance similar to the work of Glassner [4] or Gargantini [6]. The fundamental difference between their approaches is that Glassner uses a linear octree, while Gargantini uses a hierarchical pointer octree. The approach in this article is different from their approaches in two aspects: (1) no explicit intersections between rays and octants neither between rays and objects are ever calculated; (2) the spatial recursive subdivision continues further down after the first voxel is found. This subdivision is done in the same way as the recursive voxelization, however, it is done on the fly and nothing needs to be stored, since rays and surfaces are voxelized concurrently. For each octant the eight sub-octants are sorted in the order that they might be traversed by the ray, and tested to verify if they are indeed traversed by the ray. This process is summarized in the pseudo code of Fig. 1. The *IsoValueTest* and *PartialDifferentialTest* appearing in Fig. 1 are discussed in section 2.5.

In the implementations of Glassner's [4] and Gargantini's [6] approaches explicit intersection calculation is performed when traversing the subdivided space, which cannot

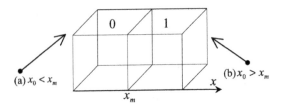

**Fig. 2.** Illustration of the BSP ordering technique

guarantee robustness. Our approach is significantly more reliable since the ray voxelization (see section 2.4) is robust and the BSP ordering scheme relies on calculations not affected by rounding errors. The traversal continues to recursively descend in the subdivided space whenever octants that satisfy the following three conditions are found:

1. The ray passes through the octant. This is tested using our new ray voxelization algorithm (see section 2.4).
2. The octant contains a part of a surface in the scene. When the traversal is within the octree, it is easily tested by verifying the data stored in the octree. When the traversal goes beyond the octree resolution, the implicit function inclusion function [13] used to voxelize the surface is reused to check this condition.
3. The octant is the nearest to $(x_0, y_0, z_0)$ satisfying conditions 1 and 2. This is ensured using the BSP ordering technique (see section 2.3).

The traversal continues until an octant is finally found after descending *NL* levels in the subdivided space, where *NL* is the preset maximum traversal level, otherwise it is assumed that there is no intersection between the ray and the scene. The intersection point is then considered to be $(x_m, y_m, z_m)$. The precision obtained is dictated by the maximum distance between this point and the real intersection, that is, half of the octant diagonal length, which is $\sqrt{3} \cdot 2^{-NL-1}$. It can be seen that better precision can be obtained by increasing *NL*. For the IEEE 754 double precision numbers, the maximum value of *NL* is 53, which is the number of bits in the significand including the hidden 1, a binary digit always set to 1 except in special cases, which is not explicitly represented.

### 2.3    Enhanced BSP Octant Ordering

The BSP ordering technique is shown in Fig. 1. It is based on the relative location of the starting point of the ray in respect to the three splitting planes. This information is stored into the variable *izyx* as illustrated in Fig. 1. The variable *i* stores the indexes of the octants as they appear in the memory, but these indexes are not in the front to back order as required in our ray traversal. As illustrated in Fig. 2, the memory voxel order is preserved in certain cases; otherwise it is reversed. Moreover, if the ray does not pass through a splitting plane, the four sub-octants at the opposite side of the splitting plane in respect to the starting point of the ray will never be traversed and are ignored. The efficiency of these two calculations comes from their extreme simplicity and reduced number of instructions.

## 2.4    Novel Robust Ray Voxelization Algorithm

**Ray Inclusion Function.** Our method to test whether a ray passes through an octant is based on the implicit form of the ray. After eliminating the $t$ variable in (1) and applying a common factor $|dx| \cdot |dy| \cdot |dz|$, the implicit representation of the ray is obtained:

$$cx \cdot (x - x_0) = cy \cdot (y - y_0) = cz \cdot (z - z_0) \tag{3}$$

where

$$
\begin{aligned}
cx &= sign(dx) \cdot |dy| \cdot |dz| \\
cy &= sign(dy) \cdot |dz| \cdot |dx| \\
cz &= sign(dz) \cdot |dx| \cdot |dy|
\end{aligned}
$$

Actually, each one of the three parts of (3) is the multiplication of $t$ in (1) by $|dx| \cdot |dy| \cdot |dz|$. The $sign(dx)$, $sign(dy)$ and $sign(dz)$ ensure that the values obtained from (3) have the same sign as the real $t$ in (1). Replacing the variables $x$, $y$ and $z$ in equation (3) by the three intervals in (2) produces one interval in each of the three parts of (3), as shown in (4).

$$
\begin{aligned}
\left[ f_x(x_c),\ f_x(X_c) \right] &= \left[ cx \cdot (x_c - x_0),\ cx \cdot (X_c - x_0) \right] \\
\left[ f_y(y_c),\ f_y(Y_c) \right] &= \left[ cy \cdot (y_c - y_0),\ cy \cdot (Y_c - y_0) \right] \\
\left[ f_z(z_c),\ f_z(Z_c) \right] &= \left[ cz \cdot (z_c - z_0),\ cz \cdot (Z_c - z_0) \right]
\end{aligned}
\tag{4}
$$

As can be seen, the ray passes through the octant only if these three intervals overlap and this condition is what is used to test if a ray passes through an octant.

**Using Correct Rounding Modes.** The approach still suffers from rounding errors caused by the multiplications. The precision of the traversal is determined by the number of subdivision levels, each further subdivision corresponding to an additional bit of precision. In experiments where subdivision levels are close to the limits of floating point double precision (e.g. 50 levels), some octants are missed by rays that pass very close to octants' edges or vertices. In this case the two bounds of the octants in each axis differ only in the last few bits of the significand. Consequently, after applying them to (3) the resulting intervals are more prone to rounding errors. To guarantee robustness of this calculation, interval arithmetic with appropriate rounding modes is applied. The calculation in (4) is done with two rounding modes, towards '$-\infty$' and '$+\infty$', to the lower and upper bounds respectively. Consequently when the three intervals obtained with correct rounding modes do not overlap, it is assured that the ray does not pass through the octant. Moreover, we consider that there is no overlap when a lower bound of an interval is equal to the upper bound of another, since the ray in this case passes through neither one of them. To avoid penalties in performance normally involved in frequent change of rounding modes, each bound is calculated in a different floating point unit (FPU) preset with the required rounding mode. The Pentium-4 allows executing 2 double precision operations in SSE2 FPU and one double precision operation in the normal x87 FPU with independent rounding modes.

**Optimizations.** The actual implementation of this algorithm differs from what is described in Fig. 1 due to optimization reasons. The number of operations shown in section

```
double tx[4] = { x_c, x_m, x_m, X_c };
/* calculate array tx*/
int itx = (signDir and 1)≪1;
/* but for the other two axes it would be: */
/* int itx = (signDir and 2); → y axis */
/* int itx = (signDir and 4)≫1; → z axis */
tx[itx] = f_x−(tx[itx]);
tx[itx+1] = f_x−(tx[itx+1]);
tx[itx xor 2] = f_x+(tx[itx xor 2]);
tx[(itx xor 2)+1] = f_x+(tx[(itx xor 2)+1]);
/* retrieving corresponding interval from tx */
/* idx is the index of the sub-octant as defined in Fig. 1 */
int ix = idx and 1;
/* but for the other two axes it would be: */
/* int ix = (idx and 2)≫1; → y axis */
/* int ix = (idx and 4)≫2; → z axis */
double lbx = tx[ix+itx];
double ubx = tx[ix+(itx xor 2)];
```

**Fig. 3.** Storing and retrieving the interval bounds with correct rounding modes without testing to avoid stalling the processor pipeline

2.6 takes these optimizations into account. In Fig. 1, the test to verify if a ray passes through an octant is carried out once for the octant itself and once for each of its sub-octants, thus giving rise to 9 tests involving the evaluation of 27 intervals. Since they share bounds with each other, only 3 intervals along each axis are really considered for testing all the 9 octants, so only 9 of the 27 intervals are indeed calculated. To avoid these repetitions all the eight sub-octants and their parent octant are tested together. What remains to be elucidated is the selection of the correct rounding mode for each bound of an interval. For the $x$ coordinate, the following three intervals, $[x_c, X_c]$, $[x_c, x_m]$ and $[x_m, X_c]$ will be applied to the implicit ray equation (3), and three resulting intervals will be obtained. However, the proper rounding modes to guarantee numerical robustness depend on the direction of the ray along the x axis ($dx$). In the case that $dx > 0$, the resulting intervals are

$$[f_{x-}(x_c), f_{x+}(X_c)], [f_{x-}(x_c), f_{x+}(x_m)], [f_{x-}(x_m), f_{x+}(X_c)]$$

where $f_{x-}()$ and $f_{x+}()$ represent the calculation of $f_x()$ in (4) using rounding modes towards '$-\infty$' and '$+\infty$' respectively. When $dx < 0$, the resulting intervals will be

$$[f_{x-}(X_c), f_{x+}(x_c)], [f_{x-}(X_c), f_{x+}(x_m)], [f_{x-}(x_m), f_{x+}(x_c)]$$

Evidently $x_c$ and $X_c$ are calculated only once under different rounding modes, and $x_m$ is always calculated twice, each one under a different rounding mode respectively. Therefore, an array of four elements $tx[4]$ is used to represent the three resulting intervals. When $dx > 0$, $tx[0]$ and $tx[1]$ are calculated with rounding mode 'towards $-\infty$', and $tx[2]$ and $tx[3]$ are calculated with rounding mode 'towards $+\infty$'; or vice versa when $dx < 0$. The same logic is used along y and z axis. To avoid tests and branch instructions that may stall the processor pipeline, the indices of the elements in the arrays to store the values calculated under each rounding mode are automatically selected. A similar scheme is used to retrieve the correct lower and upper bounds stored in $tx$ array for each sub-octant. The procedure is shown in Fig. 3.

## 2.5    Partial Differential Test

The robustness of interval arithmetic can guarantee that no parts of the surfaces were omitted during voxelization/traversal. On the other hand, however, it cannot guarantee that each leaf voxel really contains a part of a surface, thus resulting in overestimations during the voxelization/traversal process. To evaluate and reduce these overestimations, two algorithms as proposed in [18] are applied on the final octant and the surface contained in it. The first one calculates the eight iso-values of the function corresponding to the surface at the eight corners of the octant, and checks if there is a change in sign. This algorithm may suffer from rounding errors as suggested in [18], since it directly uses the surface equation. In some cases, these rounding errors will possibly be added to the half of the octant diagonal length error described at the end of section 2.2. Even though it is possible to solve the problem, the reliability at the leaf level is already quite high; therefore, the use of this algorithm is acceptable.

```
bool PartialDifferentialTest (Surface, Octant) {
     f(x, y, z) : implicit function corresponding to Surface;
     DFX(X), DFY(Y), DFZ(Z) :
        inclusion functions of ∂f/∂x, ∂f/∂y and ∂f/∂z;
     X, Y, Z : the three intervals corresponding to Octant;
     DX, DY, DZ : three intervals;
     DX = DFX(X);   DY = DFY(Y);   DZ = DFZ(Z);
     return   ((0 ∈ DX) || (0 ∈ DY) || (0 ∈ DZ))
}
```

**Fig. 4.** Partial differential test of an octant

   If there is a change in sign, the octant is considered to be correct at the given precision, otherwise the second algorithm will be applied, which examines the monotonicity of the implicit function as shown in Fig. 4. If the partial differential test returns true, the function is monotonic within the $x$, $y$, and $z$ range of the octant, thus the function has no zeroes within the octant, so the octant is ignored and the traversal continues. If the test returns false, the traversal stops, despite that in very special cases, this octant may not contain a part of a surface. This case does not occur in any surface ray traced in this paper; however, it needs to be considered in future work.

## 2.6    Comparison with Gargantini's Algorithm

Octree traversal algorithms, such as Gargantini's [6], calculate intersections between the ray and the octants it traverses. Therefore, the traversal algorithm in section 2.4 is compared with Gargantini's. The rest of the program is exactly the same.

   Gargantini's algorithm exploits the fact that a ray can pass through at most four sub-octants in an octant. Assuming that the entry and exit points of a ray into an octant are known, intersections between the ray and all three subdivision mid-planes are then calculated using (1). Only the $m$ ($m \leq 3$) intersected points that are within the octant are retained and sorted in ascending order according to their corresponding $t$ values, resulting in $m + 1$ ray segments. The lower and upper bounds of each segment correspond to the entry and exit points of the ray into one sub-octant, thus the sub-octants traversed by the ray are obtained in the correct order. The technique above was described in [6]. Due to differences between the octants' structures, the implementation here has some variations in respect to [6]. Here, an octant only contains an array of pointers to its sub-octants. After the index of a sub-octant is known, its coordinates and then its middle point need to be calculated. In [6] only the indices of the sub-octants are needed, since their coordinates are not used, and the octant structure already contains the middle point coordinates. Because of the variations in design, the number of operations of Gargantini's algorithm contains more floating point additions and bit operations than what was claimed in [6]. The extra data cannot be stored in our case since our traversal is also done on the fly as well as in the octree, while in [6] it is limited to the octree.

The advantage of the traversal algorithm in [6] is that it eliminates all the sub-octants that are not traversed by the ray. Comparatively, the BSP ordering and auxiliary techniques in our algorithm can only partially eliminate them, whereas the remaining ones are still tested by the ray voxelization algorithm in section 2.4. However, as can be seen from the results in section 3, the traversal algorithm in this article is a bit faster than Gargantini's, and is just slightly slower after performing the calculations using robust interval arithmetic with different rounding modes. This is due to the high efficiency of our BSP ordering and ray voxelization techniques.

If the precision/resolution is not too high (e.g. 30 levels), the traversals in both Gargantini's and our approach are exactly the same for the scenes tested. However, when subdivision levels become close to the limits of the floating point precision the two traversals start to mismatch. To verify which traversal is correct, a 128 bits precision binary floating point arithmetic package was used. A verification program using this package calculates the intersections between a ray and each sub-octant of a traversed octant, giving the sub-octants indices in the order they are traversed. In all the scenes tested our approach applying interval arithmetic with correct rounding modes (with SSE2) always exhibited traversals identical to the ones obtained by the verification program. Gargantini pointed out in [6] the two cases when the errors may occur, and described a method to avoid the choice of incorrect octant under certain conditions restricted to ray casting. However, the method is not a complete solution for all cases, and can only eliminate a part of the errors. Comparatively, after applying interval arithmetic with correct rounding modes, the algorithm described in section 2.4 can guarantee that an octant will never be missed without much effect in the performance, thus solving the problem in Gargantini's algorithm.

## 3    Results

Table 1 shows the times for generating $512 \times 512$ images. Six kinds of surfaces are ray traced using a PC with a Pentium-4 2.4GHz processor and 512Mbytes of main memory. Their equations can be seen in Table 1. For each kind of surface, images were generated for both ray casting and ray tracing, using Gargantini's algorithm, our algorithm with and without proper rounding modes using SSE2 instructions respectively, see Fig. 5.

(a) Spheres

(b) Holesrot

(c) Quartdef

(d) Tanglecube

(e) Asterisk(m=3, n=4)

(f) Asterisk(m=9, n=18)

**Fig. 5.** Ray tracing results with 5 level reflections

**Table 1.** Ray tracing times (sec) for $512\times512$ images, 30 bits precision, 2 light sources and 5 levels of reflections (except for ray casting)

| Primitive | Equations | Gar. | Ours | with SSE2 |
|---|---|---|---|---|
| spheres | | 21.672 | 20.687 | 27.062 |
| ray casting | | 15.516 | 15.235 | 19.860 |
| quartdef | | 37.609 | 35.140 | 45.281 |
| ray casting | $(x^2-1)^2+(y^2-1)^2+(z^2-1)^2-1=0$ | 26.109 | 24.719 | 31.891 |
| tanglecube | | 46.437 | 43.015 | 56.078 |
| ray casting | $x^4-5x^2+y^4-5y^2+z^4-5z^2+11.8=0$ | 30.719 | 28.813 | 37.531 |
| holesrot | | 43.719 | 41.313 | 53.500 |
| ray casting | $x^3+y^3+z^3-x-y-z=0$ | 30.469 | 29.313 | 37.937 |
| asterisk(3,4) | | 204.327 | 193.233 | 209.608 |
| ray casting | $\sin(3\theta)\sin(4\phi)-R=0$ | 113.875 | 109.546 | 118.750 |
| asterisk(9,18) | | 435.638 | 410.841 | 445.982 |
| ray casting | $\sin(9\theta)\sin(18\phi)-R=0$ | 135.390 | 129.312 | 140.562 |

The surfaces are voxelized at octree level 9 ($512^3$ resolution), with $NL=30$. Only images generated by our algorithm with SSE2 are shown since the images have no visible differences in comparison to those generated by Gargantini's.

It can be seen from Table 1 that our algorithms exhibit similar performance as Gargantini's. The one using interval arithmetic under proper rounding modes (with SSE2 instructions) is slightly slower, whereas the one without rounding (Ours) is several seconds faster (see discussion in section 2.6). Gargantini's algorithm was compared with Samet's in [6], and the results showed that the time in Gargantini's approach is nearly half of the time of Samet's. Thus our approach is also fairly twice faster than Samet's.

Analyzing the performance based on the number of rays per second (rays/sec), one concludes that the surfaces with tight inclusion functions exhibit roughly the same performance, thus suggesting the algorithm is insensitive to the surfaces' complexity. The best performance was with spheres, since they have quite simple equations and also the tightest inclusion functions. Naively applying interval arithmetic in the tanglecube and the holesrot did not originally provide tight inclusion functions. Cleverly decomposing their expressions led to quite tight inclusion functions significantly enhancing the performance. With the optimizations (shown in the times in the Table 1), the performance of the holesrot almost doubles, whereas the tanglecube is almost 4 times faster.

## 4    Conclusion

A new algorithm is here shown to ray trace implicit surfaces without explicit intersection. The intersection estimation converges in $O(\log_8 N)$, where $N$ is the number of voxels of the discrete space ($2^{3NL}$). It works by voxelizing rays and objects by recursively subdividing the space and using interval arithmetic to discard regions not crossed by a ray or a surface. A novel BSP octant ordering technique is used to efficiently traverse the rays; it is robust since the values involved are exact. Both the ray and object voxelizations are also robust, thus guaranteeing the reliability. A partial differential test algorithm is sometimes applied to eliminate overestimations of interval arithmetic. The results show that the algorithm is insensitive to the surfaces' complexity but quite sensitive to the inclusion functions tightness, since different surfaces with similar inclusion function tightness exhibit similar rays/sec performance.

# References

1. Whitted, T.: An Improved Ilumination Model for Shaded Display. Communications of the ACM **23** (1980) 343–349
2. Rubin, S.M., Whitted, T.: A 3-Dimensional Representation for Fast Rendering Complex Scenes. Computer Graphics **14** (1980) 110–116
3. Kay, T., Kajiya, J.: Ray Tracing Complex Scenes. Computer Graphics **20** (1986) 269–278
4. Glassner, A.S.: Space Subdivision for Fast Ray Tracing. IEEE - CGA **10** (1984) 15–22
5. Jevans, D., Wyvill, B.: Adaptative Voxel Subdivision for Ray Tracing. In: Proceedings of Graphics Interface '89, Toronto, Ontario, Canadian Information Processing Society (1989) 164–172
6. Gargantini, I.: Ray tracing an Octree: Numerical Evaluation of the First Intersection. Computer Graphics forum **12** (1993) 199–210
7. Endl, R., Sommer, M.: Classification of Ray-Generators in Uniform Subdivisions and Octrees for Ray Tracing. Computer Graphics *forum* **13** (1994) 3–19
8. Fujimoto, A., Tanaka, T., Iwata, K.: ARTS: Accelerated Ray Tracing System. IEEE - CGA **6** (1986) 16–26
9. Stolte, N., Caubet, R.: Discrete Ray-Tracing of Huge Voxel Spaces. Computer Graphics Forum **14** (1995) 383–394
10. Yagel, R., Cohen, D., Kaufman, A.: Discrete Ray Tracing. IEEE - CGA **12** (1992) 19–28
11. Moore, R.E.: Interval Analysis. Prentice-Hall, Englewood Cliffs, NJ (1966)
12. Moore, R.E.: Methods and Application of Interval Analysis. Society for Industrial and Applied Mathematics (SIAM), Philadelphia (1979)
13. Snyder, J.M.: Interval Analysis For Computer Graphics. Computer Graphics **26** (1992) 121–130
14. Duff, T.: Interval Arithmetic and Recursive Subdivision for Implicit Functions and Constructive Solid Geometry. Computer Graphics **26** (1992) 131–138
15. Goldberg, D.: What Every Computer Scientist Should Know About Floating-Point Arithmetic. ACM Computing Surveys **23** (1991) 5–48
16. de Cusatis Junior, A., de Figueiredo, L.H., Gattass, M.: Interval Methods for Ray Casting Implicit Surfaces with Affine Arithmetic. SIBGRAPHI (1999) 17–20
17. Kalra, D., Barr, A.: Guaranteed Ray Intersections with Implicit Surfaces. Computer Graphics **23** (1989) 297–306
18. Stolte, N.: High Resolution Discrete Spaces: A New Approach for Modeling and Realistic Rendering (Espaces Discrets de Haute Résolutions: Une Nouvelle Approche pour la Modelisation et le Rendu d'Images Réalistes). PhD thesis, Université Paul Sabatier - Toulouse - France (1996)
19. Kaufman, A.: An Algorithm for 3D Scan-Conversion of Polygons. In: Eurographics'87, Amsterdam, North Holand (1987) 197–208
20. Kaufman, A.: Efficient Algorithms for 3D Scan-Conversion of Parametric Curves, Surfaces, and Volumes. Computer Graphics **21** (1987) 171–179
21. Greene, N.: Voxel Space Automata: Modeling with Stochastic Growth Processes in Voxel Space. Computer Graphics **23** (1989) 175–184
22. Taubin, G.: Rasterizing Algebraic Curves and Surfaces. IEEE - CGA (1994) 14–23

23. Stolte, N., Caubet, R.: Comparison between different Rasterization Methods for Implicit Surfaces. In Rae Earnshaw, John A. Vince and How Jones, ed.: Visualization and Modeling. Academic Press (1997) 191–201
24. Stolte, N., Kaufman, A.: Novel Techniques for Robust Voxelization and Visualization of Implicit Surfaces. Graphical Models **63** (2001) 387–412
25. Bidasaria, H.B.: Defining and Rendering of Textured Objects through The Use of Exponential Functions. Graphical Models and Image Processing **54** (1992) 97–102

# Author Index

Alayrangues, Sylvie    195
Alpers, Andreas    92
Andrès, Éric    23

Balázs, Péter    104
Bernard, Thierry M.    1
Berthé, Valerie    276
Bertrand, Gilles    172
Borgefors, Gunilla    68
Braquelaire, Achille    404
Brimkov, Valentin E.    287
Brun, Luc    34
Brunetti, Sara    92
Buzer, Lilian    299

Castiglione, Giusi    115
Celasun, Işil    45
Chassery, Jean-Marc    347
Chatelier, Pierre Y.    392
Ciria, Jose C.    161
Coeurjolly, David    311
Couprie, Michel    172, 216
Crespo, Jose    206

Damiand, Guillaume    56
Dantchev, Stefan S.    287
de Miguel, Angel    161
de Vieilleville, François    240
Debled-Rennesson, Isabelle    371
Domínguez, Eladio    161
Dupont, Florent    347

Feschet, Fabien    126, 371
Fiorio, Christophe    276
Francés, Angel R.    161
Frosini, Andrea    115
Fuchs, Laurent    195

Gérard, Yan    126
Geniet, Dominique    23
Gies, Valentin    1
Grasset-Simon, Carine    56
Guédon, JeanPierre    79, 153

Hamanaka, Masatoshi    323

Ikonen, Leena    228

Jamet, Damien    276

Köthe, Ullrich    383
Kenmochi, Yukiko    323, 335
Kerautret, Bertrand    404
Kingston, Andrew    136
Klette, Reinhard    183
Kuba, Attila    148

Lachaud, Jacques-Olivier    195, 240
Lakaemper, Rolf    11
Largeteau, Gaëlle    23
Latecki, Longin Jan    11
Li, Fajie    183
Lienhardt, Pascal    56
Lindblad, Joakim    252

Malgouyres, Rémy    392
Maojo, Victor    206
Melkisetoğlu, Rupen    45
Meyer, Fernand    34
Mokhtari, Myriam    34

Najman, Laurent    172
Nomura, Yusuke    335
Normand, Nicolas    79, 153

Peltier, Samuel    195
Prasad, Lakshman    263

Quintero, Antonio    161

Restivo, Antonio    115
Rinaldi, Simone    115
Rouyer-Degli, Jocelyne    371

Servières, Myriam    153
Sivignon, Isabelle    347
Stelldinger, Peer    383
Stolte, Nilo    414
Strand, Robin    68

Sugimoto, Akihiro    323
Sun, Xinyu    11
Svalbe, Imants    136

Tekalp, A. Murat    45

Veelaert, Peter    359

Vialard, Anne    240
Vidal, Javier    206

Woeginger, Gerhard J.    148
Wolter, Diedrich    11

Zrour, Rita    216

# Lecture Notes in Computer Science

For information about Vols. 1–3352

please contact your bookseller or Springer

Vol. 3456: H. Rust, Operational Semantics for Timed Systems. XII, 223 pages. 2005.

Vol. 3455: H. Treharne, S. King, M. Henson, S. Schneider (Eds.), ZB 2005: Formal Specification and Development in Z and B. XV, 493 pages. 2005.

Vol. 3453: L. Zhou, B.C. Ooi, X. Meng (Eds.), Database Systems for Advanced Applications. XXVII, 929 pages. 2005.

Vol. 3452: F. Baader, A. Voronkov (Eds.), Logic for Programming, Artificial Intelligence, and Reasoning. XI, 562 pages. 2005. (Subseries LNAI).

Vol. 3450: D. Hutter, M. Ullmann (Eds.), Security in Pervasive Computing. XI, 239 pages. 2005.

Vol. 3449: F. Rothlauf, J. Branke, S. Cagnoni, D.W. Corne, R. Drechsler, Y. Jin, P. Machado, E. Marchiori, J. Romero, G.D. Smith, G. Squillero (Eds.), Applications on Evolutionary Computing. XX, 631 pages. 2005.

Vol. 3448: G.R. Raidl, J. Gottlieb (Eds.), Evolutionary Computation in Combinatorial Optimization. XI, 271 pages. 2005.

Vol. 3447: M. Keijzer, A. Tettamanzi, P. Collet, J.v. Hemert, M. Tomassini (Eds.), Genetic Programming. XIII, 382 pages. 2005.

Vol. 3444: M. Sagiv (Ed.), Programming Languages and Systems. XIII, 439 pages. 2005.

Vol. 3443: R. Bodik (Ed.), Compiler Construction. XI, 305 pages. 2005.

Vol. 3442: M. Cerioli (Ed.), Fundamental Approaches to Software Engineering. XIII, 373 pages. 2005.

Vol. 3441: V. Sassone (Ed.), Foundations of Software Science and Computational Structures. XVIII, 521 pages. 2005.

Vol. 3440: N. Halbwachs, L.D. Zuck (Eds.), Tools and Algorithms for the Construction and Analysis of Systems. XVII, 588 pages. 2005.

Vol. 3439: R.H. Deng, F. Bao, H. Pang, J. Zhou (Eds.), Information Security Practice and Experience. XII, 424 pages. 2005.

Vol. 3436: B. Bouyssounouse, J. Sifakis (Eds.), Embedded Systems Design. XV, 492 pages. 2005.

Vol. 3434: L. Brun, M. Vento (Eds.), Graph-Based Representations in Pattern Recognition. XII, 384 pages. 2005.

Vol. 3433: S. Bhalla (Ed.), Databases in Networked Information Systems. VII, 319 pages. 2005.

Vol. 3432: M. Beigl, P. Lukowicz (Eds.), Systems Aspects in Organic and Pervasive Computing - ARCS 2005. X, 265 pages. 2005.

Vol. 3431: C. Dovrolis (Ed.), Passive and Active Network Measurement. XII, 374 pages. 2005.

Vol. 3429: E. Andres, G. Damiand, P. Lienhardt (Eds.), Discrete Geometry for Computer Imagery. X, 428 pages. 2005.

Vol. 3427: G. Kotsis, O. Spaniol, Wireless Systems and Mobility in Next Generation Internet. VIII, 249 pages. 2005.

Vol. 3423: J.L. Fiadeiro, P.D. Mosses, F. Orejas (Eds.), Recent Trends in Algebraic Development Techniques. VIII, 271 pages. 2005.

Vol. 3422: R.T. Mittermeir (Ed.), From Computer Literacy to Informatics Fundamentals. X, 203 pages. 2005.

Vol. 3421: P. Lorenz, P. Dini (Eds.), Networking - ICN 2005, Part II. XXXV, 1153 pages. 2005.

Vol. 3420: P. Lorenz, P. Dini (Eds.), Networking - ICN 2005, Part I. XXXV, 933 pages. 2005.

Vol. 3419: B. Faltings, A. Petcu, F. Fages, F. Rossi (Eds.), Constraint Satisfaction and Constraint Logic Programming. X, 217 pages. 2005. (Subseries LNAI).

Vol. 3418: U. Brandes, T. Erlebach (Eds.), Network Analysis. XII, 471 pages. 2005.

Vol. 3416: M. Böhlen, J. Gamper, W. Polasek, M.A. Wimmer (Eds.), E-Government: Towards Electronic Democracy. XIII, 311 pages. 2005. (Subseries LNAI).

Vol. 3415: P. Davidsson, B. Logan, K. Takadama (Eds.), Multi-Agent and Multi-Agent-Based Simulation. X, 265 pages. 2005. (Subseries LNAI).

Vol. 3414: M. Morari, L. Thiele (Eds.), Hybrid Systems: Computation and Control. XII, 684 pages. 2005.

Vol. 3412: X. Franch, D. Port (Eds.), COTS-Based Software Systems. XVI, 312 pages. 2005.

Vol. 3411: S.H. Myaeng, M. Zhou, K.-F. Wong, H.-J. Zhang (Eds.), Information Retrieval Technology. XIII, 337 pages. 2005.

Vol. 3410: C.A. Coello Coello, A. Hernández Aguirre, E. Zitzler (Eds.), Evolutionary Multi-Criterion Optimization. XVI, 912 pages. 2005.

Vol. 3409: N. Guelfi, G. Reggio, A. Romanovsky (Eds.), Scientific Engineering of Distributed Java Applications. X, 127 pages. 2005.

Vol. 3408: D.E. Losada, J.M. Fernández-Luna (Eds.), Advances in Information Retrieval. XVII, 572 pages. 2005.

Vol. 3407: Z. Liu, K. Araki (Eds.), Theoretical Aspects of Computing - ICTAC 2004. XIV, 562 pages. 2005.

Vol. 3406: A. Gelbukh (Ed.), Computational Linguistics and Intelligent Text Processing. XVII, 829 pages. 2005.

Vol. 3404: V. Diekert, B. Durand (Eds.), STACS 2005. XVI, 706 pages. 2005.

Vol. 3403: B. Ganter, R. Godin (Eds.), Formal Concept Analysis. XI, 419 pages. 2005. (Subseries LNAI).

Vol. 3401: Z. Li, L.G. Vulkov, J. Waśniewski (Eds.), Numerical Analysis and Its Applications. XIII, 630 pages. 2005.

Vol. 3399: Y. Zhang, K. Tanaka, J.X. Yu, S. Wang, M. Li (Eds.), Web Technologies Research and Development - APWeb 2005. XXII, 1082 pages. 2005.

Vol. 3398: D.-K. Baik (Ed.), Systems Modeling and Simulation: Theory and Applications. XIV, 733 pages. 2005. (Subseries LNAI).

Vol. 3397: T.G. Kim (Ed.), Artificial Intelligence and Simulation. XV, 711 pages. 2005. (Subseries LNAI).

Vol. 3396: R.M. van Eijk, M.-P. Huget, F. Dignum (Eds.), Agent Communication. X, 261 pages. 2005. (Subseries LNAI).

Vol. 3395: J. Grabowski, B. Nielsen (Eds.), Formal Approaches to Software Testing. X, 225 pages. 2005.

Vol. 3394: D. Kudenko, D. Kazakov, E. Alonso (Eds.), Adaptive Agents and Multi-Agent Systems III. VIII, 313 pages. 2005. (Subseries LNAI).

Vol. 3393: H.-J. Kreowski, U. Montanari, F. Orejas, G. Rozenberg, G. Taentzer (Eds.), Formal Methods in Software and Systems Modeling. XXVII, 413 pages. 2005.

Vol. 3392: D. Seipel, M. Hanus, U. Geske, O. Bartenstein (Eds.), Applications of Declarative Programming and Knowledge Management. X, 309 pages. 2005. (Subseries LNAI).

Vol. 3391: C. Kim (Ed.), Information Networking. XVII, 936 pages. 2005.

Vol. 3390: R. Choren, A. Garcia, C. Lucena, A. Romanovsky (Eds.), Software Engineering for Multi-Agent Systems III. XII, 291 pages. 2005.

Vol. 3389: P. Van Roy (Ed.), Multiparadigm Programming in Mozart/OZ. XV, 329 pages. 2005.

Vol. 3388: J. Lagergren (Ed.), Comparative Genomics. VII, 133 pages. 2005. (Subseries LNBI).

Vol. 3387: J. Cardoso, A. Sheth (Eds.), Semantic Web Services and Web Process Composition. VIII, 147 pages. 2005.

Vol. 3386: S. Vaudenay (Ed.), Public Key Cryptography - PKC 2005. IX, 436 pages. 2005.

Vol. 3385: R. Cousot (Ed.), Verification, Model Checking, and Abstract Interpretation. XII, 483 pages. 2005.

Vol. 3383: J. Pach (Ed.), Graph Drawing. XII, 536 pages. 2005.

Vol. 3382: J. Odell, P. Giorgini, J.P. Müller (Eds.), Agent-Oriented Software Engineering V. X, 239 pages. 2005.

Vol. 3381: P. Vojtáš, M. Bieliková, B. Charron-Bost, O. Sýkora (Eds.), SOFSEM 2005: Theory and Practice of Computer Science. XV, 448 pages. 2005.

Vol. 3380: C. Priami, Transactions on Computational Systems Biology I. IX, 111 pages. 2005. (Subseries LNBI).

Vol. 3379: M. Hemmje, C. Niederee, T. Risse (Eds.), From Integrated Publication and Information Systems to Information and Knowledge Environments. XXIV, 321 pages. 2005.

Vol. 3378: J. Kilian (Ed.), Theory of Cryptography. XII, 621 pages. 2005.

Vol. 3377: B. Goethals, A. Siebes (Eds.), Knowledge Discovery in Inductive Databases. VII, 190 pages. 2005.

Vol. 3376: A. Menezes (Ed.), Topics in Cryptology – CT-RSA 2005. X, 385 pages. 2005.

Vol. 3375: M.A. Marsan, G. Bianchi, M. Listanti, M. Meo (Eds.), Quality of Service in Multiservice IP Networks. XIII, 656 pages. 2005.

Vol. 3374: D. Weyns, H.V.D. Parunak, F. Michel (Eds.), Environments for Multi-Agent Systems. X, 279 pages. 2005. (Subseries LNAI).

Vol. 3372: C. Bussler, V. Tannen, I. Fundulaki (Eds.), Semantic Web and Databases. X, 227 pages. 2005.

Vol. 3371: M.W. Barley, N. Kasabov (Eds.), Intelligent Agents and Multi-Agent Systems. X, 329 pages. 2005. (Subseries LNAI).

Vol. 3370: A. Konagaya, K. Satou (Eds.), Grid Computing in Life Science. X, 188 pages. 2005. (Subseries LNBI).

Vol. 3369: V.R. Benjamins, P. Casanovas, J. Breuker, A. Gangemi (Eds.), Law and the Semantic Web. XII, 249 pages. 2005. (Subseries LNAI).

Vol. 3368: L. Paletta, J.K. Tsotsos, E. Rome, G.W. Humphreys (Eds.), Attention and Performance in Computational Vision. VIII, 231 pages. 2005.

Vol. 3367: W.S. Ng, B.C. Ooi, A. Ouksel, C. Sartori (Eds.), Databases, Information Systems, and Peer-to-Peer Computing. X, 231 pages. 2005.

Vol. 3366: I. Rahwan, P. Moraitis, C. Reed (Eds.), Argumentation in Multi-Agent Systems. XII, 263 pages. 2005. (Subseries LNAI).

Vol. 3365: G. Mauri, G. Păun, M.J. Pérez-Jiménez, G. Rozenberg, A. Salomaa (Eds.), Membrane Computing. IX, 415 pages. 2005.

Vol. 3363: T. Eiter, L. Libkin (Eds.), Database Theory - ICDT 2005. XI, 413 pages. 2004.

Vol. 3362: G. Barthe, L. Burdy, M. Huisman, J.-L. Lanet, T. Muntean (Eds.), Construction and Analysis of Safe, Secure, and Interoperable Smart Devices. IX, 257 pages. 2005.

Vol. 3361: S. Bengio, H. Bourlard (Eds.), Machine Learning for Multimodal Interaction. XII, 362 pages. 2005.

Vol. 3360: S. Spaccapietra, E. Bertino, S. Jajodia, R. King, D. McLeod, M.E. Orlowska, L. Strous (Eds.), Journal on Data Semantics II. XI, 223 pages. 2005.

Vol. 3359: G. Grieser, Y. Tanaka (Eds.), Intuitive Human Interfaces for Organizing and Accessing Intellectual Assets. XIV, 257 pages. 2005. (Subseries LNAI).

Vol. 3358: J. Cao, L.T. Yang, M. Guo, F. Lau (Eds.), Parallel and Distributed Processing and Applications. XXIV, 1058 pages. 2004.

Vol. 3357: H. Handschuh, M.A. Hasan (Eds.), Selected Areas in Cryptography. XI, 354 pages. 2004.

Vol. 3356: G. Das, V.P. Gulati (Eds.), Intelligent Information Technology. XII, 428 pages. 2004.

Vol. 3355: R. Murray-Smith, R. Shorten (Eds.), Switching and Learning in Feedback Systems. X, 343 pages. 2005.

Vol. 3354: M. Margenstern (Ed.), Machines, Computations, and Universality. VIII, 329 pages. 2005.

Vol. 3353: J. Hromkovič, M. Nagl, B. Westfechtel (Eds.), Graph-Theoretic Concepts in Computer Science. XI, 404 pages. 2004.